Bluetongue

Biology of Animal Infections

Series Editor
Paul-Pierre Pastoret, World Organisation for Animal Health, Publications
Department, Paris, France

PUBLISHED VOLUMES

Marek's Disease: *An Evolving Problem*
Edited by Fred Davison and Venugopal Nair
ISBN: 978-0-12-0883790

Rinderpest and Peste des Petits Ruminants: *Virus Plagues of Large and Small Ruminants*
Edited by Thomas Barrett, Paul-Pierre Pastoret and William P. Taylor
ISBN: 978-0-12-088385-1

Bluetongue

Edited by
Philip S. Mellor, Matthew Baylis
and Peter P.C. Mertens

AMSTERDAM • BOSTON • HEIDELBERG • LONDON • NEW YORK • OXFORD
PARIS • SAN DIEGO • SAN FRANCISCO • SINGAPORE • SYDNEY • TOKYO

ELSEVIER

Academic Press is an imprint of Elsevier

Academic Press is an imprint of Elsevier
32 Jamestown Road, London NW1 7BY, UK
Radarweg 29, PO Box 211, 1000 AE Amsterdam, The Netherlands
525 B Street, Suite 1900, San Diego, CA 92101-4495, USA
30 Corporate Drive, Suite 400, Burlington, MA 01803, USA

First edition 2009

Library of Congress Cataloging-in-Publication Data
A catalog record for this book is available from the Library of Congress

British Library Cataloguing in Publication Data
A catalogue record for this book is available from the British Library

ISBN: 978-0-12-369368-6

For information on all Academic Press publications
visit our web site at books.elsevier.com

Printed and bound by CPI Group (UK) Ltd, Croydon, CR0 4YY
Transferred to Digital Print 2011

Working together to grow
libraries in developing countries

www.elsevier.com | www.bookaid.org | www.sabre.org

ELSEVIER   BOOK AID
              International   Sabre Foundation

Contents

Colour Plate Section

Series Introduction: Biology of Animal Infections

The Institute for Animal Health (IAH) in the United Kingdom is one of the institutes sponsored by the Biotechnology and Biological Sciences Research Council (BBSRC). It is a leading world institute in multidisciplinary research on infectious diseases of farm animals. Its remit is to understand the processes of infectious diseases and from that knowledge improve the efficiency and sustainability of livestock farming, enhance animal health and welfare, safeguard the safety and supply of food and protect the environment.

The IAH also undertakes disease surveillance for the Department of Environment, Food and Rural Affairs (DEFRA) in the United Kingdom and provides reference laboratory services for DEFRA, the Office International des Epizooties (OIE), in effect the World Organisation for Animal Health, the United Nations Food and Agriculture Organisation (FAO) and the European Union (EU).

The IAH was formed following the merger of three previously independent institutes.

One of these was the Houghton Laboratory, formerly the Poultry Research Station, situated near Cambridge that specialised in infectious and parasitic diseases of poultry. This institute was moved to Compton in 1992 and as a result the IAH has laboratories located on two sites, at Compton in Berkshire and Pirbright in Surrey.

The laboratory at Compton concentrates on immunology and enzootic diseases, including some work on transmissible spongiform encephalopathies (TSEs). The laboratory at Pirbright specialises in exotic viral diseases.

The IAH carries out research on the most important exotic diseases, including foot-and-mouth disease, rinderpest, peste des petits ruminants, African horse sickness, bluetongue, African swine fever, classical swine fever, swine vesicular disease, lumpy skin disease, sheep and goat pox and highly pathogenic avian influenza. As far as enzootic diseases are concerned, the IAH works on some of the most important ones such as Marek's disease, bovine tuberculosis, avian infectious bronchitis and avian coccidiosis. Work is also carried out on bacteria responsible for food poisoning in humans. The IAH's research is supported by a strong expertise in immunology, virology, genetics, molecular biology, microbiology and epidemiology.

The IAH decided to use its longstanding expertise, experience and international contacts acquired through its work in these areas to produce a series of monographs on some of the diseases that are currently considered to be of the highest importance for animal health.

Each monograph is devoted to a single disease and produced along similar lines, which includes historical background, disease description, pathology, aetiology, molecular biology of host–pathogen interactions, immune responses, epidemiology and transmission, in addition to more practical topics, such as prophylaxis and vaccination.

The aim is to facilitate access to the most relevant of the current literature for each disease in order to constitute a practical and valuable source of information. The monographs are aimed at readers in veterinary faculties and specialists in veterinary research institutes worldwide.

They should also be of value to final year undergraduate students pursuing immunological, microbiological, virological, epidemiological and veterinary or medical studies.

The title of the series is Biology of Animal Infections. We use the term *infection* rather than disease since infection of an animal does not always result in disease; the outcome of an infection depends on many factors including genetics of the host and the longstanding evolutionary relationship between the host and its parasites. Also many infections may result in persistence of the so-called pathogens in the host, leading to the establishment of sub-clinical carriers of the infectious agent, which may be of crucial importance in the transmission and the epidemiology of the associated disease. The term *biology* refers to the fact that in the series we wish to focus on the biology, the mechanism of the infection rather than a mere descriptive approach to studying disease, though this is also important.

The editors of the monographs are convinced that to study infections (or diseases) in the actual target (host) is the best way to gain an understanding of health. The name 'Institute for Animal Health' is a true reflection of this concept. The first monograph was devoted to Marek's disease. The reason for this choice was the fundamental scientific significance of this disease, which crosses a number of biological disciplines. The next one was devoted to Rinderpest and Peste des petits ruminants; Rinderpest is the major plague of cattle and, hopefully, will be the first animal virus disease to be eradicated and only the second following the eradication of smallpox.

The present monograph is devoted to Bluetongue, which can be considered as an emerging infection, extending its range, notably in Europe. This infection is extremely complex, depending on a number of factors, including global warming and other major changes in our environment. Both the editors and I hope that this monograph will be a basic contribution for the understanding of this fascinating infection.

Paul-Pierre Pastoret
Series Editor

Preface

Robert F. Sellers
4, Pewley Way, Guildford, Surrey

A century ago bluetongue was a disease of sheep in southern Africa. Today bluetongue is known worldwide. Evidence of its presence through disease, isolation of virus or development of antibodies has been found north and south of the equator in Africa and southern Europe, Middle East and Asia, Australia and North and South America. The fear of its presence whether in the form of disease, virus or antibody, influences actions by countries free of bluetongue to prevent its introduction. Investigations over the years initiated either unintentionally through the import of susceptible sheep to countries where virus persists without causing disease or intentionally through research have led to an understanding of the disease, the nature of the virus, its life cycle between ruminants and *Culicoides* and its distribution.

Over the years advances have been made in understanding the structure and function of the virus and its components, the biology of the *Culicoides* species able to transmit it and the persistence of virus in *Culicoides* and the mammalian host. Recently the introduction of bluetongue virus into southern Europe from Africa and the Near East has emphasised the need to understand more fully how the virus spreads. The effect of climate change in enabling the colonisation of new habitats and the development of virus within additional species of *Culicoides* are of prime importance. All this has required field and laboratory investigations to carry out virus isolation, antibody tests as well as investigations on the pathogenesis of bluetongue. In addition, methods of control for *Culicoides* and for the mammalian host, such as vaccines have merited attention.

The editors, all very knowledgeable in their field, have brought together a group of contributors with wide experience of the various facets of bluetongue. Their chapters present the latest findings on the situation of bluetongue in the world, the latest research on all aspects of the disease, virus, host and vector and the likely developments in the future. Such a volume should be of interest to a wide selection of readers. Teachers and researchers in veterinary departments of universities, workers in research institutes and government animal health departments and international agencies dealing with diagnosis, control and international trade of animals and animal products will find in the monograph topics of special interest as well as the general background of bluetongue. Those whose concern is insect-borne disease will find many parallels with the diseases they investigate, and molecular biologists will be interested in the investigations on the structure and function of a double-stranded RNA virus. In short as an authoritative text the monograph appeals to veterinarians, entomologists, molecular biologists and diagnosticians as well as those with a general interest in science and disease control.

Acknowledgement

The editors would like to thank not only Maureen Twaig and Lisa Tickner at Elsevier for their efficiency and patience in guiding this volume through to publication but also the contributors themselves for sticking to the task over a very long road, and the many colleagues and friends around the world who have helped us immeasurably in discussion and by sharing their ideas and thoughts about Bluetongue. We are also grateful to Susan Baker at Liverpool who ploughed her way through every chapter in the Monograph to compile the Glossary, with great efficiency.

Finally, grateful thanks are due to Elsevier, who with great perception have agreed to publish this volume on Bluetongue at precisely the time when the disease is achieving the most intense world-wide attention for over 50 years, because of the recent and remarkable changes in its distribution and epidemiology.

Introduction 1

PHILIP S. MELLOR, MATTHEW BAYLIS AND PETER P.C. MERTENS

Institute for Animal Health, Pirbright Laboratory, Pirbright, Woking, Surrey, UK

Background

Bluetongue (BT), as a disease in cattle and sheep, was first described in the late eighteenth century as 'tong-sikte' by a French biologist Francois de Vaillant during his travels in the Cape of Good Hope between 1781 and 1784 (Gutsche, 1979). About 40 years later, Duncan Hutcheon, the Chief Veterinary Officer of the Cape Colony, recorded some clinical features of the disease in his annual report for 1880 and, in 1902, wrote of 'malarial catarrhal fever' in sheep, in the Veterinary Record, where he postulated that the disease agent was insect-transmitted (Hutcheon, 1902).

In 1902, Spreull first reported on the results of some of his experiments with malarial catarrhal fever in sheep (Spreull, 1902). Then in 1905, he described a typical case of the disease, the onset of which was marked by high fever, and after 7–10 days, distinctive lesions appeared in the mouth and the tongue turned dark blue. In consequence of these lesions, he suggested that the name of the disease should be changed from malarial catarrhal fever to bluetongue (BT) (Spreull, 1905).

Robertson and Theiler, quoted in Spreull (1905), first showed that the agent of BT was filterable and then Spreull himself showed that it was transmissible to cattle and goats but that infection in these ruminant species was apparently subclinical (Spreull, 1905).

Most of the early findings relating to BT aetiology, epidemiology and transmission (by *Culicoides* biting midges), multi-typic nature and control,

ISBN-13: 978-0-12-369368-6

and many of the later findings too were led by scientists working in southern Africa. Their major contributions in these fields have been well summarized in Chapter 2.

Some of the more startling and many of the most recent advances in understanding of BT virus (BTV) have come about by the use of powerful new molecular tools and through a synthesis of structural, biochemical and cell biology data, which have been used to elucidate how the virus infects the host cells and evades both the innate and adaptive immune responses (see Chapters 4–7). Part of this work has resulted in the determination of the structure of the core particle of BTV, by X-ray crystallography, to a resolution approaching 3.5 Å (Grimes *et al.*, 1998). At the time, the BTV core particle represented the largest molecular structure determined in such detail and 10 years later this is still the case.

Bluetongue, an emerging disease

Bluetongue was regarded as an African disease until 1943 when an outbreak was recorded in Cyprus (Gambles, 1949), although it was believed retrospectively that outbreaks had been occurring on the island since 1924 (Polydorou, 1985). The severity of the disease in Cyprus in 1943–1944 when around 2500 sheep died with some flocks experiencing 70% mortality focussed international attention on this 'new disease' coming out of Africa. Further outbreaks of BT, in Palestine in 1943, in Turkey in 1944, 1946 and 1947 and in Israel in 1949, served to reinforce the idea that this new disease was 'emerging' (Gambles, 1949; Komarov and Goldsmit 1951), and this concept was strengthened by the realization that 'soremuzzle', a disease of sheep in California and Texas in North America, was caused by BTV serotype 10, a type of BTV previously identified in South Africa (Hardy and Price, 1952; McKercher *et al.*, 1953). The identification of further North American serotypes of BTV (11, 13, 17 and 2) followed over succeeding years (Barber, 1979; Gorman 1990, Chapters 9 and 13).

At more or less the same time that the North American BT story was unfolding, a major outbreak of disease caused by serotype 10 of the virus began in Portugal in 1956, which rapidly expanded to Spain in 1957 (Manso-Ribiero *et al* 1957) and continued in the Iberian Peninsula until 1960 (Manso-Ribiero *et al.*, 1957; Campano Lopez and Sanchez Botija, 1958). The outbreak was extremely severe with over 179 000 sheep dying and with a mortality rate of around 75% being recorded in affected sheep. The outbreak was eventually controlled by imposing animal movement restrictions, slaughter of infected individuals and compulsory annual vaccination with a live attenuated vaccine. Subsequently, BTV was not recorded in Europe for almost 20 years until an outbreak of BTV-4 was recorded in the Greek islands of Lesbos and Rhodes in 1979–1980 (Vassalos, 1980; Dragonas 1981, Chapter 11).

Bluetongue was first reported in West Pakistan in 1958 in previously vaccinated Rambouillet sheep imported from the USA. The virus was identified as BTV-16, a serotype that had previously been isolated in Israel in 1966. The virus apparently caused severe disease in local sheep and goats (Sapre, 1964). This and more recent outbreaks of BTV on the Indian subcontinent are described in Chapter 8.

In 1977, BTV was isolated in Australia for the first time, from *Culicoides* captured two years earlier in the Northern Territory (St George *et al.*, 1978). The virus was the first identification of BTV-20. Over the next nine years, until 1986, a further seven BTV serotypes were identified in the country making eight serotypes in all (1, 3, 9, 15, 16, 20, 21 and 23) (Kirkland, 2004). Despite the presence of these serotypes, Australia has remained free of disease, although disruption in trade has had a significant financial impact (Kirkland, 2004). The absence of disease in Australia is at least in part because sheep are rarely raised in the regions of northern and eastern Australia where BTV occurs, and the viruses present in regions adjacent to sheep-rearing areas are considered to be non-pathogenic. However, disease has been observed when small groups of sheep were moved into the tropical part of Northern Territory. Overall, the available evidence suggests that there is a graduation of virulence of BTV types and strains across SE Asia from India, where significant disease is experienced in both local and exotic sheep breeds (see Chapter 8), to Australia, where disease is rarely or never observed (see Chapter 10).

This situation of the apparent expansion of BTV into N. America, Australia, Europe and the Indian subcontinent during the 1940s and 1980s was a cause of major international concern and the reason for the worldwide perception of BT as an emerging disease of animals (Howell, 1963). It was also a time when increasing restrictions were placed on the movement of animals and their products from 'BTV-infected' countries and regions in attempts to halt the spread of the virus and when BT was added to List A of the OIE International Animal Health Code. However, in the early 1990s, in the absence of any recent major clinical losses due to BT, there was a reassessment of its significance as an emergent disease.

In a review in 1994, Gibbs and Greiner asked whether BT should still be considered an emerging disease and wondered whether it justified its continued inclusion as an OIE List A disease. These authors wrote, '. . . has BT in the period between the 1960s and the 1990s reached its potential to emerge and spread or have we simply learned enough about the virus to recognize that our earlier assessments of the calamity that could befall the sheep and cattle industries of the world presented a doomsday scenario?' The authors argued that BTV occurs around the world in relatively stable ecosystems that are defined by the competence of indigenous vector *Culicoides*, and they postulated that climatic and geographical barriers restrict movements of both vectors and hosts to each ecosystem. Viruses that are introduced into a different ecosystem were supposed to 'die out' due to the lack of an efficient vector. Gibbs and Greiner endorsed the proposal that BT should be removed from OIE's List A and concluded that 'the designation of BT as an emerging disease is now of historic interest only. With hindsight, such a designation in the 1960s was hasty .... and counterproductive to reasoned argument'.

Unfortunately, hindsight is the poor relation of foresight. Just 4 years after the publication of their review, outbreaks of BT occurred in Europe, starting on four Greek islands in 1998 and spreading over the next seven years to include at least 16 Mediterranean countries including nine (France, Italy, Bulgaria, Macedonia, Yugoslavia, Bosnia, Albania, Croatia and Tunisia) that had never previously experienced the disease, with the virus spreading as far north as 44°50′ (see Chapter 11). This was the furthest north, by several hundred kilometres, that BTV had ever reached in Europe. By 2005, the outbreak had already killed well over a million sheep, which made this the largest epidemic of BT ever seen (Purse et al., 2005). Sadly, this was not the end of the matter, and in 2006, BTV exhibited a further massive northwards leap, into northern Europe, by a so far unidentified route and invaded a series of countries in the NW of the continent (northern France, Belgium, The Netherlands, Luxembourg and Germany). Even more astonishingly, the virus succeeded in overwintering in this northern region and re-emerged in 2007 in each of the affected countries with massively increased vigour, afflicting livestock on over 45 000 farms and expanding into the UK, Denmark and the Czech Republic, and spreading as far north as 53° N, the furthest north BTV has ever been recorded anywhere in the world (see Chapter 11).

The BTV incursions into Europe between 1998 and 2005 seem to be due at least in part to northerly extensions in the range of the major Old World vector, *Culicoides imicola,* which now occupies most of the northern part of the Mediterranean Basin. These range changes of *C. imicola*, which is an Afro-Asiatic species, have been linked to ongoing climate change (see Chapter 16 and Purse et al., 2005). Furthermore, in many areas of eastern Europe, BTV has expanded beyond the new range limits of *C. imicola* and for the first time in recorded history, indigenous, northern Palearctic species have been implicated in its transmission in non-*imicola* areas of Europe. The involvement of these northern Palearctic species has been linked to the recent increase in temperature being experienced in many parts of Europe, as BTV is transmitted by *Culicoides* vectors with increasing efficiency as the temperatures rise (see Chapter 14).

At the same time that these unprecedented incursions of BTV into Europe were occurring, investigations in N. America showed that over a 7-year period from 1999, a further eight BTV serotypes (1, 3, 5, 6, 14, 19, 22 and 24) had entered the southern USA (Florida, Mississippi and Louisiana) (see http://www.reoviridae.org/dsRNA_virus_proteins/ReoID/BTV-isolates.htm, Mertens, Personal Communication). In Australia, after a 21-year period of apparent stability, a ninth serotype of BTV, BTV-7, was identified from sentinel cattle in the Northern Territory in 2007. As BTV-7 occurs in countries to the north of Australia, it is supposed that infected vectors were blown into the country on monsoonal winds (OIE, 2008).

These current and ongoing changes in the range of many BTV serotypes in North America, Australia and Europe, and the involvement of novel vector species in Europe, with increasing virulence also being reported from that continent (see Chapter 12) suggest that major and worldwide changes are underway in BT epidemiology. Far from the concept of BT as an emerging disease being of historic interest only, it is now a matter of intense international concern to scientists, farmers, veterinary services and legislators. At the time of writing, much effort has been spent in attempting to predict the next stage in the epidemiology of this disease and many have been wondering whether BT is merely providing a foretaste of other diseases yet to come. The information contained within the chapters of this monograph provides a wealth of information on BT, ranging from the atomic structure of the virus particle through diagnostics and control to transmission, distribution and clinical signs. We believe that the book makes a major contribution to our understanding of this most important disease.

References

Barber, T.L. (1979) Temporal appearance and species of origin of bluetongue virus serotypes in the United States. *Am. J. Vet. Res.* **40**, 1654–6.

Campano Lopez, A. and Sanchez Botija, C. (1958) L'epizootie de fievre catarrhale ovine en Espangne (Blue Tongue). *Bull. Off. Int. Epizoot.* **50**, 65–93.

Dragonas, P.N. (1981) Evolution of bluetongue in Greece. *Off. Int. Epizoot. Monthly Epizoot. Circ.* **9**, 10–11.

Gambles, R.M. (1949) Bluetongue of sheep in Cyprus. *J. Comp. Pathol.* **59**, 176–90.

Gibbs, E.P.J. and Greiner, E.C. (1994) The epidemiology of bluetongue. *Comp. Immunol. Microbiol. Infect. Dis.* **17**, 207–20.

Gorman, B.M. (1990) The bluetongue viruses. *Curr. Top. Microbiol. Immunol.* **162**, 1–19.

Grimes, J.M., J.Burroughs, N.J., Gouet, P., Diprose, J.M., Malby, R., Ziéntara, S., Mertens, P.P.C. and Stuart, D.I. (1998) The atomic structure of the bluetongue virus core. *Nature* **395**, 470–8.

Gutsche, T. (1979) *There was a Man*. Cape Town: Timmins, p. 4.

Hardy, W.T. and Price, D.A. (1952) Soremuzzle of sheep. *J. Am. Vet. Med. Assoc.* **120**, 23–5.

Howell, P.G. (1963) Bluetongue. In: Emerging Diseases of Animals. FAO Agricultural Studies No. 61, Rome, pp. 109–53.

Hutcheon, D. (1902) Malarial catarrhal fever of sheep. *Vet. Rec.* **14**, 629–33.

Kirkland, P.D. (2004) Bluetongue viruses, vectors and surveillance in Australia – the current situation and unique features. *Vet. Ital.* **40**, 47–50.

Komarov, A. and Goldsmit, L. (1951) A disease similar to bluetongue in cattle and sheep in Israel. *Refu. Vet.* **8**, 96–100.

Manso-Ribiero, J., Rosa-Azevedo, J.A., Noronha, F.O., Braco-Forte-Junior, M.C., Grave-Periera, C. and Vasco-Fernandes, M. (1957) Fievre catarrhale du mouton (blue-tongue). *Bull. Off. Int. Epizoot.* **48**, 350–67.

McKercher, D.G., McGowan, B., Howarth, J.A. and Saito, J.K. (1953) A preliminary report on the isolation and identification of the bluetongue virus from sheep in California. *J. Am. Vet. Med. Assoc.* **122**, 300–301.

OIE (2008) Bluetongue in Australia. In: WAHID Interface; World Animal Health Information Database. Information received on 14 April 2008, pp. 3.

Polydorou, K. (1985) Bluetongue in Cyprus. *Prog. Clin. Biol. Res.* **178**, 539–44.

Purse, B.V., Mellor, P.S., Rogers, D.J., Samuel, A.R., Mertens, P.P.C. and Baylis, M. (2005) Climate change and the recent emergence of bluetongue in Europe. *Nat. Rev. Microbiol.* **3**, 171–81.

Sapre, S.N. (1964) An outbreak of bluetongue in goats and sheep in Maharashtra State. *Indian Vet. Rev.* **15**, 69–71.

Spreull, J. (1902) Report from Veterinary Surgeon Spreull on the result of his experiments with malarial catarrhal fever of sheep. *Agric. J. Cape Good Hope* **20**, 469–77.

Spreull, J. (1905) Malarial catarrhal fever (bluetongue) of sheep in South Africa. *J. Comp. Pathol. Ther.* **18**, 321–37.

St George, T.D., Standfast, H.A., Cybinski, D.H., Dyce, A.L., Muller, M.J., Doherty, R.L., Carley, J.G., Filippich, C. and Frazier, C.L. (1978) The isolation of bluetongue virus from *Culicoides* collected in the Northern Territory of Australia. *Aust. Vet. J.* **54**, 153–4.

Vassalos, M. (1980) Cas de fievre catarrhale du mouton dans l'Ille de Lesbos (Grece). *Bull. Off. Int. Epizoot.* **92**, 547–55.

The history of bluetongue

2

BALTUS J. ERASMUS* AND A. CHRISTIAAN POTGIETER†

* Deltamune (Pty.) Ltd., P.O. Box 14167, Lyttelton, 0140, South Africa. (Formerly: Head, Onderstepoort Biological Products).

† Virology Division, Agricultural Research Council, Onderstepoort Veterinary Institute, P O Box X05, Onderstepoort, 0110, South Africa.

Introduction

Research on bluetongue (BT) and more particularly on BT virus (BTV) has made spectacular progress in recent years as will become clear in the various chapters of this book. However, it may be appropriate to acknowledge that this progress has been made possible by the solid foundation laid by the pioneers of BT research. Although they lacked sophisticated equipments and techniques, they compensated for this by keen powers of observation, logical reasoning, ingenuity and persistence.

It is not intended to give a chronological review of BT research but rather to highlight certain milestones that contributed significantly to a better understanding of the disease and its aetiological agent. Since most of the pioneering research work was done at the Onderstepoort Veterinary Institute in South Africa, the authors admit to a certain degree of bias in their approach to this contribution, emphasising the African perspective.

Geographical distribution

Bluetongue had already been known to sheep farmers in South Africa during the early part of the nineteenth century, most likely soon after the introduction of Merino sheep. However, the first detailed scientific descriptions of the

ISBN-13: 978-0-12-369368-6

disease in sheep were given by Hutcheon (1902) and Spreull (1902, 1905). The latter publication is still remarkable for the detailed and accurate description of the clinical signs and gross pathology.

Bluetongue is historically an African disease, and although it has not been reported from all African countries, it is probably enzootic throughout the entire continent. Clinical recognition of the disease depends largely on the presence of highly susceptible European sheep breeds, which invariably act as indicators.

The first confirmed outbreak of BT outside Africa occurred in Cyprus in 1943 (Gambles, 1949). However, it is believed that periodic unconfirmed outbreaks had occurred in Cyprus since 1924 (Polydorou, 1985). The severity of the 1943 outbreak served to focus attention on BT as a serious disease, and this increased level of alertness facilitated recognition of subsequent outbreaks in this region, e.g. in Israel in 1949 (Komarov and Goldsmit, 1951), in Pakistan in 1959 (Sarwar, 1962) and in India in 1963 (Sapre, 1964). In all these instances, the confirmation of the diagnosis and the eventual serotyping of the causal BTV were conducted at Onderstepoort.

In 1952, a condition termed 'sore muzzle' was described in California, and although its aetiology could not be established, its similarity to BT was noted (Hardy and Price, 1952). South African scientists offered their assistance and BTV was isolated and identified from Californian sheep samples submitted to Onderstepoort (Alexander, 1959; Callis, 1985). Early in 1952, Dr R. A. Alexander, Director of the Onderstepoort Veterinary Institute, was invited to the United States to give expert advice. The presence of several breaks on the hooves of recovered sheep enabled Alexander to postulate the existence of two or more serotypes of BTV in the area (Callis, 1985). The details of the California BT experience have been published by McKercher et al. (1953).

The California isolates were later identified by Howell (Howell and Verwoerd, 1971) as Type 10. Several isolates were made in subsequent years, and there were indications that some of these isolates belonged to serotypes other than Type 10. In 1975, representative isolates were submitted to the OIE World Reference Centre for Bluetongue at Onderstepoort where Erasmus serotyped these as Types 10, 11, 13 and 17 (Barber, 1979). The presence of BTV Type 2 in Florida and the Caribbean was reported later by Gibbs et al. (1983).

The first occurrence of BT in Europe, outside of Cyprus, was in 1956 when the disease was diagnosed in southern Portugal (Ribeiro et al., 1957) and shortly thereafter also in Spain (Lopez and Botija, 1958). The virus was isolated at Onderstepoort and later identified as BTV-10 (Howell, 1960). Surprisingly, the infection apparently did not become established in the Iberian Peninsula from where it eventually disappeared after about 4 years.

The extension of BTV infection into the Middle East and Asia and eventually into Australia and Oceania has been well summarised by Parsonson and Snowdon (1985), Hassan (1992) and Doyle (1992) as well as in several chapters of this book.

Host range

Historically BT has been regarded as a disease of sheep. All breeds are susceptible in the virological sense, but there are marked differences in the clinical manifestation of disease between different breeds. Indigenous African breeds (particularly in southern Africa) generally show subclinical infections. The Merino is more resistant than the European mutton breeds such as the Dorset Horn (Spreull, 1905; Neitz, 1948), which are generally regarded as highly susceptible. However, there is also a marked variation in susceptibility among individuals within a specific breed. This, together with the marked variation in pathogenicity of different BTV strains, is responsible for the unpredictable and variable mortality encountered in both natural and experimental infections.

The severity of BT in sheep is greatly influenced by environmental conditions. Neitz and Riemerschmid (1944) convincingly proved that solar radiation aggravates the course of the disease, particularly in recently shorn animals. Hyperaemia of the skin is usually very severe in those areas not covered by wool but often involves the entire body (Erasmus, 1975). This dermatitis contributes to the 'break' of wool fibres with occasional casting of the entire fleece 3–6 weeks later (Verwoerd and Erasmus, 1994). The mortality rate can be very high if infected sheep are exposed to cold, wet conditions as often happens in late autumn when the morbidity rate may be at its peak (Erasmus, 1990).

Other domestic ruminants have also been shown to be susceptible. Spreull (1905) demonstrated the susceptibility of calves to experimental BTV infection and Bekker et al. (1934) described natural outbreaks in cattle. The latter publication described very severe clinical signs and lesions in both naturally and experimentally infected cattle. In some cases, the mouth and foot lesions reported were much more severe than those ever manifested by sheep and were reminiscent of foot-and-mouth disease. De Kock et al. (1937) and Mason and Neitz (1940) had recovered BTV from naturally exposed cattle but did not record any clinical signs or mouth or foot lesions in any of their cases. Nor could they produce clinical diseases in any of their experimental cases. The generally held belief in South Africa is that Bekker et al. (1934) were probably dealing with a dual infection of BT and some other viral condition.

The susceptibility of cattle was subsequently confirmed in many countries (Komarov and Goldsmit, 1951; Hardy and Price, 1952; Lopez and Botija, 1958). The disease in cattle is summarised by Hourrigan and Klingsporn (1975).

The pathogenesis of BT in cattle possibly differs from that in sheep. There is evidence of a rapid accumulation of IgE antibodies directed against the virus after infection, resulting in immediate hypersensitivity reactions involving the release of mediators such as histamine, prostaglandins and thromboxane A2 (Anderson et al., 1975; Emau et al., 1984). The development of overt clinical

BT, in cattle therefore, appears to be due to BTV infection of animals previously sensitised by BTV or BTV-related viruses (Anderson *et al.*, 1975), though this may not always the case, as evidenced by the clinical signs exhibited by some BTV-8-infected cattle in northern Europe in 2006 and 2007, in what was previously a completely naïve population.

The role of cattle in the epizootiology of BT has been stressed by du Toit (1962), Owen *et al.*, (1965) and subsequently in various other publications.

The susceptibility of goats to BTV infection was first recorded by Spreull (1905), who commented on the inapparent nature of infection under natural conditions. Artificial infection of Angora goats caused no clinical signs although blood collected on the fifth and twentieth days after infection proved infective for sheep. During the extensive outbreak in Israel, clinical signs suggestive of BT were observed in Saanen goats (Komarov and Goldsmit, 1951). Sapre (1964) reported clinical signs in goats that were regarded as characteristic of BT and some goats even succumbed to infection. Barzilai and Tadmor (1971) recorded a mild febrile reaction and slight hyperaemia of the conjunctival and nasal mucous membranes in some artificially infected goats. Viraemia was demonstrated in some goats for 19 days but the level of viraemia was about a hundred-fold lower than in sheep. Luedke and Anakwenze (1972) infected Saanen goats with American strains of BTV. All the goats became infected as evidenced by leucopaenia, viraemia and the development of precipitating antibodies. Febrile responses were variable and viraemia was about 10-fold lower than in sheep.

Neitz (1933) demonstrated the susceptibility of blesbuck (*Damaliscus albifrons*) to artificial infection with BTV. No clinical signs were evident but viraemia between days 8 and 17 post-infection could be demonstrated. The significance of this seemingly insignificant experiment was that it focused attention to wild ruminants as potential hosts of BTV. Virological and serological evidence of BTV infection has been found in many wild ruminants (Hourrigan and Klingsporn, 1975; Hoff and Hoff, 1976; Jessup, 1985). African antelope do not develop clinical disease, whereas in the USA, white-tailed deer (*Odocoileus virginianus*), pronghorn (*Antilocapra americana*) and desert bighorn sheep (*Ovis canadensis*) may develop severe clinical disease (Hoff and Hoff, 1976).

It can be hypothesised that, historically, the primary epizootiological cycle of BTV involved species of African antelope and *Culicoides* midges. With the agricultural development of large parts of Africa, which led to wild ruminants being replaced by large numbers of domestic ruminants, the traditional epizootiological role of wild animals has largely been taken over by cattle. In regions such as southern Africa, the infection seems to progress in the cattle–midge cycle during spring and early summer, and once a certain level of infection is reached, it spills over to sheep. This generally occurs in late summer or in autumn and is often characterised by simultaneous outbreaks of BT in sheep in various regions of the country. Sheep, therefore, seem to be involved in a secondary epizootiological cycle (Erasmus, 1980, 1990).

Aetiology

Theiler (1906) established the filterable nature of the causal agent of BT and considered it to be a virus closely associated with the blood but not exclusively with the red corpuscles.

Progress in the study of BTV was greatly hampered by the fact that for a long time, sheep was the only host system available for diagnostic and research purposes. The successful propagation of BTV in embryonated eggs by Mason et al. (1940) and more particularly the improvement in cultivation achieved at optimal temperatures of incubation (Alexander, 1947) facilitated diagnostic and research work. A further major milestone was the demonstration by Goldsmit and Barzilai (1965) that the intravenous route of inoculating embryonated eggs resulted in a hundred- to a thousand-fold greater sensitivity than other routes.

However, the most significant breakthrough was the successful cultivation of five strains of egg-adapted BTV in primary cultures of lamb kidney cells (Haig et al., 1956). Subsequent to this, the virus was adapted to a wide range of primary cells as well as cell lines (Fernandes, 1959a, b). Research work was further facilitated by the development of a plaque technique (Howell et al., 1967) for the accurate assay of BTV concentrations and the other applications that emanated from this finding.

Early electron microscopic studies (Owen and Munz, 1966; Studdert et al., 1966) suggested a Reovirus-like structure. More extensive morphological studies were later conducted by Els and Verwoerd (1969) and Murphy et al. (1971).

Bluetongue virus is characterised by the existence of multiple serotypes. Howell (1960) described the first 12 and later (Howell, 1970) a further 4 serotypes (BT13-16). Bluetongue virus serotypes 17–24 were serotyped by Erasmus (Unpublished). BTV-17 was originally isolated in the United States (Barber, 1979), while BTV-18 and BTV-19 were isolated from clinical cases of BT in South Africa in 1976 (Erasmus, Unpublished). BTV-20 and BTV-21 were first isolated in Australia (St. George et al., 1978, 1980).

BTV-22 and BTV-23 were first isolated from wild-caught Culicoides in South Africa (Nevill et al., 1992) and later BTV-23 was also isolated from sentinel cattle in Australia (Gard et al., 1987). BTV-24 was isolated from sick sheep (Erasmus, Unpublished) as well as from wild-caught Culicoides in South Africa (Nevill et al., 1992).

The large-scale cultivation and purification of the virus was first shown by Verwoerd (1969). A major breakthrough in BTV molecular research came soon after the extraction and purification of the viral genetic material. It was shown unequivocally that BTV possesses a segmented, double-stranded ribonucleic acid (RNA) genome (Verwoerd *et al.*, 1970), similar to that of Reoviruses (Bellamy *et al.*, 1967). Distinct differences were also noticed between Reoviruses and BTV, which led to the recognition of a new group of dsRNA viruses termed diplornaviruses (Verwoerd, 1970). This group later became the orbiviruses.

The molecular studies on BTV continued at Onderstepoort and 3 years later a landmark paper was published on the structure of the BTV capsid (Verwoerd *et al.*, 1972). Not only were the seven structural proteins of BTV identified but also important structural features of complete virions, infectious subviral particles and cores shown. Combination of these data with electron micrographs of the same particles allowed the identification of the outer- and inner-layer proteins. The molecular weights of the viral proteins and nucleic acid segments estimated from polyacrylamide gel electrophoresis allowed the gene assignments of seven of the ten genome segments of BTV. In addition, the relative copy number of each viral structural protein per virus particle was determined. Infectivity of core particles could be partially restored by recombination with the outer-layer proteins (VP2 and VP5). This was the first paper on the molecular characterisation of the aetiological agent of BT. The two major non-structural proteins of BTV (NS1 and NS2) were identified by Huismans (1979). The third (NS3) was detected by Gorman *et al.* (1981). NS1 tubules were shown in electron micrographs of BTV-infected cells (Lecatsas, 1968) and were later characterised further by Huismans (1979) and Huismans and Els (1979), who showed that similar tubules were found in cells infected by other orbiviruses as well. The segments that encode these proteins were later identified by Van Dijk and Huismans (1988).

Another milestone in BT molecular biology was evident on the genetic relatedness of different BTV serotypes and isolates based on the homology of individual dsRNA segments. Homology was determined using the hybridisation of mRNA and dsRNA from different virus isolates (Huismans and Howell, 1973). In this study, the relative abundance of the 10 mRNA transcripts in BTV-infected cells was also determined. Cloned cDNA from the 10 BTV genomic segments were first used as probes in DNA/RNA hybridisation to show the genetic relation amongst several BTV serotypes and field isolates by Huismans and Cloete (1987) and Huismans *et al.* (1987). Importantly, it was already evident that BTV S2, which encodes VP2, was the most variable segment among BTV serotypes. The complete gene assignment of all 10 segments of BTV was shown for the first time in 1988 by virtue of the in vitro transcription and translation of the mRNA from all 10 BTV genomic segments (Van Dijk and Huismans, 1988). The full nucleotide sequence of all 10 segments of BTV-10 was published by Roy (1989). These initial studies were the basis for the cloning and sequencing of many BTV gene segments as well as for the expression of their respective recombinant proteins. The sequencing and hybridisation studies were the basis for molecular epidemiological studies and also RT-PCR techniques (Dangler *et al.*, 1990; Wade-Evans *et al.*, 1990), which are used today for routine diagnostic detection of BTV nucleic acids.

The *in vitro* activation of BTV transcription was first published in 1980 (Van Dijk and Huismans). Purified BTV selectively stripped of the outer capsid proteins was shown to have transcription activity in vitro. The conditions that influenced transcription were also determined. This was a very important step towards the understanding of the uncoating and replication of the virus during the infection cycle. The recent rescue of BTV from purified ssRNA transcripts (Boyce and Roy, 2007), an extremely important step to further characterise the virus, is based on the in vitro transcription from the former study.

The work mentioned here formed the cornerstone for most molecular biological studies on BTV that followed. Today, BT molecular biology continues to improve our understanding of the aetiological agent in terms of its structure, protein function, replication and interaction with the host and the vector. It allows studies on the molecular epidemiology of the disease, the improvement of diagnostic tests and finally the development of safe and effective vaccines to combat the disease.

Transmission

It was common knowledge in South Africa more than a century ago that BT was most prevalent during the late summer months, especially following wet seasons; also that the disease was more common in low-lying pastures and valleys than on high-lying ground and that sheep stabled in sheds during summer nights escaped infection. There was, therefore, very strong circumstantial evidence to suggest that BT was an insect-transmitted disease. This view was expressed by Hutcheon (1902) and Spreull (1902, 1905). These workers further noticed the sudden cessation of BT outbreaks soon after the first heavy frosts of the season.

Mosquitoes were initially suspected as the most likely vectors of BTV. Several transmission attempts involving *Aedes* mosquitoes were carried out by Nieschultz *et al.* (1934). Although BTV was apparently recovered on three occasions from mosquitoes previously fed on infected sheep, the evidence was not convincing.

The disappointing results obtained with mosquitoes compelled du Toit to consider other haematophagous insects as potential vectors of BTV. Amongst these were *Culicoides* midges, a group of insects that was hitherto not viewed as a potential vector of any pathogen. Their relative abundance in areas and during seasons when BTV was highly prevalent contributed to the decision to investigate their vector potential.

To the surprise of local as well as international entomologists, du Toit (1944) succeeded in transmitting BTV either by injecting emulsions of wild-caught *Culicoides* into sheep or by allowing *Culicoides pallidipennis (imicola)*, previously fed on infected sheep, to bite susceptible sheep 10 days later. This remarkable finding was confirmed by Hardy and Price (1952) in the United States. with *Culicoides variipennis*. The biological transmission of BTV by *Culicoides* was proven beyond any doubt by Foster *et al.* (1963) with colonised *C. variipennis*. The vectorial role of *Culicoides* has since been corroborated in all areas in which BTV occurs as well as for many other orbiviruses.

For decades, *C. imicola* was regarded as the main if not sole vector of BTV in South Africa. The presence of BTV in regions where *C. imicola* was either absent or present in very low numbers led to a reappraisal of this tenet. Work done by Venter *et al.* (1998) identified *C. bolitinos* as an additional vector of BTV, and Paweska *et al.* (2002) proved that *C. bolitinos* had a significantly higher transmission potential for BTV-1 than did *C. imicola*.

The significant role of various *Culicoides* species in the transmission of BTV in different regions of the world is well documented in several other chapters of this book.

Control

In enzootic situations such as South Africa, it was realised more than a century ago that prophylactic vaccination of sheep would offer the most practical control measure. The earliest recorded attempt at immunisation was by Spreull (1905), who simultaneously administered virulent sheep blood and hyperimmune serum to sheep. Theiler (1908) believed that BTV became attenuated after serial passage of the virus in sheep. This 'attenuated' virus was used for the production of a 'vaccine' composed of the blood of infected sheep. Despite the variable results obtained, this 'vaccine' was used until 1946, during which time more than 50 million doses were issued. It made sheep farming possible in areas of South Africa where it was previously uneconomical. The results were indeed better than what would normally be expected of a monovalent vaccine. This can perhaps be ascribed to the fact that the Theiler virus strain, which was later identified as BTV-4 (Howell, 1960), is a serotype that is serologically related to most other BTV serotypes (Erasmus, 1990) and, therefore, had the potential of cross-protection against many serotypes.

Although some of the earlier researchers surmised the existence of a multiplicity of virus strains, it was Neitz (1948) who proved conclusively by cross-protection tests in sheep that several antigenically different BTVs existed in nature. Furthermore, he proved that BTV could not be attenuated by serial passage in sheep. These discoveries revealed the shortcomings of the Theiler blood vaccine and stressed the need for alternate host systems for the isolation, propagation and attenuation of BTV.

Alexander *et al.* (1947) greatly facilitated the cultivation of BTV in embryonated eggs by showing that the optimal virus replication occurred at an incubation temperature of 33.6°C. A further outcome of this finding was the attenuation through serial passage in eggs of a few virus isolates. This eventually led to the development of a quadrivalent avianised vaccine, which, on account of the great demand for the vaccine, is still produced in sheep according to the Theiler method (Alexander and Haig, 1951). Once a sufficient number of embryonated eggs became available, routine vaccine production was done in this host system but the macerated embryo material was still suspended in normal sheep blood. For some reason this practice solved some of the problems of contamination and poor shelf life. Around 1951, antibiotics became more readily available and the normal sheep blood was replaced with penicillin, streptomycin and crystal violet to control contamination.

In 1953, it was discovered that buffered lactose peptone was an excellent stabiliser of the virus, which also allowed successful lyophilisation of the vaccine. Subsequently, vaccine could be produced all year round and stockpiled for periods of greater demand. By 1958, annual production had increased to more than 24 million doses compared to only 2 million in 1946.

The cultivation of egg-adapted BTV in lamb kidney cell cultures and the demonstration that cytopathic effects could be inhibited by homologous antibody in sheep serum were major milestones (Haig *et al.*, 1956). This allowed the performance of in vitro neutralisation tests, which enabled Howell (1960) to recognise 12 serologically distinct serotypes of BTV.

A further outcome of this milestone was the adaptation of the egg-attenuated BTV vaccine strains to cell culture and the eventual routine production of the vaccine in such cultures (Howell, 1963). This greatly facilitated the production of large volumes of vaccine. A further major improvement was instituted in 1968 when production switched from stationary lamb kidney cell cultures to BHK_{21} cells in roller cultures.

Despite the major improvements in BTV vaccine production, problems were still encountered with its efficacy. This was caused by various factors. The most significant of these was undoubtedly the multiplicity of BTV serotypes. As more serotypes were identified in the field, more were included in the polyvalent vaccine. Around 1968, the vaccine contained 14 serotypes, which was ludicrous if one considers that these were all live attenuated vaccine strains, and good immunity was dependent upon replication of each of these vaccine strains in recipient sheep. It, therefore, soon became clear that administration of a polyvalent live attenuated vaccine does not automatically lead to complete polyvalent immunity. The problem of poor immunity created by the competition to replicate was further aggravated by the low immunogenicity of certain vaccine strains. As an interim measure, some of the poorly immunogenic strains were omitted, and between 1969 and 1975, the vaccine contained only nine serotypes.

In 1976, BT was very prevalent in South Africa, affecting both unvaccinated as well as previously vaccinated sheep. Two new BTV serotypes (Types 18 and 19) were identified from field samples. Type 19 was the most prevalent one and showed a minor relationship to other BTV serotypes, thus explaining the fact that previously vaccinated sheep appeared completely susceptible (Erasmus, Unpublished data). It thus became imperative to have Type 19 included in the BTV vaccine to prevent future losses.

Attenuation of BTVs in embryonated eggs requires an average of 70–100 serial passages. In view of the urgency associated with Type 19, a modified approach was used to expedite the process of attenuation. Once the Type 19 isolate had undergone 29 serial passages in embryonated eggs, it was argued that the proportion of avirulent to virulent virus particles should have changed sufficiently to allow the selection of avirulent virus by plaque selection. Unfortunately no genetic markers that correlated with loss of pathogenicity could be found. Despite this drawback, 10 plaques were selected at random at the dilution end point in Vero cells. These were grown up on BHK-21 cells and screened in BT-susceptible sheep for immunogenicity and lack of pathogenicity. The three most promising candidates were then tested more extensively and eventually the best performing of these was selected as the potential vaccine candidate and subjected to testing in large numbers of sheep under laboratory and field conditions.

This approach proved very successful and it was decided to apply it to all the BT vaccine strains. The lowest egg passage levels that were preserved (usually 40 or higher) were used (Erasmus and Weiss, Unpublished data).

With the addition of Type 19, it meant that the polyvalent vaccine should be composed of 15 different serotypes. Research work done by Erasmus (1980) had indicated that irrespective of the number of live attenuated serotypes contained in a BTV vaccine, sheep will, on average, respond to three strains only (not necessarily the same three strains in every instance). The most practical option was, therefore, to issue three pentavalent vaccines that had to be administered at intervals of at least 3 weeks. To encourage correct usage, the three pentavalent vaccines are packed and sold as a unit and the order in which they should be administered is clearly indicated (bottle A contains serotypes 1, 4, 6, 12 and 14; bottle B contains 3, 8, 9, 10 and 11 and C contains 2, 5, 7, 13 and 19). When used correctly, this vaccine resulted in demonstrable antibody levels against nine serotypes on average (after a single series of vaccinations). Annual boosters broaden the scope of immunity to the extent that properly vaccinated sheep were extremely well protected (Erasmus and Weiss, Unpublished data).

A monovalent egg-attenuated Type 10 BTV vaccine was used in the United States on a relatively small scale until 1970 (Walton, 1992). A BTV-10 vaccine attenuated by serial passage in foetal bovine kidney cell cultures has been licensed for use in sheep but is apparently not in great demand (Walton, 1992). A quadrivalent-modified live BTV vaccine has been used experimentally in several wildlife species with promising results (McConnell et al., 1985).

Until recently little work has been conducted on the development of inactivated BTV vaccines. Parker *et al.* (1985) obtained good results in sheep with a bivalent BTV vaccine inactivated with beta-propiolactone. Campbell *et al.* (1985) demonstrated the production of protective immunity in sheep with a BTV vaccine inactivated by gamma irradiation.

Once it was demonstrated that the outer capsid protein VP2 of BTV is the major serotype-specific protein and that it can induce protective immunity (Huismans and Erasmus, 1981; Huismans *et al.*, 1987), the potential production of recombinant vaccines became a reality. VP2 and VP5 proteins derived from single baculovirus expression vectors and BTV-like particles derived from multiple baculovirus expression vectors have been administered to sheep. Neutralising antibody responses and clinical protection were demonstrated (results summarised by Roy and Erasmus, 1992).

The continuous development of very powerful techniques (many of which are described elsewhere in this book) will certainly contribute to novel approaches regarding diagnostic aids and vaccine development in the near future, which should greatly facilitate the control of BT.

References

Alexander, R.A. (1947). The propagation of bluetongue virus in the developing chick embryo with particular reference to the temperature of incubation. *Onderstepoort J. Vet. Sci. Anim. Ind.* **22**, 7–26.

Alexander, R.A. (1959) Blue tongue as an international problem. *Bull. Off. Int. Epizoot.* **51**, 432–9.

Alexander, R.A. and Haig, D.A. (1951) The use of egg attenuated bluetongue virus in the production of a polyvalent vaccine for sheep. A. – Propagation of the virus in sheep. *Onderstepoort J. Vet. Res.* **25**, 3–15.

Alexander, R.A., Haig, D.A. and Adelaar, T.F. (1947) The attenuation of bluetongue virus by serial passage through fertile eggs. *Onderstepoort J. Vet. Sci. Anim. Ind.* **21**, 231–41.

Anderson, G.A., Stott, J.L., Gershwin, L.J. and Osburn, B.I. (1975) Subclinical and clinical bluetongue disease in cattle: Clinical, pathological and pathogenic considerations. In: T.L. Barber and M.M. Jochim (eds), *Bluetongue and Related Orbiviruses*. New York: Alan R. Liss, Inc., 103–107.

Barber, T.L. (1979) Temporal appearance, geographic distribution, and species of origin of bluetongue virus serotypes in the United States. *Am. J. Vet. Res.* **40**, 1654–6.

Barzilai, E. and Tadmor, A. (1971) Multiplication of bluetongue virus in goats following experimental infection. *Refu. Vet.* **28**, 11–20.

Bekker, J.G., de Kock, G.v.d.W. and Quinlan, J.B. (1934) The occurrence and identification of bluetongue in cattle – the so-called pseudo-foot and mouth disease in South Africa. *Onderstepoort J. Vet. Sc. Anim. Ind.* **2**, 393–507.

Bellamy, A.R., Shapiro, L., August, J.T. and Joklik, W.K. (1967) Studies on reovirus I. RNA characterization of reovirus genome RNA. *J. Mol. Biol.* **29**, 1–17.

Boyce, M. and Roy, P. (2007) Recovery of infectious bluetongue virus from RNA. *J. Virol.* **8**, 2179–86.

Callis, J. (1985) Bluetongue in the United States. In: T.L. Barber and M.M. Jochim (eds), *Bluetongue and Related Orbiviruses*. New York: Alan R. Liss, Inc., pp. 37–42.

Campbell, C.H., Barber, T.L., Knudsen, R.C. and Swaney, L.M. (1985) Immune responses of mice and sheep to bluetongue virus inactivated by gamma irradiation. In: T.L. Barber and M.M. Jochim (eds), *Bluetongue and Related Orbiviruses*. New York: Alan R. Liss, Inc., pp. 639–47.

Dangler, C.A., de Mattos, C.A., de Mattos, C.C. and Osburn, B.I. (1990) Identifying bluetongue virus ribonucleic acid sequences by the polymerase chain reaction. *J. Virol. Methods*. **28**, 283–92.

De Kock, G., Du Toit, R. and Neitz, W.O. (1937) Observations on blue tongue in cattle and sheep. *Onderstepoort J. Vet. Sci. Anim. Ind.* **8**, 129–80.

Doyle, K.A. (1992) An overview and perspective on orbivirus disease prevalence and occurrence of vectors in Australia and Oceania. In: T.L. Barber and M.M. Jochim (eds), *Bluetongue, African Horse Sickness, and Related Orbiviruses*. Boca Raton: CRC Press, Inc., pp. 44–57.

Du Toit, R.M. (1944) The transmission of bluetongue and horse sickness by *Culicoides*. *Onderstepoort J. Vet. Sc. Anim. Ind.* **7**, 16–390.

Du Toit, R. (1962) The role played by bovines in the transmission of bluetongue in sheep. *J. S. Afr. Vet. Med. Assoc.* **33**, 483–90.

Els, H.J. and Verwoerd, D.W. (1969) Morphology of the bluetongue virus. *Virology* **38**, 213–19.

Emau, P., Giri, S.N., Anderson, G.A., Stott, J.L. and Osburn, B.I. (1984) Function of prostaglandins, thromboxane A2 and histamine in hypersensitivity reaction to experimental bluetongue disease in calves. *Am. J. Vet. Res.* **45**, 1852–7.

Erasmus, B.J. (1975) Bluetongue in sheep and goats. *Aust. Vet. J.* **51**, 165–70.

Erasmus, B.J. (1980) The epidemiology and control of bluetongue in South Africa. *Bull. Off. Int. Epizoot.* **92**, 461–7.

Erasmus, B.J. (1990) Bluetongue virus. In: Z. Dinter and B. Morein (eds), *Virus Infections of Ruminants*, Vol. 3. Amsterdam: Elsevier, pp. 227–37.

Fernandes, M.V. (1959a) Isolation and propagation of bluetongue virus in tissue culture. *Am. J. Vet. Res.* **20**, 408.

Fernandes, M.V. (1959b) Cytopathogenic effects of bluetongue virus on lamb tissues. *in vitro. Tex. Rep. Biol. Med.* **17**, 94–105.

Foster, N.M., Jones, R.H. and McCory, B.R. (1963) Preliminary investigations on insect transmission of bluetongue virus in sheep. *Am. J. Vet. Res.* **24**, 1195–200.

Gambles, R.M. (1949) Bluetongue of sheep on Cyprus. *J. Comp. Path.* **59**, 176–90.

Gard, G.P., Shorthose, J.E., Weir, R.P. and Erasmus, B.J. (1987) The isolation of a bluetongue serotype new to Australia. *Aust. Vet. J.* **64**, 87.

Gibbs, E.P.J., Greiner, E.C., Taylor, W.P., Barber, T.L., House, J.A. and Pearson, J.E. (1983) Isolation of bluetongue virus serotype 2 from cattle in Florida: Serotype of bluetongue virus hitherto unrecognized in the western hemisphere. *Am. J. Vet. Res.* **44**, 2226–8.

Goldsmit, L. and Barzilai, E. (1965) Isolation and propagation of a bluetongue virus strain in embryonating chicken eggs by the intravenous route of inoculation – preliminary report. *Refu. Vet.* **22**, 285–379.

Gorman, B.M., Taylor, J., Walker, P.J., Davidson, W.L. and Brown, F. (1981) Comparisons of bluetongue type 20 with certain viruses of the bluetongue and Eubenangee serological groups of orbiviruses. *J. Gen. Virol.* **57**, 251–61.

Haig, D.A., McKercher, D.G. and Alexander, R.A. (1956) The cytopathogenic action of bluetongue virus on tissue cultures and its application to the detection of antibodies in the serum of sheep. *Onderstepoort J. Vet. Res.* **27**, 171–7.

Hardy, W.T. and Price, D.A. (1952) Sore muzzle of sheep. *J. Am. Vet. Med. Assoc.* **120**, 23–5.

Hassan, A. (1992) Status of bluetongue in the Middle East and Asia. In: T.E. Walton and B.I. Osburn (eds), *Bluetongue, African Horse Sickness, and Related Orbiviruses*. Boca Raton: CRC Press, Inc., pp. 38–43.

Hoff, G.L. and Hoff, D.M. (1976) Bluetongue and epizootic hemorrhagic disease: A review of these diseases in non-domestic artiodactyles. *J. Zoo Anim. Med.* **7**, 26–30.

Hourrigan, J.L. and Klingsporn, A.L. (1975) Bluetongue: The disease in cattle. *Aust. Vet. J.* **51**, 170–4.

Howell, P.G. (1960) A preliminary antigenic classification of strains of bluetongue virus. *Onderstepoort J. Vet. Res.* **28**, 357–63.

Howell, P.G. (1963) *Bluetongue. Emerging Diseases of Animals*. Rome: FAO, pp. 111–53.

Howell, P.G. (1970) The antigenic classification and distribution of naturally occurring strains of bluetongue virus. *J. S. Afr. Vet. Med. Assoc.* **41**, 215–23.

Howell, P.G. and Verwoerd, D.W. (1971) Bluetongue virus. In: S. Gard, C. Hallauer and K.F. Meyer (eds), *Virology Monographs*, Vol. 9. Springer-Verlag, Vienna, pp. 35–74.

Howell, P.G., Verwoerd, D.W. and Oellermann, R.A. (1967) Plaque formation by bluetongue virus. *Onderstepoort J. Vet. Res.* **34**, 317–32.

Huismans, H. (1979) Protein synthesis in bluetongue virus infected cells. *Virology* **92**, 385–96.

Huismans, H. and Cloete, M. (1987) A comparison of different cloned bluetongue virus segments as probes for the detection of virus-specified RNA. *Virology* **158**, 373–80.

Huismans, H. and Els, H.J. (1979) Characterization of the tubules associated with the replication of three different orbiviruses. *Virology* **92**, 397–406.

Huismans, H. and Erasmus, B.J. (1981) Identification of the serotype-specific and group-specific antigens of bluetongue virus. *Onderstepoort J. Vet. Res.* **48**, 51–8.

Huismans, H. and Howell, P.G. (1973) Molecular hybridization studies on the relationships between different serotypes of bluetongue virus and on the difference between virulent and attenuated strains of the same serotype. *Onderstepoort J. Vet. Res.* **40**, 93–104.

Huismans, H., Van der Walt, N.T., Cloete, M. and Erasmus, B.J. (1987) Isolation of a capsid protein of bluetongue virus that induces a protective immune response in sheep. *Virology* **157**, 172–9.

Hutcheon, D. (1902) Malarial catarrhal fever of sheep. *Vet. Rec.* **14**, 629–33.

Jessup, D.A. (1985) Epidemiology of two orbiviruses in California's native wild ruminants; Preliminary report. In: T.L. Barber and M.M. Jochim (eds), *Bluetongue and Related Orbiviruses*. New York: Alan R. Liss, Inc., pp. 53–65.

Komarov, A. and Goldsmit, L. (1951) A disease similar to bluetongue in cattle and sheep in Israel. *Refu. Vet.* **8**, 96–100.

Lecatsas, G. (1968) Electron microscopic study of the formation of bluetongue virus. *Onderstepoort J. Vet. Res.* **35**, 139–49.

Lopez, A.C. and Botija, C.S. (1958) Epizootie de fiévre catarrhale ovine en Espagne (bluetongue). *Bull. Off. Int. Epizoot.* **50**, 65–93.

Luedke, A.J. and Anakwenze, E.I. (1972) Bluetongue virus in goats. *Am. J. Vet. Res.* **33**, 1739–45.

Mason, J.H., Coles, J.D.W.A. and Alexander, R.A. (1940) Cultivation of bluetongue virus in fertile eggs produced on a vitamin deficient diet. *Nature* **145**, 1022–3.

Mason, J.H. and Neitz, W.O. (1940) The susceptibility of cattle to the virus of bluetongue. *Onderstepoort J. Vet. Sci.* **15**, 149–57.

McConnell, S., Morrill, J.C. and Livingston, C.W. (1985) Use of a quadrivalent modified-live bluetongue virus vaccine in wildlife species. In: T.L. Barber and M.M. Jochim (eds), *Bluetongue and Related Orbiviruses*. New York: Alan R. Liss, Inc., pp. 631–8.

McKercher, D.G., McGowan, B., Howarth, J.A. and Saito, J.K. (1953) A preliminary report on the isolation and identification of the bluetongue virus from sheep in California. *J. Am. Vet. Med. Assoc.* **122**, 300–301.

Murphy, F.A., Borden, E.C., Shope, R.E. and Harrison, A. (1971) Physicochemical and morphological relationships of some arthropod-borne viruses to bluetongue virus – a new taxonomic group. Electron Microscope Studies. *J. Gen. Virol.* **13**, 273–88.

Neitz, W.O. (1933) The Blesbuck (*Damaliscus albifrons*) as a carrier of heartwater and blue tongue. *J. S. Afr. Vet. Med. Assoc.* **4**, 24–7.

Neitz, W.O. (1948) Immunological studies on bluetongue in sheep. *Onderstepoort J. Vet. Sci. Anim. Ind.* **23**, 93–136.

Neitz, W.O. and Riemerschmid, G. (1944) The influence of sunlight on the course of bluetongue. *Onderstepoort J. Vet. Sci. Anim. Ind.* **19**, 69–70.

Nevill, E.M., Erasmus, B.J. and Venter, G.J. (1992) A six-year survey of viruses associated with *Culicoides* biting midges throughout South Africa (Diptera: Ceratopogonidae). In: T.E. Walton and B.I. Osburn (eds), *Bluetongue, African Horse Sickness and Related Orbiviruses. Proceedings of the 2nd International Symposium*. Boca Raton: FL: CRC Press, Inc., pp. 314–19.

Nieschultz, O., Bedford, G.A.H. and du Toit, R.M. (1934) Investigations into the transmission of bluetongue in sheep during the season 1931–1932. *Onderstepoort J. Vet. Sci.* **2**, 509–62.

Owen, N.C., du Toit, R.M. and Howell, P.G. (1965) Bluetongue in cattle: Typing of viruses isolated from cattle exposed to natural infections. *Onderstepoort J. Vet. Res.* **32**, 3–6.

Owen, N.C. and Munz, E.K. (1966) Observations on a strain of bluetongue virus by electron microscopy. *Onderstepoort J. Vet. Res.* **33**, 9–14.

Parker, J., Herniman, K.A.J., Gibbs, E.P.J. and Sellers, R.F. (1985) An experimental inactivated vaccine against bluetongue. *Vet. Rec.* **96**, 284–7.

Parsonson, I.M. and Snowdon, W.A. (1985) Bluetongue, epizootic haemorrhagic disease of deer and related viruses: Current situation in Australia. In: T.L. Barber and M.M. Jochim (eds), *Bluetongue and Related Orbiviruses*. New York: Alan R. Liss, Inc., pp. 27–35.

Paweska, J.T., Venter, G.J. and Mellor, P.S. (2002) Vector competence of South African *Culicoides* species for bluetongue virus serotype 1 (BTV-1) with special reference to the effect of temperature on the rate of virus replication in *C. imicola and C. bolitinos*. *Med. Vet. Entomol.* **16**, 10–21.

Polydorou, K. (1985) Bluetongue in Cyprus. In: T.L. Barber and M.M. Jochim (eds), *Bluetongue and Related Orbiviruses*. New York: Alan R. Liss, Inc., pp. 539–44.

Ribeiro, J.M., Rosa-Azevedo, J.A., Noronha, F.M.O., Bracoforte, M.C., Grave-Pereira, C. and Vasco-Fernandes, M. (1957) Bluetongue in Portugal. *Bull. Off. Int. Epizoot.* **48**, 350–67.

Roy, P. (1989) Bluetongue virus genetics and genome structure. *Virus Res.* **13**, 179–206.

Roy, P. and Erasmus, B.J. (1992) Second generation candidate vaccine for bluetongue disease. In: T.L. Barber and M.M. Jochim (eds), *Bluetongue and Related Orbiviruses*. New York: Alan R. Liss, Inc., 856–67.

Sapre, S.N. (1964) An outbreak of bluetongue in goats and sheep in Maharashtra State, India. *Vet. Rec.* **15**, 69–71.

Sarwar, M.M. (1962) A note on bluetongue in sheep in West Pakistan. *Pak. J. Anim. Sci.* **1**, 1–2.

Spreull, J. (1902) Report from veterinary surgeon Spreull on the result of his experiments with the malarial catarrhal fever of sheep. *Agric. J. Cape of Good Hope.* **20**, 469–77.

Spreull, J. (1905) Malarial catarrhal fever (bluetongue) of sheep in South Africa. *J. Comp. Pathol.* **18**, 321–37.

St. George, T.D., Cybinski, D.H., Della-Porta, A.J., McPhee, D.A., Wark, M.C. and Bainbridge, M.A. (1980) The isolation of two bluetongue viruses from healthy cattle in Australia. *Aust. Vet. J.* **56**, 562–3.

St. George, T.D., Standfast, H.A., Cybinski, D.H., Dyce, A.L., Muller, M.J., Doherty, R.L., Carley, J.G., Filippich, C. and Frazier, C.L. (1978) The isolation of a bluetongue virus from *Culicoides* in the NT of Australia. *Aust. Vet. J.* **54**, 153–4.

Studdert, M.J., Pangborn, J. and Addison, R.B. (1966) Bluetongue virus structure. *Virology* **29**, 509–11.

Theiler, A. (1906) Bluetongue in sheep. *Ann. Rep. Dir. Agric. Tvl.* **1904–05**, 110–21.

Theiler, A. (1908) The inoculation of sheep against bluetongue and the results in practice. *Vet. J.* **64**, 600–607.

Van Dijk, A.A. and Huismans, H. (1980) The *in vitro* activation and further characterization of the bluetongue virus associated transcriptase. *Virology* **104**, 347–56.

Van Dijk, A.A. and Huismans, H. (1988) *In vitro* transcription and translation of bluetongue virus mRNA. *J. Gen. Virol.* **69**, 573–81.

Venter, G.J., Paweska, J.T., Van Dijk, A.A., Mellor, P.S. and Tabachnick, W.J. (1998) Vector competence of *Culicoides bolitinos* and *C. imicola* (Diptera: Ceratopogonidae) for South African bluetongue virus serotypes 1, 3 and 4. *Med. Vet. Entomol.* **12**, 101–108.

Verwoerd, D.W. (1969) Purification and characterisation of bluetongue virus. *Virology* **38**, 203–12.

Verwoerd, D.W. (1970) Diplornaviruses: A newly recognized group of double-stranded RNA viruses. *Progr. Med. Virol.* **12**, 192–210.

Verwoerd, D.W. and Erasmus, B.J. (1994) Bluetongue. In: J.A.W. Coetzer, G.R. Thomson, and R.C. Tustin (eds), *Infectious Diseases of Livestock* Cape Town: Oxford University Press, pp. 443–59.

Verwoerd, D.W., Louw, H. and Oellermann, R.A. (1970) Characterization of bluetongue virus ribonucleic acid. *J. Virol.* **5**, 1–7.

Verwoerd, D.W., Els, H.J., De Villiers, E.-M. and Huismans, H. (1972) Structure of the bluetongue virus capsid. *J. Virol.* **10**, 783–94.

Wade-Evans, A.M., Mertens, P.P. and Bostock, C.J. (1990) Development of the polymerase chain reaction for the detection of bluetongue virus in tissue samples. *J. Virol. Methods.* **30**, 15–24.

Walton, T.E. (1992) Attenuated and inactivated orbiviral vaccines. In: T.L. Barber and M.M. Jochim (eds), *Bluetongue and Related Orbiviruses*. New York: Alan R. Liss, Inc., pp. 851–5.

Bluetongue virus, other orbiviruses and other reoviruses: Their relationships and taxonomy

3

HOUSSAM ATTOUI, SUSHILA MAAN, SIMON J. ANTHONY, AND PETER P.C. MERTENS

Institute for Animal Health, Pirbright, Woking, Surrey, GU24 0NF, UK

Introduction

The icosahedral viruses that contain genomes composed of multiple (9, 10, 11 or 12) linear segments of dsRNA are all classified within the family *Reoviridae* and comprises a total of 15 virus genera, found in a wide variety of ecological niches (Mertens, 2004; Mertens *et al.*, 2005a). The distinct genera of reoviruses (a term used here to include any member of the family) have now been recognized by the International Committee for the Taxonomy of viruses (ICTV) (Mertens *et al.*, 2005a; Table 3.1), along with several other unclassified viruses of insects, arachnids and crustacea. Individual reoviruses can infect a wide variety of terrestrial and non-terrestrial vertebrates, terrestrial and non-terrestrial invertebrates, plants and fungi. The genus *Orbivirus* is the largest of the genera of reoviruses and contains 22 virus species as well as 10 unclassified 'orbiviruses', some of which may represent still further species (Table 3.2; Mertens *et al.*, 2005b). *Bluetongue virus* (BTV), which has been studied in the greatest depth and is the 'type' species of the genus *Orbivirus*,

ISBN-13: 978-0-12-369368-6

Table 3.1 The genera of the family *Reoviridae*

Genus	Structure	No. of genome segments (genome size)	Number of species (+ unassigned or proposed species)	Hosts (vectors)
Aquareovirus	Non-turreted	11 (~23.7 kbp)	6 (+5)	Molluscs, finfish, Crustacea
Cardoreovirus	Non-turreted	12 (~20.15 kbp)	1 (+2)	Crabs
Coltivirus	Non-turreted	12 (~29 kbp)	2 (+1)	Mammals (including humans) [ticks]
Cypovirus	Turreted	10 (~24.8 to 33.3 kbp)[a]	21 (+2)	Insects (Lepidoptera Diptera and Hymenoptera), a single isolate was reported from a freshwater daphnid
Fijivirus	Turreted	10 (~28.7 kbp)	5	Plants (Graminae, Liliaceae) (delphacid planthoppers)
Dinovernavirus	Turreted	9 (~23.35 kbp)	1	Mosquitoes
Idnoreovirus	Turreted	10 +1[b] (~25.1 kbp)	5 (+1)	Insects (Hymenoptera)
Mimoreovirus	Non-turreted	11 (~25.56 kbp)	1	Phytoplankton
Mycoreovirus	Turreted	11 or 12 (~24.4 kbp)	3	Fungi
Orbivirus	Non-turreted	10 (~19.2 kbp)	22 (+10)	Mammals (including humans), birds [*Culicoides*, mosquitoes, phlebotomines, ticks]
Orthoreovirus	Turreted	10 (~23.5 kbp)	5	Birds, reptiles, mammals
Oryzavirus	Turreted	10 (~26.1 kbp)	2	Plants (Graminae)) [delphacid planthoppers]
Phytoreovirus	Non-turreted	12 (~25.1 kbp)	3 (+1)	Plants (cicidellid leafhoppers)
Rotavirus	Non-turreted	11 (~18.5 kbp)	5 (+2)	Birds, mammals
Seadornavirus	Non-turreted	12 (~21 kbp)	3	Mammals (including humans) [mosquitoes]
Unassigned reoviruses (potentially representing new genera)		10, 11 or 12	5	

Source: Data primarily from Mertens *et al.* (2005a).

[a] Individual idnoreovirus particles may contain a 11th genome segment depending on the sex and ploidy of the host wasp from which they are derived.

[b] In some cases, the size of the genome size for different *Cypovirus* species was estimated only by electrophoretic analyses.

Table 3.2 Species in the genus *Orbivirus*

Virus species	Number of serotypes/strains	Vector species	Host species
African horse sickness virus (AHSV)	9 numbered serotypes (AHSV-1 to AHSV-9)	*Culicoides* spp.	Equids, dogs, elephants, camels, cattle, sheep, goats, humans (in special circumstances), predatory carnivores (by eating infected meat)
Bluetongue virus (BTV, Orbivirus type-species)	24 numbered serotypes (BTV-1 to BTV-24)	*Culicoides* spp.	All ruminants, camelids and predatory carnivores (by eating infected meat)
Changuinola virus (CGLV)	12 named serotypes	Phlebotomine sand flies, culicine mosquitoes	Humans, rodents, sloths
Chenuda virus (CNUV)	7 named serotypes	Ticks	Seabirds
Chobar Gorge virus (CGV)	2 named serotypes	Ticks	Bats
Corriparta virus (CORV)	6 named serotypes/strains[a]	Culicine mosquitoes	Humans, rodents
Epizootic haemorrhagic disease virus (EHDV)	7 numbered or named serotypes/strains[a] (EHDV-1 and EHDV-3 to EHDV-8 (EHDV-1 and 3 can be grouped together as one 'type'))	*Culicoides* spp.	Cattle, sheep, deer, camels, llamas, wild ruminants, marsupials
Equine encephalosis virus (EEV)	7 numbered serotypes (EEV-1 to EEV-7)	*Culicoides* spp.	Equids
Eubenangee virus (EUBV)	4 named serotypes	*Culicoides* spp., anopheline and culicine mosquitoes	Unknown hosts

(*Continued*)

Table 3.2 (*Continued*)

Virus species	Number of serotypes/strains	Vector species	Host species
Ieri virus (IERIV)	3 named serotypes	Mosquitoes	Birds
Great Island virus (GIV)	36 named serotypes/strains[a]	*Argas, Ornithodoros, Ixodes* ticks	Seabirds, rodents and humans
Lebombo virus (LEBV)	1 numbered serotype (LEBV-1)	Culicine mosquitoes	Humans, rodents
Orungo virus (ORV)	4 numbered serotypes (ORUV-1 to ORUV-4)	Culicine mosquitoes	Humans, camels, cattle, goats, sheep, monkeys
Palyam virus (PALV)	13 named serotypes/strains[a]	*Culicoides* spp., culicine mosquitoes	Cattle, sheep
Peruvian horse sickness virus (PHSV)	1 numbered serotype (PHSV-1)	Mosquitoes	Horses
St. Croix River virus (SCRV)	1 numbered serotype: (SCRV-1)	Ticks	Hosts unknown
Umatilla virus (UMAV)	4 named serotypes	Culicine mosquitoes	Birds
Wad Medani virus (WMV)	2 named serotypes	*Boophilus, Rhipicephalus, Hyalomma, Argas* ticks	Domesticated animals
Wallal virus (WALV)	3 serotypes/strains	*Culicoides* spp.	Marsupials
Warrego virus (WARV)	3 serotypes/strains	*Culicoides* spp., anopheline and culicine mosquitoes	Marsupials
Wongorr virus (WGRV)	8 serotypes/strains	*Culicoides* spp., mosquitoes	Cattle, macropods

Yunnan orbivirus (YUOV)	*Culex tritaeniorhyncus, Ochlerotatus scapularis*	Cattle, equids
2 serotypes		
Tentative species		
Andasibe virus (ANDV)	Mosquitoes	Unknown hosts
Codajas virus (COV)	Mosquitoes	Rodents
Ife virus (IFEV)	Mosquitoes	Rodents, birds, ruminants
Itupiranga virus (ITUV)	Mosquitoes	Unknown hosts
Japanaut virus (JAPV)	Mosquitoes	Unknown hosts
Kammavanpettai virus (KMPV)	Unknown vectors	Birds
Lake Clarendon virus (LCV)	Ticks	Birds
Matucare virus (MATV)	Ticks	Unknown hosts
Tembe virus (TMEV)	Mosquitoes	Unknown hosts
Tracambe virus (TRV)	Mosquitoes	Unknown hosts

[a]In some species, the serological relationships between strains have not been fully determined. For more information concerning individual named types, see Mertens et al. (2005b).

includes 24 distinct serotypes, as well as many different topotypes (groups reflecting the geographical origins of each virus isolate) and nucleotypes (lineages), although these have no 'formal' taxonomic status within the virus species itself (Mertens *et al.*, 2005b; Maan *et al.*, 2007a).

The genera of the family *Reoviridae*

The linear dsRNA molecules, which make up the reovirus genome (see Figure 10.5) are packaged as exactly one copy of each segment within each icosahedral virus particle, ~60–90 nm in diameter (see Chapter 6). Several of the genera of reoviruses include arboviruses (viruses that are transmitted by arthropods, in which they can also replicate). These include the plant reoviruses, classified within the genera *Fijivirus*, *Oryzavirus* and *Phytoreovirus*, and the vertebrate reoviruses (including certain viruses of humans) that are classified within the genera *Coltivirus*, *Orbivirus* and *Seadornavirus* (see Table 3.1).

Clear overall similarities exist in the structure and replication strategies of the different reoviruses, which are reflected in the phylogenetic relationships of their genome segments. These relationships suggest that all of the reoviruses have evolved from a single common ancestor, in parallel with their different hosts (possibly originally from a marine species), a process known as co-speciation (Attoui *et al.*, 2002). Subsequent gene deletions or duplications, followed by gene rearrangements, have generated changes in both the number of genome segments and the proteins they encode (Mohd Jaafar *et al.*, 2005; Mohd Jaafar *et al.*, 2008). For example, although the rotaviruses and seadornaviruses are two of the more closely related genera of reoviruses, they have different numbers of genome segments (11 and 12, respectively). The aquareoviruses and coltiviruses are also 'related' and again have 11 or 12 genome segments, respectively (Mohd Jaafar *et al.*, 2008).

Reovirus particles are composed of up to three concentric shells of virus structural proteins (depending on the genus to which they belong) surrounding the segments of the virus genome (see Chapter 6). Although there are similarities in the innermost sub-core shell structure and protein (the T2 protein) (Chapter 6), the structure of the outer capsid and outer core layers varies significantly between members of different genera within the family (see Section "Relationships with other reoviruses").

The orbiviruses (including BTV) have a three-layered icosahedral protein capsid (see Figure 6.1 and colour plates 1, 2 and 3). However, they can also acquire an outer membrane layer by budding through the cell membrane during cell exit, forming 'membrane-enveloped virus particles' (MEVP) (Chapter 6). This membrane appears to be relatively unstable and is not required for infectivity (Mertens *et al.*, 1987a; 1996).

The genus *Orbivirus*

The orbivirus genome is composed of 10 segments of linear dsRNA, which code for the seven structural proteins (VP1–VP7) that make up the virus particle and three distinct non-structural proteins (NS1–NS3) (see Figure 5.10 and colour plate 4). The genus *Orbivirus* is the largest of the genera within the family *Reoviridae,* containing 22 distinct virus species and 10 further 'unassigned' viruses that have not yet been fully characterized in terms of either their serological or phylogenetic relationships (see Table 3.1). The orbiviruses are primarily transmitted between their vertebrate hosts by arthropod vectors – which (depending on virus species) include ticks and/or haematophagous insects [such as mosquitoes and biting midges (*Culicoides* spp.)]. As the most economically important and 'type' species, BTV has been studied more extensively than the other orbiviruses, providing a useful paradigm for their structural properties and replication. The transmission, maintenance, pathogenesis, pathology and structural biology of BTV are described in other chapters in this book.

Each of the different *Orbivirus* species contains a number of distinct virus serotypes, the identity of which is determined by the specificity of reactions between the neutralizing antibodies that are generated during infection of the vertebrate host and the protein components of the outer capsid (Gould and Eaton, 1990, Mertens, 1999). There are 24 serotypes of BTV, determined primarily by the larger of the two outer capsid proteins, VP2 [encoded by genome segment 2 (Seg-2)] (see Figure 5.10) (Maan *et al.*, 2007a). However, the smaller outer capsid component (VP5, encoded by Seg-6) can also have some influence on BTV serotype, possibly by affecting the structure or conformation of VP2 (Huismans and Erasmus, 1981; Cowley and Gorman, 1989; Mertens *et al.*, 1989; DeMaula *et al.*, 1993; Singh *et al.*, 2005). In some other orbiviruses [e.g. Great Island virus GIV)], serotype can be determined by the smaller outer capsid proteins (Moss *et al.*, 1987).

The different serotypes of BTV and other orbiviruses largely fail to cross-neutralize or cross-protect, and these are therefore both biologically and epidemiological significant. Their identification is also essential in the design and implementation of appropriate disease control measures to ensure the use of the correct vaccine serotype (Chapters 18 and 19). Although there are no reports of genome segment exchange/reassortment between different species of the family *Reoviridae* (Mertens, 2004), the different strains (including the different serotypes and different topotypes) within the same *Orbivirus* species can reassort, potentially generating enormous numbers of different progeny virus strains (Nuttall *et. al.*, 1990, 1992; O'Hara *et. al.*, 1998). As a consequence, the different types and strains in each species are not individually recognized within the taxonomic classification of these viruses (Mertens *et al.*, 2005a,b; see below).

Classification and differentiation of the *Orbivirus* species

Closely related reoviruses that infect the same cell can exchange genome segments by a process known as reassortment, generating new progeny virus strains. This ability to reassort is regarded as a primary indication that individual virus strains belong to the same virus species, within each of the genera of the *Reoviridae*. However, in the absence of data concerning the ability of specific reovirus strains to reassort (particularly for new virus isolates), other parameters can be used (either singly or in combination) to identify the members of the same virus species. This provides the polythetic definition of virus species recommended by ICTV. The parameters used to identify members of different *Orbivirus* species (reviewed by Mertens *et al.*, 2005b) include the following:

1. High-level serological cross-reaction [detected by ELISA or other serological assays, such as complement fixation (CF), or agar gel immunodiffusion (AGID)] using either polyclonal sera or monoclonal antibodies against conserved antigens such as VP7(T13) (Chapter 17). For example, in the competition ELISA at a test serum dilution of 1/5, a positive serum will show >50% inhibition of colour formation, while a negative control serum or serum that is specific for a different species will normally produce <25% inhibition of colour compared to a no antibody control. Related species may show a lower-level serological cross-reaction, which may be only one way.

2. High levels of RNA sequence similarity in conserved genome segments. For example, viruses belonging to distinct *Orbivirus* species (serogroups) will normally show 21.4–72% amino acid (aa) sequence identity in genome segment 3 (Seg-3) encoding the major sub-core structural protein VP3(T2).

3. Relatively efficient cross-hybridization of conserved genome segments (those not encoding outer capsid components or other variable proteins) under high stringency conditions (>70% homology). For example, as detected by Northern or dot blots with probes made from viral RNA or cDNA (Mertens *et al.*, 1987b).

4. Real-time or conventional reverse transcription polymerase chain reaction (RT-PCR) assays using primers (and probes) targeting conserved genome segments or regions. For example, assays targeting segments such as Seg-1 (Toussaint *et al.*, 2007; Shaw *et al.*, 2007), Seg-3 (McColl and Gould, 1991), Seg-5 (Aradaib *et al.*, 2003; Toussaint *et al.*, 2006) and Seg-7 (Anthony *et al.*, 2007). The assays can be coupled with cross-hybridization analysis (Northern or dot blots).

5. Identification of the virus serotype, as a virus type already classified within a specific *Orbivirus* species. None of the serotypes from different species has been shown to cross-neutralize. For example, virus serotypes

can be identified by conventional serum neutralization tests (SNT), virus neutralization tests (VNT) or by serotype-specific RT-PCR assays and sequence analyses of genome segment 2 (seg-2) (Maan *et al.*, 2007b; Mertens *et al.*, 2007 – Chapters 7 and 17).

6. Analysis of the viral RNA electropherotype by agarose gel electrophoresis (AGE) but not by PAGE. Viruses within a single *Orbivirus* species (e.g. 24 reference strains of BTV) show a relatively uniform AGE electropherotype (Maan *et al.*, 2007a). However, a major deletion or insertion event can result in two distinct electropherotypes within a single species; as observed with epizootic haemorrhagic disease virus (EHDV) (Eaton and Gould, 1987; Anthony, 2007), similarities can exist between the more closely related species, for example, BTV and EHDV.

7. Conserved upstream and downstream terminal regions (hexanucleotides) on the different genome segments (www.reoviridae.org/dsrna_virus_proteins//CPV-RNA-Termin.htm). However, the terminal sequences of some *Orbivirus* are not conserved on all of their genome segments. The positive strand of the BTV genome segments usually has the following conserved termini: 5'-GUUAAA ... ACUUAC-3'. Closely related species can also have identical terminal sequences on at least some segments (Table 3.3).

8. Identification of common vector or host species and the clinical signs produced. For example, BTVs are transmitted only by certain *Culicoides* species and will infect cattle, sheep and other ruminants including camelids, producing clinical signs of varying severity but are not thought to infect horses. The reverse is true of African horse sickness virus (AHSV).

Table 3.3 Conserved terminal sequences of orbivirus genome segments

Virus isolate	Conserved RNA termini (+ strand)
Bluetongue virus	5'-GUUAAA.........................UUAC-3'
African horse sickness virus	5'-GUU $^A/_U$ A$^A/_U$.....................AC$^A/_U$UAC-3'
Epizootic haemorrhagic disease virus	5'-GUUAAA.........................C$^C/_U$ $^C/_U$ AC-3'
Great Island virus	5'-GUAAAA.........................A$^A/_G$GAUAC-3'
Palyam virus	5'-GU $^A/_U$ AAA.........................$^A/_G$CUUAC-3'
Equine encephalosis virus[a]	5'-GUUAAG.........................UGUUAC-3'
St Croix River virus	5'-$^A/_G$UAAU$^G/_{A/U}$...........$^G/_{A/U}$ $^C/_U$ $^C/_A$ UAC-3'
Peruvian horse sickness virus	5'-GUUAAAA.........................$^A/_G$ $^C/_G$ $^A/_G$UAC-3'
Yunnan orbivirus	5'-GUUAAAA.........................$^A/_G$UAC-3'

Source: Based on: www.reoviridae.org/dsrna_virus_proteins//CPV-RNA-Termin.htm.
[a] Based on genome segment 10 (only) of the seven different serotypes.

The economically important orbiviruses

Some orbiviruses can cause severe and economically important diseases of domesticated and wild animals, including members of the virus species such as *AHSV, BTV, EHDV* and *equine encephalosis virus* (EEV). The introduction of these orbiviruses into areas that are usually free of the disease, and which therefore contain immunologically naive populations of susceptible host species, can cause high levels of morbidity and mortality, as seen during the outbreak caused by BTV serotype 8, which started in northern Europe during 2006 (Elbers *et al.*, 2007; Wilson *et al.*, 2007). Even in endemic areas, BTV can cause significant reductions in the overall productivity of domesticated animals (Tabachnick *et al.*, 1996; Yadin *et al.*, 2008; Chapters 8 and 9). Infectious BTV has been detected in bull semen, and early experiments (although apparently not reproducible) have indicated the possibility of long-term persistence of the virus, coupled with immunotolerance, in cattle that were naturally infected in utero (Bowen *et al.*, 1985; Howard *et al.*, 1985). As a consequence, there are restrictions on the international import and export of both animals and germ line materials (semen and ova) from infected areas, although the exact regulations, involving quarantine periods or even a complete import ban, differ from country to country (OIE, 2007). These barriers to trade in themselves represent a cause of major financial losses associated with outbreaks of bluetongue (BT) and African horse sickness (AHS).

The orbiviruses that infect wildlife species (e.g. deer, seabirds etc) may also have an impact on the sustainability of wild animal populations, particularly if they emerge into a new geographical area containing susceptible host species, and they can therefore pose a significant threat to the conservation of endangered species.

Bluetongue

Any serotype of BTV has the potential to cause BT, a disease of ruminants that can not only result in a high rate of mortality in sheep but can also cause significant economic losses in other ruminant species, including cattle and goats (McLaughlin *et al.*, 2003; Mertens *et al.*, 2006; Darpel *et al.*, 2007; Wilson *et al.*, 2007). The history of BT is reviewed in Chapters 2 and 8–11.

African horse sickness

African horse sickness was initially recognized as early as 1780, during the early days of European colonization and importation of horses into southern Africa, resulting in epizootics with high levels of mortality in infected animals. There are nine serotypes of AHSV, which are all considered to be endemic in sub-Saharan Africa. There have also been major epizootics caused by AHSV-9 in the Middle East, the Indian subcontinent, North Africa and the Iberian Peninsula (1959/1960) (Mellor and Boorman, 1995). An outbreak of AHS was also caused by the importation of a zebra infected with AHSV-4 in 1987 into Spain, which lasted for 4 years spreading to Portugal and Morocco (Sánchez-Vizcaíno, 2004). Recent changes in climate have been linked to the spread of BTV and outbreaks of disease in Europe (Purse *et al.*, 2005), suggesting that the risk of other arbovirus and particularly other orbiviruses that are transmitted by the same insect species (e.g. AHSV, and EHDV) has also increased. Indeed, outbreaks of AHS occurred during 2007, caused by AHSV-2 and AHSV-7 in Senegal and Mauritania and by AHSV-4 in Kenya, suggesting that the distribution of other orbiviruses is also changing (www.reoviridae.org/dsrna_virus_proteins//outbreaks.htm#AHS-Outbreaks).

Epizootic haemorrhagic disease

Within the genus Orbivirus, BTV is most closely related to EHDV, which was formally identified in 1955 following reports of a fatal epizootic in white-tailed deer (*Odocoileus virginianus*) in New Jersey (USA) (Shope *et al.*, 1955). However, older records suggest that EHDV has caused disease in wild ruminants throughout the south-eastern USA since 1890, where it was known among woodsmen and hunters as 'blacktongue', reflecting one of the more severe clinical signs of disease in deer (Shope *et al.*, 1960).

Initially SNTs and VNTs were used to compare different EHDV isolates. However, serotype identification has been hampered by the lack of well-characterized reference strains and antisera for the different types that could be used for the identification of new strains from around the world. Although eight different serotypes of EHDV are described in the literature, phylogenetic comparisons of the full genome from reference strains (particularly genome segment 2 – encoding outer capsid protein VP2) have identified only seven distinct types (Anthony *et al.*, 2007). These data indicate that the unclassified strain '318' belongs to type 6, with a close phylogenetic and serological relationship between types 6 and 8. They also showed that types 1 and 3 are the same, indicating that there are only seven EHDV types so far, types 1(+3), 2, 4, 5, 6, 7 and 8 (Anthony, 2007; Anthony *et al.*, 2008). These results were subsequently confirmed by serological analyses using VNT (Anthony, 2007; Anthony *et al.*, 2008).

The New Jersey and South Dakota strains of EHDV-1 were isolated in 1955 and 1956, respectively, and a second serotype was isolated in the Canadian province of Alberta in 1962 (Shope *et al.*, 1955; Fay *et al.*, 1956; Shope, *et al.*, 1960; Karstad *et al.*, 1961; Ditchfield *et al.*, 1964; Trainer, 1964; Nettles *et al.*, 1992; Stallknecht *et al.*, 1995; Stallknecht *et al.*, 1996). In recent years, there has been an increase in EHDV outbreaks recorded in the United States with >200 virus isolations during 2007. During 2006, EHDV-6 was also isolated for the first time in North America, from white-tailed deer in Illinois and

Indiana, and then again in Missouri during 2007 (Pittman, 2007; Stallknecht, 2007a; Stallknecht, 2007b; Stallknecht, 2008).

Epizootic haemorrhagic disease virus has been isolated from a range of animal species (including cattle), which may represent 'reservoir' hosts (Anthony *et al.*, 2007). Although cattle are usually considered to be asymptomatic for EHDV infection, there have been reports of significant levels of morbidity and mortality following both experimental and natural infection (Anthony *et al.*, 2007). An outbreak in Japan during 1959, caused by Ibaraki virus (a strain of EHDV-2), resulted in the deaths of >39,000 cattle (Omori *et al.*, 1969a,b). During 2006 and 2007, EHDV (type 2) was isolated for the first time in North America, also from clinically diseased cattle (Anthony *et al.*, 2007). Outbreaks of disease caused by EHDV-6 and -7 also occurred in cattle in North Africa and Israel, respectively (Yadin *et al.*, 2008). Serological evidence indicates the presence of EHDV-1 in cattle in the Caribbean (Jamaica, Antigua, Barbados, Grenada, Trinidad and Tobago) and South America (Guyana and Surinam) and of EHDV-2 in South America (Guyana) and Trinidad and Tobago. EHDV-1 and -2 have also been isolated in Australia and Japan (Ibaraki virus) (Omori *et al.*, 1969a,bb; Cambell, *et al.*, 1978; St. George *et al.*, 1983; Weir *et al.*, 1997) and EHDV-2 has been isolated in Africa (Mohammed and Mellor, 1990).

The reclassification of the Nigerian isolate of type 3 as EHDV-1 means that serotype 1 is also present in Africa (Lee, 1979; Anthony, 2007; Anthony *et al.*, 2008). EHDV-4 has been recorded only in Africa (Lee, 1979), with serotypes 5 and 8 only in Australia (Cambell *et al.*, 1978; St. George *et al.*, 1983). EHDV-6 has also been isolated in Australia (Cambell *et al.*, 1978; St. George *et al.*, 1983) and recently in the United States (Stallknecht, 2007a; Stallknecht, 2007b; Stallknecht, 2008). EHDV-6 (strain 318) has been isolated in Bahrain (1983), the Sudan, the Sultanate of Oman and Algeria (2006) and Morocco (2004, 2006) (Mohammed and Mellor, 1990; Mohammed *et al.*, 1996; Benazzou, 2004, 2006; Mertens *et al.*, 2006; Yadin *et al.*, 2008). EHDV-7 was initially detected only in Australia (St. George *et al.*, 1983) but has recently caused disease outbreaks in Israel during 2006 (Chaimovitz, 2006; Mertens *et al.*, 2006; Yadin *et al.*, 2008). Epizootic haemorrhagic disease virus has also been detected (either by the presence of antibody or by the isolation of the virus), but not yet fully characterized, in Cyprus and Greece (personal communication: Dr. K. Nomikou), India (personal communication: Dr. S. Maan), the Island of Reunion (Breard *et al.*, 2004), Taiwan and Indonesia.

Equine encephalosis virus

Equine encephalosis virus is associated with disease of horses in southern Africa. The virus was first identified in 1967 from horses that died from an unknown peracute illness. Serological investigations revealed that widespread EEV infections of horses had occurred during the summer of 1967 but that infection had not occurred in South Africa to any appreciable extent in the preceding 10 years. In some of the years following 1967, equine encephalosis (EE) took on epidemic proportions. Seasonal outbreaks of the disease have been coupled to the presence of the suspected insect vector *Culicoides* (Theodoridis *et al.*, 1979). Outbreaks of EE have also been associated with equine abortion during the first 5–6 months of gestation. An early undetected abortion could therefore be misdiagnosed as infertility in a mare.

Six different EEV serotypes have been identified in southern Africa of which Kyalami, Bryanston and Cascara are the best known. Complement fixation tests have failed to demonstrate relatedness between any of the EEV serotypes and other orbiviruses such as BTV, AHSV and EHDV.

Emerging orbiviruses

The orbiviruses can infect a wide range of host species including ruminants, camelids, equids, marsupials, bats, seabirds and in some cases humans. New orbiviruses have been isolated from both insect and tick vectors, as well as their mammalian hosts (Attoui *et al.*, 2001). Some of these viruses can cause clinical diseases that have not been previously reported. For example, in 1997, a new orbivirus was isolated from an outbreak of disease in horses from Peru (Attoui *et al.*, 2008). This virus is a member of a new species of *Orbivirus* that was named as '*Peruvian horse sickness virus*' (PHSV). Elsey station virus, isolated from sick horses in Australia, was subsequently shown to be a second isolate of PHSV (Attoui *et al.*, 2008).

In 2005, a new orbivirus was reported from *Culex tritaeniorhynchus* mosquitoes collected in Yunnan province of China. This virus was also shown to belong to a new species of *Orbivirus*, subsequently named as *Yunnan orbivirus* (YUOV) (Attoui *et al.*, 2005). Multiple isolates belonging to the same species were subsequently identified in lysates of wild-caught mosquitoes collected in 1999 from various provinces in China (Attoui *et al.*, 2005). At the time of isolation, the virus was not associated with a specific mammalian host but was shown to replicate in experimental animals, particularly mice. Very recently, YUOV was shown to be closely related to the Rioja virus (RIOV), which was isolated along with PHSV in Peru during 1997 (Attoui *et al.*, 2008). Rioja virus was originally isolated from sick cattle and donkeys and from *Ochlerotatus scapularis* mosquitoes in 1997. Rioja virus shows 98% aa identity in the sub-core shell protein VP2(T2) with YUOV, indicating that both viruses belong to the same species (Attoui *et al.*, 2008). In 2007, Middle Point orbivirus (MPOV) was isolated from cattle in Australia (Cowled *et al.*, 2007) and shown to be closely related to YUOV, with 96% aa identity in VP2(T2), indicating that it belongs to the same species.

Yunnan orbivirus and PHSV have been implicated in disease of equids and cattle and are transmitted by various mosquito species. Their widespread distribution (South America, China and Australia) indicates that these viruses have spread, although their origin is still unidentified. Amino acid identity values between the homologous proteins of PHSV and BTV ranged from 17% (in the serotype-determining protein) to 45% (in the polymerase). Identity values between the homologous proteins of the YUOV and BTV ranged from 15% (in the NS3 protein) to 36% (in the polymerase).

Orbiviruses that can infect humans

Several orbiviruses have been isolated from humans. Antibodies indicative of orbivirus infection have also been detected in human sera. Changuinola virus (CGLV) (one of the twelve named serotypes within the *Changinola virus* species) was isolated in Panama from a human with a brief febrile illness (Karabatsos, 1985). The virus was also isolated from phlebotomine sand flies, and antibodies were detected in rodents. Changuinola virus replicates in a mosquito cell line (C6/36 cells) without producing cytopathic effect (CPE) (in the same way as BTV replicates in C6/36 cells) and in a *Culicoides sonorensis* cell line (KC cells) without producing CPE and is pathogenic for newborn mice or hamsters following intracerebral inoculation (Karabatsos, 1985).

Kemerovo, Lipovnik and Tribec are among 36 serotypes of tick-borne viruses belonging to the *GIV* species. These viruses have been implicated as the cause of non-specific fever or neurological infection, in the former USSR (Kemerovo) and Central Europe (Lipovnik and Tribec) (Libikova *et al.*, 1970). More than 20 strains of Kemerovo virus were isolated during 1962 in the Kemerovo region of Russia from *Ixodes persulcatus* ticks and from human patients with meningoencephalitis. The virus, which was also isolated from birds, can infect Vero or BHK-21 cells. Lipovnik and Tribec viruses have been implicated in Central European encephalitis (CEE), with >50% of CEE patients having antibodies to Lipovnik virus (Libikova *et al.*, 1970). The virus is also suspected in the aetiology of some chronic neurological diseases, including polyradiculoneuritis and multiple sclerosis. Antibodies against a Kemerovo-related virus have been detected in Oklahoma and Texas, in patients with Oklahoma tick fever. Sixgun City virus is one of the seven tick-borne serotypes of the *Chenuda virus* species that have been isolated from birds. Several Oklahoma tick fever patients also had antibodies to Sixgun City virus, although no virus was isolated (Fields *et al.*, 1985).

Lebombo virus-1 (LBV-1) (the only known serotype of the *Lebombo virus* species) was isolated in Ibadan and Nigeria, in 1968, from a child with fever (Fields *et al.*, 1985). Lebombo virus-1 replicates in C6/36 cells without CPE but causes lysis in Vero cells and LLC-MK2 cells (Rhesus monkey kidney cells). It is also pathogenic for suckling mice and has been isolated from both rodents and mosquitoes (*Mansonia* and *Aedes* species) in Africa (Fields *et al.*, 1985).

The species *Orungo virus* (ORV) contains four distinct serotypes (ORV-1 to ORV-4), which are transmitted by *Anopheles*, *Aedes* and *Culex* mosquitoes

(Fields *et al.*, 1985). Orungo virus is widely distributed in tropical Africa where it has been isolated from humans, camels, cattle, goats, sheep, monkeys and mosquitoes. Orungo virus was first isolated in Uganda during 1959 from the blood of a human patient with fever and diarrhoea who developed weakness of the legs and generalized convulsions. The weakness eventually progressed to flaccid paralysis. Despite a high prevalence of virus infection and three deaths in Uganda, only a few clinical cases of human disease (involving fever, headache, myalgia, nausea and vomiting) have been reported (Fields *et al.*, 1985). Orungo virus causes lethal encephalitis in suckling mice and hamsters and replicates in adult *Aedes aegypti* mosquitoes after intra-thoracic inoculation (Karabatsos, 1985). It also causes CPE and forms plaques in Vero and BHK-21 cells (Karabatsos, 1985). High rates of co-infection with yellow fever and ORVs have been reported, reflecting their similar geographical distribution and transmission by *Aedes* mosquitoes, as principal vectors.

The genome coding assignments of BTV (see Figure 5.10) and of GIV are equivalent for genome segments 1, 7, 8, 9 and 10. However, in all of the tick-borne orbiviruses that have been characterized, the sub-core shell 'T2' protein is VP2, which is equivalent to VP3 of BTV. The GIV genome segments 3, 4, 5 and 6 are homologous to segments 4, 6, 2 and 5, respectively, of BTV. Amino acid identity values between GIV and BTV in homologous proteins range from 23% (in segments 7 or 9) to 46% (in the polymerase).

Relationships with other reoviruses

The cores of some reoviruses have 12 icosahedrally arranged surface projections, known as 'turrets' or 'spikes', while others have a relatively smooth (non-turreted) surface appearance (Mertens *et al.*, 2005a; Attoui *et al.*, 2006). It has been proposed (to ICTV) that the family *Reoviridae* should be separated into two sub-families, containing 'turreted' and 'non-turreted' viruses. The turreted viruses have 12 surface projections, located on the surface of the sub-core 'T2' layer, at the five-fold vertices of the icosahedral particle. Each of these 'turrets' (or spikes) is composed of a pentamer of the viral capping enzyme. This group includes members of genera *Cypovirus*, *Orthoreovirus* and *Aquareovirus*. Core particles of the 'non-turreted' viruses have a relatively smooth external surface, with their capping enzyme contained within the sub-core itself (Mertens *et al.*, 2005a). The cores of the 'non-turreted' viruses are composed of two protein layers. The innermost sub-core layer (composed of 120 copies of the 'T2' protein) act as a scaffolding layer for attachment of the outer layer of the core (composed of 780 copies of the T13 protein) (Mertens *et al.*, 2005b). The non-turreted reoviruses include members of the genera *Orbivirus*, *Rotavirus* and *Seadornavirus*. The magnitude of differences in virus structure, protein function and aa sequence makes meaningful phylogenetic comparisons of the majority of reovirus proteins difficult or impossible between different genera. However, phylogenetic analysis of the aa sequences of the viral RNA-dependent RNA polymerase (RdRp) shows that the turreted and non-turreted viruses form separate phylogenetic clusters (Figure 3.1). The orbiviruses also form a distinct cluster, showing 13–15% aa identity to members of the other non-turreted genera but less than 8% identity to the turreted viruses (Attoui, unpublished data).

Several distinct protein functions and structural roles can be identified for the proteins of viruses belonging to the different genera of reoviruses. However, these roles are not necessarily carried out by direct homologous proteins in each case (see: www.reoviridae.org/dsrna_virus_proteins//protein-comparison.htm).

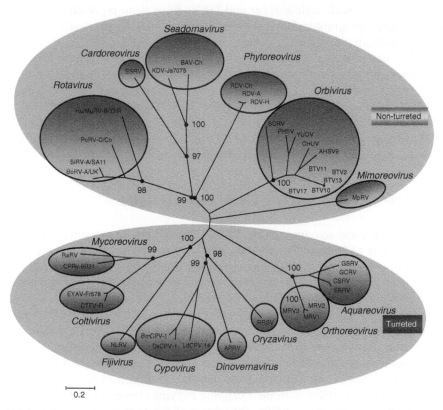

Figure 3.1 Radial neighbour-joining phylogenetic tree constructed using the amino acid (aa) sequences of RNA-dependent RNA polymerases of 14 genera of the family *Reoviridae*. The bootstrap values (500 replications) supporting the validity of the branching are indicated at the nodes. The bar represents the number of substitutions per site. Turreted and non-turreted viruses cluster separately into two broad phylogenetic groups. The tree clearly shows orbiviruses as members of distinct phylogenetic cluster among other non-turreted viruses.

For example, the viral inclusion body matrix protein (VIP) of the orbiviruses (including BTV) is a non-structural protein 2 (NS2). This function appears to be carried by two separate proteins in both the orthoreoviruses (Sigma-NS and Mu-NS) and the rotaviruses (NSP2 and NSP5) (Taraporewala and Patton, 2004). The helicase activity carried by VP6 of BTV (a minor structural protein and a component of the replication complex thought to be located at the 12 five-fold axes of the particle; Grimes *et al.*, 1998) is provided by the sub-core 'T2' protein layer of the orthoreoviruses. This function appears to be missing entirely from the rotaviruses (even though they have a more similar structural organization to that of the orbiviruses – including BTV).

The reovirus outer capsid proteins are involved in cell attachment and penetration. They, therefore, determine the specificity of interactions with different cell types from a wide variety of hosts that are targeted by the different genera and species of viruses. As a consequence, these outer capsid components are also the most variable proteins, showing little or no conservation between the orbiviruses and other reoviruses.

Phylogenetic relationships of the orbivirus proteins

There are clear overall similarities in the protein structure and composition of the different orbiviruses. There are also serological cross-reactions between the members of some different *Orbivirus* species and significant levels of aa or nucleotide sequence identity in the more conserved proteins or the genes from which they are translated. These similarities demonstrate that the different orbiviruses have a common ancestry.

In contrast, the outermost capsid layer interacts with cellular components of the mammalian immune system and with the neutralizing antibodies that are generated in response to orbivirus infections within the vertebrate host. Sequence variations in the outer capsid proteins (and the genome segments from which they are translated) therefore determine the specificity of interactions with these neutralizing antibodies. As a consequence, they also determine the identity of the different virus serotypes within each *Orbivirus* species.

The outer capsid components are also subjected to selective pressure to change and avoid recognition by the host's immune system. In BTV, AHSV or EHDV, it is the larger of the two outer capsid proteins (VP2 – encoded by genome segment 2) that mediates cell attachment and is the major target for neutralizing antibodies (Mertens *et al.*, 2005b). VP2 shows up to 27% aa sequence variation within a single serotype, while different BTV types can show as much as 73% variation in the sequence of this protein (Maan *et al.*, 2007a). Variations in the aa sequence of VP2 can also reflect the serological relationships (low-level cross-reactions) that exist between different serotypes

(Erasmus, 1990; Anthony, 2007) (Figures 3.2 and 3.3). The smaller orbivirus outer-coat protein (VP5) can also influence virus serotype, and variations in VP5 of BTV show a partial correlation with virus type (Maan *et al.*, 2008). In some of the other orbiviruses (e.g. the GIVs), the smaller outer capsid protein exerts a greater influence over serotype (Moss *et al.*, 1987).

Within a single *Orbivirus* species, individual genome segments can display sequence variations that reflect the geographical origin of the virus isolate (topotypes) (Pritchard *et al.*, 1995; 2004; Maan *et al.*, 2008; Mertens *et al.*, 2008 – Chapter 7). For example, BTV and EHDV isolates can be divided into eastern and western groups, based on sequence comparisons of the majority of their genome segments/proteins (i.e. 'eastern' viruses from Australia, India and Asia and 'western' viruses from Africa and America) (Maan, 2004; Anthony, 2007; Mertens *et al.*, 2007; Maan *et al.*, 2008). Genome segments 2 and 6 of BTV and EHDV (which encode the outer capsid proteins VP2 and VP5, respectively) also show variations within each type that reflect the geographical origins of the virus lineage (Maan *et al.*, 2007a, 2008).

Since 1998, both 'eastern' and 'western' strains of BTV have invaded Europe, providing a unique opportunity for these viruses to co-infect the same host and exchange (reassort) genome segments derived from a number of different geographical origins. This has the potential to generate reassortant viruses that may have novel biological characteristics (Batten *et al.*, 2008; Maan *et al.*, 2008). Genome segments 7 and 10 (encoding VP7 and NS3 of both BTV and EHDV) also show variations that divide them into a number of clades that have as yet unknown biological significance (Wilson *et al.*, 2000, 2007; Anthony, 2007; Balasuriya *et al.*, 2008; Maan *et al.*, 2008). In the absence of outer capsid proteins, VP7 can mediate cell attachment and pene-tration during infection of *Culicoides* cells by BTV core particles (Mertens *et al.*, 1987a). NS3 can also mediate cell exit from insect cells in the absence of cell lysis (Hyatt *et al.*, 1989). It has therefore been suggested that variations in these genome segments may be related to the identity of the insect species that act as vectors for the virus in different geographical regions.

The variations that have been detected in VP2/Seg-2 and VP5/Seg-6 within individual serotypes indicate that the different BTV types emerged as an early step in the evolution of the virus species. Strains of each serotype subsequently became geographically dispersed and then acquired the further point mutations that distinguish the different topotypes of each serotype. Reverse transcription polymerase chain reaction assays and sequence analyses of BTV Seg2/VP2 provide rapid and reliable diagnostic methods to determine both the serotype and the origin of individual isolates (molecular epidemiology) (Mertens *et al.*, 2007; Maan *et al.*, 2008). These molecular assays are currently replacing serological methods for BTV-type identification within reference laboratories although VNT and SNT remain the gold standard for identification of virus serotype (Mertens *et al.*, 2008; Chapter 17).

The Seg-2/VP2 sequences of the reference strains of different BTV sero-types cluster as 10 distinct evolutionary lineages, identified as nucleotypes A–J (see Figure 3.2). The grouping of the different BTV types within the 10 nucleotypes also reflects the serological relationships (low-level cross-reactions between different BTV serotypes (see Figure 3.3). The open reading frame (ORF) and non-coding region (NCRs) of Seg-2, from the different serotypes within each nucleotype of BTV, are uniform in length. However,

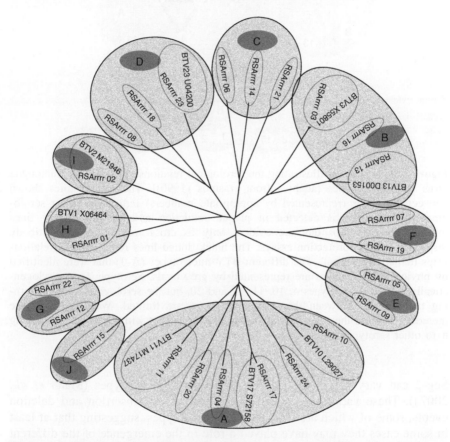

Figure 3.2 Phylogenetic relationships of bluetongue virus (BTV) outer capsid protein VP2. Unrooted neighbour-joining tree showing the relationships between deduced amino acid (aa) sequences of VP2 (encoded by genome segment 2) from the 24 BTV serotypes. This neighbour-joining tree was constructed using MEGA program, version 3.1 and using the *p*-distance algorithm and the full-length VP2 sequences of the 24 BTV types. The different serotypes (indicated by dotted circles) are distinct but show some relationships (grey bubbles) that mirror the serological relatedness (cross-reactions) that are known to exist between different serotypes. These groupings are reflected in the nucleotide sequences of genome segment 2 and 10 distinct groups have previously been identified as nucleotypes A–J.

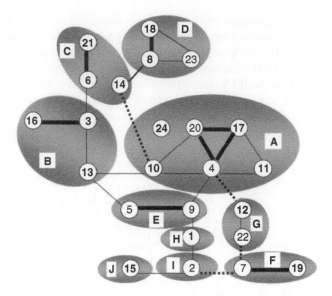

Figure 3.3 Schematic illustrating the serological relationships between bluetongue virus (BTV) serotypes [adapted from erasmus (1990)]. The thicker lines shown between serotypes (represented by appropriate numbers) indicate stronger serological relationships, as detected in plaque reduction assays. The thinner lines represent relationships that are evident only as cross- or heterotypic antibody responses in cross-protection assays. The black dotted lines represent interrelationships that are very weak. The different BTV nucleotypes (A–J) that were identified by phylogenetic analyses are represented by grey ovals. Serotype 24 is phylogenetically related to serotypes 4, 10, 11, 17 and 20, but the serological data concerning relationships between this and other types are limited. We have therefore included it within the oval shown for nucleotype 'a' but without a line connecting it to other serotypes.

Seg-2 can vary in length between different BTV serotypes (Maan *et al.*, 2007a). These variations appear to be the result of insertion and deletion events, some of which are unique to individual types, suggesting that at least in some cases they may have played a role in the emergence of the different nucleotypes and serotypes. For example, BTV-8 shows an insertion of 12 nt at position 122–134 when aligned to Seg-2 of the reference strain of BTV-1 (RSArrrr/01 – position 112–113) and an insertion of 6 nt at position 830–835 when aligned with Seg-2 sequence of BTV-19 (position 773–774 with reference to RSArrrr/01). Seg-2 of types 12, 15 and 22 contains several longer insertions (e.g. at nt 360–369) as well as deletions (between positions 1888 and 1926) in their aligned sequences compared to Seg-2 of other serotypes. This reflects the relatively close genetic relationship detected between types 12 and 22 (nucleotype G) and type 15 (nucleotype J) even though they are classified within separate nucleotypes. There are also several areas of Seg-2 from the 24 serotypes where no insertions or deletions were evident (e.g. nt 370–424, 900–1250, 1580–1850 in the aligned sequences).

Seg-2 encodes the VP2 protein, which ranges in size from 950 aa (2852 bp ORF) in BTV-12 to 962 aa (2888 bp ORF) in BTV-19. VP2 shows between 22.7% (BTV-4 and BTV-20) and 72.9% (BTV-6 and BTV-22) aa variation between BTV types. Only the carboxy-terminus (aa 946–961) and residues at positions 338–379 appeared to be relatively conserved. The amino-terminus of VP2 is one of the least conserved regions (Maan *et al.*, 2007a).

Comparisons of VP2s from the members of different *Orbivirus* species (AHSV, EHDV, Broadhaven virus (BRDV), SCRV and Chuzan virus) showed only low levels of aa sequence identity. Phylogenetic analyses confirmed that Seg-2s/VP2s of these orbiviruses exist as distinct mono-phyletic groups, despite their overlapping global distributions. This provides further evidence that the members of distinct *Orbivirus* species do not exchange genetic information by reassortment. It also confirms previous serological and sequencing studies, which indicate that BTVs are more closely related to EHDVs than to AHSVs or Palyam viruses (PALVs) (Maan, 2004; Anthony, 2007).

Phylogenetic relationships between the RNA-dependent RNA polymerase (Pol – RdRP) of the different orbiviruses

The RdRp aa sequence has been determined for members of several *Orbivirus* species, including *BTV, AHSV, PALV* (Chuzan strain) (PALV), YUOV, *PHSV, St. Croix River virus* (SCRV) and *EHDV*. Phylogenetic analyses have confirmed that the RdRps of these species are more closely related than their outer capsid proteins [e.g. PHSV and YUOV: aa identity in the VP3 (outer capsid and serotype defining) is 27% while aa identity in the polymerase is 69%]. A phylogenetic tree (Figure 3.4) built with the RdRps of these viruses shows that SCRV roots the various other viruses, suggesting that it is closest to the ancestral virus from which all others orbiviruses originated.

Comparisons of VP1(Pol) from the members of different *Orbivirus* species showed a close relationship within a single species. For example, the aa identity of the RdRp of different BTVs ranged from 96 to 98% and of different EHDVs ranged from 90 to 99%. However, lower levels of aa identity were detected between the different *Orbivirus* species, ranging from 32 to 70%. In each case, the lowest level of identity was detected between SCRV and the other orbiviruses. This reflects its more distant relationship with other members of the genus.

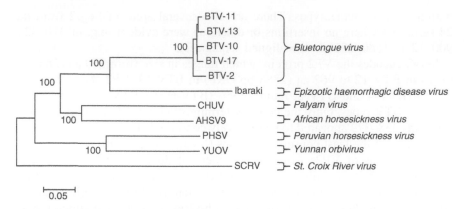

Figure 3.4 Phylogenetic tree depicting the relationship between various orbiviruses, based on the amino acid (aa) sequences of their polymerases. The bootstrap values (500 replications) supporting the validity of the branching is indicated at the nodes. The bar represents the number of substitutions per site. The tree clearly show St. Croix River virus (SCRV) rooting all other orbiviruses.

Phylogenetic relationships based on the sub-core shell T2 protein of bluetongue virus and other orbiviruses

The sub-core of BTV and the other orbiviruses is composed of 120 copies of a single 'T2' protein, identified as VP3 of BTV, AHSV, EHDV and EEV (www.reoviridae.org/dsRNA_virus_proteins/BTV.htm). In some orbiviruses, (e.g. EEV) the sub-cores are relatively stable and can be purified (Mertens and Diprose, 2004; Chapter 6). The T2 protein of BTV can also self-assemble when expressed in isolation (e.g. by recombinant baculovirus) to form sub-core-like particles (Chapter 4). These observations suggest that formation of the sub-core is an initial step in virus morphogenesis. The aa sequence of T2, which determines its conformation and structure, also determines its assembly into sub-cores, and therefore both the overall size and the morphology and organization of the virus capsid (Chapter 6). The T2 protein also interacts with the dsRNA of the virus genome segments, helping to determine their organization within the central space of the sub-core shell (Diprose *et al.*, 2002). These important and complex functional roles are reflected in very high levels of aa sequence identity of the T2 protein (>91%) within a single *Orbivirus* species (serogroup), which could be used for the identification orbivirus species (Attoui *et al.*, 2001; Anthony, 2007).

In some of the *Orbivirus* species [including GIV, SCRV, PHSV, YUOV, Wongorr virus (WGRV) and Corriparta virus (CORV)], the larger of the two outer capsid proteins is smaller than the sub-core T2 protein, which is

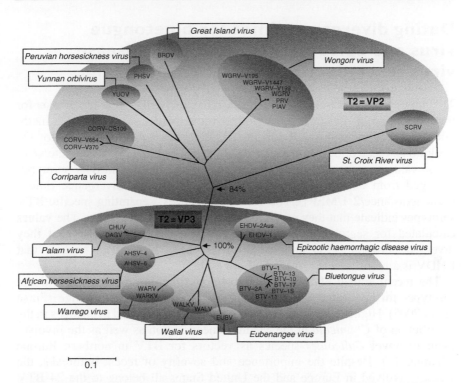

Figure 3.5 Phylogenetic comparison of sub-core shell (T2) proteins of different orbiviruses. Unrooted neighbour-joining tree showing relationships between the deduced amino acid (aa) sequences of the sub-core shell protein (T2) of different orbivirus species. This neighbour-joining tree was constructed using MEGA program, version 3.1 and the *p*-distance algorithm is based on partial sequences for VP3/VP2, as indicated (aa 393–548 relative to BTV-10 sequence). Similar trees were obtained with the Poisson correction and the gamma distance. Two major clusters of virus species were detected, one group with T2 proteins (VP2) encoded by genome segment 2 (supported by bootstrap values of > 84%) and a second group with T2 protein (VP3) encoded by genome segment 3 (e.g. BTV and AHSV). These two groups appear to represent distinct evolutionary lineages. St. Croix River virus (SCRV) is the most divergent virus from either insect-borne or the other tick-borne orbiviruses and forms a separate small cluster.

therefore identified as VP2 rather than as VP3. The VP3(T2) and VP2(T2) groups form two distinct phylogenetic lineages (Figure 3.5). Within the group of the VP3(T2) (which are all insect-borne), aa identity values between species varied from 57.3 to 78.7%. Within the group of the VP2(T2), aa identity values between insect-borne species varied from 55.3 to 60.5%. Within the group of the VP2(T2), aa identity values between tick-borne species varied from 23.4 to 24.9%.

Dating divergence times for bluetongue virus and epizootic haemorrhagic disease virus serotypes

The nucleotide sequences of Seg-2 from BTV and EHDV provide a basis for the calculation of molecular evolutionary rates (MER) using Bayesian methodologies (Drummond and Rambaut, 2007). Upper and lower limits for the evolutionary rate of Seg-2 were estimated at 10^{-4} and 10^{-5} changes/site/year. These values were used to calculate the time at which different BTV serotypes diverged from a common ancestor (using the formula: divergence time = [geneticdistance/2*1/MER]). The 'genetic distances' separating specific BTV serotypes indicate that they separated $\sim$1500 to 72 000 years ago. The values calculated for segment 2 of different EHDV serotypes indicate that they diverged between $\sim$500 and 40 000 years ago. These data suggest that EHDV is a more 'recent' virus (Anthony, 2007).

The recent spread of BTV into Europe and the emergence of additional serotypes in the United States have been linked to climatic change (Purse et al., 2005). Higher temperatures within Europe are reflected in changes in the distribution of C. imicola in the Mediterranean region as well as the involvement of novel Culicoides species as vectors for BTV in northern Europe (Chapter 11). Despite the importance and severity of recent outbreaks, the viruses involved in Europe and the United States all belong to the 24 BTV serotypes that have already been identified. With the increasing refinement and speed of sequencing and PCR-based assay methods (Mertens et al., 2007; Maan et al., 2007b), new isolates of different orbiviruses can be identified and characterized more rapidly than ever before. This not only helps to identify new species within the genus but also provides valuable information concerning their relationships, distribution and transmission, as well as their involvement as the causes of clinical disease (Attoui et al., 2001, 2005, 2008).

References

Anthony, S.J. (2007) Genetic studies of epizootic haemorrhagic disease virus (EHDV). Department of Medical Sciences and the Wellcome Trust Centre for Human Genetics D. Phil Thesis: Linacre College, University of Oxford.

Anthony, S., Jones, H., Darpel, K.E., Elliott, H., Maan, S., Samuel, A., Mellor, P.S. and Mertens, P.P. (May 2007) A duplex RT-PCR assay for detection of genome segment 7 (VP7 gene) from 24 BTV serotypes. J. Virol. Methods. **141**(2), 188–97. Epub 2007 Jan 22. http://www.ncbi.nlm.nih.gov/pubmed/17241676?ordinalpos=7&itool=Entrez System2. PEntrez.Pubmed. Pubmed_ResultsPanel.Pubmed_RVDocSum

Anthony, S.J., Maan, S., Maan, N., Kgosana, L., Bachanek-Bankowska, K., Batten, C.A., Darpel, K.E., Attoui, H. and Mertens, P.P.C. (2008) Genetic and phylogenetic analysis of the outer-coat proteins VP2 and VP5 of epizootic haemorrhagic disease virus (EHDV):

Comparison of genetic and serological data to characterise the EHDV serogroup. Submitted

Aradaib, I.E., Smith, W.L., Osburn, B.I. and Cullor, J.S. (2003) A multiplex PCR for simultaneous detection and differentiation of North American serotypes of bluetongue and epizootic hemorrhagic disease viruses. *Comp. Immunol. Microbiol. Infect. Dis.* **26**, 77–87.

Attoui, H., Fang, Q., Mohd Jaafar, F., Cantaloube, J.F., Biaginim, P., de Micco, P. and de Lamballerie, X. (2002) Common evolutionary origin of aquareoviruses and orthoreoviruses revealed by genome characterization of golden shiner reovirus, grass carp reovirus, striped bass reovirus and golden Ide reovirus (genus *Aquareovirus*, family Reoviridae). *J. Gen. Virol.* **83**, 1941–51.

Attoui, H., Mendez-Lopez, M.R., Rao, S., Hurtado-Alendes, A., Samuel, A.R., Pritchard, L.I., Melville, L., Weir, R., Davis, S., Lunt R., Calisher, C.H., Tesh, R.B., Fujita, R. and Mertens, P.P.C. (2008) Peruvian horse sickness virus and Yunnan orbivirus isolated from equine, bovine, ovine and mosquito species in Peru and Australia. *Virology* Submitted.

Attoui, H., Mohd Jaafar, F., Belhouchet, M., Aldrovandi, N., Tao, S., Chen, B., Liang, G., Tesh, R.B., de Micco, P. and de Lamballerie, X. (2005) Yunnan orbivirus, a new orbivirus species isolated from *Culex tritaeniorhynchus* mosquitoes in china. *J. Gen. Virol.* **86**, 3409–17.

Attoui, H., Mohd Jaafar, F., Belhouchet, M., de Micco, P., de Lamballerie, X. and Brussaard, C.P.D. (2006) The *Micromonas pusilla* reovirus: A member of family *Reoviridae* with an extra outer coat. Assignment to a novel proposed genus (Mimoreovirus). *J. Gen. Virol.* **87**, 199–208.

Attoui, H., Stirling, J.M., Munderloh, U.G., Billoir, F., Brookes, S.M., Burroughs, J.N., de Micco, P., Mertens, P.P.C. and de Lamballerie, X. (2001) Complete sequence characterisation of the genome of the St. Croix River virus, a new orbivirus isolated from *Ixodes scapularis* cells. *J. Gen. Virol.* **82**, 795–804.

Balasuriya, U.B., Nadler, S.A., Wilson, W.C., Pritchard, L.I., Smythe, A.B., Savini, G., Monaco, F., De Santis, P., Zhang, N., Tabachnick, W.J. and Maclachlan, N.J. (January 1, 2008) The NS3 proteins of global strains of bluetongue virus evolve into regional topotypes through negative (purifying) selection. *Vet. Microbiol.* **126**(1–3), 91–100. Epub 2007 Jul 10. http://www.ncbi.nlm.nih.gov/pubmed/17706379?ordinalpos=2&itool= EntrezSystem2.PEntrez.Pubmed.Pubmed_ResultsPanel.Pubmed_RVDocSum

Batten, C.A., Maan, S., Shaw, A.E., Maan, N.S. and Mertens, P.P. (July 28, 2008) A European field strain of bluetongue virus derived from two parental vaccine strains by genome segment reassortment. *Virus Res.* [Epub ahead of print]. http://www.ncbi. nlm.nih.gov/pubmed/18598726?ordinalpos=1&itool=EntrezSystem2.PEntrez.Pubmed. Pubmed_ResultsPanel.Pubmed _RVDocSum

Benazzou, H. (2004) *OIE Disease Information: Bluetongue in Morocco,* Vol. 17, pp. 273.

Benazzou, H. (2006) Outbreak: Epizootic haemorrhagic disease in cattle, Morocco. International Society for Infectious Disease (ProMED) Archive Number: 20061010.2906.

Bowen, R.A., Howard, T.H., Elsden, R.P. and Seidel, G.E. (1985) Bluetongue virus and embryo transfer in cattle. *Prog. Clin. Biol. Res.* **178**, 85–9. http://www.ncbi.nlm.nih.gov/ pubmed/2989931?ordinalpos=3&itool=EntrezSystem2.PEntrez.Pubmed.Pubmed_Results Panel.Pubmed_RVDocSum

Breard, E., Sailleau, C., Hamblin, C., Graham, S., Gourreau, J. and Zientara, S. (2004) Outbreak of epizootic haemorrhagic disease on the island of reunion. *Vet. Rec.* **14**, 422–3.

Cambell, C.H., Barber, T.L. and Jochim, M.M. (1978) Antigenic relationship of Ibaraki, bluetongue, and epizootic haemorrhagic disease viruses. *Vet. Microbiol.* **3**, 15–22.

Chaimovitz, M. (2006) Outbreak of epizootic haemorrhagic disease in cattle, Israel. International Society for Infectious Diseases (ProMED) Archive Number: 20060925.2739.

Cowled, C., Melville, L., Weir, R., Walsh, S., Hyatt, A., Van Driel, R., Davis, S., Gubala, A. and Boyle, D. (2007) Genetic and epidemiological characterization of middle point orbivirus, a novel virus isolated from sentinel cattle in northern Australia. *J. Gen. Virol.* **88**, 3413–22.

Cowley, J.A. and Gorman, B.M. (1989) Cross-neutralization of genetic reassortants of bluetongue virus serotypes 20 and 21. *Vet. Microbiol.* **19**, 37–51.

Darpel, K.E., Batten, C.A., Veronesi, E., Shaw, A.E., Anthony, S., Bachanek-Bankowska, K., Kgosana, L., bin-Tarif, A., Carpenter, S., Müller-Doblies, U.U., Takamatsu, H.H., Mellor, P.S., Mertens, P.P.C. and Oura, C.A. (2007) Clinical signs and pathology shown by British sheep and cattle infected with bluetongue virus serotype 8 derived from the 2006 outbreak in northern Europe. *Vet. Rec.* **161**, 253–61.

DeMaula, C.D., Heidner, H.W., Rossitto, P.V., Pierce, C.M. and MacLachlan, N.J. (1993) Neutralization determinants of united states bluetongue virus serotype ten. *Virology* **195**, 292–6.

Diprose, J.M., Grimes, J.M., Sutton, G.C., Burroughs, J.N., Meyer, A., Maan, S., Mertens, P.P.C. and Stuart, D.I. (2002) The core of bluetongue virus binds double-stranded RNA. *J. Virol.* **76**, 9533–6.

Ditchfield, J., Debbie, J.G. and Karstad, L. (1964) The virus of epizootic hemorrhagic disease of deer. *Trans. North Am. Wildl. Nat. Resour. Conf.* **29**, 196–9.

Drummond, A.J. and Rambaut, A. (2007) BEAST: Bayesian evolutionary analysis by sampling trees. *BMC Evol. Biol.* **7**, 214.

Eaton, B.T. and Gould, A.R. (1987) Isolation and characterization of orbivirus genotypic variants. *Virus Res.* **6**, 363–82.

Elbers, A.R., Backx, A., Ekker, H.M., van der Spek, A.N. and van Rijn, P.A. (2007) Performance of clinical signs to detect bluetongue virus serotype 8 outbreaks in cattle and sheep during the 2006-epidemic in the Netherlands. *Vet. Microbiol.* Nov 17, [Epub ahead of print].

Erasmus, B.J. (1990) Bluetongue virus. In: Z. Dinter and B. Morein, (eds.), *Virus infections of ruminants*, vol. 3, New York: Elsevier Science Publishers, pp. 227–37.

Fay, L.D., Boyce, A.P. and Youatt, W.G. (1956) Twenty-First American Wildlife Conference.

Fields, B.N., Knipe, D.M. and Howley, P.M. (1985) *Fields Virology*, 3rd edn. Philadelphia, Lippincott-Raven.

Gould, A.R. and Eaton, B.T. (1990) The amino acid sequence of the outer coat protein VP2 of neutralizing monoclonal antibody-resistant, virulent and attenuated bluetongue viruses. *Virus Res.* **17**, 161–72.

Grimes, J.M., Burroughs, J.N., Gouet, P., Diprose, J.M., Malby, R., Ziéntara, S., Mertens, P.P.C. and Stuart, D.I. (1998) The atomic structure of the bluetongue virus core. *Nature.* **395**, 470–8.

Howard, T.H., Bowen, R.A. and Pickett, B.W. (1985) Isolation of bluetongue virus from bull semen. *Prog. Clin. Biol. Res.* **178**, 127–34. http://www.ncbi.nlm.nih.gov/pubmed/2989846?ordinalpos=4&itool=EntrezSystem2.PEntrez.Pubmed.Pubmed_ResultsPanel.Pubmed_RVDocSum

Huismans, H. and Erasmus, B.J. (1981) Identification of the serotype-specific and group-specific antigens of bluetongue virus. *Onderstepoort J. Vet. Res.* **48**, 51–8.

Hyatt, A.D., Eaton, B.T. and Brookes, S.M. (November 1989) The release of bluetongue virus from infected cells and their superinfection by progeny virus. *Virology* **173**(1), 21–34. http://www.ncbi.nlm.nih.gov/pubmed/2554570?ordinalpos=9&itool=Entrez System2.PEntrez.Pubmed.Pubmed_ResultsPanel.Pubmed_RVDocSum

Karabatsos, N. (1985) International catalogue of arboviruses including certain other viruses of vertebrates. *The Subcommittee on Information Exchange of the American Committee on Arthropod-Borne Viruses,* 3rd edn. San Antonio, TX: American Society for Tropical Medicine and Hygiene.

Karstad, L., Winter, A. and Trainer, D.O. (1961) Pathology of epizootic hemorrhagic disease of deer. *Am. J. Vet. Res.,* 227–35.

Lee, V.H. (1979) Isolations of viruses from field populations of *Culicoides* (Diptera: Ceratopogonidae) in Nigeria. *J. Med. Entomol.* **16**, 76–9.

Libikova, H., Tesarova, J. and Rajcani, J. (1970) Experimental infection of monkeys with Kemerovo virus. *Acta Virol.* **14**, 64–9.

Maan, S. (2004) Complete nucleotide sequence analyses of genome segment 2 from the twenty-four serotypes of bluetongue virus: development of nucleic acid based typing methods and molecular epidemiology. PhD thesis, University of London.

Maan, S., Maan, N.S., Ross-smith, N., Batten, C.A., Shaw, A.E., Anthony, S.J., Samuel, A.R., Darpel, K.E., Veronesi, E., Oura, C.A., Singh, K.P., Nomikou, K., Potgieter, A.C., Attoui, H., van Rooij, E., van Rijn, P., De Clercq, K., Vandenbussche, F., Zientara, S., Bréard, E., Sailleau, C., Beer, M., Hoffman, B., Mellor, P.S. and Mertens, P.P. (August 1, 2008) Sequence analysis of bluetongue virus serotype 8 from the Netherlands 2006 and comparison to other European strains. *Virology* **377**(2), 308–18. http://www.ncbi. nlm.nih.gov/pubmed/18570969?ordinalpos=2&itool=EntrezSystem2.PEntrez.Pubmed. Pubmed_ResultsPanel.Pubmed_RVDocSum

Maan, S., Maan, N.S., Samuel, A.R., Rao, S., Attoui, H. and Mertens, P.P.C. (2007a) Analysis and phylogenetic comparisons of full-length VP2 genes of the 24 bluetongue virus serotypes. *J. Gen. Virol.* **88**, 621–30.

Maan, S., Rao, S., Maan, N.S., Anthony, S.J., Attoui, H., Samuel, A.R. and Mertens, P.P.C. (2007b) Rapid cDNA synthesis and sequencing techniques for the genetic study of bluetongue and other dsRNA viruses. *J. Virol. Methods.* **143**, 132–9.

McColl, K.A. and Gould, A.R. (September 1991) Detection and characterisation of blue-tongue virus using the polymerase chain reaction. . *Virus Res.* **21**(1), 19–34. http:// www.ncbi.nlm.nih.gov/pubmed/1660214?ordinalpos=5&itool=EntrezSystem2.PEntrez. Pubmed.Pubmed_ResultsPanel.Pubmed_RVDocSum

McLaughlin, B.E., DeMaula, C.D., Wilson, W.C., Boyce, W.M. and MacLachlan, N.J. (2003) Replication of bluetongue virus and epizootic hemorrhagic disease virus in pulmonary artery endothelial cells obtained from cattle, sheep, and deer. *Am. J. Vet. Res.* **64**, 860–5.

Mellor, P.S. and Boorman, J. (February 1995) The transmission and geographical spread of African horse sickness and bluetongue viruses. *Ann. Trop. Med. Parasitol.* **89**(1), 1–15. Review. http://www.ncbi.nlm.nih.gov/pubmed/7741589?ordinalpos=6&itool=Entrez System2.PEntrez.Pubmed.Pubmed_ResultsPanel.Pubmed_RVDocSum

Mertens, P.P.C. (1999) Orbiviruses and coltiviruses – general features. In: R.G. Webster and A. Granoff, (eds.), *Encyclopaedia of Virology,* 2nd edn. London: Academic Press, pp. 1043–61.

Mertens, P.P.C. (2004) The dsRNA viruses. *Virus Res.* **101**, 3–13. http://www.ncbi.nlm. nih.gov/pubmed/15010213?ordinalpos=10&itool=EntrezSystem2.PEntrez.Pubmed.Pubmed_ ResultsPanel.Pubmed_RVDocSum

Mertens, P.P.C., Anthony, S., Kgosana, L., Bankowska, K., Batten, C.A., El Harrak, M., Madani, H. and Yadin, H. (2006) Outbreak: Epizootic haemorrhagic disease in cattle – Morocco, Algeria and Israel. Serotyped. International Society for Infectious Disease (ProMED) Archive Number: 20061214.3513.

Mertens, P.P.C., Attoui, H., Duncan, R. and Dermody, T.S. (2005a). Reoviridae. In: C.M. Fauquet, M.A. Mayo, J. Maniloff, U. Desselberger, and L.A. Ball, (eds.), *Virus Taxonomy. Eighth Report of the International Committee on Taxonomy of Viruses*. London: Elsevier/Academic Press, pp. 447–54.

Mertens, P.P.C., Burroughs, J.N. and Anderson, J. (1987a) Purification and properties of virus particles, infectious subviral particles, and cores of bluetongue virus serotypes 1 and 4. *Virology*. **157**, 375–86.

Mertens, P.P.C., Burroughs, J.N., Walton, A., Wellby, M.P., Fu, H., O'Hara, R.S., Brookes, S.M. and Mellor, P.S., (1996) Enhanced infectivity of modified bluetongue virus particles for two insect cell lines and for two *Culicoides* vector species. *Virology*. **217**, 582–93.

Mertens, P.P.C. and Diprose, J. (2004) The bluetongue virus core: A nano-scale transcription machine. *Virus Res*. **101**, 29–43.

Mertens, P.P.C., Maan, N.S., Prasad, G., Samuel, A.R., Shaw, A.E., Potgieter, A.C., Anthony, S.J. and Maan, S. (2007) Design of primers and use of RT-PCR assays for typing European bluetongue virus isolates: Differentiation of field and vaccine strains. *J. Gen. Virol*. **88**, 2811–23.

Mertens, P.P.C., Maan, S., Samuel, A. and Attoui, H., (2005b). Orbivirus, Reoviridae. In: C.M. Fauquet, M.A. Mayo, J. Maniloff, U. Desselberger and L.A. Ball, (Eds), *Virus Taxonomy, VIIIth Report of the ICTV*. Elsevier/Academic Press, London, pp. 466–83.

Mertens, P.P.C., Pedley, S., Cowley, J. and Burroughs, J.N. (1987b) A comparison of six different bluetongue virus isolates by cross-hybridization of the dsRNA genome segments. *Virology* **161**, 438–47.

Mertens, P.P.C., Pedley, S., Cowley, J., Burroughs, J.N., Corteyn, A.H., Jeggo, M.H., Jennings, D.M. and Gorman, B.M. (1989) Analysis of the roles of bluetongue virus outer capsid proteins VP2 and VP5 in determination of virus serotype. *Virology* **170**, 561–5.

Mohammed, M.E.H., Aradaib, I.E., Mukhtar, M.M., Ghalib, H.W., Riemann, H.P., Oyejide, A. and Osburn, B.I. (1996) Application of molecular biological techniques for detection of epizootic hemorrhagic disease virus (EHDV-318) recovered from a sentinel calf in central Sudan. *Vet. Microbiol* **52**, 201–8.

Mohammed, M.E.H. and Mellor, P.S. (1990) Further studies on bluetongue and bluetongue-related orbiviruses in the Sudan. *Epidemiol. Infect*. **105**, 619–32.

Mohd Jaafar, F., Attoui, H., Mertens, P., de Micco, P. and de Lamballerie, X. (2005) Structural organisation of a human encephalitic isolate of Banna virus (genus *Seadornavirus*, family *Reoviridae*). *J. Gen. Virol*. **86**, 1141–6.

Mohd Jaafar, F., Goodwin, A.E., Belhouchet, M., Merry, G., Fang, Q., Cantaloube, J.F., Biagini P., de Micco, P., Mertens, P.P.C. and Attoui, H. (2008) Complete characterisation of the american grass carp reovirus genome (genus *Aquareovirus*: Family *Reoviridae*) reveals an evolutionary link between aquareoviruses and coltiviruses. *Virology* **373**, 310-21. Epub 2008 Jan.

Moss, S.R., Ayres, C.M. and Nuttall, P.A. (1987) Assignment of the genome segment coding for the neutralizing epitope(s) of orbiviruses in the great island subgroup (kemerovo serogroup). *Virology* **157**, 137–44.

Nettles, V.F., Hylton, S.A. and Stallknecht, D.E. (1992) Bluetongue, African Horse Sickness and Related Orbiviruses: Proceedings of the Second International Symposium, Paris.

Nuttall, P.A., Jacobs, S.C., Jones, L.D., Carey, D. and Moss, S.R. (1992) Enhanced neurovirulence of tick-borne orbiviruses resulting from genetic modulation. *Virology* **187**, 407–12.

Nuttall, P.A., Moss, S.R., Carey, D., Jones, L.D. and Jacobs, S.C. (1990) Genetic determinants modulating the pathogenic phenotype of tick-borne orbiviruses. *Virology* **174**, 430–5.

O'Hara, R.S., Meyer, A.J., Burroughs, J.N., Pullen, L., Martin, L.A. and Mertens, P.P.C. (1998) Development of a mouse model system, coding assignments and identification of the genome segments controlling virulence of African horse sickness virus serotypes 3 and 8. *Arch. Virol. suppl.* **14**, 259–79.

OIE (2007) Terrestrial animal health code, Chapter 2.2.13. *Bluetongue,* pp. 131–6.

Omori, T., Inaba, Y., Morimoto, T., Tanaka, Y., Ishitani, R., Kurogi, H. and Matsumoto, M. (1969a) Ibaraki virus, an agent of epizootic hemorrhagic disease of cattle, resembling bluetongue. I. *Jpn. J. Vet. Microbiol.* **13**, 139–57.

Omori, T., Inaba, Y., Morimoto, T., Tanaka, Y., Kono, M., Kurogi, H. and Matsumoto, M. (1969b) Ibaraki virus, an agent of epizootic hemorrhagic disease of cattle resembling bluetongue. II. *Jpn. J. Vet. Microbiol.* **13**, 159–68.

Pittman, L. (2007) Epizootic hemorrhagic disease in bovines, USA. Promed News, 22-NOV-2007.

Pritchard, L.I., Gould, A.R., Wilson, W.C., Thompson, L., Mertens, P.P. and Wade-Evans, A.M. (March 1995) Complete nucleotide sequence of RNA segment 3 of bluetongue virus serotype 2 (Ona-A). Phylogenetic analyses reveal the probable origin and relationship with other orbiviruses. *Virus Res.* **35**(3), 247–61. http://www.ncbi.nlm.nih. gov/pubmed/7785314?ordinalpos=5&itool=EntrezSystem2.PEntrez.Pubmed.Pubmed_ ResultsPanel.Pubmed_RVDocSum

Pritchard, L.I., Sendow, I., Lunt, R., Hassan, S.H., Kattenbelt, J., Gould, A.R., Daniels, P.W. and Eaton, B.T. (May 2004) Genetic diversity of bluetongue viruses in south east Asia. *Virus Res.* **101**(2), 193–201. http://www.ncbi.nlm.nih.gov/pubmed/15041187?ordinalpos= 6&itool=EntrezSystem2.PEntrez.Pubmed.Pubmed_ResultsPanel.Pubmed_RVDocSum

Purse, B.V., Mellor, P.S., Rogers, D.J., Samuel, A.R., Mertens, P.P. and Baylis, M (2005) Climate change and the recent emergence of bluetongue in Europe. *Nat. Rev. Microbiol.* **3**, 171–81.

Sánchez-Vizcaíno, J.M. (2004) Control and eradication of African horse sickness with vaccine. *Dev. Biol. (Basel)* **119**, 255–8.

Shaw, A.E., Monaghan, P., Alpar, H.O., Anthony, S., Darpel, K.E., Batten, C.A., Guercio, A., Alimena, G., Vitale, M., Bankowska, K., Carpenter, S., Jones, H., Oura, C.A., King, D.P., Elliott, H., Mellor, P.S. and Mertens, P.P.C. (2007) Development and initial evaluation of a real-time RT-PCR assay to detect bluetongue virus genome segment 1. *J. Virol. Methods* **145**, 115–26.

Shope, R.E., MacNamara, L.G. and Mangold, R. (1955) *Epizootic hemorragic disease of deer. New Jersey Outdoors* (November), 16–21.

Shope, R.E., MacNamara, L.G. and Mangold, R. (1960) A virus induced epizootic hemorrhagic disease of the Virginia white-tailed deer (*Odocoileus virginianus*). *J. Exp. Med.* **111**, 155–70.

Singh, K.P., Maan, S., Samuel, A.R., Rao, S., Meyer, A. and Mertens, P.P.C. (2005) Phylogenetic analysis of bluetongue virus genome segment 6 (encoding VP5) from different serotypes. *Vet. Ital.* **40**, 479–83.

St. George, T.D., Cybinski, D.H., Standfast, H.A., Gard, G.P. and Della-Porta, A.J. (1983) The isolation of five different viruses of the epizootic haemorrhagic disease of deer serogroup. *Aust. Vet. J.* **60**, 216–7.

Stallknecht, D.E. (2007a) Haemorrhagic disease in 2007. Southeastern Cooperative Wildlife Disease Study: College of Veterinary Medicine, University of Georgia, **23**(October), pp.1–3.

Stallknecht, D.E. (2007b) hemorrhagic disease events in 2006. Southeastern Cooperative Wildlife Disease Study: College of Veterinary Medicine, University of Georgia, **23**(2) pp. 3–4.

Stallknecht, D.E. (2008) Orbiviruses new and old – what do we need to know? Southeastern Cooperative Wildlife Disease Study: College of Veterinary Medicine, University of Georgia, **23**(4) pp. 1–3.

Stallknecht, D.E., Luttrell, M.P., Smith, K.E. and Nettles, V.F. (1996) Hemorrhagic disease in white-tailed deer in Texas: A case for enzootic stability. *J. Wildl. Dis.* **32**, 695–700.

Stallknecht, D.E., Nettles, V.F., Rollor, III E.A. and Howerth, E.W. (1995) Epizootic hemorrhagic disease virus and bluetongue virus serotype distribution in white-tailed deer, in Georgia. *J. Wildl. Dis.* **31**, 331–8.

Tabachnick, W.J., MacLachlan, N.J., Thompson, L.H., Hunt, G.J. and Patton, J.F. (1996) Susceptibility of *Culicoides variipennis sonorensis* to infection by polymerase chain reaction-detectable bluetongue virus in cattle blood. *Am. J. Trop. Med. Hyg.* **54**, 481–5.

Taraporewala, Z.F. and Patton, J.T. (2004) Nonstructural proteins involved in genome packaging and replication of rotaviruses and other members of the Reoviridae. *Virus Res.* **101**, 57–66.

Theodoridis, A., Nevill, E.M., Els, H.J. and Boshoff, S.T. (1979) Viruses isolated from *Culicoides* midges in South Africa during unsuccessful attempts to isolate bovine ephemeral fever virus. *Onderstepoort J. Vet. Res.* **46**, 191–8.

Toussaint, J.F., Sailleau, C., Mast, J., Houdart, P., Czaplicki, G., Demeestere, L., Vanden-Bussche, F., van Dessel, W., Goris, N., Bréard, E., Bounaadja, L., Etienne, T., Zientara, S. and De Clercq, K. (2007) Bluetongue in Belgium, 2006. *Emerg. Infect. Dis.* **13**, 614–6.

Trainer, D.O. (1964) Epizootic hemorrhagic disease of deer. *J. Wildl. Manage.* **28**, 377–87.

Weir, R.P., Harmsen, M.B., Hunt, N.T., Blacksell, S.D., Lunt, R.A., Pritchard, L.I., Newberry, K.M., Hyatt, A.D., Gould, A.R. and Melville, L.F. (1997) EHDV. *Vet. Microbiol.* **1**, 135–43.

Wilson, A., Carpenter, S., Gloster, J. and Mellor, P. (2007) Re-emergence of bluetongue in northern europe in 2007. *Vet. Rec.* **161**, 487–9.

Wilson, W.C., Ma, H.C., Venter, E.H., van Djik, A.A., Seal, B.S. and Mecham, J.O. (April 2000) Phylogenetic relationships of bluetongue viruses based on gene S7. *Virus Res.* **67**(2), 141–51. erratum in: March 2001 *Virus Res.* **73**(2), 201–2. http://www.ncbi.nlm.nih.gov/pubmed/10867193?ordinalpos=3&itool=EntrezSystem2.PEntrez.Pubmed.Pubmed_ResultsPanel.Pubmed_ RVDoc Sum

Yadin, H., Brenner, J., Gelman, B., Bumbrov, V., Oved, Z., Stram, Y., Galon, N., Klement, E., Perl, S., Anthony, S.J., Maan, S., Batten, C.A. and Mertens, P.P.C. (2008) Epizootic haemorrhagic disease virus type 7 infection in cattle in Israel. *Vet. Rec.* **162**, 53–6.

Bluetongue virus replication and assembly

4

ROB NOAD AND POLLY ROY

Department of Infectious and Tropical Diseases, London School of Hygiene and Tropical Medicine, Keppel Street, London, WC1E 7HT, UK

Introduction

Chapter 6 will discuss in detail the arrangement of proteins in the bluetongue virus (BTV) core. This chapter will therefore focus on reviewing insights from experimental studies into the molecular events that occur during BTV replication and assembly. As it has only recently been possible to introduce directed mutations into BTV genes and recover infectious virus (Boyce *et al.*, 2008), much of the progress in BTV research has relied on direct observations of laboratory strains of the virus or the expression of viral proteins in heterologous systems. However, the latter approach has allowed for the production of mutant proteins and structures that would not be viable in replicating virus and provided valuable data on the molecular strategies for BTV replication and assembly.

Bluetongue disease is currently an emerging disease in Europe but was first described as a disease of wild and domestic ruminants in Africa in the late eighteenth century. The causative agent of the disease is a virus that has been subsequently isolated from many tropical, subtropical and temperate zones. Bluetongue virus is the type member of the arthropod-transmitted genus *Orbivirus* within the family *Reoviridae* and has the characteristic segmented dsRNA genome and multilayered capsid of all viruses within this family (Verwoerd, 1969; Verwoerd *et al.*, 1970; Verwoerd *et al.*, 1972) (Figure 4.1). In sheep, the bluetongue disease is acute and mortality is often accordingly high. Due to its economic significance, BTV has been the subject of extensive molecular, genetic and structural study. As a consequence, it now represents one of the best characterized viruses.

ISBN-13: 978-0-12-369368-6

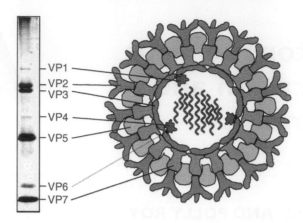

Figure 4.1 The architecture of the bluetongue virus (BTV) particle. (Left) Silver-stained gel showing all of the structural proteins of BTV. Centre cartoon showing the location of the major structural proteins VP3, VP7, VP2 and VP5 and the enzymatic proteins VP1, VP4 and VP6. The position of VP6 is assumed to be complexed with RNA at the VP1/VP4 transcriptase complex. (Right) Cryo-EM reconstruction of intact BTV.

Arrangement of proteins in the virus outer capsid and their role in virus entry

Compared to rotavirus and reovirus, BTV virions (550S) are relatively fragile, and infectivity is lost under mildly acidic conditions (Verwoerd, 1969; Verwoerd *et al.*, 1972). The virions are architecturally complex structures composed of seven discrete proteins that are organized into two concentric shells, the inner and outer capsids (Verwoerd, 1969; Verwoerd *et al.*, 1972; Mertens *et al.*, 1987; Hewat *et al.*, 1992) (see Figure 4.1). Although the structure of the enzymatically active inner capsid (core) particle has been solved to atomic resolution (Grimes *et al.*, 1998), the arrangement of the proteins in the virus outer capsid has, until recently, been relatively uncharacterized. However, cryo-EM studies of the virus particle have recently shown that the outer capsid contains two structures, a triskelion-shaped structure that protrudes from the surface of the virus and a globular structure (Hewat *et al.*, 1992; Nason *et al.*, 2004). When expressed as a recombinant protein (Hassan and Roy, 1999), VP2 oligomerizes into a trimer, which has a volume that matches that of the triskelion structures observed on the virus surface (Nason *et al.*, 2004). Therefore the globular protein detected on the virus surface must be the other viral outer capsid protein VP5.

VP2 is the most variable BTV protein (see reviews by Roy, 1992; Mertens *et al.*, 2005) and elicits virus-neutralizing antibodies, specifies the serotype and is the viral haemagglutinin (Huismans and Erasmus, 1981; Kahlon *et al.*, 1983; French and Roy, 1990; Roy *et al.*, 1990). Enzymatic treatments of mammalian cells permissive for BTV infection suggest that the cellular receptor molecule for BTV is a glycoprotein, although it is likely that additional co-receptors are also required (Eaton and Crameri, 1989; Hassan and Roy, 1999). Soluble VP2 trimers also have a strong affinity for binding to glycophorin A and are taken up into mammalian cells that are susceptible to virus infection (Hassan and Roy, 1999) (Figure 4.2). These data are all consistent with the assignment of VP2 as the spiky, outermost viral capsid protein.

During BTV entry into susceptible mammalian cells, VP2 mediates virus attachment to the cell and the virus is internalized in clathrin-coated vesicles by receptor-mediated endocytosis. Both VP2 and VP5 can be detected in the early endosome, and the addition of compounds that prevent endosomal acidification prevents the release of the viral core particle into the cytoplasm (Eaton *et al.*, 1990). Late endosomes are not involved in BTV entry (Eaton *et al.*, 1990; Hassan and Roy, 1999; Forzan *et al.*, 2004). Furthermore, inhibition of the clathrin entry pathway using pharmacological and siRNA reagents has been found to dramatically inhibit virus replication (Forzan *et al.*, 2007). During the entry process, both VP2 and VP5 are lost from the surface of the core. Although the role of VP2 is clearly in virus attachment, VP5 is critical in the pH-dependant release of viral cores from the endosomal compartment. VP5 has structural features also observed in the fusion proteins of enveloped viruses, including a membrane-inserting hydrophobic region followed by a long heptad repeat region (mediating oligomerization) at its N-terminus. The predicted structure of VP5 includes a carboxyl-terminal globular domain separated from a coiled-coil amino terminus by a flexible hinge region (Figure 4.3). VP5 is a cytotoxic protein when over-expressed separately from other viral proteins and has membrane-permeabilizing activity *in vitro* (Hassan *et al.*, 2001; Forzan *et al.*, 2004). This activity is dependent on N-terminal amphipathic helices (Hassan *et al.*, 2001) and can functionally mimic the fusion activity of enveloped virus when VP5 is expressed on the cell surface (Forzan *et al.*, 2004). In addition, fusion in these experiments was found to be dependent on a low-pH incubation to trigger the activity. Deletion

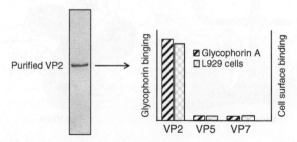

Figure 4.2 Purified recombinant protein was used to identify which of the surface-exposed viral proteins was responsible for the binding of bluetongue virus (BTV) to mammalian cells. Only VP2 demonstrated significant binding to cells or glycophorin A.

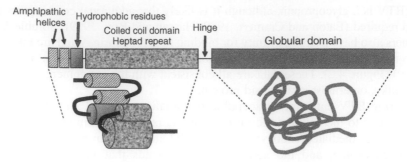

Figure 4.3 Predicted domain structure of VP5. Two amphipathic helices are present at the N-terminus of the protein, followed by a coiled-coil domain, a linker region and a globular domain.

of the amphipathic helices completely prevented fusion activity (Forzan *et al.*, 2004). There is no evidence that BTV VP5 requires the proteolytic activation step necessary for rotavirus entry. This difference is probably related to the normal route of transmission of the different viruses.

In the cryo-EM reconstruction of BTV particles, the triskelion protein (VP2) is positioned on the top surface of the VP7 trimers, while the globular protein (VP5) makes contact mainly with the sides of the VP7 trimers (Figure 4.4).

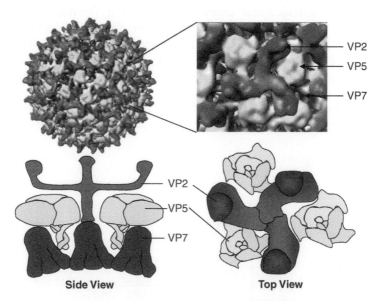

Figure 4.4 Arrangement of the proteins in the outer capsid of bluetongue virus (BTV). (Top) Cryo-EM reconstructions highlighting the relative position of the tris-kelion protein (VP2) and the globular protein (VP5). (Below) Cartoon showing the arrangement of VP2, VP5 and VP7 relative to each other (See colour plate 1).

Overall, the principal protein–protein contacts of the globular and triskelion proteins of the BTV outer capsid are with the underlying VP7 layer rather than with each other. This may be related to the sequential attachment and endosomal penetration steps mediated by each protein during virus entry into uninfected cells. Combining the structural and biochemical data, a plausible hypothesis would be that the low pH inside endosomes induces a rearrangement and conformational change in VP5, loosening the interactions of VP5 and VP7 and simultaneously allowing VP5 to form a protein layer with intrinsic outside-in curvature. Proteolytic removal or low-pH-induced conformational change of VP2 then allows the VP5 amphipathic helices to freely interact with membranes and initiate the permeabilization process.

Functional dissection of the enzymatic core proteins

In the same way that the function of the outer capsid proteins VP2 and VP5 has been elucidated by studying their activity when expressed and purified in the baculovirus expression system, the enzymatic functions of all three minor core enzymes have been demonstrated *in vitro* using the same system.

Upon release into the cytoplasm, BTV core particles initiate the transcription of the viral genome, and newly synthesized viral messenger-sense RNAs (mRNAs) that are capped but not polyadenylated are extruded into the cytoplasm (Verwoerd and Huismans, 1972). The two major proteins of the core, VP3 and VP7, account for the overall morphology of the particle, while the three minor proteins, VP1, VP4 and VP6, are responsible for the transcription of mRNA. The core particle marks an end point in virus disassembly and thus protects the viral dsRNA genome from cellular antiviral surveillance mechanisms. When intact cores isolated from virus particles are activated *in vitro* in the presence of magnesium ions and NTP substrates, distinct conformational changes can be seen around the five-fold axis of the BTV core (Gouet *et al.*, 1999). This is interpreted as outward movement of VP3 and VP7, allowing opening of a pore in the VP3 layer at the five-fold axes. The mRNAs are extruded through these pores and leave the core particle through the central space in the pentameric rings of VP7 trimers. The detailed structure of the transcriptase complex positioned inside the core at the five-fold axis was not resolved in the X-ray structure. However, cryo-EM reconstructions of core-like particle (CLP) formed with and without the minor enzymatic proteins VP1 and VP4 have revealed that they form a flower-like structure that protrudes inwards from the underside of the VP3 layer (Figure 4.5). Although the structural integrity of the core particle appears to be essential for the maintenance of efficient transcriptional activity, it has been possible to use *in vitro*

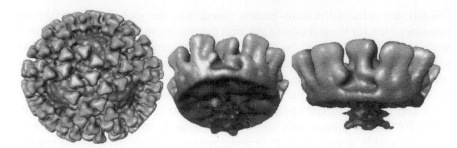

Figure 4.5 Cryo-EM reconstruction of core-like particles (CLPs) incorporating VP1 (polymerase) and VP4 (capping enzyme). (Left) CLPs in which the VP7 trimer at the five-fold axis has been lost. (Centre and right) Position and shape of the transcriptase complex (blue) relative to VP3 (pink) and VP7 (green) at the five-fold axes (See colour plate 2).

assays with three minor proteins of the core to delineate specific roles for each within the transcription complex.

Since transcription of the dsRNA genome of members of the *Reoviridae* occurs by a fully conservative process (Bannerjee and Shatkin, 1970), it is logical that this process involves a helicase activity, either to unwind the dsRNA ahead of the transcriptase protein or to separate the parental and newly synthesized RNAs following transcription. For BTV, VP6, which is rich in basic amino acids (Arg and Lys), has a strong binding affinity for ssRNAs and dsRNAs and has ATP binding, ATP hydrolysis and RNA unwinding activities *in vitro* (Stauber *et al.*, 1997). Mutation studies have also functionally mapped a typical conserved ATP-binding domain of helicases ($AXXGXGK_{110}V$) and the RNA-binding site (R_{205}) within VP6 (Kar and Roy, 2003). Furthermore, purified VP6 forms stable oligomers with a ring-like morphology in the presence of nucleic acids similar to those observed for other helicases (Kar and Roy, 2003) (Figure 4.6).

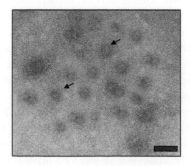

Figure 4.6 Electron micrograph of the ring-like structures that are formed when the bluetongue virus (BTV) helicase, VP6, is incubated with viral RNA.

The replication of viral dsRNA occurs in two distinct steps. First, as just described, plus-strand RNAs (mRNAs) are transcribed, using the negative strand of dsRNA segments as templates, and extruded from the core particle (Bannerjee and Shatkin, 1970). Second, the plus-strand RNAs serve as templates for the synthesis of new minus-strand RNAs at an as yet undefined stage during the assembly of new virus core particles. It is not known whether all the *de novo* dsRNA molecules are synthesized simultaneously or consecutively. *In vitro*, the largest core protein VP1 (150 kDa) has the ability to both initiate and elongate minus strand synthesis *de novo* in the absence of other core proteins or structures (Boyce *et al.*, 2004). This does not require a specific secondary structure present at the 3′ end of the plus-strand template. Studies of replicase activity in other members of the *Reoviridae* have used very short templates (Tao *et al.*, 2002) or have required a particulate replicase in which the catalytic subunit is proposed to possess replicase activity only in the context of a subviral particle (Patton *et al.*, 1997). Since the replicase activity of isolated recombinant BTV VP1 *in vitro* is relatively low, it is possible that the activity of VP1 is enhanced by other viral proteins present in the assembling core particle. Although a detailed structure of the equivalent protein for another member of the *Reoviridae* has been solved (Tao *et al.*, 2002), there is currently no structure for BTV VP1. Comparisons of the sequences of VP1 and the reovirus polymerase *in silico* have revealed that there is most homology between the proteins in both the N-terminal and C-terminal domains. The polymerase domain of VP1 itself more closely resembles that of the RNA-dependent RNA polymerase of two positive-sense RNA viruses: rabbit haemorrhagic disease virus and polio virus (Wehrfritz *et al.*, 2007). From these comparisons it has been possible to assign a putative structure for domains within VP1 (Figure 4.7). This structural assignment is supported by reconstitution experiments where each of the three domains is expressed separately and subsequently mixed to recover the enzymatic activity of the protein (see Figure 4.7).

Although VP1 has been shown to be active as the viral RNA-dependent RNA polymerase, it is not sufficient for the synthesis of the methylated cap structure found at the 5′ end of BTV mRNAs. Recombinant, purified VP4 (76.4 kDa) can synthesize type 1-like 'cap' structures *in vitro* that are identical to those found on authentic BTV mRNA. VP4 possesses guanylyltransferase, RNA triphosphatase and two methyltransferase activities that are necessary for the capsid structure, and this does not require any other viral or cellular proteins to cap viral RNA (Martinez Costas *et al.*, 1998; Ramadevi and Roy, 1998; Ramadevi *et al.*, 1998) (Figure 4.8). This is notably different from other viral capping enzymes, e.g. those of vaccinia virus, where completion of capping is dependent on a complex of three proteins (Venkatesan *et al.*, 1980). Most recently, it has been possible to generate an atomic structure for this enzyme at 2.5 Å resolution (Sutton *et al.*, 2007). This is the first example of the structural determination of an uncomplexed, single protein from any

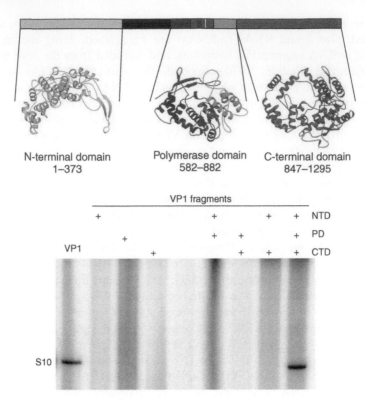

Figure 4.7 (Top) Predicted structure of domains of VP1 based on homology with other viral RNA-dependent RNA polymerases. (Bottom) The assignment of domains of VP1 was validated by the demonstration that polymerase activity could be reconstituted by mixing expressed N-terminal domain (NTD), polymerase domain (PD) and C-terminal domain (CTD) in a standard polymerase assay for the intact VP1 protein (See colour plate 3).

source that has two methyltransferase [guanine-N7-methyltransferase (N7MTase) and nucleoside-2′-*O*-methyltransferase activity (2′OMTase)] as well as the phosphotransferase and guanylyltransferase activities and has provided tantalizing data on how the enzyme activities associated with this protein are arranged in its structure. Currently, assignment of functional domains for this enzyme has been made based on structural homology with enzymes that share some of the activities of VP4 and using crystallographic studies in which crystals are incubated with enzyme cofactors treatments. Future studies on this protein from these structural studies may well lead to exciting new insights into how the enzymatic activities are achieved.

In summary, each of the three minor proteins of BTV core has the ability to function on its own. Together they constitute a molecular motor that can

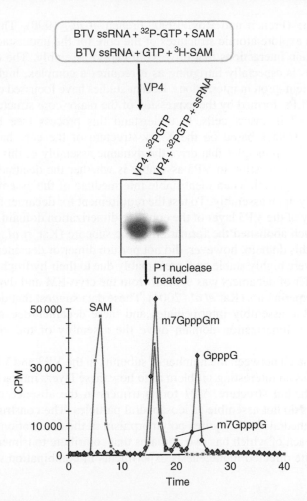

Figure 4.8 HPLC analysis of the reaction products from incubation with purified BTV VP4 with single-stranded RNA (ssRNA) and either Ado[methyl-3H]Met (SAM) or 32P-GTP, followed by P1 nuclease digestion reveals the presence of the guanylyl-transferase and two methyltransferase activities necessary for the production of the type I cap structure present in viral RNA.

unwind RNAs, synthesize ssRNAs of both polarities and modify the 5′ termini of the newly synthesized mRNA molecules.

Assembly of the viral core

Bluetongue virus structural proteins, with the possible exception of VP6, have inherent capacity to self-assemble into virus-like particles (VLPs) that lack the viral genome (French and Roy, 1990; French *et al.*, 1990). This has been exploited to explore atomic structures of the core for the understanding of the protein–protein interactions that drive viral capsid assembly. The assembly of BTV capsids is especially intriguing as it requires a complex, highly ordered, series of protein–protein interactions. Recent studies have focussed on the use of VLPs and CLPs, formed by the expression of the major core structural proteins VP3 and VP7 in insect cells, to understand this process (see Roy, 2005). Mutagenesis studies based on the X-ray structure of the core have revealed some of the key principles that drive the dynamic assembly of this structure.

An important question in VP3 assembly is whether the decamer present in the assembled particle is an identifiable intermediate of the assembly process or arises only upon assembly. To test the requirement for decamer formation in the assembly of the VP3 layer of the core, the dimerization domain of VP3 was deleted, which abolished the formation of the subcore (Kar *et al.*, 2004). The deletion of this domain, however, did not perturb dimer or decamer formation. Decamers were highly stable and, presumably due to their hydrophobic nature, a higher order of decamers was evident from the cryo-EM and dynamic light-scattering experiments (Kar *et al.*, 2004). These data suggest that decamers are the first stable assembly intermediates and these decamer–decamer interactions via the dimerization domain drive the assembly of the viral subcore (Figure 4.9).

The mismatch between the number of subunits in the VP3 and VP7 layers of the core poses an interesting problem as to how these layers interact to form an intact icosahedral structure. VP7 forms trimers in the absence of VP3, but these trimers do not assemble as icosahedral particles. The construction of the $T = 13$ icosahedral shell requires polymorphism in the association of the VP7 monomers each of which has two domains that contribute to trimer formation. Extensive site-directed mutagenesis experiments in combination with the use

Figure 4.9 Steps in the assembly of the VP3 layer (subcore) of bluetongue virus (BTV). VP3 adopts two different conformations in the mature core structure. Mutagenesis studies have revealed that the first stable structural intermediate in the assembly of the subcore is a decamer and that these decamers are the building blocks of subcore assembly (See colour plate 4).

of various assembly assay systems have suggested the order of the assembly pathway of VP7 onto the subcore (VP3) layer during stable core formation.

Mutations in VP7 directed at perturbing monomer–monomer, trimer–trimer and VP7–VP3 interactions have been generated. The effects of these mutations on VP7 solubility, ability to trimerize and form CLPs in the presence of VP3 were investigated (Limn *et al.*, 2000; Limn and Roy, 2003). Residues critical for the formation of the VP7 layer were identified, and it was shown that core assembly depends on trimer formation, the precise 'shape' of the trimers being sufficient to drive the formation of the tight lattice of 260 VP7 trimers on the core surface. Mutagenesis data also suggest that initially multiple sheets of VP7 form around different nucleation sites instead of a cascade of trimer associations originating from a single nucleation site as was postulated from the initial structural studies (Grimes *et al.*, 1998). A likely pathway of core assembly is therefore that a number of strong VP7 trimer–VP3 contacts act as multiple equivalent initiation sites and that a second set of weaker interactions then 'fills the gaps' to complete the outer layer of the core (Figure 4.10). This model is possible because not all of the VP7 trimers have equal contacts with the VP3 layer. There is a clear sequential order of trimer attachment onto the VP3 scaffold. The 'T' trimers (of the P, Q, R, S and T trimers), which are at the three-fold axes of symmetry, act as nucleation, while 'P' trimers that are furthest from the three-fold axis and closer to the five-fold axis, are the last to attach (Limn *et al.*, 2000; Limn and Roy, 2003).

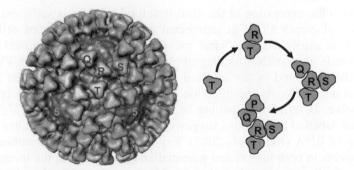

Figure 4.10 Combined mutagenesis and cryo-EM studies have revealed the order of assembly of the VP7 layer onto the VP3 subcore. If VP7 trimers in core-like particle (CLPs) are labelled P, Q, R, S, T with the P timer positioned at the five-fold axis and the T trimer at the three-fold axis of the VP3 layer, the P trimer has the poorest contacts with the VP3 layer and is easily removed and the T trimer has the most favourable contacts. (Left) Cryo-EM reconstruction of CLP in which the VP7 P trimer has been removed in a CsCl gradient and the other trimers and the visible five-fold axes are labelled. (Right) Cartoon showing order of assembly of VP7 trimers on the VP3 subcore. The T trimer nucleates assembly at the three-fold axis and other trimers are sequentially added from this point (See colour plate 5).

In addition to the production of CLPs that contain two major structural proteins VP3 and VP7, it is also possible to form CLPs that also contain the minor enzymatic proteins VP1 and VP4 proteins. Indeed both of these proteins independently associate with the VP3 layer, with VP4 having more direct contact with the undersurface of VP3 at the five-fold axes of the particle (Loudon and Roy, 1992; Nason et al., 2004). A VP1/VP4 complex also directly interacts with the VP3 decamer in solution (Kar et al., 2004). By contrast, BTV RNAs, while interacting with the intact VP3 in vitro very efficiently, fail to associate with VP3 decamers under same conditions. The interaction of VP1 and VP4 with VP3 decamers suggests that assembly of the BTV core may initiate with the complex formed by these two proteins and the VP3 decamers, and these assembly intermediates subsequently recruit the viral RNA and VP6 complexes prior to completion of the assembly of the VP3 subcore.

Unlike VP1 and VP4, it has not been possible to confirm the location of VP6 in the core, although it is likely that VP6 is also located within the five-fold axes of the VP3 layer directly beneath the decamer together with VP1 and VP4. VP6 forms defined hexamers in the presence of BTV transcripts and subsequently assembles into distinct ring-like structures that can be isolated by glycerol gradient centrifugation (Kar and Roy, 2003). It is not known whether such structures are present within the core during intracellular morphogenesis and whether these are incorporated either together with VP1 and VP4 into the VP3 decamer intermediates stage or independently during the assembly of the 12 VP3 decamers.

Although the expression of the viral structural proteins is sufficient for the formation of empty CLPs in uninfected cells, in virus-infected cells, core assembly is associated with viral inclusion bodies (VIBs) formed largely from a single viral ssRNA-binding phosphoprotein, NS2. During BTV infection, VIBs are found in the infected cells, predominantly near the nucleus (Eaton et al., 1990). As the infection progresses, VIBs increase in both size and number. In addition, labelling of new RNA synthesis in BTV-infected cells with labelled nucleotides suggests that VIBs are principal sites for the synthesis of RNA (Kar et al., 2007). Expression of BTV NS2 without other viral proteins in both insects and mammalian cells results in the formation of inclusion bodies that are indistinguishable from the VIBs found in virus-infected cells (Thomas et al., 1990; Modrof et al., 2005). Co-expression of core structural proteins has shown that while VP1, VP3, VP4 and VP6 all co-localize with NS2 inclusions without other viral proteins, VP7 requires the co-expression of VP3 to be recruited to these structures. This may be an early indication that partially assembled core structures in which VP3 and VP7 are already interacting are the building blocks of core formation.

In addition to binding to viral core proteins, NS2 also binds to viral ssRNA. Purified recombinant NS2 has strong ssRNA binding activity in vitro, and BTV RNAs are preferentially bound over non-specific RNAs (Theron and Nel,

1997; Lymperopoulos *et al.*, 2003; Markotter *et al.*, 2004). These observations support the suggestion that NS2 may have a role in the recruitment of mRNA for replication. The amino terminus of NS2 is essential for ssRNA binding, but neither the amino- nor the carboxy-terminus of the protein is required for oligomerization (Zhao *et al.*, 1994). These observations are significant since the mechanism by which 10 dsRNA segments that make up the viral genome are selectively recruited and packaged into newly assembling virus particles is unresolved and represents one of the most enduring questions in the field. Electrophoretic mobility shift competition assays *in vitro* have suggested that NS2 has a high affinity for specific BTV RNA structures that are unique in each RNA segment (Lymperopoulos *et al.*, 2003; Lymperopoulos *et al.*, 2006). The sequences that represent the putative binding partners for NS2 are neither the conserved 5′ terminal octanucleotides nor 3′ terminal hexanucleotides of all BTV RNA segments or the potential panhandle structures (formed by the partial complimentary sequences of the 5′ and 3′ termini). Instead they are distributed throughout the coding and non-coding regions of the different genome segments (Lymperopoulos *et al.*, 2003). Chemical and enzymatic structure probing of regions bound preferentially by NS2 has shown that the NS2 binds BTV RNA transcripts at unique hairpin-loop secondary structures. The NS2–hairpin interaction has been further confirmed by using hairpin mutants, which alter the sequence or the structure of the RNA independently (Lymperopoulos *et al.*, 2006). No other RNA-binding protein of any other member of *Reoviridae* so far has been shown to have similar RNA structural specificity. However, rotavirus NSP3 appears to bind a linear sequence found at the 3′ end of all rotavirus RNA segments (Poncet *et al.*, 1993), and rotavirus VP1 also binds to the 3′ end of rotavirus RNA (Patton, 1996). The significance of these interactions for genome packaging is, at this stage, unclear and it is worth pointing out that, other than BTV, there are several systems where viral proteins recognize packaging signals that are contained in hairpin structures with no apparent sequence homology (Bae *et al.*, 2001; Beasley and Hu, 2002). Although the RNA binding activity of NS2 can explain how BTV mRNAs are selected from the pool of cellular messages for incorporation into assembling virus particles, it remains unclear how only a single copy of each genome segment is included in newly formed core particles. NS2 could bring the viral RNAs into close proximity, and inter-segment RNA interactions allow the formation of an RNA complex that is the basis of core assembly. Alternately, different RNA subsets that have already interacted may be bound by NS2. The latter hypothesis is consistent with the observations that NS2 may form decameric complexes (Butan *et al.*, 2004) and that each NS2 protein subunit may have several RNA-binding domains (Fillmore *et al.*, 2002).

Although inclusion bodies have obvious advantages in providing a site for core assembly and genome packaging as well as in preventing premature shutoff of core transcription by preventing the addition of viral outer capsid proteins, they may also represent a 'trap' for newly assembled core particles. In order for newly assembled core to be released from inclusions, it is necessary that these structures must be dynamic and easily disassembled as well as assembled. NS2 is the only BTV-encoded phosphoprotein. Recent data suggest that protein kinase casein kinase II (CKII) interacts with NS2 and is able to phosphorylate it *in vitro* (Modrof *et al.*, 2005). Mutagenesis of two normally phosphorylated serine residues (aa 249 and 259) to alanine in NS2 abrogates the formation of distinct VIBs, whilst engineered mutation of the same residues to aspartic acid, mimicking a constitutively phosphorylated state, results in the formation of normal inclusion body structures (Modrof *et al.*, 2005). Furthermore, the expression of the non-phosphorylated mutant of NS2 was sufficient to disrupt the formation of VIBs in virus-infected cells, demonstrating the potential of the mutant to act in a dominant negative fashion and the control of the degree of phosphorylation over BTV VIB stability (Modrof *et al.*, 2005). The sequence context of the phosphorylated serine 249 ($_{248}$LSDDDDQ$_{254}$) in NS2 conforms to a conserved sequence, which generally stabilizes helix unfolding (Zetina, 2001), suggesting that phosphorylation of NS2 in this position could be involved in stabilizing protein folding within its C-terminal domain. Thus, phosphorylation may act to control the propensity of NS2 to form large VIBs and thus provide a mechanism by which newly assembled core particles can be released from VIBs.

Assembly of the viral outer capsid

Although our understanding of the physical contacts between the outer layer of the core, VP7 and the viral outer capsid proteins, VP2 and VP5, has increased recently (see above, Nason *et al.*, 2004), our understanding of how this process is controlled is less clear. Since the addition of VP2 and VP5 abolishes the transcription activity of the core, addition of the outer capsid is likely to be a highly regulated process to prevent premature shutoff of transcription. Unlike rotavirus, there is no evidence that assembly of the outer capsid is associated with the ER. Instead, within BTV-infected cells, virus particles are found associated with the vimentin intermediate filaments (Eaton *et al.*, 1990). Furthermore, the presence of both VP2 and VP5 was found to be necessary to direct VLPs to the cytoskeleton of insect cells (Hyatt *et al.*, 1993). By contrast, VP2 alone associates with vimentin intermediate filaments, and disruption of these filaments drastically reduces the normal release of virus particles from infected cells (Bhattacharya *et al.*, 2007). Given the importance of these structures for the assembly and egress of many other viruses (Arcangeletti *et al.*, 1997; Nedellec *et al.*, 1998), it may be that intermediate filaments are used to control the addition of the VP2 and possibly VP5 to the assembled core.

Release of progeny virions from infected cells

For BTV, release of newly assembled virus particles varies between host cell types, with insect cells allowing a non-lytic release of virus, while the majority of virus particles in mammalian cells remain cell associated (Hyatt *et al.*, 1989). However, even in mammalian cells that show substantial cytopathic effect in response to BTV infection, the titre of virus in the culture supernatant increases significantly before the onset of dramatic CPE. Thus it is clear that there are defined mechanisms that traffic newly formed virus particles out of infected cells. Virus particles have been observed to leave infected cells in one of two ways. First, particularly early in infection, particles can be seen to bud through the cell membrane and acquire, at least temporarily, an envelope, although evidence for the presence of viral antigen on the outside of such particles is lacking (Bowne, 1967; Eaton *et al.*, 1990). Second, egress of virus particles from infected cells is accomplished by a process of extrusion in which individual or groups of virus particles move through a local disruption of plasma membrane (Hyatt *et al.*, 1989). This appears to happen without a significant effect on host cell viability, and both types of particles have been observed to be released from some cell surfaces (Hyatt *et al.*, 1989).

Recent evidence has implicated the viral non-structural proteins NS1 and NS3 in this process and has revealed intriguing parallels between the egress of non-enveloped BTV and enveloped viruses (Beaton *et al.*, 2002; Owens and Roy, 2004). NS3 (229 aa) and its shorter form NS3A (216 aa) are the only membrane proteins encoded by BTV (Lee and Roy, 1986; French *et al.*, 1989). Both proteins can be found associated with smooth-surfaced, intracellular vesicles (Hyatt *et al.*, 1993) but do not form part of the stable structure of the mature virus. NS3/NS3A proteins have long N-terminal and a shorter C-terminal cytoplasmic domains connected by two transmembrane domains and a short extracellular domain (Wu *et al.*, 1992; Bansal *et al.*, 1998; Beaton *et al.*, 2002). A single glycosylation site is present in the extracellular domain of BTV NS3 but is missing in NS3 of AHSV, a closely related orbivirus, and therefore does not seem to be essential for the function of the protein (van Staden and Huismans, 1991; Bansal *et al.*, 1998). The NS3/NS3A proteins accumulate to only very low levels in BTV-infected mammalian cells, but in invertebrate cells, the expression levels of these proteins are high (French *et al.*, 1989; Guirakhoo *et al.*, 1995). The correlation between high NS3/NS3A expression and non-lytic virus release has suggested a significant functional role for NS3 in virus egress from invertebrate cells, and NS3 and NS3A facilitate the release of baculovirus-expressed VLPs (acting as surrogates for authentic virions) in heterologous insect cells. The NS3 protein is localized at the site of the membrane where VLPs are released (Hyatt *et al.*, 1993), confirming that the integration of NS3/NS3A into the plasma membrane may trigger the release of mature virions. The NS3 protein of BTV and AHSV exhibit cytotoxicity when expressed separately in mammalian or insect cells (French *et al.*, 1989; van Staden *et al.*, 1995). The cytotoxicity of NS3 requires membrane association of the protein and depends on the presence of the N-terminal transmembrane domain – suggesting that NS3 might function as a viroporin and facilitate virus release by inducing membrane permeabilization (Han and Harty, 2004). It is possible that this permeabilization activity causes local disruption of the plasma membrane allowing virus particles to be extruded through a membrane pore without acquiring a lipid envelope.

The first 13 amino acids of NS3 that are absent in NS3A have the potential to form an amphipathic helix. This cytoplasmic region of NS3 also interacts with the calpactin light chain (p11) of the cellular annexin II complex (Beaton *et al.*, 2002), a complex that has been implicated in membrane-related events along the endocytic and regulated secretory pathways including the trafficking of vesicles (for reviews see Creutz, 1992; Raynal and Pollard, 1994). The interaction between NS3 with p11 is highly specific, and an NS3 peptide (a mimic of the sequences of the p11-binding domain) inhibits the progeny virions from BTV-infected insect cells. Although the exact physiological role of this interaction is still unknown, it is likely that interaction of p11 with NS3 may direct NS3 to sites of active cellular exocytosis or that NS3 could become part of an active extrusion process. There are some indications that cytoskeletal material is released at sites of BTV egress (Hyatt *et al.*, 1991), which may be annexin II being drawn through the membrane during the extrusion process, while it is still associated with NS3. The significance of this interaction to BTV egress becomes more apparent in the light of the observation that the other cytoplasmic domain of the protein, situated at the C-terminal end, interacts specifically with the BTV outer capsid protein VP2 (Beaton *et al.*, 2002).

In addition to its interaction with the p11 component of the annexin II complex, NS3 is also capable of interaction with Tsg101 (Figure 4.11), a cellular protein implicated in the intracellular trafficking and release of a number of enveloped viruses (Freed, 2004). Tsg101 specifically interacts with a PTAP motif that is present in the late domain of the retroviral Gag protein. Other motifs present in this Gag include the YPDL motif, interacting with the protein Alix that functions downstream of Tsg101, and the PPXY motif that plays a role in recruiting host ubiquitin ligases. The PTAP and PPXY motif have also been identified in proteins of other enveloped viruses, where they exhibit an equivalent function as the late domain motifs of retro-viruses (Freed, 2002, 2004). Similar motifs are also present within the NS3 of BTV and certain other orbiviruses (Wirblich *et al.*, 2006). Recent findings showed that NS3 and NS3A of both BTV and AHSV bind *in vitro* with similar affinity to human Tsg101 and also to its ortholog from *Drosophila melanogaster*. This interaction is mediated by the conserved PSAP motif in NS3 of BTV and seems to play a role in virus release as knockdown of Tsg101 with siRNA inhibits release of BTV from mammalian cells (Wirblich *et al.*, 2006).

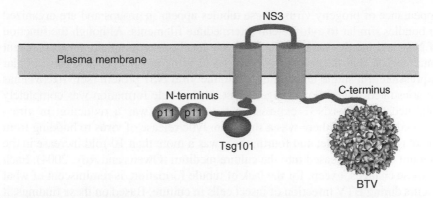

Figure 4.11 Bluetongue virus (BTV) NS3 is an integral membrane protein that interacts at its N-terminus with cellular release factors calpactin p11 and Tsg101 and at its C-terminus with the BTV outer capsid protein VP2. The interactions with calpactin and p11 are important for proper release of newly assembled virus particles from infected cells.

Like most other viral proteins that recruit Tsg101, NS3 also binds *in vitro* NEDD4-like ubiquitin ligase via a resident PPXY late domain motif. However, the late domain motifs in NS3 do not function as effectively as the late domains of other enveloped viruses (Wirblich *et al.*, 2006). This appears to be mainly due to the presence of a unique arginine at position 3 of the PPXY motif in NS3. The low activity of the NS3 late domain motifs can be reversed by converting the PPRY motif of NS3 into a more universal PPPY, rendering NS3 as effective as Gag in facilitating retroviral VLP release as the late domains of enveloped viruses.

Thus, orbivirus NS3 recruits the cellular protein Tsg101 to facilitate virus release from mammalian cells and presumably insect cells as well. The ability to usurp the vacuolar protein-sorting pathway is likely to be more important in insect hosts as orbiviruses establish persistent infections in insect cells without causing significant cytopathic effect. Although full clarification of this issue will have to await the availability of a reverse genetics system for BTV, it should be possible to identify insect proteins that interact with NS3 and to shed more light on the question of whether NS3 is better adapted to engage insect proteins, thus facilitating improved virus release.

In addition to the inclusion body protein NS2, and NS3/NS3A, BTV produces a third non-structural protein in virus-infected cells. This protein NS1 is highly conserved between viral serotypes and is synthesized in large amounts, forming up to 25% of total BTV proteins in virus-infected cells. NS1 also forms the tubule structures that are formed in infected cells and are characteristic of orbivirus infections (Huismans and Els, 1979). Tubules can be seen throughout the infection cycle and as early as 2–4 h post-infection, long before the appearance of progeny virus. These tubules appear in groups and are organized in bundles similar to cytoskeletal intermediate filaments. Although the function of NS1 and tubules in the BTV life cycle has been a complete mystery, recent data suggest that the protein is involved in virus morphogenesis. Intracellular expression of a single-chain antibody to the viral NS1 protein (scFv-αNS1) has demonstrated four major changes. First, NS1 tubule formation was completely inhibited by scFv-αNS1 expression; second, there was a reduction in virus-induced CPE; third, there was a shift from lytic release of virus to budding from the plasma membrane; and fourth, there was a more than 10-fold increase in the amount of virus released into the culture medium (Owens and Roy, 2004). Each of these changes, except for the lack of tubule formation, is reminiscent of what occurs during BTV infection of insect cells in culture. Based on these findings, it has been proposed that NS1 is a major determinant of pathogenicity of BTV in the vertebrate host and that its mechanism of action is the augmentation of virus cell association (but not transport of virus to the cell surface), ultimately leading to the lysis of the infected cell.

Differential virus release in different cells also suggests the involvement of host proteins. Indeed a yeast two-hybrid analysis have identified NS1 interaction with the cellular protein SUMO-1, which is responsible for the induction of post-translational modification of proteins involved in intracellular trafficking (Noad and Roy, unpublished data). In addition, sumoylation of cellular proteins is highly enhanced in BTV-infected cells. Upregulation of SUMO-1 has recently been shown to regulate dynamin-dependent protein trafficking within the cell (Mishra et al., 2004). Thus, it is plausible that the role of NS1 is to control the trafficking of immature or mature virus particles in infected cells through the interaction with cellular proteins such as SUMO-1.

In summary, it is likely that both NS1/tubules and NS3 are involved in progeny virus maturation and trafficking by acting to facilitate interaction between virus and cellular components (Figure 4.11).

Recent development of reverse genetics system for BTV that allows direct introduction of site-directed mutagenesis in viral genome.

Recently it has been possible to demonstrate that viral ssRNA transcripts synthesed *in vitro* from purified BTV cores and introduced into mammalian cells led to the recovery of infectious virus (Boyce and Roy, 2007). The ability to recover infectious BTV wholly from ssRNA also suggested a means for establishing a helper virus-independent reverse genetics system for BTV. Subsequently it was found to be possible to replace one or more BTV transcripts in this system with transcripts derived *in vitro* from cDNA downstream of a T7 RNA polymerase promoter (Boyce et al., 2008). Further, when all 10 BTV transcripts, synthesized *in vitro* from cDNA copies of the 10 segments of the BTV genome, were provided it was possible to also recover infectious virus (Boyce et al., 2008). The recovered infectious virus showed no difference from either the native virus or the infectious virus recovered from *in vitro* core-derived transcripts. As has been sufficiently demonstrated for other viruses, a reverse genetics system for BTV should contribute to the further understanding of the functions of viral proteins in replicating virus and allow the corroboration of the enzymatic or structural functions already assigned. Further, the ability to recover specific mutations in the genome of BTV for the first time not only provides a novel tool for the molecular dissection of BTV and related orbiviruses, but also the opportunity to develop specifically attenuated vaccines to these viruses.

In addition reverse genetic will allow the mapping of the *cis*-acting RNA sequences that control the replication, packaging, and expression of orbivirus genomes that are still need to be defined. The replacement of outer capsid proteins can be used to generate vaccine strains with different serotypes based on a common genetic background and it may be possible to identify the determinants of pathogenicity of BTV and related orbiviruses such that strains with varying levels of attenuation could be generated. In summary such system has now opened up a window of opportunity both for fundamental basic science as well as for targeted designed vaccines for BTV and other orbiviruses.

Concluding remarks

Understanding of the assembly and replication of BTV and other orbiviruses has benefited greatly from detailed structural understanding of the viral core. However, a complete understanding of a virus replication cycle is possible only through the synthesis of structural, biochemical and cell biology data to build an overall model for how the virus completes all of the necessary steps to invade the host and evade innate and adaptive immune responses. By combining these approaches, it has been possible to provide a detailed understanding of the role of each of the viral proteins in virus replication and assembly (Figure 4.12). However, significant questions such as how the packaging of a complete set of 10 different genome segments into each assembling core is controlled and the role of tubules in virus replication remain unanswered. Also, recent work has revealed unexpected and striking parallels between the entry and release pathways of non-enveloped BTV and pathways involved in the entry and release of enveloped viruses. These parallels may be because BTV evolved from an enveloped ancestor virus, or because there simply are a limited number of cellular pathways that can be used for egress of large protein complexes from cells. It is notable that the NS3 glycoprotein of BTV is an integral membrane protein that is functionally involved in virus egress by bridging between the outer capsid protein VP2 and the cellular export machinery. Although no cell-free enveloped form of BTV has been isolated, a number of EM studies have shown structures that may be BTV particles apparently budding with envelope from infected cells at the plasma membrane. The resolution of these questions regarding virus structure, assembly and interaction with the host cell will undoubtedly generate exciting and unexpected findings in the future. Indeed,

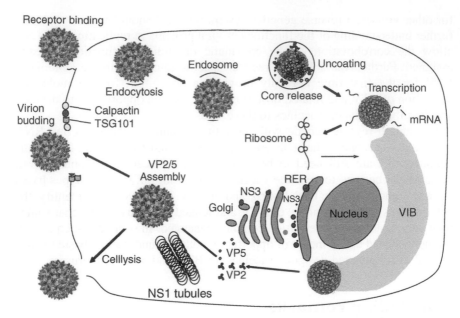

Figure 4.12 Summary of our current understanding of the intracellular replication cycle of BTV. Virus attaches to the cell surface via contacts made by the triskelion protein VP2 and is internalised by receptor-mediated endocytosis. The acidic pH of the endosome activates the membrane-permeabilizing activity of VP5, which mediates the release of cores into the cytoplasm. The transcriptase activity of the cores is activated and VP1 (polymerase), VP4 (capping enzyme) and VP6 (helicase) act together to synthesize viral messenger sense RNA. Following protein synthesis, new cores are assembled first from the interaction of decamers of VP3, then the ordered assembly of VP7 trimers. NS2 plays a role in bringing together proteins and RNA to facilitate this assembly process. Newly synthesized cores are released from inclusions, possibly by dephosphorylation of the NS2 protein, and the outer capsid proteins VP2 and VP5 are added before newly synthesized virus particles are trafficked out of cells by the activity of NS3 and its interaction with Tsg101 and calpactin p11.

a recent system that allows introduction of defined mutations into replicating virus is likely to catalyse research in the field.

References

Arcangeletti, M.C., Pinardi, F., Missorini, S., De Conto, F., Conti, G., Portincasa, P., Scherrer, K. and Chezzi, C. (1997) Modification of cytoskeleton and prosome networks in relation to protein synthesis in influenza A virus-infected LLC-MK2 cells. *Virus Res.* **51**, 19–34.

Bae, S.H., Cheong, H.K., Lee, J.H., Cheong, C., Kainosho, M. and Choi, B.S. (2001) Structural features of an influenza virus promoter and their implications for viral RNA synthesis. *Proc. Natl. Acad. Sci. USA.* **98**, 10602–7.

Bannerjee, A.K. and Shatkin, A.J. (1970) Transcription *in vitro* by reovirus associated ribonucleic acid-dependent polymerase. *J. Virol.* **6**, 1–11.

Bansal, O.B., Stokes, A., Bansal, A., Bishop, D.H.L. and Roy, P. (1998) Membrane organization of bluetongue virus non-structural glycoprotein NS3. *J. Virol.* **72**, 3362–9.

Beasley, B.E. and Hu, W.S. (2002) Cis-acting elements important for retroviral RNA packaging specificity. *J. Virol.* **76**, 4950–60.

Beaton, A.R., Rodriguez, J., Reddy, Y.K. and Roy, P. (2002) The membrane trafficking protein calpactin forms a complex with bluetongue virus protein NS3 and mediates virus release. *Proc. Natl. Acad. Sci. USA.* **99**, 13154–9.

Bhattacharya, B., Noad, R.J. and Roy, P. (2007) Interaction between bluetongue virus outer capsid protein VP2 and vimentin is necessary for virus egress. *Virol. J.* Jan 15; **4**, 7.

Bowne, J.G. and Jochim, M.M. (1967) Cytopathologic changes and development of inclusion bodies in cultured cells infected with bluetongue virus. *Am. J. Vet. Res.* **28**, 1091–1105.

Boyce, M., Celma, C.C. and Roy, P. (2008) A Reverse genetics system for Bluetongue virus. *J. Virol.* **82**, 8339–48.

Boyce, M. and Roy, P. (2007) Recovery of infectious bluetongue virus from RNA. *J. Virol.* **81**, 2179–86.

Boyce, M., Wehrfritz, J., Noad, R. and Roy, P. (2004) Purified recombinant bluetongue virus VP1 exhibits RNA replicase activity. *J. Virol.* **78**, 3994–4002.

Butan, C., Van Der Zandt, H. and Tucker, P.A. (2004) Structure and assembly of the RNA binding domain of bluetongue virus non-structural protein 2. *J. Biol. Chem.* **279**, 37613–21.

Creutz, C.E. (1992) The annexins and exocytosis. *Science* **258**, 924–31.

Eaton, B.T. and Crameri, G.S. (1989) The site of bluetongue virus attachment to glycophorins from a number of animal erythrocytes. *J. Gen. Virol.* **70**, 3347–53.

Eaton, B.T., Hyatt, A.D. and Brookes, S.M. (1990) The replication of bluetongue virus. *Curr. Top. Microbiol. Immunol.* **162**, 89–118.

Fillmore, G.C., Lin, H. and Li, J.K.K. (2002) Localization of the single-stranded RNA-binding domains of bluetongue virus nonstructural protein NS2. *J. Virol.* **76**, 499–506.

Forzan, M., Marsh, M. and Roy, P. (2007) Bluetongue virus entry into cells. *J Virol.* **81**, 4819–27.

Forzan, M., Wirblich, C. and Roy, P. (2004) A capsid protein of nonenveloped bluetongue virus exhibits membrane fusion activity. *Proc. Natl. Acad. Sci. USA.* **101**, 2100–105.

Freed, E.O. (2002) Viral late domains. *J. Virol.* **76**, 4679–87.

Freed, E.O. (2004) Mechanisms of enveloped virus release. *Virus Res.* **106**, 85–6.

French, T.J., Inumaru, S. and Roy, P. (1989) Expression of two related nonstructural proteins of bluetongue virus (BTV) type 10 in insect cells by a recombinant baculovirus: production of polyclonal ascitic fluid and characterization of the gene product in BTV-infected BHK cells. *J. Virol.* **63**, 3270–8.

French, T.J., Marshall, J.J. and Roy, P. (1990) Assembly of double-shelled, virus-like particles of bluetongue virus by the simultaneous expression of four structural proteins. *J. Virol.* **64**, 5695–700.

French, T.J. and Roy, P. (1990) Synthesis of bluetongue virus (BTV) corelike particles by a recombinant baculovirus expressing the two major structural core proteins of BTV. *J. Virol.* **64**, 1530–6.

Gouet, P., Diprose, J.M., Grimes, J.M., Malby, R., Burroughs, J.N., Zientara, S., Stuart, D.I. and Mertens, P.P. (1999) The highly ordered double-stranded RNA genome of bluetongue virus revealed by crystallography. *Cell* **97**, 481–90.

Grimes, J.M., Burroughs, J.N., Gouet, P., Diprose, J.M., Malby, R., Zientara, S., Mertens, P.P.C. and Stuart, D.I. (1998) The atomic structure of the bluetongue virus core. *Nature* **395**, 470–8.

Guirakhoo, F., Catalan, J.A. and Monath, T.P. (1995) Adaptation of bluetongue virus in mosquito cells results in overexpression of NS3 proteins and release of virus particles. *Arch. Virol.* **140**, 967–74.

Han, Z. and Harty, R.N. (2004) The NS3 protein of bluetongue virus exhibits viroporin-like properties. *J. Biol. Chem.* **279**, 43092–7.

Hassan, S.H. and Roy, P. (1999) Expression and functional characterization of bluetongue virus VP2 protein: role in cell entry. *J. Virol.* **73**, 9832–42.

Hassan, S.H., Wirblich, C., Forzan, M. and Roy, P. (2001) Expression and functional characterization of bluetongue virus VP5 protein: role in cellular permeabilization. *J. Virol.* **75**, 8356–67.

Hewat, E.A., Booth, T.F. and Roy, P. (1992) Structure of bluetongue virus particles by cryoelectron microscopy. *J. Struct. Biol.* **109**, 61–9.

Huismans, H. and Els, H.J. (1979) Characterization of the tubules associated with the replication of three different orbiviruses. *Virology* **92**, 397–406.

Huismans, H. and Erasmus, B.J. (1981) Identification of the serotype-specific and group-specific antigens of bluetongue virus. *Onderstepoort J. Vet. Res.* **48**, 51–8.

Hyatt, A.D., Eaton, B.T. and Brookes, S.M. (1989) The release of bluetongue virus from infected cells and their superinfection by progeny virus. *Virology* **173**, 21–34.

Hyatt, A.D., Gould, A.R., Coupar, B. and Eaton, B.T. (1991) Localization of the non-structural protein NS3 in bluetongue virus-infected cells. *J. Gen. Virol.* **72**, 2263–7.

Hyatt, A.D., Zhao, Y. and Roy, P. (1993) Release of bluetongue virus-like particles from insect cells is mediated by BTV nonstructural protein NS3/NS3A. *Virology* **193**, 592–603.

Kahlon, J., Sugiyama, K. and Roy, P. (1983) Molecular basis of bluetongue virus neutralization. *J. Virol.* **48**, 627–32.

Kar, A.K., Bhattacharya, B. and Roy, P. (2007) Bluetongue virus RNA binding protein NS2 is a modulator of viral replication and assembly. *BMC Mol. Biol.* **8**, 4.

Kar, A.K., Ghosh, M. and Roy, P. (2004) Mapping the assembly of bluetongue virus scaffolding protein VP3. *Virology* **324**, 387–99.

Kar, A.K. and Roy, P. (2003) Defining the structure-function relationships of bluetongue virus helicase protein VP6. *J. Virol.* **77**, 11347–56.

Lee, J.W. and Roy, P. (1986) Nucleotide sequence of a cDNA clone of RNA segment 10 of bluetongue virus (serotype 10). *J. Gen. Virol.* **67**, 2833–7.

Limn, C.K. and Roy, P. (2003) Intermolecular interactions in a two-layered viral capsid that requires a complex symmetry mismatch. *J. Virol.* **77**, 1111–24.

Limn, C.H., Staeuber, N., Monastyrskaya, K., Gouet, P. and Roy, P. (2000) Functional dissection of the major structural protein of bluetongue virus: identification of key residues within VP7 essential for capsid assembly. *J. Virol.* **74**, 8658–69.

Loudon, P.T. and Roy, P. (1992) Interaction of nucleic acids with core-like and subcore-like particles of bluetongue virus. *Virology* **191**, 231–6.

Lymperopoulos, K., Noad, R., Tosi, S., Nethisinghe, S., Brierley, I. and Roy, P. (2006) Specific binding of bluetongue virus NS2 to different viral plus-strand RNAs. *Virology* **353**, 17–26.

Lymperopoulos, K., Wirblich, C., Brierley, I. and Roy, P. (2003) Sequence specificity in the interaction of bluetongue virus non-structural protein 2 (NS2) with viral RNA. *J. Biol. Chem.* **278**, 31722–30.

Markotter, W., Theron, J. and Nel, L.H. (2004) Segment specific inverted repeat sequences in bluetongue virus mRNA are required for interaction with the virus non structural protein NS2. *Virus Res.* **105**, 1–9.

Martinez Costas, J., Sutton, G., Ramadevi, N. and Roy, P. (1998) Guanylyltransferase and RNA 5′-triphosphatase activities of the purified expressed VP4 protein of bluetongue virus. *J. Mol. Biol.* **280**, 859–66.

Mertens, P.P., Burroughs, J.N. and Anderson, J. (1987) Purification and properties of virus particles, infectious subviral particles, and cores of bluetongue virus serotypes 1 and 4. *Virology* **157**, 375–86.

Mertens, P.P.C., Maan, S., Samuel, A. and Attoui, H. (2005) Orbivirus, Reoviridae. In: Fauquet, C.M., Mayo, M.A., Maniloff, J., Desselberger, U. and Ball, L.A. (eds) *Virus Taxonomy, VIIIth Report of the ICTV*. London, Elsevier/Acadamic Press 466–83.

Mishra, R.K., Jatiani, S.S., Kumar, A., Simhadri, V.R., Hosur, R.V. and Mittal, R. (2004) Dynamin interacts with members of the sumoylation machinery. *J. Biol. Chem.* **279**, 31445–54.

Modrof, J., Lymperopoulos, K. and Roy, P. (2005) Phosphorylation of bluetongue virus nonstructural protein 2 is essential for formation of viral inclusion bodies. *J. Virol.* **79**, 10023–31.

Nason, E., Rothnagel, R., Muknerge, S.K., Kar, A.K., Forzan, M., Prasad, B.V.V. and Roy, P. (2004) Interactions between the inner and outer capsids of bluetongue virus. *J. Virol.* **78**, 8059–67.

Nedellec, P., Vicart, P., Laurent-Winter, C., Martinat, C., Prevost, M.C. and Brahic, M. (1998) Interaction of Theiler's virus with intermediate filaments of infected cells. *J. Virol.* **72**, 9553–60.

Owens, R. and Roy, P. (2004) Role of an arbovirus nonstructural protein in cellular pathogenesis and virus release. *J. Virol.* **78**, 6649–56.

Patton, J.T. (1996) Rotavirus VP1 alone specifically binds to the 3′ end of viral mRNA, but the interaction is not sufficient to initiate minus-strand synthesis. *J. Virol.* **70**, 7940–7.

Patton, J.T., Jones, M.T., Kalbach, A.N., He, Y.W. and Xiaobo, J. (1997) Rotavirus RNA polymerase requires the core shell protein to synthesize the double-stranded RNA genome. *J. Virol.* **71**, 9618–26.

Poncet, D., Aponte, C. and Cohen, J. (1993) Rotavirus protein NSP3 (NS34) is bound to the 3′ end consensus sequence of viral mRNAs in infected cells. *J. Virol.* **67**, 3159–65.

Ramadevi, N., Burroughs, J.N., Mertens, P.P.C., Jones, I.M. and Roy, P. (1998) Capping and methylation of mRNA by purified recombinant VP4 protein of bluetongue virus. *Proc. Natl. Acad. Sci. USA.* **95**, 13537–42.

Ramadevi, N. and Roy, P. (1998) Bluetongue virus core protein VP4 has nucleoside triphosphate phosphohydrolase activity. *J. Gen. Virol.* **79**, 2475–80.

Raynal, P. and Pollard, H.B. (1994) Annexins: the problem of assessing the biological role for a gene family of multifunctional calcium- and phospholipid-binding proteins. *Biochim. Biophys. Acta.* **1197**, 63–93.

Roy, P. (1992) Bluetongue virus proteins. *J. Gen. Virol.* **73**, 3051–64.

Roy, P. (2005). Bluetongue virus proteins and their role in virus entry, assembly and release. In: P. Roy, (ed), *Virus Structure and Assembly,* . USA: Elsevier Academic Press, USA: Elsevier Academic Press.

Roy, P., Marshall, J.J. and French, T.J. (1990) Structure of the bluetongue virus genome and its encoded proteins. *Curr. Top. Microbiol. Immunol.* **162**, 43–87.

Stauber, N., Martinez-Costas, J., Sutton, G., Monastyrskaya, K. and Roy, P. (1997) Blue-tongue virus VP6 protein binds ATP and exhibits an RNA-dependent ATPase function and a helicase activity that catalyze the unwinding of double-stranded RNA substrates. *J. Virol.* **71**, 7220–6.

Sutton, G., Grimes, J.M., Stuart, D.I. and Roy, P. (2007) Bluetongue virus VP4 is an RNA-capping assembly line. *Nat. Struct. Mol. Biol.* **14**, 449–51.

Tao, Y., Farsetta, D.L., Nibert, M.L. and Harrison, S.C. (2002) RNA synthesis in a cage-structural studies of reovirus polymerase lambda3. *Cell* **111**, 733–45.

Theron, J. and Nel, L.H. (1997) Stable protein-RNA interaction involves the terminal domains of bluetongue virus mRNA, but not the terminally conserved sequences. *Virology* **229**, 134–42.

Thomas, C.P., Booth, T.F. and Roy, P. (1990) Synthesis of bluetongue virus-encoded phosphoprotein and formation of inclusion bodies by recombinant baculovirus in insect cells: it binds the single-stranded RNA species. *J. Gen. Virol.* **71**, 2073–83.

van Staden, V. and Huismans, H. (1991) A comparison of the genes which encode non-structural protein NS3 of different orbiviruses. *J. Gen. Virol.* **72**, 1073–9.

van Staden, V., Stoltz, M.A. and Huismans, H. (1995) Expression of nonstructural protein NS3 of African horse sickness virus (AHSV): evidence for a cytotoxic effect of NS3 in insect cells, and characterization of the gene products in AHSV infected vero cells. *Arch. Virol.* **140**, 289–306.

Venkatesan, S., Gershowitz, A. and Moss, B. (1980) Modification of the 5' end of mRNA. Association of RNA triphosphatase with the RNA guanylyltransferase-RNA (guanine-7-) methyltransferase complex from vaccinia virus. *J. Biol. Chem.* **255**, 903–8.

Verwoerd, D.W. (1969) Purification and characterization of bluetongue virus. *Virology* **38**, 203–12.

Verwoerd, D.W., Els, H.J., De Villiers, E.M. and Huismans, H. (1972) Structure of the bluetongue virus capsid. *J. Virol.* **10**, 783–94.

Verwoerd, D.W. and Huismans, H. (1972) Studies on the *in vitro* and the *in vivo* transcription of the bluetongue virus genome. *Onderstepoort J. Vet. Res.* **39**, 185–91.

Verwoerd, D.W., Louw, H. and Oellermann, R.A. (1970) Characterization of bluetongue virus ribonucleic acid. *J. Virol.* **5**, 1–7.

Wehrfritz, J.M., Boyce, M., Mirza, S. and Roy, P. (2007) Reconstitution of bluetongue virus polymerase activity from isolated domains based on a three-dimensional structural model. *Biopolymers* **86**, 83–94.

Wirblich, C., Bhattacharya, B. and Roy, P. (2006) Nonstructural protein 3 of blue-tongue virus assists virus release by recruiting the ESCRT-I protein Tsg101. *J. Virol.* **86**, 460–73.

Wu, X., Chen, S.Y., Iwata, H., Compans, R.W. and Roy, P. (1992) Multiple glycoproteins synthesized by the smallest RNA segment (S10) of bluetongue virus. *J. Virol.* **66**, 7104–12.

Zetina, C.R. (2001) A conserved helix-unfolding motif in the naturally unfolded proteins. *Proteins* **44**, 479–83.

Zhao, Y., Thomas, C., Bremer, C. and Roy, P. (1994) Deletion and mutational analyses of bluetongue virus NS2 protein indicate that the amino but not the carboxy terminus of the protein is critical for RNA-protein interactions. *J. Virol.* **68**, 2179–85.

Bluetongue virus: cell biology

5

NATALIE ROSS-SMITH, KARIN E. DARPEL, PAUL MONAGHAN AND PETER P.C. MERTENS

Institute for Animal Health, Pirbright, Ash Road GU24 0NF, Surrey, UK

Introduction

Bluetongue virus (BTV) can infect and replicate in a wide range of mammalian and insect cell types, *in vivo* and *in vitro* (Wechsler and McHolland, 1988). The production of virus-specific RNAs, proteins and structures, and their interaction with the cellular components involved in assembly and release of progeny virus particles can have far reaching consequences for the viability of the cell. However, different cell types can respond in very different ways to BTV infection, showing variations not only between mammalian and insect cells but also between different mammalian cell types.

Interactions between BTV and the host cell have been studied in mammalian epithelial cells (including BHK cells, BSR cells (a BHK-derived cell clone) and Vero cells; Huismans and Howell, 1973; Mertens *et al.*, 1984; Hyatt *et al.*, 1989; Hyatt *et al.*, 1991; Boyce and Roy 2007), endothelial cells (of bovine, ovine and cervine origins; Russell *et al.*, 1996; DeMaula, 2002a, 2002b; McLaughlin *et al.*, 2003) as well as in insect cells derived from *Culicoides*, mosquitoes and certain species of Lepidoptera (Hyatt *et al.*, 1993; Fu, 1995; Guirakhoo *et al.*, 1995; Mertens *et al.*, 1996; Xu *et al.*, 1997).

Although this chapter summarises current knowledge of BTV interactions with the host cell *in vitro*, many questions concerning differences observed between host cell types and tissues remain unanswered. In areas where our knowledge of BTV cell biology is limited, the mechanisms used by other

ISBN-13: 978-0-12-369368-6

related viruses within the family *Reoviridae* are discussed, mainly those used by orthoreoviruses (ORVs) and rotaviruses (RVs). Bluetongue virus replication and assembly are also described in Chapter 4.

Bluetongue virus entry into the cellular host

Many viruses initiate infection via a complex, multi-step process, interacting in a sequential manner with several different receptors (López and Arias, 2004, 2006; Pelkmans and Helenius, 2003). Interactions with the cell surface are often initiated via binding of an 'attachment protein' on the virus, with a generalised cellular receptor. This is usually followed by further interactions and binding to a specific receptor for the virus, which may be sufficient for virus internalisation to occur, although other 'factors' may also be required.

Four distinct mechanisms are thought to be used by viruses to enter cells (reviewed by Pelkmans and Helenius, 2003). These include the following: (1) clathrin-mediated endocytosis, which is receptor-specific and can be identified by the presence of virions within coated vesicles that subsequently lose their clathrin and are then processed from early to late endosomes. The reduced pH within the endosomes may cause viral uncoating. This method of entry is thought to be used by ORV. (2) Endocytosis can also occur via caveolae in a non-constitutive process that requires cell activation. This mechanism bypasses the endosomal pathway, so there is no pH drop within the cellular compartment. (3) Pinocytosis is used by viruses that bind sialic acid but is not receptor-specific. This method of entry involves lipid rafts and the formation of pinocytotic vesicles that have a distinctive appearance when viewed by electron microscopy (EM). This mechanism is thought to be used by RV. (4) Non-clathrin, non-caveolae-dependent endocytosis is less well defined but can be distinguished by the presence of uncoated vesicles as observed by EM.

Binding of intact BTV particles to mammalian cells appears to be mediated primarily by VP2 (Huismans *et al.*, 1983; Hassan and Roy, 1999). Early EM studies suggested that adsorption and penetration of the cell occurred within 10 min of infection, possibly via pinocytotic vesicles (Lecatsas, 1968). Although a specific cell surface receptor has not yet been identified, the site to which the virus binds seems to be characterised by the presence of clathrin. The clathrin-coated membrane invaginates and eventually detaches from the cell membrane, generating a coated vesicle (Hyatt *et al.*, 1989; Eaton *et al.*, 1990). Sialoglycoproteins may also be involved in the process of BTV adsorption, at least when binding to erythrocytes (Eaton and Hyatt, 1989; Hassan and Roy, 1999). After internalisation, the clathrin coat is rapidly lost and vesicle fusion results in the formation of an endocytic vesicle (early endosome) within which the outer virus coat is lost due to the low pH within the vesicle. The changes that occur in the outer capsid release amphipathic helices on VP5, which interact with the endosomal membrane to initiate permeabilisation, thus releasing the BTV core into the host cell cytoplasm (Hassan *et al.*, 2001). Most BTV particles appear to be converted to core particles and appear in the cell cytoplasm within 1-h post-infection (Huismans *et al.*, 1987; Eaton *et al.*, 1990; Forzan *et al.*, 2007).

The initial stages of cell attachment and virus entry into insect cells show significant differences from those in mammalian cell systems. The specific infectivity of the BTV particle for adult *Culicoides* and *Culicoides* cell lines (but not mammalian cells) is very significantly enhanced by proteolytic cleavage of the outer capsid and cell attachment protein VP2 (as mediated by trypsin or chymotrypsin) (Mertens *et al.*, 1996). Cleavage of VP2 generates modified 'infectious subviral particles' (ISVP) that can be readily purified (Mertens *et al.*, 1987). Monoclonal antibodies to the core surface protein VP7 are unable to bind to purified ISVP in solution, indicating that the core surface is not exposed by this protease treatment (Hutchinson, 1999). The infectivity of ISVP for mammalian cells, their enhanced infectivity for *Culicoides* cells and their ability to raise neutralising antibodies (to both intact virus and ISVP) in rabbits and guinea pigs suggests that the VP2 cleavage products and VP5 on the ISVP surface can still mediate BTV cell attachment and particle entry. However, cleavage of VP2 does inhibit binding to erythrocytes (and haemagglutination), indicating that there are certain other differences in the cell attachment and infection mechanisms used by BTV in mammalian or insect systems. The enhanced infectivity of ISVP for adult *Culicoides* suggests that proteolytic cleavage of the virus capsid proteins, possibly by proteases in insect saliva (Langner *et al.*, 2007) or in the insect gut (Mertens *et al.*, 1996), may play an important role in the infection and transmission pathway between the mammalian and insect hosts.

The BTV core particle is also infectious for adult *Culicoides*, insect cell cultures and some mammalian cell types, in the complete absence of the outer capsid proteins (Mertens *et al.*, 1996). Bluetongue virus cores have a much lower specific infectivity than does intact virus or ISVP for mammalian cells (Fu, 1995; Mertens *et al.*, 1996), suggesting that the outer capsid proteins are particularly important during infection of the mammalian host. However, cores can infect insect cells with the same efficiency as intact virus particles, indicating the existence of a distinct cell attachment and penetration mechanism that is mediated by the core surface protein VP7 rather than components of the outer capsid layer (Xu *et al.*, 1997). This is supported by observations that some polyclonal and monoclonal antibodies to VP7 can neutralise the BTV core particle infectivity for either insect or mammalian cell cultures but do not alter the infectivity of virus or ISVP (Hutchinson, 1999). The infectivity of BTV cores is also independent of the reduced pH in endosomes (Hutchinson, 1999).

Bluetongue virus cores can bind efficiently to glycosaminoglycans on the cell surface (Hutchinson, 1999), although it appears likely that a further 'specific' cellular receptor is also involved. Bluetongue virus core binding to *Culicoides* cells may be facilitated through an arginine–glycin–aspartate tripeptide (known as an 'RGD motif') present on the outer surface of each monomer of the core surface protein VP7, which is therefore repeated 780 times across the outside of the core particle (Basak *et al.*, 1996; Grimes *et al.*, 1998; Tan *et al.*, 2001). The RGD peptide is frequently involved in binding of viruses to cell surface integrins (Jackson *et al.*, 1997).

The higher infectivity of BTV cores for adult *Culicoides* and for insect cell lines, taken together with the absence of outer capsid proteins from the cypoviruses, members of a distinct genus of the reoviruses that infect only insects (reviewed by Mertens *et al.*, 2005), suggests that the orbiviruses may have acquired their outer capsid during evolution and adaptation to the infection of mammalian cells. Removal of the BTV outer capsid proteins is essential for activation of the virus transcriptase, allowing the core particles that are released into the host cell cytoplasm to synthesise viral mRNAs and initiate replication (Van Dijk and Huismans 1980; Huismans *et al.*, 1987) (Figure 5.1).

Other members of the family *Reoviridae* also use a multi-step process for virus entry, which can also involve modification of outer capsid components by proteases. Rotaviruses require both integrins ($\alpha 2\beta 1$, $\alpha v\beta 3$, $\alpha x\beta 2$ and $\alpha 4\beta 1$) and heat-shock protein 70 (hsc70) (Guerrero *et al.*, 2000; Ciarlet *et al.*, 2002; López and Arias, 2004). Mammalian reoviruses have been shown to use junction-adhesion molecule 1 (JAM-1) as a receptor for binding and $\beta 1$ integrin for virus internalisation (Guglielmi *et al.*, 2006; Maginnis *et al.*, 2006).

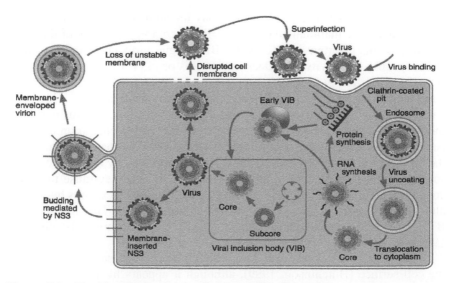

Figure 5.1 The bluetongue virus replication cycle. (See colour plate 6).

Viral structures and their functions

Electron microscopy has also been used to explore the function and location of individual BTV proteins, as well as the assembly of progeny virus particles within the infected cell (Hyatt and Eaton, 1988; Hyatt *et al.*, 1991). Several specific virus-induced structures have been discovered, including non-structural protein 1 (NS1) tubules (Figures 5.2 and 5.3); viral

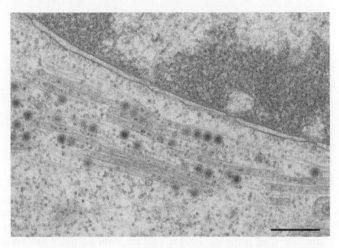

Figure 5.2   Electron micrograph of BHK-21 cells infected with bluetongue virus (BTV) and fixed at 18 hpi showing a portion of an infected cell with NS1 tubules running alongside the nucleus. There are a small number of BT virions associated with the tubules. Scale bar = 250 nm.

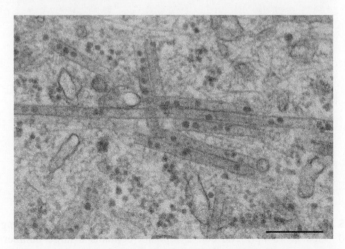

Figure 5.3   Higher power image of cell prepared as in Figure 5.1. The NS1 tubules clearly contain small particles of unknown composition. Scale bar = 200 nm.

inclusion bodies (VIBs) (Figure 5.4) and virus particles (Figure 5.5) (Owen, 1966; Lecatsas, 1968; Murphy *et al.*, 1971; Thomas *et al.*, 1990). These structures have subsequently been identified in every BTV-infected mammalian and insect cell type that has been studied, including primary mammalian cells such as

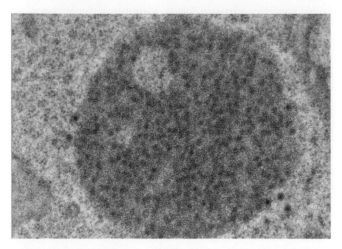

Figure 5.4 Electron micrograph of BHK-21 cell infected with bluetongue virus (BTV) and fixed at 18 hpi showing a viral inclusion body (VIB), which contains a number of immature viral particles. Larger mature viral particles are present in the cytoplasm at the periphery of the VIB.

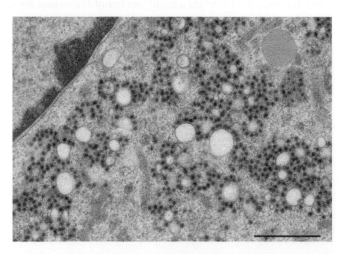

Figure 5.5 Electron micrograph of BHK-21 cells infected with bluetongue virus (BTV) and fixed at 18 hpi. The cell cytoplasm contains a large number of viral particles attached to vesicular membranes. Scale bar = 1 μm.

monocytes and lymphocytes (undergoing blastogenesis) (Whetter *et al.*, 1989; Fu, 1995; Darpel, 2007).

Viral inclusion bodies

Dense cytoplasmic granular aggregates, known as viroplasms or 'VIBs', are one of the most striking features of cells infected with BTV and several other reoviruses (see Figure 5.4). Non-structural protein 2 (NS2) is a major component of VIBs, although they also contain structural protein components of virus core and viral RNAs (Thomas *et al.*, 1990; Brookes *et al.*, 1993). Non-structural protein 2 is the only BTV phosphoprotein (Theron *et al.*, 1994; Taraporewala *et al.*, 2001). It is expressed by segment 8 (Mertens *et al* 1984) and is highly conserved, with a molecular weight of 41 kDa protein. It is also one of the first viral proteins detected in BTV-infected cells (Thomas *et al.*, 1990). Non-structural protein 2 binds ssRNA but not dsRNA or DNA (Huismans *et al.*, 1987; Thomas *et al.*, 1990; Fillmore *et al.*, 2002), although it is unclear exactly what effect phosphorylation has on this binding.

Expression of NS2, or a GFP-NS2 fusion protein, results in the formation of protein aggregates resembling VIBs within the cell cytoplasm, suggesting that NS2 can form VIBs in the absence of other viral factors. In contrast, both ORV and RV require at least two viral proteins to form similar structures (Fabbretti *et al.*, 1999; Touris-Otero *et al.*, 2004; Patton *et al.*, 2006). Mutation of two NS2 serines (aa 249 and 259) completely inhibits phosphorylation and oligomerisation of the expressed protein, which remains dispersed throughout the cytoplasm (Modrof *et al.*, 2005). The ability of NS2 to bind ssRNA as well as its role in the formation of VIBs (which are thought to be the site of BTV particle assembly) suggests that it plays a significant role in BTV genome assembly, replication and packaging (see Figure 5.1). Phosphorylation of NS2 by casein kinase II (Modrof *et al.*, 2005) may play some part of regulating these processes within the infected cell (Huismans *et al.*, 1987; Theron *et al.*, 1994).

Non-structural protein 2 is initially dispersed throughout the cytoplasm of BTV-infected BHK-21 cells but rapidly forms into a large number of small discrete cytoplasmic inclusions. Confocal microscopy of BTV-infected cells has shown co-localisation of microtubules and NS2 within early or pre-VIBs (Ross-Smith, 2008) (Figure 5.6). As BTV infection progresses, the VIBs appear to move along tubules to a perinuclear location, becoming larger but fewer in number. Tubulin has previously been shown to be involved in the formation of ORV VIBs (Parker *et al.*, 2002). Nocodazole, which disrupts tubulin, reduced the number of VIBs in BTV-infected BHK-21 cells. In addition, the VIBs that were produced after such treatment were also randomly distributed within the cell, and none of the larger 'mature' VIBs was observed (Ross-Smith, 2008). Nascent viral particles can often be observed within the

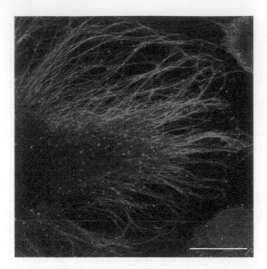

Figure 5.6 CPAE cell infected with bluetongue virus (BTV) and fixed for fluorescence microscopy at 20 hpi. The microtubules were labelled with mouse anti-tubulin detected with species-specific Alexa488 conjugate (green) and NS2 was labelled with specific antibody detected with species-specific Alexa568 conjugate (red). The majority of small viral inclusion antibodies (VIBs) are associated with the microtubules. Nuclei were stained with DAPI (blue). Scale bar = 10 μm (See colour plate 7).

matrix of VIBs by EM and fully formed particles are found at their periphery (see Figure 5.4) (Brookes *et al.*, 1993; Fu, 1995).

Fusion of BTV VIBs is believed to occur as their size increases, in some cases trapping areas of cytosol that contain ribosome-like structures and intact BTV virus particles (Fu, 1995). Early EM studies have indicated the formation of VIBs in association with each infecting 'parental' virus core particle (Eaton *et al.*, 1988; Hyatt *et al.*, 1989; Eaton *et al.*, 1990). The exchange of genome segments between different BTV strains that infect the same cell may therefore depend on the fusion of distinct VIBs (Mertens, 2004; Silvestri *et al.*, 2004).

Culicoides varipennis cells (KC cells) can be persistently infected by BTV. At a late stage of infection (>14 d.p.i.), each infected cell can contain a single massive VIB, indicating that a high level of VIB fusion has occurred (Fu, 1995). However, at these late time points, very few, if any, virus particles or subparticles are present within the large VIBs, or within the insect cell itself (Fu, 1995). This suggests that there are intracellular mechanisms (such as RNA silencing) that can eventually control or suppress BTV replication in insect cells.

Initial immuno-gold labelling studies have demonstrated that in addition to NS2, the dense fibrillar structures of the VIB matrix also contain VP1, VP3, VP4, VP6 with VP5 towards the periphery and VP2 only at its surface. After their release from VIBs, intact progeny virus particles can accumulate in the host cell cytoplasm or may associate with cellular membranes and vesicles. These structures might be involved in the transport of virus particles to the cell surface (see Figure 5.5). These observations support the hypothesis that VIBs are the primary site of virus replication and assembly (Huismans and Els, 1979; Eaton *et al.*, 1988; Hyatt and Eaton, 1988; Hyatt *et al.*, 1991; Brookes *et al.*, 1993; Kar *et al.*, 2005). However, confocal microscopy studies have shown a more complex situation, with NS2 present only around the periphery of the larger VIBs (Figure 5.7), suggesting that it may be involved in transporting viral components between the VIBs and cytoplasm, leading to its accumulation near the VIB surface (Ross-Smith, 2008).

Viral inclusion bodies also contain large amounts of RNA (Browne and Jochim, 1967). RNA-silencing studies with RV have shown that although translation of the mRNA can be inhibited, there is a population of RNAs within the viral inclusion (for packaging within progeny virus particles) that are unaffected by silencing, (Silvestri *et al.*, 2004). Bluetongue virus mRNAs are synthesised exclusively within the viral core (Mertens and Diprose, 2004) and can be directed either into assembling progeny virions within the VIB or into the cytoplasm for translation (see Figure 5.1). Bluetongue virus NS2 binds to ssRNA and has a greater affinity for BTV ssRNA, as compared to cellular mRNAs (Lymperopoulos *et al.*, 2003). Non-structural protein 2 also interacts with the transcriptase complex, VP1 (Modrof *et al.*, 2005), suggesting that interactions with NS2 may be

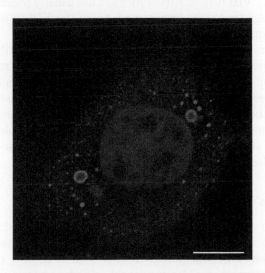

Figure 5.7 BHK-21 cells were infected with bluetongue virus (BTV) and fixed for fluorescence microscopy at 18 hpi. Cells were labelled with anti-NS2 antibody, which was detected with species-specific Alexa568 conjugate (red). Nuclei were stained with DAPI (blue). Scale bar = 10 μm (See colour plate 8).

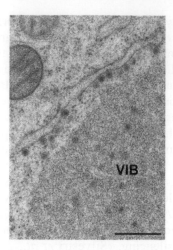

Figure 5.8   Higher magnification of cell prepared as in Figure 5.4 showing part of a viral inclusion body (VIB) and closely associated rough endoplasmic reticulum (RER). A number of viral particles are attached to RER membrane. Scale bar = 500 nm.

involved in controlling the fate of BTV mRNAs. Structures resembling ribosomes have occasionally been visualised within VIBs by EM, although they may be present only as a result of the incorporation of cytoplasmic inclusions during VIB fusion. Ribosomes are not usually present within VIBs and BTV mRNAs need to be transported out of the VIB to be translated by ribosomes within the host cell cytoplasm.

Confocal microscopy and staining of BHK cells with an antibody against S6 (one of the ribosomal small subunit proteins) indicated a redistribution of ribosomes during BTV infection. Although a dramatic shift in the distribution of free ribosomes has not been detected by EM, approximately one-third of VIBs appear to be surrounded by ribosome-covered membranes (Figure 5.8; rough endoplasmic reticulum (RER) predominantly responsible for translation within the cell) (Ross-Smith, 2008). Eaton *et al.* (1987) also reported ribosomes forming large clusters around VIBs.

NSI Tubules

The functions of 'tubules' formed from the NS1, which are characteristically observed within BTV-infected cells, has not yet been determined (Eaton *et al.*, 1988). Non-structural protein 1 is a 64-kDa protein expressed from BTV genome segment 5 (Mertens *et al.*, 1984; Anderson *et al.*, 1993). Structures resembling these tubules are formed within cells expressing NS1 in the absence of other BTV proteins, indicating that no other viral factors are required for their assembly (Urakawa and Roy, 1988; Hewat *et al.*, 1992). The

BTV tubules have a helical configuration with an average diameter of 52.3 nm and can assemble up to 1000 nm in length. Similar tubular structures are also seen within the cytoplasm of cells infected with other orbiviruses (Huismans and Els, 1979).

Electron microscopic studies have indicated an association between NS1 tubules and VIBs (Eaton *et al.*, 1988), although this appears to be a rare occurrence. It was suggested that the mature virus particles that are released from VIBs associate with the NS1 tubules that are frequently detected in the vicinity of the VIBs, and that these tubules are in some way involved in transporting virus particles to the cell's cytoskeleton.

However, NS1 tubules also accumulate around the nucleus of BTV-infected mammalian cells or in clusters throughout the cytoplasm but do not form a pattern that logically suggests involvement in virion transport from the VIB to the cell surface (Ross-Smith, 2008). Antibodies against NS1 have been shown to react with emerging virions and with fibrillar structures seen in proximity to VIBs, in cytoskeletal extracts (Eaton *et al.*, 1988), leading to suggestions that the NS1 might be necessary for virion assembly and that the NS1 tubes are a repository of 'used' NS1. However, distinct particles can clearly be seen within the hollow centre of the tubules, indicating that the assembled structure may itself have a specific functional role associated with virus replication (see Figure 5.3).

It has also been suggested that NS1 may be involved in the mechanism of transport of virus particle to the plasma membrane and release from infected cells (Owens *et al.*, 2004) (see below).

Interaction of bluetongue virus with the cytoskeleton

Interactions of BTV with the cytoskeleton and elucidation of the role it potentially plays in the assembly of progeny virions and their exit from the cell were initially studied by EM (Eaton *et al.*, 1987; Eaton *et al.*, 1988; Hyatt and Eaton, 1988; Hyatt *et al.*, 1989; Hyatt *et al.*, 1991; Hyatt *et al.*, 1993). The cytoskeleton is a network of protein fibres primarily consisting of microfilaments, intermediate filaments and microtubules within the cell cytoplasm. It gives shape and organisation to the cell and is involved in cell movement, as well as in holding and moving organelles and proteins throughout the cell. Bluetongue virus is not alone in using the cytoskeleton to move its own proteins to the sites of virion assembly within the cell and to transport the assembled virion to the cell membrane for virus egress. Other known examples include herpes virus, adenovirus, vaccinia virus and African swine fever virus (reviewed by Smith and Enquist, 2002).

Eaton *et al.* (1987) demonstrated the presence of all of the BTV structural proteins in purified cytoskeletal fractions, although it is not known if this is due to associations with fully assembled virions or individual proteins. Although virus particles were seen associated with filamentous structures, it is not clear if the cytoskeleton is directly involved in shuttling BTV proteins to the VIBs for assembly or for the transport of the virion through the cytoplasm or both.

Drugs that disrupt microtubules and microfilaments within the infected cell cause little effect on virus replication, leading to the suggestion that BTV-cytoskeleton interactions are predominantly with intermediate filaments, including vimentin. Bhattacharya et al. (2007) reported that BTV outer capsid protein VP2 interacts with vimentin, suggesting that this process was important for virus egress. There are also indications that at least a portion of the NS1 tubules and most VIBs interact with the cytoskeleton (Eaton et al., 1987; Eaton et al., 1988). Both ORV (Sharpe et al., 1982; Mora et al., 1987) and RV use vimentin for virion transport (Musalem and Espejo, 1985; Weclewicz et al., 1994; Brunet et al., 2000). However, associations with actin and/or tubulin have also been reported (Gardet et al., 2006; Berkova et al., 2007; Gardet et al., 2007).

Involvement of the cellular translation machinery during bluetongue virus infection

Capped positive-sense mRNAs that are released from the BTV core into the cytoplasm use the host cells' translational machinery to synthesise viral proteins. The BTV mRNAs are capped by VP4 (the BTV core-associated guanylyltransferase and transmethylase), suggesting that they are translated in a cap-dependent manner (Ramadevi et al., 1998). However, little else is known about the mechanism of translation. Most of the mRNA segments are translated from the AUG start codon nearest to their 5′ terminus, although some translation of mRNAs 9 and 10, encoding VP6 and NS3, respectively, also occurs from a downstream in-frame AUG, in each case resulting in the synthesis of two related proteins of slightly different lengths (Figure 5.9). Although BTV mRNAs are capped, they do not have a poly(A) tail, which is normally required to circularise cellular mRNAs for efficient translation. An interaction occurs between the 3′ poly(A) tail and the poly(A)-binding protein (PABP) located at the 5′ terminus through its association with the cap-binding initiation complex eIF-4F.

Bluetongue virus infection of certain mammalian cells (e.g. BHK cells) induces 'shut-off' of host cell translation, which is usually complete by 12 h p.i. with optimal expression of BTV proteins at ~16 h p.i. (Mertens et al., 1984). The mechanisms by which BTV mRNAs are efficiently translated, while cellular protein synthesis is inhibited, are currently unknown. Like BTV, the mRNAs of RV do not have a poly(A) tail. However, circularisation of RV mRNAs, which allows efficient translation, is achieved by binding to one of the RV non-structural proteins (NSP3) at the 3′ end of the mRNAs and to the 5′ initiation complex (Poncet et al., 1993, 1996).

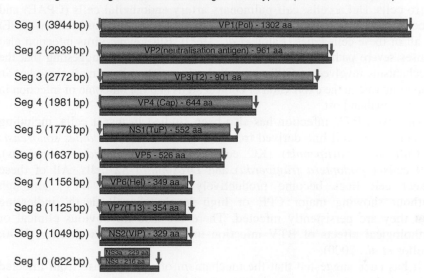

Figure 5.9 The genome organisation of bluetongue virus type 8 (BTV-8 NET2006/04).

During ORV infection, there is a switch from cap-dependent to cap-independent translation, which is accompanied by the synthesis of uncapped viral mRNAs by the progeny virus cores (Zarbl *et al.*, 1980; Skup *et al.*, 1981). A similar switch to non-cap-dependent translation has been observed in BHK cells infected with BTV (Stirling, 1996). However, unlike ORV, the BTV-capping enzyme (VP4) appears to remain active throughout infection, and there is no evidence that the progeny core particles of BTV can make uncapped RNA.

Bluetongue virus persistently infects *Culicoides* cells (e.g. KC cells) but does not cause shut-off of host cell protein synthesis (see below). The virus can also persistently infect some mammalian cells (e.g. γδ T cells; Takamatsu *et al.*, 2003), suggesting that host cell shut-off is also circumvented in these cells.

Fate of bluetongue virus-infected cells and virus egress

The ability of BTV to infect both mammalian and insect cells lines reflects the situation *in vivo* where both insect vectors (*Culicoides* species) and mammalian (ruminant) hosts are infected. Most BTV infection studies have been carried out in mammalian epithelial, endothelial or fibroblastic-transformed cell lines, particularly BHK cells (or BHK-derived clones like BSR cells), Vero cells, HeLa cells, calf pulmonary artery endothelial cells (CPAE) and L-cells (mouse fibroblasts) causing significant or severe cytopathic effects (CPE) in all of these cells types (Eaton *et al.*, 1990). Bluetongue virus infection also causes severe pathology in some of its mammalian hosts, suggesting that the mechanisms involved in CPE and cell death in cell culture may also play an important role in the expression of clinical signs and the outcome of infection in the mammalian host.

However, BTV infection has also been studied in insect cells, including an embryonic cell line derived from the BTV vector *Culicoides sonorensis* (=*Culicoides variipennis*) (KC cells), C6/36 cells (*Aedes albopictus*), Sf9 cells (*Spodoptera frugiperda*) and *Drosophila* S2 cells. All of these insect cell lines become productively infected with BTV, although without showing major CPE or high levels of cell death, indicating that they are persistently infected. There are also no obvious clinical or pathological effects of BTV infection in adult *Culicoides* (Mellor, 1990; Mellor *et al.*, 2000).

It has been suggested that the mechanism of virus release from infected cells can help to influence whether a cell becomes lytically or persistently infected. Bluetongue virus particles appear to cause damage to the mammalian cell membrane as they leave the cell possibly culminating in cell lysis (Fu, 1995; Owens *et al.*, 2004). In contrast, the progeny BTV particles can bud out of infected insect cells that are persistently infected, leaving the cell membrane intact. However, the release of progeny BTV particles from mammalian cells does not rely on a single 'exit' mechanism but appears to change over the course of infection. Virus particles can be released from infected BHK cells as early as 4–5-h post-infection, while up to 80% of the cells are still intact at 24-h post-infection (Ross-Smith, 2008). Lysis of the cell membrane is therefore not the major release mechanism during the early stages of infection in mammalian cells (Hyatt *et al.*, 1989; Eaton *et al.*, 1990).

Three different mechanisms of virus release have been observed in EM studies of BTV-infected mammalian cells. These include the following: (1) budding, which seems to be a major route of virus release prior to host cell shut-off in the early to mid-stage of infection (8–16 h) (Figure 5.10). (2) At later stages of post-infection (16–28 h), the virus is extruded from the cells and appears to make holes through the cell membrane. This appears to cause significant damage, with some release of cell contents, although penetration of the membrane does not appear to immediately destroy the overall cell integrity. However, the damage that is caused may be cumulative eventually leading to cell death and lysis. (3) At around 24–28-h post–infection, extensive CPE is increasingly visible. Virus is then released as a result of cell lysis but may remain associated with the cell debris and membranes (Hyatt *et al.*, 1989; Eaton *et al.*, 1990).

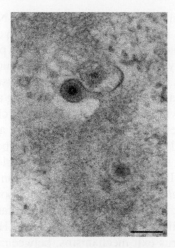

Figure 5.10   Electron micrograph of cells prepared as in Figure 5.4 showing a mature virus particle budding through the cell membrane. Scale bar = 125 nm.

In contrast, the release of progeny virus particles from KC cell cultures, at early stages of post-infection, was observed only by budding through the cell membrane (Fu, 1995). Virus particles that bud through the cell membrane are released initially as membrane-enveloped virus particles (MEVP) (see Figure 5.10) with the lipid membrane envelope acquired from the cellular plasma membrane of the infected cell. Significant and increasing levels of infectivity are detected in the supernatants from BTV-infected KC cell cultures, suggesting that MEVP may also be infectious without further modification. Although the stability of the MEVP membrane envelope has not been examined directly, previous studies have suggested that it may be relatively unstable (Hyatt et al., 1991; Hyatt et al., 1993).

Segment 10 encodes NS3, the only BTV glycoprotein or membrane protein. Seg-10 mRNA has two in-frame initiation codons, resulting in the production of NS3 (229 aa) and a slightly shorter protein identified as NS3a (216 aa). Both NS3 proteins have two cytoplasmic domains linked by two hydrophobic regions, which span the cell membrane. Newly synthesised NS3 and NS3a proteins are transported to the golgi for glycosylation prior to transport to the cell membrane for incorporation, although both proteins can exist in glycosylated and non-glycosylated forms (Bansal et al., 1998). The difference in roles of the glycosylated and non-glycosylated forms of NS3 and NS3a is not known, although loss of the amino-terminal domain from NS3a may affect its ability to interact with components of the mammalian cell (P11) (Beaton et al., 2002). The glycosylation of NS3/NS3a is likely to result in a highly variable molecular weight and smearing during SDS PAGE, making it even more difficult to detect these proteins in mammalian cells.

Labelling of infected mammalian cells with an anti-NS3 antibody or expression of GFP-NS3 results in the detection of the protein around the periphery of the cell. The native protein is associated with BTV particles that are in the process of extrusion (Hyatt *et al.*, 1991). Non-structural protein 3/NS3a can also facilitate the release of virus-like particles from insect cells and is found at the location of their release (Hyatt *et al.*, 1993). Using a dihybrid system, NS3 was found to interact with p11, the calpactin light chain of the annexin II complex, which has been implicated in endocytosis and regulated secretary events. These studies also indicated an interaction between NS3 and BTV outer capsid protein VP2, suggesting that NS3 forms a bridge between virions and p11 (part of the annexin II complex) directing the virion to the cellular exocytic machinery (Beaton *et al.*, 2002). Another study reported that NS3 exhibits viroporin-like properties, suggesting that it may also facilitate virus release by destabilising or permeabilising the cellular membrane (Han and Harty, 2004).

The differences in virus exit mechanisms, between insect and mammalian host cells, have been attributed to the amount of NS3 present within these different cell types. Although NS3 is produced in detectable amounts early post-infection in mammalian cells, synthesis drops to non-detectable levels as infection continues. However, in insect cells, NS3 is produced throughout infection, resulting in far higher expression levels compared to some of the mammalian cell types (Guirakhoo *et al.*, 1995). Reducing the amount of available NS1 in the cell by inducible expression of a recombinant, single-chain antibody fragment (scFv) derived from an NS1-specific monoclonal antibody increased virus budding and inhibited cell lysis (Owens *et al.*, 2004). It was suggested that although NS3 is required for budding of BTV through the cell membrane, the ratio of NS3 to NS1 may also be important (Owens *et al.*, 2004) (see Chapter 4).

Bluetongue virus replicates primarily in the endothelial cells and agranular leukocytes of the ruminant host (see Chapters 12 and 13). Primary cell cultures of ovine and bovine endothelial cells, monocytes, lymphocytes undergoing blastogenesis and skin fibroblasts can also be infected with BTV *in vivo* producing progeny virus particles. Although endothelial cells, fibroblasts and monocytes all develop CPE, they often do so much more slowly than established cell lines (Takamatsu and Jeggo, 1989; Barratt-Boyes *et al.*, 1992; DeMaula, 2001; Darpel, 2007).

Although lymphocytes can become infected with BTV when they undergo blastogenesis (Stott *et al.*, 1990; Barratt-Boyes *et al.*, 1992), the fate of infected lymphocyte cultures is uncertain as both lytic and non-lytic (persistent) infections have been reported in different lymphocyte subsets (Stott *et al.*, 1990; Barratt-Boyes *et al.*, 1992). Lytic infection of CD4[+] cultures, chronic infection for CD8[+] and persistent infection of 'null cell' cultures were all demonstrated in one investigation (Stott *et al.*, 1990). The null cells are now thought to be γδ T cells, persistent infection of which has been reported elsewhere (Takamatsu *et al.*, 2003; Darpel, 2007). Several other studies have demonstrated that leukocytes and/or peripheral blood mononuclear cells (PBMC) collected ex vivo from infected ruminants, then cultured after stimulation with interleukin-2 (IL-2) and concanavalin A (Con A) can be maintained in an infected state for long periods (Takamatsu *et al.*, 2003; Lunt *et al.*, 2006). These observations suggest that some mammalian cell populations can become persistently infected, with significant implications for the long-term persistence of BTV in the mammalian hosts (Takamatsu *et al.*, 2003).

Endothelial cells are capable of an inflammatory response and represent a major cellular target for BTV infection *in vivo*. The responses of these cells to BTV infection have been studied *in vitro* (DeMaula *et al.*, 2001, 2002a, 2002b; McLaughlin *et al.*, 2003). These authors were able to demonstrate 'upregulation' of certain cytokines. They also showed a difference between endothelial cells collected from ovine and bovine origins. Although IL-1, IL-6, IL-8, COX-2 and NO mRNA were upregulated in bovine endothelial cell cultures, the upregulation of IL-6, COX-2 and NO was much lower in ovine endothelial cell cultures. As a consequence, ovine microvascular endothelial cells would produce much less prostacyclin, which could lead to a different antithrombatic response in sheep, as compared to cattle. It was suggested that the differences observed help to explain why a much higher proportion of sheep display severe haemorrhagic disease as a result of BTV infection. In contrast, infected cattle remain mostly subclinical (DeMaula *et al.*, 2001, 2002a, 2002b; McLaughlin *et al.*, 2003).

These studies also detected differences in the response of endothelial cells from different tissue locations within a single species (e.g. endothelial cells collected from the umbilical cord, as compared to lung endothelium) (Russell *et al.*, 1996; DeMaula *et al.*, 2001, 2002a, 2002b; McLaughlin *et al.*, 2003). Other immune-competent cells including monocytes and lymphocytes may also become infected with BTV and respond with antiviral and inflammatory responses.

Bluetongue virus can induce apoptosis in mammalian cells but not in insect cells, which may help explain why CPE is not observed in insect cell cultures (Mortola *et al.*, 2004). It has been reported that the addition of either virus or purified VP2 and VP5 can trigger apoptosis in some mammalian cells, although addition of either protein alone does not cause this effect (Mortola *et al.*, 2004). This indicates that BTV can induce apoptosis in certain cells by activating the apoptotic pathway during virus entry, before replication has occurred. It was demonstrated independently that the development of CPE could be prevented by cloning the gene for the *A. albopictus* inhibitor of apoptosis 1 into the BSR cell genome (Li *et al.*, 2007).

Hu *et al.* (2007) infected primary human and mouse embryo fibroblast, as well as human- and mouse-transformed cell lines, with BTV-10 and reported that only the transformed cell lines became infected with BTV and displayed CPE. Primary ovine and bovine lung microvascular endothelial cells and bovine pulmonary artery endothelial cells that were infected with partially purified virus showed necrosis rather than apoptosis. Only ovine pulmonary artery endothelial cells displayed a mix of apoptotic and necrotic cell death (DeMaula *et al.*, 2001). However, the mechanism of cell death shifted to mainly apoptosis in these cell cultures when a cell lysate of BTV-infected endothelial cells was used for infection. This finding could be recreated using partially purified BTV and IL-1, suggesting that the inflammatory mediators in the original cell lysate may have an important influence on the pathway of cell death (DeMaula *et al.*, 2001). These studies also demonstrate difficulties in identifying the precise mechanism of death in different cells after infection with BTV *in vitro* and in drawing valid conclusions concerning the mechanisms of cell death and tissue damage *in vivo*.

Conclusions

Our knowledge concerning the molecular mechanisms involved in cellular and tissue damage is rapidly increasing, providing a better understanding of the responses of different species and tissues to BTV infection. However, further research is required to clarify the mechanisms leading to the death of certain mammalian cell types while allowing insect cells and possibly some mammalian cell (sub) types to become persistently infected in response to BTV infection. Only with a detailed understanding of the molecular interactions and mechanisms involved, we will be able to fully understand the processes of BTV infection and transmission.

References

Anderson, J., Mertens, P.P.C. and Herniman, K.A. (1993) A competitive ELISA for the detection of anti-tubule antibodies using a monoclonal antibody against bluetongue virus non-structural protein NS1. *J. Virol. Methods* **43**, 167–75.

Bansal, O.B., Stokes, A., Bansal, A., Bishop, D. and Roy, P. (1998) Membrane organization of bluetongue virus nonstructural glycoprotein NS3. *J. Virol* **72**, 3362–9.

Barratt-Boyes, S.M., Rossitto, P.V., Stott, J.L. and MacLachlan, N.J. (1992) Flow cytometric analysis of *in vitro* bluetongue virus infection of bovine blood mononuclear cells. *J. Gen. Virol.* **73**, 1953–60.

Basak, A.K., Gouet, P., Grimes, J., Roy, P. and Stuart, D. (1996) Crystal structure of the top domain of African horse sickness virus VP7: comparisons with bluetongue virus VP7. *J. Virol.* **70**, 3797–806.

Beaton, A.R., Rodriguez, J., Reddy, Y.K. and Roy, P. (2002) The membrane trafficking protein calpactin forms a complex with bluetongue virus protein NS3 and mediates virus release. *Proc. Natl. Acad. Sci. U.S.A.* **99**, 13154–59.

Berkova, Z., Crawfordm, S.E., Blutt, S.E., Morris, A.P. and Estes, M.K. (2007) Expression of rotavirus NSP4 alters the actin network organization through the actin remodeling protein cofilin. *J. Virol.* **81**, 3545–53.

Bhattacharya, B., Noad, R.J. and Roy, P. (2007) Interaction between bluetongue virus outer capsid protein VP2 and vimentin is necessary for virus egress. *Virol. J.* **4**, 7.

Boyce, M. and Roy, P. (2007) Recovery of infectious bluetongue virus from RNA. *J. Virol.* **81**, 2179–86.

Brookes, S.M., Hyatt, A.D. and Eaton, B.T. (1993) Characterization of virus inclusion bodies in bluetongue virus-infected cells. *J. Gen. Virol.* **74**, 525–30.

Browne, J.G. and Jochim, M.M. (1967) Cytopathologic changes and development of inclusion bodies in cultured cells infected with bluetongue virus. *Am. J. Vet. Res.* **28**, 1091–105.

Brunet, J.P., Jourdan, N., Cotte-Laffitte, J., Linxe, C., Géniteau-Legendre, M., Servin, A. and Quéro, A.M. (2000) Rotavirus infection induces cytoskeleton disorganization in human intestinal epithelial cells: implication of an increase in intracellular calcium concentration. *J. Virol.* **74**, 10801–806.

Ciarlet, M., Crawford, S.E., Cheng, E., Blutt, S.E., Rice, D.A., Bergelson, J.M. and Estes, M.K. (2002) VLA-2 (alpha2beta1) integrin promotes rotavirus entry into cells but is not necessary for rotavirus attachment. *J. Virol.* **76**, 1109–23.

Darpel, K.E. (2007) The bluetongue virus 'ruminant host – insect vector' transmission cycle; the role of *Culicoides* saliva proteins in infection. A thesis submitted in partial fulfilment of the requirements of the University of London for the degree of Doctor of Philosophy.

Darpel, K.E., Batten, C.A., Veronesi, E., Shaw, A.E., Anthony, S., Bachanek-Bankowska, K., Kgosana, L., bin-Tarif, A., Carpenter, S., Müller-Doblies, U.U., Takamatsu, H.H., Mellor, P.S., Mertens, P.P.C. and Oura, C.A. (2007) Clinical signs and pathology shown by British sheep and cattle infected with bluetongue virus serotype 8 derived from the 2006 outbreak in Northern Europe. *Vet. Rec.* **161**, 253–61.

DeMaula, C.D., Jutila, M.A., Wilson, D.W. and MacLachlan, N.J. (2001) Infection kinetics, prostacyclin release and cytokine-mediated modulation of the mechanism of cell death during bluetongue virus infection of cultured ovine and bovine pulmonary artery and lung microvascular endothelial cells. *J. Gen. Virol.* **82**, 787–94.

DeMaula, C.D., Leutenegger, C.M., Bonneau, K.R. and MacLachlan, N.J. (2002a) The role of endothelial cell-derived inflammatory and vasoactive mediators in the pathogenesis of bluetongue. *Virology* **296**, 330–7.

DeMaula, C.D., Leutenegger, C.M., Jutila, M.A. and MacLachlan, N.J. (2002b) Bluetongue virus-induced activation of primary bovine lung microvascular endothelial cells. *Vet. Immunol. Immunopathol* **86**, 147–57.

Eaton, B.T. and Hyatt, A.D. (1989) Association of bluetongue virus with the cytoskeleton. *Subcell. Biochem.* **15**, 233–73.

Eaton, B.T., Hyatt, A.D. and Brookes, S.M. (1990) The replication of bluetongue virus. *Curr. Top. Microbiol. Immunol.* **162**, 89–118.

Eaton, B.T., Hyatt, A.D. and White, J.R. (1987) Association of bluetongue virus with the cytoskeleton. *Virology* **157**, 107–16.

Eaton, B.T., Hyatt, A.D. and White, J.R. (1988) Localization of the nonstructural protein NS1 in bluetongue virus-infected cells and its presence in virus particles. *Virology* **163**, 527–37.

Fabbretti, E., Afrikanova, I., Vascotto, F. and Burrone, O.R. (1999) Two non-structural rotavirus proteins, NSP2 and NSP5, form viroplasm-like structures *in vivo*. *J. Gen. Virol.* **80**, 333–9.

Fillmore, G.C., Lin, H. and Li, J.K. (2002) Localization of the single-stranded RNA-binding domains of bluetongue virus nonstructural protein NS2. *J. Virol.* **76**, 499–506.

Forzan, M., Marsh, M. and Roy, P. (2007) Bluetongue virus entry into cells. *J. Virol.* **8**, 4819–27.

Fu, H. 1995. Mechanisms controlling the infection of Culicoides biting midges with bluetongue virus. PhD Thesis, University of Hertfordshire, p. 154.

Gardet, A., Breton, M., Fontanges, P., Trugnan, G. and Chwetzoff, S. (2006) Rotavirus spike protein VP4 binds to and remodels actin bundles of the epithelial brush border into actin bodies. *J. Virol.* **80**, 3947–56.

Gardet, A., Breton, M., Trugnan, G. and Chwetzoff, S. (2007) Role for actin in the polarized release of rotavirus. *J. Virol.* **81**, 4892–4.

Grimes, J.M., Burroughs, J.N., Gouet, P., Diprose, J.M., Malby, R., Zientara, S., Mertens, P.P.C. and Stuart, D.I. (1998) The atomic structure of the bluetongue virus core. *Nature* **395**, 470–8.

Guerrero, C.A., Méndez, E., Zárate, S., Isa, P., López, S. and Arias, C.F. (2000) Integrin alpha(v)beta(3) mediates rotavirus cell entry. *Proc. Natl. Acad. Sci. U.S.A.* **97**, 14644–9.

Guglielmi, K.M., Johnson, E.M., Stehle, T. and Dermody, T.S. (2006) Attachment and cell entry of mammalian orthoreovirus. *Curr. Top. Microbiol. Immunol.* **309**, 1–38.

Guirakhoo, F., Catalan, J.A. and Monath, T.P. (1995) Adaptation of bluetongue virus in mosquito cells results in overexpression of NS3 proteins and release of virus particles. *Arch. Virol.* **140**, 967–74.

Han, Z. and Harty, R.N. (2004) The NS3 protein of bluetongue virus exhibits viroporin-like properties. *J. Biol. Chem.* **279**, 43092–7.

Hassan, S.S. and Roy, P. (1999) Expression and functional characterization of bluetongue virus VP2 protein: role in cell entry. *J. Virol.* **73**, 9832–42.

Hassan, S.H., Wirblich, C., Forzan, M. and Roy, P. (2001) Expression and functional characterization of bluetongue virus VP5 protein: role in cellular permeabilization. *J. Virol.* **75**, 8356–67.

Hewat, E.A., Booth, T.F., Wade, R.H. and Roy, P. (1992) 3-D reconstruction of bluetongue virus tubules using cryoelectron microscopy. *J. Struct. Biol.* **108**, 35–48.

Hu, J., Dong, C.Y., Li, J.K., Chen, D.E., Liang, K. and Liu, J. (2007) Selective *in vitro* cytotoxic effect of human cancer cells by bluetongue virus-10. *Acta Oncol.* **5**, 1–11.

Huismans, H. and Els, H.J. (1979) Characterization of the tubules associated with the replication of three different orbiviruses. *Virology* **92**, 397–406.

Huismans, H. and Howell, P.G. (1973) Molecular hybridization studies on the relationships between different serotypes of bluetongue virus and on the difference between the virulent and attenuated strains of the same serotype. *Onderstepoort J. Vet. Res.* **40**, 93–104.

Huismans, H., van der Walt, N.T., Cloete, M. and Erasmus, B.J. (1983) The biochemical and immunological characterization of bluetongue virus outer capsid polypeptides. In: Compans, R.W. and Bishop, D.H.L. (eds), *Double-Stranded RNA Viruses*. New York: Elsevier, pp. 165–72.

Huismans, H., Van Dijk, A.A. and Els, H.J. (1987) Uncoating of parental bluetongue virus to core and subcore particles in infected L cells. *Virology* **157**, 180–8.

Hutchinson, I.R. 1999. The role of VP7(T13) in initiation of infection of bluetongue virus. PhD Thesis, University of Hertfordshire, p. 195.

Hyatt, A.D. and Eaton, B.T. (1988) Ultrastructural distribution of the major capsid proteins within bluetongue virus and infected cells. *J. Gen. Virol.* **69**, 805–15.

Hyatt, A.D., Eaton, B.T. and Brookes, S.M. (1989) The release of bluetongue virus from infected cells and their superinfection by progeny virus. *Virology* **173**, 21–34.

Hyatt, A.D., Gould, A.R., Coupar, B. and Eaton, B.T. (1991) Localization of the non-structural protein NS3 in bluetongue virus-infected cells. *J. Gen. Virol.* **72**, 2263–7.

Hyatt, A.D., Zhao, Y. and Roy, P. (1993) Release of bluetongue virus-like particles from insect cells is mediated by BTV nonstructural protein NS3/NS3a. *Virology* **193**, 592–603.

Jackson, T., Sharma, A., Ghazaleh, R.A., Blakemore, W.E., Ellard, F.M., Simmons, D.L., Newman, J.W., Stuart, D.I. and King, A.M. (1997) Arginine-glycine-aspartic acid-specific binding by foot-and-mouth disease viruses to the purified integrin alpha(v)beta3 *in vitro. J. Virol.* **71**, 8357–61.

Kar, A.K., Iwatani, N. and Roy, P. (2005) Assembly and intracellular localization of the bluetongue virus core protein VP3. *J. Virol.* **79**, 11487–91145.

Langner, K.F., Darpel, K.E., Denison, E., Drolet, B.S., Leibold, W., Mellor, P.S., Mertens, P.P.C., Nimtz, M. and Greiser-Wilke, I. (2007) Collection and analysis of salivary proteins from the biting midge *Culicoides nubeculosus* (Diptera: Ceratopogonidae). *J. Med. Entomol.* **44**, 238–48.

Lecatsas, G. (1968) Electron microscopic study of the formation of bluetongue virus. *Onderstepoort J. Vet. Res.* **35**, 139–49.

Li, Q., Li, H., Blitvich, B.J. and Zhang, J. (2007) The *Aedes albopictus* inhibitor of apoptosis 1 gene protects vertebrate cells from bluetongue virus-induced apoptosis. *Insect Mol. Biol.* **16**, 93–105.

López, S. and Arias, C.F. (2004) Multistep entry of rotavirus into cells: a Versaillesque, dance. *Trends Microbiol.* **12**, 271–8.

López, S. and Arias, C.F. (2006) Early steps in rotavirus cell entry. *Curr. Top. Microbiol. Immunol.* **309**, 39–66.

Lunt, R.A., Melville, L., Hunt, N., Davis, S., Rootes, C.L., Newberry, K.M., Pritchard, L.I., Middleton, D., Bingham, J., Daniels, P.W. and Eaton, B.T. (2006) Cultured skin fibroblast cells derived from bluetongue virus-inoculated sheep and field-infected cattle are not a source of late and protracted recoverable virus. *J. Gen. Virol.* **87**, 3661–6.

Lymperopoulos, K., Wirblich, C., Brierley, I. and Roy, P. (2003) Sequence specificity in the interaction of bluetongue virus non-structural protein. *J. Biol. Chem.* **2**, 31722–30.

Maginnis, M.S., Forrest, J.C., Kopecky-Bromberg, S.A., Dickeson, S.K., Santoro, S.A., Zutter, M.M., Nemerow, G.R., Bergelson, J.M. and Dermody, T.S. (2006) Beta1 integrin mediates internalization of mammalian reovirus. *J. Virol.* **80**, 2760–70.

McLaughlin, B.E., DeMaula, C.D., Wilson, W.C., Boyce, W.M. and MacLachlan, N.J. (2003) Replication of bluetongue virus and epizootic hemorrhagic disease virus in pulmonary artery endothelial cells obtained from cattle, sheep, and deer. *Am. J. Vet. Res.* **64**, 860–5.

Mellor, P.S. (1990) The replication of bluetongue virus in *Culicoides* vectors. *Curr. Top. Microbiol. Immunol.* **162**, 143–61.

Mellor, P.S., Boorman, J. and Baylis, M. (2000) *Culicoides* biting midges: their role as arbovirus vectors. *Annu. Rev. Entomol.* **45**, 307–40.

Mertens, P.P.C. (2004) The dsRNA viruses. *Virus Res.* **101**, 3–13.

Mertens, P.P.C., Brown, F. and Sangar, D.V. (1984) Assignment of the genome segments of bluetongue virus type 1 to the proteins which they encode. *Virology* **135**, 207–17.

Mertens, P.P.C., Burroughs, J.N. and Anderson, J. (1987) Purification and properties of virus particles, infectious subviral particles, and cores of bluetongue virus serotypes 1 and 4. *Virology* **157**, 375–86.

Mertens, P.P.C., Burroughs, J.N., Walton, A., Wellby, M.P., Fu, H., O'Hara, R.S., Brookes, S.M. and Mellor, P.S. (1996) Enhanced infectivity of modified bluetongue virus particles for two insect cell lines and for two *Culicoides* vector species. *Virology* **217**, 582–93.

Mertens, P.P.C. and Diprose, J. (2004) The bluetongue virus core: a nano-scale transcription machine. *Virus Res.* **101**, 29–43.

Mertens, P.P.C., Maan, S., Samuel, A. and Attoui, H. (2005) Orbivirus, Reoviridae. In: Fauquet, C.M., Mayo, M.A., Maniloff, J., Desselberger, U. and Ball, L.A. (eds), *Virus Taxonomy, VIIIth Report of the ICTV*. London: Elsevier/Academic Press, pp. 466–83.

Modrof, J., Lymperopoulos, K. and Roy, P. (2005) Phosphorylation of bluetongue virus nonstructural protein 2 is essential for formation of viral inclusion bodies. *J. Virol.* **79**, 10023–31.

Mora, M., Partin, K., Bhatia, M., Partin, J. and Carter, C. (1987) Association of reovirus proteins with the structural matrix of infected cells. *Virology* **159**, 226–77.

Mortola, E., Noad, R. and Roy, P. (2004) Bluetongue virus outer capsid proteins are sufficient to trigger apoptosis in mammalian cells. *J. Virol.* **78**, 2875–83.

Murphy, F.A., Borden, E.C., Shope, R.E. and Harrison, A. (1971) Physicochemical and morphological relationships of some arthropod-borne viruses to bluetongue virus a new taxonomic group. Electron microscopic studies. *J. Gen. Virol.* **13**, 273–88.

Musalem, C. and Espejo, R.T. (1985) Release of progeny virus from cells infected with simian rotavirus SA11. *J. Gen. Virol.* **66**, 2715–24.

Owen, N.C. (1966) Investigation into the pH stability of bluetongue virus by electron microscopy. *Onderstepoort J. Vet. Res.* **33**, 9–14.

Owens, R.J., Limn, C. and Roy, P. (2004) Role of an arbovirus nonstructural protein in cellular pathogenesis and virus release. *J. Virol.* **78**, 6649–56.

Parker, J.S., Broering, T.J., Kim, J., Higgins, D.E. and Nibert, M.L. (2002) Reovirus core protein mu2 determines the filamentous morphology of viral inclusion bodies by interacting with and stabilizing microtubules. *J. Virol.* **76**, 4483–96.

Patton, J.T., Silvestri, L.S., Tortorici, M.A., Vasquez-Del Carpio, R. and Taraporewala, Z.F. (2006) Rotavirus genome replication and morphogenesis: role of the viroplasm. *Curr. Top. Microbiol. Immunol.* **309**, 169–87.

Pelkmans, L. and Helenius, A. (2003) Insider information: what viruses tell us about endocytosis. *Curr. Opin. Cell Biol.* **15**, 414–22.

Poncet, D., Aponte, C. and Cohen, J. (1993) Rotavirus protein NSP3 (NS34) is bound to the 3' end consensus sequence of viral mRNAs in infected cells. *J. Virol.* **67**, 3159–65.

Poncet, D., Aponte, C. and Cohen, J. (1996) Structure and function of rotavirus nonstructural protein NSP3. *Arch. Virol. Suppl.* **12**, 29–35.

Ramadevi, N., Burroughs, N.J., Mertens, P.P.C., Jones, I.M. and Roy, P. (1998) Capping and methylation of mRNA by purified recombinant VP4 protein of bluetongue virus. *Proc. Natl. Acad.Sci. U.S.A.* **95**, 13537–42.

Ross-Smith, N. 2008. Elucidating the role of non-structural protein 2 (ViP) in bluetongue virus infection. PhD Thesis, University of Oxford.

Russell, H., O'Toole, D.T., Bardsley, K., Davis, W.C. and Ellis, J.A. (1996) Comparative effects of bluetongue virus infection of ovine and bovine endothelial cells. *Vet. Pathol.* **33**, 319–31.

Sharpe, A.H., Chen, L.B. and Fields, B.N. (1982) The interaction of mammalian reoviruses with the cytoskeleton of monkey kidney CV-1 cells. *Virology* **120**, 399–411.

Silvestri, L.S., Taraporewala, Z.F. and Paton, J.T. (2004) Rotavirus replication: plus-sense templates for double-stranded RNA synthesis are made in viroplasms. *J. Virol.* **78**, 7763–74.

Skup, D., Zarbl, H. and Millward, S. (1981) Regulation of translation in L-cells infected with reovirus. *J. Mol. Biol.* **151**, 35–55.

Smith, G.A. and Enquist, L.W. (2002) Break ins and break outs: viral interactions with the cytoskeleton of mammalian cells. *Annu. Rev. Cell Dev. Biol.* **18**, 135–61.

Stirling, J. 1996. Studies on the replication and assembly of bluetongue virus. PhD Thesis, University of Reading, p. 237.

Stott, J.L., Blanchard-Channell, M., Scibienski, R.J. and Stott, M.L. (1990) Interaction of bluetongue virus with bovine lymphocytes. *J. Gen. Virol.* **71**, 363–8.

Takamatsu, H. and Jeggo, M.H. (1989) Cultivation of bluetongue virus-specific ovine T cells and their cross-reactivity with different serotype viruses. *Immunology* **66**, 258–63.

Takamatsu, H.H., Mellor, P.S., Mertens, P.P.C., Kirkham, P.A., Burroughs, J.N. and Parkhouse, R.M. (2003) A possible overwintering mechanism for bluetongue virus in the absence of the insect vector. *J. Gen. Virol.* **84**, 227–35.

Tan, B.H., Nason, E., Staeuber, N., Jiang, W., Monastryrskaya, K. and Roy, P. (2001) RGD tripeptide of bluetongue virus VP7 protein is responsible for core attachment to *Culicoides* cells. *J. Virol.* **75**, 3937–47.

Taraporewala, Z.F., Chen, D. and Patton, J.T. (2001) Multimers of the bluetongue virus nonstructural protein, NS2, possess nucleotidyl phosphatase activity: similarities between NS2 and rotavirus NSP2. *Virology* **280**, 221–31.

Theron, J., Uitenweerde, J.M., Huismans, H. and Nel, L.H. (1994) Comparison of the expression and phosphorylation of the non-structural protein NS2 of three different orbiviruses: evidence for the involvement of an ubiquitous cellular kinase. *J. Gen. Virol.* **75**, 3401–11.

Thomas, C.P., Booth, T.F. and Roy, P. (1990) Synthesis of bluetongue virus-encoded phosphoprotein and formation of inclusion bodies by recombinant baculovirus in insect cells: it binds the single-stranded RNA species. *J. Gen. Virol.* **71**, 2073–83.

Touris-Otero, F.J., Martinez-Costas, Vakharia, V.N. and Benavente, J. (2004) Avian reovirus nonstructural protein micro NS forms viroplasm-like inclusions and recruits protein sigma NS to these structures. *Virology* **319**, 94–106.

Urakawa, T. and Roy, P. (1988) Bluetongue virus tubules made in insect cells by recombinant baculoviruses: expression of the NS1 gene of bluetongue virus serotype. *J. Virol.* **10**, 3919–27.

Van Dijk, A.A. and Huismans, H. (1980) The *in vitro* activation and further characterization of the bluetongue virus-associated transcriptase. *Virology* **104**, 347–56.

Wechsler, S.J. and McHolland, L.E. (1988) Susceptibilities of 14 cell lines to bluetongue virus infection. *J. Clin. Microbiol.* **26**, 2324–7.

Weclewicz, K., Kristensson, K. and Svensson, L. (1994) Rotavirus causes selective vimentin reorganization in monkey kidney CV-1 cells. *J. Gen. Virol.* **75**, 3267–71.

Whetter, L.E., Maclachlan, N.J., Gebhard, D.H., Heidner, H.W. and Moore, P.F. (1989) Bluetongue virus infection of bovine monocytes. *J. Gen. Virol.* **70**, 1663–76.

Xu, G., Wilson, W., Mecham, J., Murphy, K., Zhou, E.M. and Tabachnick, W. (1997) VP7: an attachment protein of bluetongue virus for cellular receptors in *Culicoides variipennis*. *J. Gen. Virol.* **78**, 1617–23.

Zarbl, H., Skup, D. and Millward, S. (1980) Reovirus progeny subviral particles synthesize uncapped mRNA. *J. Virol.* **34**, 497–505.

The structure of bluetongue virus core and proteins

6

PETER P.C. MERTENS,* NATALIE ROSS-SMITH,* JON DIPROSE† AND HOUSSAM ATTOUI*

*Department of Arbovirology, Institute for Animal Health, Pirbright Laboratory, Pirbright, Woking GU24 0NF, UK.
†Wellcome Trust Centre for Human Genetics, Division of Structural Biology, University of Oxford OX3 7BN, UK

Introduction

Structural studies have provided insights into the biological functions of many viruses (Wikoff *et al.*, 2000; Abrescia *et al.*, 2004; Cockburn *et al.*, 2004). In comparison to DNA or single-stranded (ss) RNA viruses, the members of the family *Reoviridae* [the 'reoviruses', which includes bluetongue virus (BTV) and the other orbiviruses] have additional constraints on their life cycle imposed by the nature of their double-stranded (ds) RNA genome (Mertens *et al.*, 2005).

The BTV particle provides a secure compartment for packaging and transporting the viral genome from one infected host cell to the next. The protein components of the outer capsid layer of the virus are directly involved in binding to the cell surface and initiating cell entry. The absence of host cell RNA polymerases that are able to use dsRNA as a template for mRNA synthesis suggests that a naked dsRNA virus genome would be transcriptionally and translationally inert within the host cell cytoplasm. It would therefore be unable to initiate the processes of replication. However, the structure of the inner core particle of BTV also provides a secure space, within which 10 genome segments can be repeatedly transcribed into ssRNA by the core-associated RNA polymerase. The positive-sense RNA copies (viral mRNAs)

ISBN-13: 978-0-12-369368-6

are synthesized simultaneously from each of the genome segments and are released from the core particle directly into the host cell cytoplasm. They can then function as templates for both translation and negative-strand viral RNA synthesis within nascent progeny virus particles (see Chapters 4 and 5). This transcription mechanism requires efficient and co-ordinated movement of the 10 genome segments through the active sites of at least six different enzyme activities within the confined space of the core structure, including NTPase, helicase, polymerase, RNA capping and two distinct transmethylase activities.

Many mammalian, insect and plant cells also have defence systems that are activated by detection of intracellular dsRNAs, which represents an indicator of viral infection (Jacobs and Langland, 1996; Gitlin and Andino, 2003; Goldbach *et al.*, 2003). Once dsRNA has been detected, the host cell can modify its translation apparatus, leading to apoptosis, interference or activation of sequence-specific RNA cleavage that has the potential to destroy viral mRNAs (RNA silencing) and prevent or limit replication. From the point of view of the virus, these cellular defences need to be avoided, and the reoviruses have evolved various strategies to circumvent them (reviewed by Stuart and Grimes, 2006). As a primary defence, the structure of the BTV core provides an effective physical barrier between the viral genome and the cell cytoplasm, preventing release of dsRNA and 'hiding it' from the host's detection apparatus (Zarbl and Millward; 1983 Eaton *et al.*, 1990).

The bluetongue virion structure

The orbiviruses have a 10-segmented dsRNA genome that is packaged within a three-layered icosahedral protein capsid (Figure 6.1, see colour plate 13). Core particles of BTV-1 and BTV-10 contain almost exactly 1000 protein molecules within each particle (Stuart *et al.*, 1998). The structure of BTV-1 and BTV-10 cores has been determined by X-ray crystallography to a resolution of 3.5 and 6.5 Å, respectively, using crystals with unit cell parameters in excess of 1000 Å (Grimes *et al.*, 1998; Gouet *et al.*, 1999; Diprose *et al.*, 2001). This demonstrated, for the first time, that it was possible to analyse crystals with a unit cell mass approaching one billion Daltons. These studies revealed the structural organization of the protein bilayer that makes up the icosahedral capsid of the BTV core, providing an important paradigm for the other orbiviruses and other reoviruses.

The innermost capsid shell of BTV (the subcore) is composed of 120 copies of the essentially triangular VP3(T2) protein (901 amino acids, 103 kDa, encoded by BTV genome segment 3). VP3(T2) is arranged with icosahedral symmetry, occupying two different positions and two distinct conformations within the subcore capsid layer, which are identified by Grimes *et al.* (1998) as A and B. The VP3(T2)A and B molecules are chemically identical but undergo

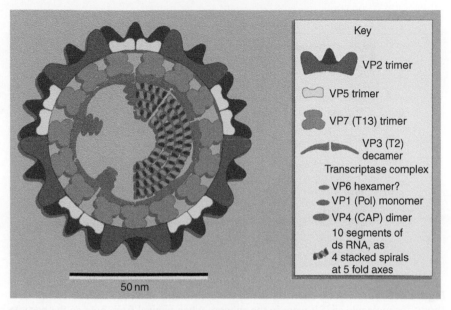

Figure 6.1 Schematic diagram of the structure of the bluetongue virus (BTV) particle based on structural data from several sources (Mertens *et al.*, 2005).

an internal conformational shift between their 'apical', 'carapace' and 'dimerization' domains, allowing them to interact and close the surface of the icosahedron (Figure 6.2). Five VP3(T2)A molecules surround each five-fold apex, leaving a small central pore. These are interspersed with five VP3(T2)B molecules, which are slightly more distant from the five-fold axis, creating a dish-shaped decamer. The subcore itself can be considered as an assembly of 12 of the VP3(T2) decamers, each of which has a 'zigzag' outer edge, which 'zips' together with its immediate neighbours to form the intact subcore layer (see Figure 6.2). This unusual method of icosahedral construction can be usefully interpreted as having a pseudo triangulation number of 2 ($T = 2$) (Caspar and Klug, 1962; Grimes *et al.*, 1998) or alternatively as a $T = 1$ arrangement of AB conformational heterodimers (Reinisch *et al.*, 2000).

The BTV subcore layer can self-assemble when VP3(T2) is synthesized separately from the other viral proteins (e.g. as expressed by a recombinant baculovirus in insect cells; P. Roy, personal communication). Subcore particles of some orbiviruses, for example, equine encephalosis virus (EEV), are also stable in the absence of the outer capsid layers (Mertens, 2004). Structure-based modifications of the VP3 molecule, which remove the dimerization domain that is responsible for inter-decamer contacts, allow decamer formation but prevent assembly of the intact subcore shell (Mertens and Diprose, 2004).

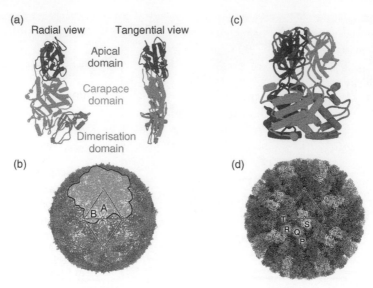

Figure 6.2 (a) Cartoon showing the domain structure of VP3(T2). The radial view shows the molecule as if it were in the subcore when viewed from outside the particle looking towards the centre. The tangential view is rotated 90° away from the radial view around the long axis of the molecule. Note how thin the molecule appears in tangential compared to radial profile showing just how thin the subcore shell is compared to the volume it encloses. (b) The BTV subcore made up of 120 copies of VP3(T2). The conformationally distinct A and B molecules have been coloured green and red, respectively. The icosahedral 5-fold, 3-fold and 2-fold symmetry axes have been marked. One decamer, a putative assembly intermediate, has been highlighted. (c) Cartoon showing the structure of the VP7 trimer. The three monomers have been coloured differently. The trimer sits on the exterior surface of the subcore, making contact through the bottom surface in this view. Inter-trimer contacts are mediated through the sides of the lower half of the molecule. (d) Cartoon showing the core structure. The icosahedral 5-fold, 3-fold and 2-fold axes are labelled and the icosahedrally independent trimers P (purple), Q (orange), R (green), S (yellow) and T (blue) are identified (Grimes *et al.*, 1998) (See colour plate 9).

The architecture of the BTV subcore shell, and even the overall shape of the VP3(T2) protein, shows remarkable similarities to the innermost capsid shell and inner capsid protein of many other dsRNA viruses (Bamford *et al.*, 2005; Mertens, 2004). This suggests that these dsRNA viruses have evolved from a common ancestor that originally developed a simple although elegant mechanism for the assembly of an inner capsid shell, which now represents an important step in the virion assembly pathway. Since the VP3(T2) capsid layer can self-assemble, the amino acid (aa) sequence must not only determine the fold and structure of the protein but also hold information that determines the overall size of the complete capsid. The assembly of VP3(T2) must therefore also dictate the organization of the other structural components of the particle that are attached or interact with it, both internally and externally.

The outer surface of the BTV core is composed of 780 copies of the VP7 protein (349 aa, 38.5 kDa, encoded by genome segment 7), arranged (as 260 trimers) with $T = 13l$ icosahedral symmetry (Grimes *et al.*, 1998). The VP7(T13) molecules are layered onto the scaffolding of the VP3(T2) subcore shell but as a result of mismatched symmetry between these layers, the base of the VP7(T13) monomer must interact with the VP3(T2) molecules of the subcore surface in 13 distinct orientations. Indeed, the individual VP7 trimers can occupy five different positions in the core surface layer, which have been identified with increasing distance from the five-fold axis, as P through Q, R and S (two-fold) to T at the three-fold axis (see Figure 6.2). The symmetries of the VP3(T2) and VP7(T13) protein layers coincide only at the three-fold icosahedral axes of the BTV core. At these points, the interactions between the VP7 and VP3 are also the most extensive, suggesting that trimers are added sequentially from the three-fold towards the five-fold axes. These interactions progressively 'curve' the VP7 layer around the particle surface forming a series of six-component, ring-shaped capsomers, finishing with rings composed of five VP7 trimers around the five-fold axes. The significance of VP3–VP7 interactions in forming the core surface layer is suggested by VP7 of African horse sickness viruses (AHSV), another *Orbivirus* species closely related to BTV, which form crystals of VP7 within the cell cytoplasm (Burroughs *et al.*, 1994). In the absence of interactions with VP3(T2) on the subcore surface, the AHSV VP7 trimers form into large 'flat hexagonal arrays' of six-membered rings (Burroughs *et al.*, 1994).

Genome packaging

The electron density maps of the core of both BTV-1 and BTV-10 that were generated by X-ray crystallography (Gouet *et al.*, 1999; Grimes *et al.*, 1998) revealed layers of density within the central space of the subcore that could not be modelled as viral proteins (Gouet *et al.*, 1999). These layers are made up of multiple strands, which in many places have a helical structure that is not only very similar for the two virus serotypes but also consistent with layers of the packaged genomic dsRNA.

From a detailed knowledge of the volume and internal contents of the BTV core (Grimes *et al.*, 1998), it has been possible to calculate the concentration of the dsRNA within the central cavity of the BTV particle. The volume of the core interior was calculated at 60.6×10^6 Å^3. The volume occupied by the protein components of the transcriptase complexes (TCs) was estimated, based on the copy numbers (VP1, 12; VP4, 24 and VP6, 72) (Stuart *et al.*, 1998) and molecular masses (VP1, 150 kDa; VP4, 76.4 kDa; VP6, 35.8 kDa) of the components (Roy, 1989), as 7.6×10^6 Å^3 (Gouet *et al.*, 1999). The residual volume for the RNA is therefore $\sim53.0 \times 10^6$ Å^3. The BTV-10 genome consists of 19 219 base pairs (bp) of dsRNA, with a molecular mass of $\sim13.1 \times 10^6$ Da (Roy, 1989). The concentration of dsRNA within the core is therefore ~410 mg/mL.

The properties of concentrated solutions of dsDNA are relatively well characterized although less is known about the properties of dsRNA. It is established that dsDNA forms liquid crystalline arrays at high concentration with the phase and helix–helix packing distance being a simple function of concentration (Livolant and Leforestier, 1996). At concentrations of ~400 mg/mL, the liquid crystalline packing arrangement for DNA has been shown to be columnar hexagonal with an inter-helix packing distance of approximately 30 Å, and it appears likely that the behaviour of dsRNA would be similar at high concentrations. This ordering of the dsRNA molecules may well be essential if they are to function effectively as templates for the core-associated transcriptase activities without becoming tangled and jamming the mechanism. One potentially relevant property of such a packing structure is that in the presence of suitable counterions, the nucleic acid chains would glide over each other with very little friction (Livolant and Leforestier, 1996). The phosphate backbone of the dsRNA carries a negative charge that would presumably be neutralized by counterions within the particle. A scanning proton microprobe has been used in an attempt to detect the presence of metal ions within the crystals of BTV core particles (Gouet et al., 1999) but failed to detect magnesium ions, which are essential for the polymerase activity (van Dijk and Huismans, 1980). Although both calcium and zinc are present, they are detected only at approximately one hundredth of the level of phosphorous. This would be insufficient to neutralize the charge on the RNA's phosphate backbone and suggests that an organic cation such as spermidine may be present, although this has not yet been confirmed.

A model, derived from X-ray crystallography studies, has been proposed for packing of the BTV genome (Gouet et al., 1999), which represents $\sim80\%$ of the 19 219 bp (a total length ~6 µm), implying that there is a particular organization of the dsRNA strands within the core. Such ordering appears to be at least partially imposed by chemically featureless grooves that form tracks for the RNA on the inside of the VP3(T2) layer (Figure 6.3). Specific RNA/protein interactions are evident at only two points in the icosahedral asymmetric unit, involving the long βJ/βK loop (residues 790 to 817), a secondary structure defined by Grimes et al. (1998) in the dimerization domains of the VP3(T2)A and VP3(T2)B molecules. This loop is partially disordered in VP3(T2)B (Grimes et al., 1998). Apart from the Lys-807 on the βJ/βK loop, there are very few basic residues on the inner surface of VP3(T2). This paucity of specific interactions may also facilitate the movement of the dsRNA segments within the core during transcription.

Figure 6.3c shows that there are some striking similarities in the interactions that the structurally distinct A and B copies of VP3(T2) make with the RNA

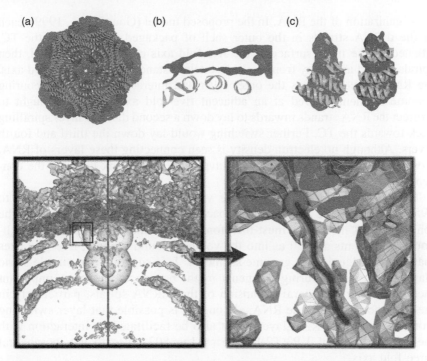

Figure 6.3 (Top) Cartoon showing dsRNA following shallow grooves on the underside of the VP3(T2) layer. (a) The top layer of dsRNA is modelled as a spiral around the 5-fold axis. The strands shown in orange are related to the strands shown in blue by icosahedral symmetry. (b) A slice through the VP3(T2) layer (molecular surface shown in red and green) showing shallow grooves on the underside of the layer. The strands appear to follow these grooves. (c) The grooves are in roughly the same place on both the A (green) and B (red) molecules. When assembled into the subcore layer, A and B are offset, and the grooves on one molecule feed into a groove further from the 5-fold on the next, giving rise to the spiral pattern (Gouet *et al.*, 1999). (Bottom) Cartoon showing the putative location of VP1(Pol) and VP4(Cap) within the subcore. VP3(T2) is shown as red and green worms, the observed electron density is shown as a grey semi-transparent surface, the icosahedral 5-fold axis is shown as a purple bar. VP1(Pol) and VP4(Cap) are thought to lie directly under the subcore layer on the icosahedral 5-fold axis, with VP1(Pol) being uppermost. The red and purple semi-transparent surfaces have been calculated to have volumes equal to those estimated for VP1(Pol) and VP4(Cap) and indicate their likely location. Shown in the right-hand panel is some weak density that may indicate that the disordered N-terminus of VP3(T2)A actually leads down into the core to make contact with VP1(Pol) and/or VP4(Cap). The possible path of the protein backbone is shown in blue (Gouet *et al.*, 1999) (See colour plate 10).

although there are also some differences that reflect the specific conformational changes between the two protein subunits. Since the two molecules lie at different radial distances from the five-fold axes, the quasi-equivalence in the protein–RNA interactions generates the spiral structure, which is observed in the organization of the RNA. In the proposed model (Gouet *et al.*, 1999), each of the dsRNA strands in the outer shell of packaged RNA leaves the TC, situated on the inner surface at the five-fold axis of the subcore shell, then spirals around and away from it. At a certain distance from the five-fold axis, the RNA effectively fills the outer layer and interacts with a neighbouring genome segment centred at an adjacent five-fold axis. This is thought to redirect the RNA strands inwards to lay down a second discrete layer spiralling back towards the TC. Further switching would lay down the third and fourth layers. Although no electron density is seen connecting these layers of RNA, icosahedral averaging would render invisible a single link between the concentric layers.

The BTV genome segments have very different lengths (from 822 to 3954 bp), and there is little free space within the particle. Some of the longer dsRNA molecules must therefore trespass into the volume of neighbouring segments as well as into the volume around the two, five-fold axes that are not occupied by genome segments. This suggests that simple steric clashes with neighbouring segments might be one of the most important factors limiting the lateral expansion of the dsRNA spirals, particularly in the innermost layer of the RNA, although it is possible that layer switching from the outermost RNA layer might also be facilitated by interaction with the flexible loop of VP3(T2)B bearing Lys-807 close to the icosahedral three-fold axis.

Enzyme functions and location

The BTV core contains a small number of TCs (10–12 copies) (Stuart *et al.*, 1998), which are activated by the removal of the outer capsid proteins. The TCs are composed of three minor structural proteins: the polymerase VP1(Pol) (Roy *et al.*, 1988; Urakawa *et al.*, 1989), the capping enzyme VP4(CaP) (Mertens *et al.*, 1992; Ramadevi *et al.*, 1998) and the helicase VP6(Hel) (Stauber *et al.*, 1997). The polymerase, which requires Mg^{2+}, transcribes positive (+ve)-sense ssRNA copies from the negative (−ve) RNA strand of the dsRNA genome segment, which acts as a template. It uses NTPs as substrates and produces pyrophosphate (PP_i) as a by-product. The resulting daughter strand is capped by VP4(CaP), which has nucleotide phosphohydrolase, guanylyltransferase (GTase) and two distinct transmethylase activities required for the synthesis of Cap 1 structures (Le Blois *et al.*, 1992; Mertens *et al.*, 1992; Martinez-Costas *et al.*, 1998; Ramadevi and Roy, 1998; Ramadevi *et al.*, 1998). The nucleotide phosphohydrolase activity removes the γ-phosphate from the 5′ G at the end of the +ve-sense RNA, releasing P_i as a by-product. The GTase activity then adds GMP derived from GDP or GTP substrates to the 5′-terminal GDP, forming GpppG as a 5′ to 5′-linked structure releasing Pi or PP_i as by-products. Finally, the methyltransferase activities use *S*-adenosyl methionine (AdoMet) as a substrate to add two methyl groups to the Cap structure, releasing *S*-adenosyl homocysteine (AdoHCy) as a further by-product.

In addition to the RNA of the viral genome, a distinct but unlayered region of electron density was also detected within the central space, immediately below the internal five-fold axes of the icosahedral BTV subcore (see Figure 6.3) (Gouet *et al.*, 1999). This is thought to represent the protein components of the TCs. These complexes appear to be attached to the inner VP3(T2) surface immediately below the pore at the five-fold axis. By analogy with the cypoviruses (Yazaki and Miura, 1980) and complying with the model for packing of BTV RNA described above (Gouet *et al.*, 1999), each TC, composed of the minor core proteins VP1(Pol), VP4(CaP) and VP6(Hel), is thought to be closely associated with a single genome segment (Urakawa *et al.*, 1989; Mertens *et al.*, 1992). These TCs transcribe the 10 genome segments, producing exactly full-length mRNA copies (Huismans and Verwoerd, 1973), which are extruded from the core surface via the pores at the five-fold axes (see Figure 6.3).

During transcription, each dsRNA genome segment must move through the active site of the polymerase enzyme. The fully conservative nature of the transcriptase imposes certain topological requirements on the process. The two strands of the parental dsRNA segments must unwind prior to transcription to allow the −ve-sense template strands to enter the polymerase active site. However, the resulting parental–daughter strand duplexes must also be separated so that the nascent RNA chains can be exported from the core particle probably through the pores in the VP3(T2) subcore shell at the five-fold axes (Grimes *et al.*, 1998; Gouet *et al.*, 1999), allowing the parental strands of the dsRNA duplex to reanneal. VP6(Hel) is a helicase and has ATPase activity using the energy released by the hydrolysis of ATP to ADP and P_i to separate dsRNA into its component strands (Stauber *et al.*, 1997). This activity may facilitate RNA strand separation before or after transcription or both.

The BTV core faces several logistical problems during transcription. It must provide entry routes and mechanisms to continuously feed the vital NTP and Ado Met substrates to the internal enzyme complexes, to maintain appropriate levels of certain metal ions and to allow the reaction by-products to escape. It must also simultaneously propel the 10 nascent mRNA molecules into the infected cell cytoplasm so that they can be translated or packaged within the next generation of virus particles. Analyses of transcribing cores of related viruses (Gillies *et al.*, 1971; Yazaki and Miura, 1980; Lawton *et al.*, 1997a) are also consistent with the TCs lying at the five-fold axes of the core and with mRNA being extruded at, or close to, these symmetry axes. In BTV, the electron density indicates that the TCs, which are non-symmetric, lie along the five-fold axes of the virus at the heart of the dsRNA spirals formed by the genome segments and immediately below the pores in the VP3(T2) layer (Grimes *et al.*, 1998), which, if slightly expanded from the resting structure, would allow the exit of the RNA (see Figure 6.3) (Diprose *et al.*, 2001). The volume of the electron density envelope assigned to proteins at each of the icosahedral five-fold axes on the inside of the VP3(T2) shell is, $\sim350 \times 10^3$ Å^3, almost sufficient to accommodate the monomeric VP1(Pol) and two subunits of VP4(CaP), which, assuming standard packing densities, would occupy a total volume of 366×10^3 Å^3.

The N-terminal 50 residues of VP3(T2)A that are situated closest to the five-fold axis appear to modulate the symmetry mismatch that exists between the five-fold symmetric VP3(T2) shell and the asymmetric TC. These N-terminal residues swing inwards, compared to their positions in the VP3(T2)B molecule where they are well ordered (Grimes *et al.*, 1998), and appear to engage the TC acting as five fingers that appear to hold at least some of the enzyme components of the TC in the correct orientation (see Figure 6.3). This domain may therefore provide an important structural basis for the TC and RNA organization, which determines the mechanics of RNA synthesis and export from the core. This is consistent with observations for orthoreovirus where the λ1 protein, the structural homologue of VP3(T2), is enlarged compared to BTV and contains a proteolytically sensitive N-terminal region, which is not completely ordered (Smith *et al.*, 1969; White and Zweerink, 1976) and which seems to be located, at least in part, close to the icosahedral five-fold axes (Dryden *et al.*, 1998).

In BTV, the dimeric protein VP4(CaP) can guanylate and methylate the 5′ end of exogenous ssRNA strands forming Cap-1 structures (Mertens *et al.*, 1992; Ramadevi *et al.*, 1998), suggesting that it lies immediately within the VP3(T2) layer (Gouet *et al.*, 1999) forming strong interactions with residues 307–328 of VP3(T2)A (see Figure 6.3). The monomeric VP1 viral transcriptase is thought to lie internal to VP4 and is probably stabilized in part by interactions with the N-terminus of VP3(T2)A (see Figure 6.3).

Substrate and product-binding sites

Bluetongue virus core particles can synthesize and cap mRNA copies of ten dsRNA genome segments. A solution consisting of 4 mM ATP, 2 mM CTP, 2 mM GTP, 2 mM UTP and 0.5 mM Ado Met in a buffer of 9 mM $MgCl_2$, 6 mM DTT and 0.1 M Tris/HCl, pH 8.0 at 30 °C is sufficient for the in vitro activity of purified core particles (Van Dijk and Huismans, 1980; Mertens *et al.*, 1987). The BTV core particles, which can be used to make crystals for X-ray diffraction studies, retain their polymerase activity even though crystal formation takes several weeks at 29°C. This activity can be demonstrated by transferring a crystal into the above solution. After approximately 30 min, the crystals develop cracks and after a few hours only tiny fragments remain (Figure 6.4). This is thought to reflect the synthesis of viral mRNA, which pushes the crystal apart.

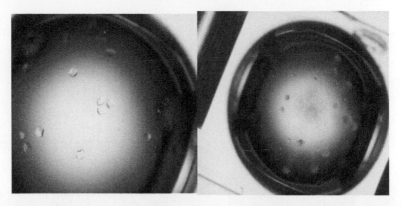

Figure 6.4 The image on the left shows native crystals of bluetongue virus (BTV)-1 cores (approximately 0.2 mm diameter) after transfer to 0.1 M Tris/HCl buffer at pH 8.0 after 4 h at 30 °C. The image on the right shows similar crystals that have shattered after transfer to a complete polymerase reaction mix (0.1 M Tris/HCl, pH 8.0, MgCl$_2$, DTT, Ado Met and NTPs) for the same period.

A quantitative measure of the BTV core transcriptase activity can be obtained by adding [^{3}H]UTP to the reaction mix and assessing the amount of label that becomes incorporated into RNA precipitated by 10% cold trichloroacetic acid (TCA). Biochemical and electron microscopic (EM) evidence from other systems suggests that all 10 genome segments are transcribed simultaneously (Gillies *et al.*, 1971; Bartlett *et al.*, 1974; Yazaki and Miura, 1980; Smith and Furuichi, 1982; Lawton *et al.*, 1997a) at a rate of up to 50 nucleotides per second per transcript (Bartlett *et al.*, 1974).

In order for these transcription reactions to occur, substrates must enter the core and the end products must leave (Figure 6.5). The BTV core structure is tightly knit, and the majority of gaps in the VP3(T2) layer are also blocked by VP7(T13) trimers. The largest of the remaining holes in the protein structure of the core that could act as entry or exit channels are situated at the icosahedral five-fold axes. These appear to function as exits for the nascent mRNA strands. However, with 10 out of the 12 five-fold axes occupied in this way, other channels must also exist to allow the entry of substrate and the exit of byproducts during transcription. The small size of the other pores that were detected during structural studies of native BTV core suggests that specific interactions between ligand (substrate or end product molecules) and channels though the capsid may be necessary for their passage. Indeed, Diprose *et al.* (2001) showed that at high concentrations, it is possible to trap and identify specific ligand molecules within these channels.

Crystals of BTV-1 core particles were soaked in solutions containing molecules of interest, prior to the collection of X-ray diffraction data, by Diprose *et al.* (2001). Various electron density maps were then calculated

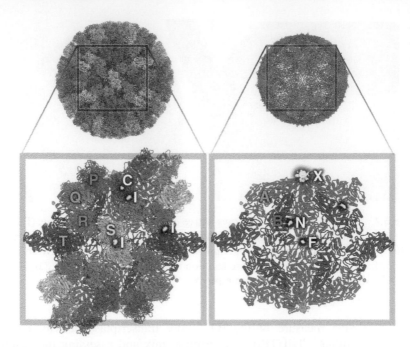

Figure 6.5 Cartoon showing substrate/product-binding sites on the bluetongue virus (BTV) core particle. The left panel shows the VP7(T13) layer (trimers P, Q, R, S, T are labelled) and associated binding sites (I, C). The right-hand panel shows the VP3(T2) layer (molecules A and B are labelled) and associated sites (X, N and F). The inter-trimer site I binds NTP and NDP and is found on the local 2-fold dyad between the top domains of trimers S–S′, P–Q and R–T. The cation-binding site C is found within all of the trimers but shown only for the P trimer for simplicity. Site X, on the icosahedral 5-fold axis, is the likely site of export of the nascent mRNA. Site N was shown to bind NTP and NDP and is a putative pore allowing NTP into the particle (see Figure 6.12 for more detail). Site F was shown to bind phosphate and may be a route out of the core for waste phosphate (Diprose *et al.*, 2001) (See colour plate 11).

and compared with reference data derived from crystals analysed in the absence of the relevant molecules. Several different nucleoside mono-, di- and triphosphates, DNA and RNA oligonucleotides, various smaller molecules (such as P_i and PP_i) and several cations were used in these analyses. The exquisite sensitivity of the 'difference maps', enhanced by difference attenuation and 30-fold real-space averaging, allowed binding to be identified at several distinct sites around the core particle. Figure 6.5 shows a composite of the results obtained, identifying a number of different binding sites on the outer surface and the inner surfaces of the core. Some of these appear to represent entrance and exit pores in the structure. The one-letter codes used to identify the different sites shown in Figure 6.5 are used below.

Transfer from the crystallization mother liquor used to grow the BTV core crystals (containing 25% saturated ammonium sulphate, 25% ethylene glycol in 0.1 M Tris/HCl, pH 8.0) to the polymerase reaction buffer caused a slight radial expansion of the core capsid that was localized to regions around the icosahedral five-fold axes. On the basis of the relatively high atomic temperature factors found during their model refinement, Grimes et al. (1998) had already hypothesized that these regions may have some flexibility. Since it is unlikely that ammonium sulphate or ethylene glycol would be excluded from the interior of the core, this would not seem to be an osmotic effect. It was suggested that a reduction in the counterions available to screen the backbone charge of the genomic RNA during core particle production and crystallization had caused an increase in strand separation, raising the pressure that the tightly packed genome exerts on the capsid. Diprose et al. (2001) point out that this radial expansion around the five-fold axes may also increase the permeability of the capsid for the substrates and products of the internal machinery.

Site X: the exit

Site X is the pore through the capsid at the icosahedral five-fold axis and is the largest of the pores identifiable from the 3.5 Å model. Lawton et al. (1997a, 1997b) used cryo-EM analysis of rotavirus, a member of a related genus within the family Reoviridae, to show that the RNA was clearly visible during transcription at the five-fold axis but only near the capsid surface. Their interpretation was that the ssRNA passed through the inner capsid layer via a site somewhat displaced from the five-fold axis and moved up the axis only once clear of the inner capsid.

Soaks with 20 base RNA oligomers demonstrated that ssRNA will bind only to the BTV core in a specific orientation at site X, exactly at the five-fold axis (Diprose et al., 2001). The absence of any other ordered association makes it unlikely that another route is available for the export of nascent RNA through the capsid. The walls of this pore are lined by 15 arginine residues, three contributed by each of the five surrounding VP3(T2) molecules, giving an area of strong positive charge that is capable of interacting with the negative charge of the RNA phosphate backbone. Site X will also bind a range of nucleosides mono-, di- and tri-phosphates and PPi. Indeed, the pore is deep enough to bind several nucleotide molecules at one time. An AMP soak revealed five bound nucleotides, taking advantage of base stacking as well as ionic interactions whilst in the pore. The aliphatic portions of the arginine side chains are also capable of interacting with the hydrophobic surfaces presented by the bases. The combination of these charged and hydrophobic sequence-independent interactions would discourage a build-up of secondary structure in the nascent RNA until it is clear of the inner capsid layer, presumably helping to prevent any blockage.

The additional ordering constraints imposed by the RNA backbone allows significantly more density to be visible, bound to the pore at the five-fold axis, in crystals soaked with RNA oligomers than with AMP monomers. However, this was still enough to account for only up to 15 bases of ssRNA. Bluetongue virus core particles can catalyse the addition of Cap 1 structures to exogenous ssRNA (Mertens *et al.*, 1992). These observations indicate that the 5′-terminus of the oligomer can reach into the active site of the capping enzyme VP4(CaP). By making a conservative estimate, the active site VP4(CaP) must be within 15 bases of site X. If the mRNA is fully extended, this could be up to 90 Å. As a comparison, the two-fold and three-fold axes pass through the capsid about 140 and 155 Å from the X site, respectively (see Figure 6.3), and the centre of the core is over 240 Å away down the five-fold axis. Unfortunately, neither internalized density was visible for the oligomer to positively locate VP4(CaP) nor any internal density was visible in soaks with GDP, which is also expected to bind to the active site of the enzyme.

Site N: entry

A difference in electron density was also observed at a second site (N) in the VP3(T2) layer after soaking with ADP, GDP or CTP but not in soaks with AMP (Figure 6.6). Site N, which is located at the interface between adjacent VP3(T2) A and B molecules about one-third of the distance from the five-fold to the two-fold axis (Diprose *et al.*, 2001), was not identified in the previous atomic model of the capsid (Grimes *et al.*, 1998). At magnesium ion concentrations that are near optimal for transcription, the site appears to represent a

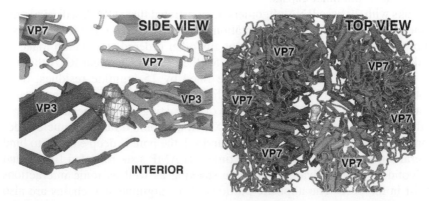

Figure 6.6   Site N, the NTP entry site through the outer core VP7(T13) and subcore VP3(T2) layers. The difference in density observed when comparing an ADP soak to a reference is shown as a semi-transparent surface. This site, accessible from both sides of the capsid, is seen to bind NTP and NDP. This suggests a role as a pore allowing NTP into the particle and waste NDP out (Diprose *et al.*, 2001) (See colour plate 12).

pore connecting the interior of the core to the outside near the VP7(T13) S-trimer. There are several arginines, lysines and aspartates around this pore, which interact to give little overall charge. Rearrangements of the arginine and lysine side chains could present either hydrophobic or positively charged surfaces suitable for binding nucleotides. The absence of density in site N in the AMP soaks suggests that this pore is incapable of accepting ssRNA and hence is not a possible export route for the nascent mRNA. However, the pore does appear to accept nucleoside di- and triphosphates and could provide a vital trafficking route both for importing NTP substrates for VP1(Pol), VP4(CaP) and VP6(Hel) and for exporting any NDP by-product (e.g. from ATP hydrolysis to ADP), keeping the molecular machinery running.

Sites I: inter-trimer

Nucleotides can also bind in several additional sites (I) that are situated between VP7(T13) trimers in the outer core layer. The I sites are very open with the nucleotide held in place between arginine-rich patches found on the surface of the VP7(T13) top domain. Each of the two neighbouring trimers contributes two arginine residues to create these sites. Not all pairs of VP7 trimers are in the correct relative orientation to create the site; out of seven unique possible combinations, they are found in only three pair wise interactions (PQ, SS and RT).

Two roles have been suggested for the binding of nucleotides in the I site. Binding is associated with a small conformational change in the capsid and could stabilize a more open conformation of the core, which may be necessary for transcription to take place. Moreover, nucleotide binding by the core surface may act as a 'substrate sink' raising the local concentration of nucleotides. Similarly, these external sites could help to draw waste NDP out of the particle. Both effects could have obvious benefits for the transcriptional activity of the particle.

Sites F and C: anion and cation

Several other small binding sites for different anions and cations have also been identified. The F sites, which are located in the VP3 layer at the icosahedral two-fold axes, can bind either phosphate or sulphate ions and have been suggested as an export route for the waste phosphate generated by the enzymatic activities of the core. The C sites are thought to bind cations with a possible structural role. No Zn^{2+} binding was found in BTV VP7(T13) trimers observed by Mathieu et al. (2001) in the equivalent rotavirus VP6(T13) structure.

Analysis of the effects of buffer changes on binding sites and pores suggests mechanisms for the transport of substrates/products in actively transcribing BTV cores. Under physiological conditions, the capsid occupies a relatively open conformation particularly around the icosahedral five-fold axes. The I sites attract NTP to the core, and this is subsequently transported into the interior of the core via the N sites. Within the core, the NTPs are converted to mRNA, NDPs, pyrophosphate and phosphate ions by the polymerase capping and helicase enzymes. The mRNA is then exported along the icosahedral five-fold axis passing through an X site as it leaves the core. Waste NDP is removed via site N and then site I, the reverse of the NTP import route, whilst waste P and Pi can escape through site F. In this way, the transcriptional machinery can be kept supplied, even though the largest and the most obvious pore is blocked by the nascent mRNA.

Binding of dsRNA to the bluetongue virus core surface

Low-resolution X-ray analyses of the crystals of native BTV cores showed that the exterior of the core particle was 'festooned' with dsRNA (Figure 6.7). The density was seen as two crystallographically related copies of three distinct intersecting sections of 412, 276 and 265 base pairs not far short of forming a continuous network throughout the crystal structure (Diprose *et al.*, 2002). Indeed, the detection of these RNA molecules may help to explain the slow formation and long stability (for over a year at 29 °C) of the crystals of BTV cores.

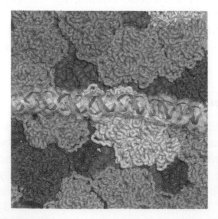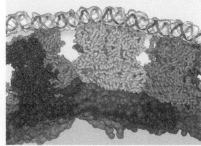

Figure 6.7 Cartoon showing bound dsRNA on the surface of core. The VP3(T2) layer is shown as balls, the VP7(T13) layer is shown as worms, the dsRNA is shown as ribbons and the observed additional electron density is shown as a semi-transparent surface. The RNA makes extensive contacts with the top surface of the VP7(T13) trimers (Diprose *et al.*, 2002) (See colour plate 13).

Each core particle is associated with ~2000 base pairs of dsRNA with a total contact area of some 84 000 Å^2. Interestingly, the majority of the contact area between neighbouring particles in the crystal is from RNA–protein interaction perhaps explaining why the RNA density is so clearly visible. However, the dsRNA binding activity of the core was also shown to be present in solution and is not just a crystallographic artefact. The strands of RNA form 39 different interactions with the VP7(T13) layer. When modelled onto a single VP7(T13) timer, these can be divided into four categories with 24 of the binding sites in the most common of these categories. The overlap of the strands when aligned in this way is strongly suggestive of specific interactions between VP7(T13) and dsRNA. However, the resolution of the model is too low to determine if the interaction is RNA sequence-specific. Solution-binding experiments indicated that longer strands of dsRNA were bound more frequently than short ones, which may indicate that binding frequency is simply a function of length suggesting that these interactions are not highly sequence-specific.

Thus it appears that the core itself is capable of soaking free dsRNA. This may provide an additional protection against the accidental release of dsRNA, helping to mask the presence of the virus from the host cell. Several dsRNA-binding proteins that can sequester free dsRNA, potentially released by any breakdown of virus particles, have also been identified in cells infected by various other reoviruses. These include sigma-3 of the orthoreoviruses; avian reovirus sigma A; NSP3 of group C rotaviruses; and pns10 of rice dwarf virus (a phytoreovirus), all of which possess dsRNA binding activity and can prevent dsRNA-activated antiviral responses in infected cells (Bergeron *et al.*, 1998; Jacobs and Langland, 1998; Martinez-Costas *et al.*, 2000; Cao *et al.*, 2005; Mérai *et al.*, 2006).

VP4: The 'capping' enzyme

The 5' 'Cap' structure stabilizes eukaryotic mRNAs and is involved in binding to ribosomes and promoting efficient translation (reviewed by Scheper and Proud, 2002). The reoviruses use structural enzyme components of the virus core to synthesize Cap structures on their mRNA molecules before they are released into the host cell cytoplasm (Furuichi and Miura, 1975; Muthukrishnan *et al.*, 1975; Shatkin, 1976; Mertens, 2004). The Cap is made in a concerted reaction involving four steps that are catalysed by a single protein, the VP4(CaP) (Ramadevi and Roy, 1998; Ramadevi *et al.*, 1998). The reaction, shown schematically in Figure 6.8, involves hydrolysis of the γ-phosphate of the 5'-terminal rGTP residue by a triphosphatase also known as an NTPase activity. GMP is then added from a GTP donor molecule to form 5'–5' triphosphate linkage (also known as pyrophosphate link) by GTase activity. The inorganic pyrophosphate generated by this step can inhibit the reaction and can be removed by an inorganic pyrophosphatase activity of VP4

(a)

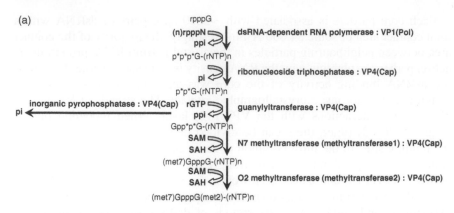

Figure 6.8 (a) The concerted reactions catalysed by the VP4(cap) resulting in the formation of a cap 1 structure. P_i = phosphate; Pp_i = pyrophosphate; SAM = S-adenosyl methionine; SAH = S-adenosyl homocysteine.

(Martinez-Costas *et al.*, 1998; Mohd Jaafar *et al.*, 2005). Two methyl groups are transferred from S-adenosyl methionine (a methyl donor usually designated as SAM or AdoMet) onto the cap structure. The first methyl group is transferred onto the N7 position by a guanine-N7-methyltransferase (N7-MTase – also known as methyltransferase-1) to give a structure identified as 'Cap 0'. Subsequent methylation to form a 'Cap 1' structure occurs on the 2'-hydroxyl of the ribose of the first nucleotide, catalysed by a nucleoside-2'-O-methyltransferase (2'-O-MTase – also known as methyltransferase-2). The by-product of the methyl group transfer reaction is S-adenosyl-L-homocysteine (AdoHcy), which can also act as an inhibitor for transmethylation.

The reoviruses include both 'turreted' and 'non-turreted' viruses (Mertens *et al.*, 2005), which have different capping strategies. The non-turreted viruses (including BTV) have their capping enzymes located inside the core, alongside the polymerase (Gouet *et al.*, 1999), while cores of the turreted viruses have their capping enzymes arranged as pentameric turrets projecting from five-fold axes on the surface of the T2 layer (Reinisch *et al.*, 2000). The capping enzymes of the non-turreted viruses are smaller and fewer copies are incorporated into each particle (probably two at each vertex).

The structure of BTV-10 VP4(CaP) has been solved at a resolution of 2.5 Å (Sutton *et al.*, 2007). The refined model includes 613 of the 644 residues. The N-terminal methionine is missing and the disordered residues 270–276, 537–549 and 601–610 are not 'visible' in the structural model. The VP4 monomer has a morphology that is reminiscent of an eccentric egg timer with dimensions 96 Å–66 Å–52 Å (Figure 6.9) composed of four domains, which include a kinase-like domain (KL; residues 1–108); methyltransferase-1 domain (N7MT; residues 109–174 and 378–509); O2-methyltransferase-2 domain (O2MT; residues 175–377) and a GTase domain (GT; residues 510–644).

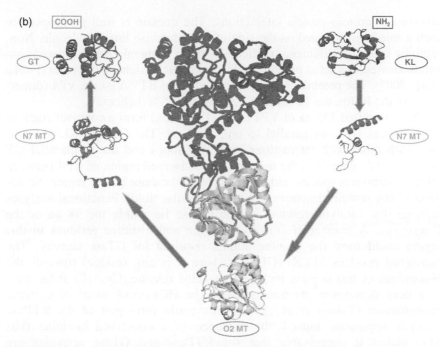

Figure 6.9 (b) The atomic structure of VP4 coloured by domain. From the N-terminus to the C-terminus: the kinase-like (KL, in red) domain, which is thought to interact with the polymerase; the N7-methyltransferase (N7MT, in cyan); the O2-methyltransferase (O2MT, in yellow) and the guanylyltransferase (GTase, in green) (See colour plate 14).

These domains are organized (from the NH_2-terminus) as KL domain linked to the amino-terminal portion of the N7MT domain; O2MT-domain; carboxy-terminal portion of the N7MT domain and (finally) the GT domain.

The $2'$-O-MT structure shows remarkable similarity to the catalytic domains of class I SAM-dependent methyltransferases and in particular to vaccinia virus VP39. A KDKE tetrad in the active site seems to be characteristic of RNA $2'$-O-MTases with the second lysine directly affecting catalysis and readily able to be superimposed on the equivalent residue of VP39. The N7-MT domain also has a class I SAM-dependent methyltransferase fold although the KDKE catalytic tetrad is absent and it is most similar to the Ecm1 mRNA cap N7-MTase.

The N-terminal 108 residues of VP4 show a kinase fold (KL domain) although this lacks the catalytic apparatus and P-loop, which is regarded as essential for kinase activity. It is the least conserved domain among the capping enzymes of different orbivirus species with only 13% identical residues (N7-MT, 22%; $2'$O-MT, 19%; GT, 27%). This domain is thought to be involved in protein–protein interactions. The domain is well positioned for such a role being situated on the side of the otherwise linear molecule. Non-catalytic kinase-like domains, particularly those of membrane-associated guanylate kinases, can act as protein-binding domains in higher eukaryotes (Funke *et al.*, 2005). The potential partners for binding of BTV VP4 are VP4 (dimerization), the RdRp, the T2 layer (VP3) and the VP6 (helicase).

The C-terminal 135 aa of VP4 (the GT domain) form a compact stack of six α-helices (five antiparallel to one another). The GT and KL domains face each other with interactions involving loops and the antiparallel GT helix. The GT domain is the most highly conserved region of VP4 between different orbivirus species, although a protein database (PDB) search did not identify any protein structures that have a similar 'fold'. Functional analyses indicate that catalytic residues of the GTase lie within the 98 aa of the C-terminus. A number of conserved lysine and histidine residues in this region could form the guanine adduct required for GTase activity. The conserved residues SLCRFxGL/IR (where x is any residue) towards the N-terminus of this domain include a cysteine residue (Cys518) at the base of a deep depression. By analogy with the HCxxxxxR motif of cysteine phosphatases (Takagi *et al.*, 1997), this could form part of the RTPase catalytic apparatus. Indeed, the presence of a conserved histidine (His 554) makes it conceivable that the RTPase and GTase activities are co-localized in this depression.

Coupling via the KL domain would place the polymerase adjacent to the GT domain with the next (N7-MTase) active site 30–40 Å along the molecule and the final O2MT 40 Å beyond that.

Bluetongue virus outer capsid proteins VP2 and VP5

The organization of the BTV outer capsid has been analysed by cryo-EM (Nason *et al.*, 2004) (Figure 6.10). Two distinct electron densities were observed representing trimers of VP2 and VP5. VP2 is the outermost BTV protein forming large triskelion motifs on the virus particle surface (Hewat *et al.*, 1992). It is encoded by genome segment 2, is 956 aa long with a molecular weight of 111 kDa and is the main cell attachment protein on the surface of the intact virus particle. The protein can be cleaved by treatment of the virus particle with proteases-generating infectious subviral particles (ISVP) that retain the cleavage products of VP2 and are highly infectious for insect cells (Mertens *et al.*, 1987; 1996). The site of protease cleavage is though to lie within the N-terminal third of the molecule. VP2 is the most variable of the BTV proteins (Maan *et al.*, 2008) containing neutralizing epitopes (Hassan and Roy, 1999) and controlling virus serotype (Maan *et al.*, 2007).

(a) (b)

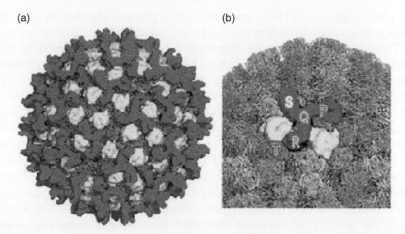

Figure 6.10 The outer capsid proteins VP2 and VP5. (a) Arrangement of outer capsid proteins VP2 and VP5 on the core surface. The 60 trimers of the VP2 triskelion are shown in *red* and the 120 copies of the globular VP5 density is shown in *yellow*. (b) The arrangement of the icosahedral asymmetric portion of VP2 and VP5 (in *grey*) on the core. The positions of the icosahedrally unique trimers of VP7(T13) are labelled (*P–T*) to help clarify the positions ofVP2 and VP5 (See colour plate 15).

VP5 is also trimeric but has a more globular structure than VP2. The VP5 monomer is 526 aa long with a molecular weight of ~59 kDa and is encoded by genome segment 6. VP5 can also influence the virus serotype (Demaula *et al.*, 2000) and has been shown to possess membrane fusion activity (Hassan *et al.*, 2001). The VP5 trimers are situated in the outer capsid above six-membered rings of VP7 trimers in the core surface layer (Hewat *et al.*, 1992; Hassan *et al.*, 2001). This assignment of the VP2 and VP5 to specific areas of electron density in the outer capsid layer was confirmed by comparisons to Broadhaven virus (Schoehn *et al.*, 1997).

More recent and higher-resolution (24 Å) cryo-EM studies (Nason *et al.*, 2004) have provided a more precise model of the BTV outer shell (see Figure 6.10). The 60 triskelions extend to give the virus a maximum diameter of approximately 880 Å. Interspersed between the triskelions lie 120 copies of the globular VP5 at a maximum diameter of approximately 820 Å. The volume of the density features indicates that both the VP2 triskelions and VP5 globules represent trimers (Stuart and Grimes, 2006). Atomic structures determined by X-ray diffraction are not yet available for either VP2 or VP5.

Bluetongue virus non-structural proteins

The three 'non-structural' BTV proteins include NS1 (552 aa with a molecular weight of ~64.5 kDa, translated from genome segment 5); NS2 (357 aa with a molecular weight of ~41 kDa, translated from genome segment 8) and NS3 (229 aa with a molecular weight of ~25.5 kDa, translated from genome segment 10). These proteins clearly play important roles in the coordination of viral assembly, virus translocation within the cell and egress through the cell membrane (Brookes *et al.*, 1993). However, until very recently, the only structural information available for them was from EM studies (Brookes *et al.*, 1993).

The first atomic-level information on a portion of NS2, derived from X-ray crystallography studies, has been provided by Butan *et al.* (2004). The NS2 protein is a major component of viral inclusion bodies (VIB) that characteristically form within the cytoplasm of BTV-infected cells. It is a phosphoprotein and can bind to ssRNA (see Chapter 5). NS2 is well conserved between different isolates of BTV, although like the other BTV proteins it can be divided into distinct eastern and western topotypes (Maan *et al.*, 2008; see Chapter 7). Despite the absence of sequence similarity, proteins that carry out similar functions have been identified from other reoviruses. Rotavirus NSP2 and NSP5 or orthoreovirus Mu-NS, which functions with Sigma-NS, binds ssRNA to form large homomultimeric complexes and can accumulate to generate VIB within the infected cell (Taraporewala *et al.*, 1999; Touris-Otero *et al.*, 2004, Zhao *et al.*, 1994). NS2 and NSP2 have NTPase activities (i.e. they hydrolyse the γ phosphodiester bonds of NTPs). NS2 also has nucleotidyl phosphotase activity (i.e. it hydrolyses the β and γ phosphodiester bonds of NTPs) (Taraporewala *et al.*, 1999, Taraporewala *et al.*, 2001). The atomic structure of NSP2 indicates that the NTPase activity is located to a cleft in the carboxy-terminal domain, which resembles the ubiquitous cellular histidine triad (HIT) group of nucleotidyl hydrolases (Vasquez-Del Carpio *et al.*, 2004; Vasquez-Del Carpio *et al.*, 2006). However, it has not yet been possible to specifically locate the active site of the nucleotidyl phosphotase activity.

NS2 is the only BTV phosphoprotein (Huismans *et al.*, 1987a, 1987b) containing two phosphorylation sites, at least one of which must be phosphorylated for VIB formation to occur. The non-phosphorylatable form of NS2 remains dispersed throughout the cytoplasm (Modrof *et al.*, 2005). In the mammalian host cell, NS2 is believed to be phosphorylated by the ubiquitous casein kinase II (CK2) (Modrof *et al.*, 2005). There is some evidence that it may also have an autokinase activity (Ross-Smith, 2008).

Apart from the requirement for phosphorylation, no other viral factors (proteins or RNA) appear to be required for the assembly of expressed BTV NS2 into large protein aggregates within uninfected cells (Thomas *et al.*, 1990; Modrof *et al.*, 2005; Chapter 6). +++Although these structures superficially resemble VIB, they do not contain BTV RNA molecules or virus structural components that include the nascent virus particles usually present in VIB. They cannot therefore be regarded as truly 'VIB-like'.

The oligomerization of BTV NS2 does not require either the N-terminal 92 aa or the C-terminus (Zhao *et al.*, 1994). However, mutagenesis/deletion studies of a consensus sequence found within the amino-terminus in EHDV

NS2 (aa 75–83) were shown to disrupt both ssRNA binding and VIB formation (Van Staden *et al.*, 1991; Theron *et al.*, 1996). This consensus sequence is present in all of the orbivirus NS2 sequences that have been analysed as well as in proteins that are thought to play a role similar to NS2 from the other reoviruses (e.g. σNS of orthoreovirus; Theron *et al.*, 1996).

NS2 selectively binds to ssRNA rather than dsRNA or DNA (Fillmore *et al.*, 2002). NS2 expressed in BTV-infected mammalian cells forms ~22S complexes with ssRNA independent of the molar ratio of NS2 to RNA (Huismans *et al.*, 1987a, 1987b) or exists as ~7S homomultimers following treatment with RNase (Uitenweerde *et al.*, 1995). Even bacterially expressed recombinant NS2 could not be detected as a monomer but was present as ~18–22S complexes with bacterial mRNA (Taraporewala *et al.*, 2001). These findings suggest that during BTV infection, NS2 is present as a multimer, usually complexed with ssRNA. The RNA-binding domain of NS2 was initially mapped to specific bases at the N-terminus of NS2 (specifically Lys^4, Arg^6 and Arg^7) (Zhao *et al.*, 1994). This region is not required for NS2 oligomerization, suggesting that RNA binding and oligomerization are separate processes. It was later shown that there are three important RNA-binding regions in NS2 (aa 2–11, 153–166 and 274–286) (Fillmore *et al.*, 2002). Cross-linking studies showed that NS2 has greater affinity for the 3′-terminus of BTV RNA than for other ssRNAs, but this does not involve the conserved terminal hexanucleotide sequences (Theron and Nel, 1997). Mutation of inverted repeat sequences adjacent to both the conserved 5′- and 3′-hexanucleotides reduced NS2 binding, with deletion of either terminal sequence having a greater effect than deletion of both. This may be due to the overall effect that deletion has on the secondary structure of the RNA rather than deletion of specific sequences (Markotter *et al.*, 2004).

A stem-loop structure has been detected within the open reading frame (ORF) of several of the BTV genome segments (Lymperopoulos *et al.*, 2003; Lymperopoulos *et al.*, 2006), which may represent specific packaging signals involved in the selection of BTV mRNAs for packaging into progeny virus particles. In view of its location within the VIB (the site of BTV genome assembly and replication) and its ability to bind ssRNA, it has been suggested that NS2 plays a central role in selection of one copy of each of the 10 BTV genome segments for packaging into progeny virus particles.

Crystals of the N-terminal domain of recombinant expressed and purified NS2 have been generated under various conditions (Figure 6.11). The atomic structure of the NS2 N-terminal domain was determined by X-ray crystallography and is composed predominantly of two β-sheets arranged as a sandwich (Stuart and Grimes, 2006). These structures form tight oligomeric complexes with two distinct sets of oligomeric interactions (Butan *et al.*, 2004). NS2 dimers can produce a crystal lattice, which is made up of broad, shallow spirals of NS2 molecules, arranged with 12 subunits per turn of the 6₁ helix (Ross-Smith, 2008) (Figure 6.12).

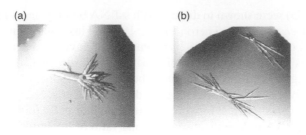

Figure 6.11 (a) Crystals formed from bacterial-expressed NS2 protein in 0.8 m sodium/potassium phosphate, pH 5.6. (b) Crystals formed from bacterial-expressed NS2 protein in 0.8 m sodium/potassium phosphate, pH 6.3.

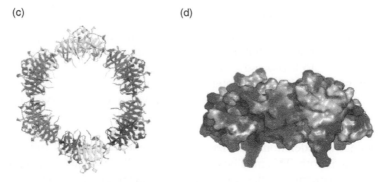

Figure 6.12 (c) Diagram showing the helical packing of the RNA-binding domains of NS2 observed in the crystals (Butan *et al.* 2004). Subunits are coloured individually and drawn as secondary structural cartoons. (d) Electrostatic potential of an NS2 dimer showing the amphipathic nature of the molecule (negative potential: *red*, positive: *blue*, on an arbitrary scale). The putative RNA-binding domain is towards the bottom of the molecule (See colour plate 16).

The motifs, which have been suggested as RNA-binding domains (Zhao *et al.*, 1994), are arranged relatively closely to each other in space so that they are well positioned to collaborate in the formation of an RNA-binding surface – the putative RNA-binding domain being towards the bottom of the molecule. The mode of binding of ssRNA is reminiscent of other ssRNA-binding proteins where the RNA-binding surface is formed by β-sheets (Handa *et al.*, 1999). The organization of the RNA-binding surface suggests that the packing of NS2 observed within the crystal structure may resemble that of biologically relevant higher-order assemblies of this molecule, e.g. within the VIB.

Bluetongue virus particle structure and cell entry mechanisms

There are several structurally distinct BTV particles that are infectious for either mammalian or insect cells, suggesting that the virus can use a number of different cell entry pathways or mechanisms to initiate infection (Mertens *et al.*, 1996). The BTV particle can bud out of infected cells (see Chapters 4 and 5) generating particles that have an outer membrane layer derived from the cell surface membrane. These membrane-enveloped virus particles (MEVP) are poorly characterized and the membrane envelope may be relatively unstable. It is uncertain if MEVP are infectious in their own right, although the BTV outer capsid protein VP5 can mediate membrane fusion (Forzan *et al.*, 2004). Insect cell cultures can be persistently infected by BTV with little sign of cytopathic effect (CPE) or cell death. The BTV particles that are released from infected insect cell cultures, presumably by budding, have a high level of infectivity, suggesting that one mechanism of infection by BTV might involve fusion of MEVP with target cells, although this has not been confirmed.

The high levels of infectivity that are associated with intact and purified BTV virus particles, or with the protease-modified ISVP (Mertens *et al.*, 1996), and their neutralization by antibodies to the outer capsid proteins (Huismans *et al.*, 1987a, 1987b) confirm that the outer capsid proteins are centrally involved in mechanisms used by BTV to bind and gain entry to mammalian cells. These particle types are also highly infectious for adult *Culicoides* and for insect cell lines, although ISVP have a very significantly enhanced infectivity ($\sim$1000-fold) for insect cells, suggesting a biologically significant difference in their infection mechanism (Mertens *et al.*, 1996).

Although the outer capsid proteins are entirely missing from purified BTV core particles, which removes their infectivity for CHO cells, they still retain a low but still significant infectivity for BHK-21 cells (Mertens *et al.*, 1987, 1996). However, BTV core particles show a similar level of infectivity to that of the intact virus particle for both adult *Culicoides* and for *Culicoides* cell cultures (KC cells). The mechanism by which cores enter cells is not known although they can bind to glycosaminoglycans on the cell surface (Hutchinson, 1999). There is also a putative integrin-binding motif (RGD) in an exposed position on VP7(T13), which is ideally placed to bind a more specific receptor molecule (Grimes *et al.*, 1995, 1998). The involvement of the RGD peptide in the initiation of BTV infection is also indicated by cell-binding assays using core-like particles, which showed reduced binding when the RGD tripeptide was mutated (Tan *et al.*, 2001).

The infectivity of BTV core particles is also independent of a reduction in pH within target cell endosomes, which is thought to be essential for the release of the outer capsid proteins from the virus particles (Hutchinson, 1999). The high infectivity associated with BTV cores for insect systems demonstrates that VP7(T13) can itself mediate cell attachment and penetration presumably by a different mechanism/receptor to that mediated by outer capsid proteins VP2 and VP5. These observations together with the single-shelled nature of some of the insect reoviruses (members of the genus *Cypovirus*) suggest that the outer capsid layer of BTV and the other orbiviruses could be an evolutionary adaptation allowing them to infect mammals.

Although there are no direct data available to address the issue of conformational changes in the core that might lead to or accompany cell entry, structural studies of the outer protein of the core VP7(T13) have demonstrated that the two domains of this molecule can flex and that its twisted architecture provides a natural mechanism for large-scale conformational change. VP7(T13)s from both BTV and AHSV can undergo proteolytic cleavage at a point between the two domains (Basak *et al.*, 1996; Basak *et al.*, 1997). This may simply reflect inherent flexibility in the molecule, which is manifested in a low-resolution structure of BTV VP7(T13) (Basak *et al.*, 1997), revealing that the α-helical domains can rotate by roughly 160° to produce a massive unwinding of the trimer. All known viral fusion machines employ large-scale conformational rearrangements of trimeric molecules. Although there is no direct evidence that such mechanisms are employed by non-enveloped viruses, a very similar mechanism has been suggested for the μ1 protein of reovirus (Liemann *et al.*, 2002), which has a fold that is similar to BTV VP7(T13).

Although the 'core' architecture of the of non-turreted reoviruses appears to be rather well conserved, the outer layers of the virus particles are much more variable (Mertens, 2004). Recent analyses have shown the folding of outer shell protein components for several reoviruses (Dormitzer *et al.*, 2002, 2004; Mohd Jaafar *et al.*, 2005). The relative positions of the outer capsid proteins of the non-turreted reoviruses, as revealed in low-resolution structures, indicates that it is useful to view certain proteins as orthologues. In particular, these analyses suggest that VP2 of BTV and VP4 of rotavirus have similarities in their structures and functions. Cryo-EM data demonstrate that the three arms of VP4 are disordered prior to proteolytic activation (Dormitzer *et al.*, 2004). VP5 is the larger of the two protein fragments that remain attached to the rotavirus particle after proteolytic activation of VP4. Although VP5 of rotavirus was initially thought to exist on the virus particle as dimers, it is now known to be trimeric, like BTV VP2 (Dormitzer *et al.*, 2004). The structure of VP5, determined by X-ray crystallography, fits into a comparable portion of BTV VP2. BTV VP2 may therefore be regarded as a relative of rotavirus VP4 in which the pre-activation state is well ordered by interactions with the virus core. In rotavirus, the VP7 outer layer proteins prevents such stabilization.

Rotavirus VP8 is the smaller cleavage product of VP4 and is involved in interactions with receptors on the cell surface. VP8 also shows some structural similarity to the trimeric receptor-binding protein VP9 of Banna virus. Banna virus belongs to the genus *Seadornavirus,* non-turreted reoviruses that represent an emerging threat to human health (Mohd Jaafar *et al.*, 2005).

Despite different copy numbers of BTV VP5 (120 trimers) and rotavirus VP7 (260 trimers), comparison of low-resolution cryo-EM reconstructions for the two molecules shows significant similarities. These analyses suggest that somewhat larger (59 kDa vs 37 kDa) VP5 subunit of BTV may consist of two domains, arranged as a twisted stack, with the lower portion resembling rotavirus VP7.

The argument for drawing speculative parallels between these different reoviruses is supported by the similar biological functions attributed to the different protein molecules. By combining data from cryo-EM analyses of BTV and rotavirus with X-ray data on the structures of rotavirus VP5 and VP8 and Banna virus VP9 (Dormitzer *et al.*, 2002, 2004; Nason *et al.*, 2004; Mohd Jaafar *et al.*, 2005), it is possible to propose a generic quasi-atomic model for these molecular systems, which appears to mediate receptor interactions and cell entry. As noted by Dormitzer *et al.* (2004), recent structural results suggest that there might also be mechanistic similarities between the cell entry mechanisms used by dsRNA viruses and membrane fusion proteins (Stuart and Grimes, 2006).

Further structural studies of BTV and other reoviruses, particularly of their more variable outer capsid proteins, will inevitably provide more information concerning their evolutionary relationships. Such studies will also help to reveal the nature of interactions between the virus and cell including mechanisms used for cell attachment and penetration, supporting the design of novel antiviral agents and improved subunit vaccines.

References

Abrescia, N.G., Cockburn, J.J., Grimes, J.M., Sutton, G.C., Diprose, J.M., Butcher, S.J., Fuller, S.D., San Martin, C., Burnett, R.M., Stuart, D.I., Bamford, D.H. and Bamford, J.K. (2004) Insights into assembly from structural analysis of bacteriophage PRD1. *Nature* **432**, 68–74.

Bamford, D.H., Grimes, J.M. and Stuart, D.I. (December 2005) What does structure tell us about virus evolution? *Curr. Opin. Struct. Biol.* **15**(6), 655–63. Epub 2005 Nov 3. Review.

Bartlett, N.M., Gillies, S.C., Bullivant, S. and Bellamy, A.R. (1974) Electron-microscopy of reovirus reaction cores. *J. Virol.* **14**, 315–26.

Basak, A.K., Gouet, P., Grimes, J., Roy, P. and Stuart, D. (1996) Crystal structure of the top domain of African horse sickness virus VP7: Comparisons with bluetongue virus VP7. *J. Virol.* **70**, 3797–806.

Basak, A.K., Grimes, J.M., Gouet, P., Roy, P. and Stuart, D.I. (1997) Structures of orbivirus VP7: Implications for the role of this protein in the viral life cycle. *Structure* **5**, 871–83.

Bergeron, J., Mabrouk, T., Garzon, S. and Lemay, G. (1998) Characterization of the thermosensitive ts453 reovirus mutant: Increased dsRNA binding of sigma 3 protein correlates with interferon resistance. *Virology* **246**, 199–210.

Brookes, S.M., Hyatt, A.D. and Eaton, B.T. (1993) Characterization of virus inclusion bodies in bluetongue virus-infected cells. *J. Gen. Virol.* **74**, 525–30.

Burroughs, J.N., O'Hara, R.S., Smale, C.J., Hamblin, C., Walton, A., Armstrong, R. and Mertens, P.P.C. (1994) Purification and properties of virus particles, infectious subviral particles, cores and VP7 crystals of African horse sickness virus serotype 9. *J. Gen. Virol.* **75**, 1849–57.

Butan, C., Van Der Zandt, H. and Tucker, P.A. (2004) Structure and assembly of the RNA binding domain of bluetongue virus non-structural protein 2. *J. Biol. Chem.* **279**, 37613–21.

Cao, X., Zhou, P., Zhang, X., Zhu, S., Zhong, X., Xiao, Q., Ding, B. and Li, Y. (2005) Identification of an RNA silencing suppressor from a plant double-stranded RNA virus. *J. Virol.* **79**, 13018–27.

Caspar, D.L.D. and Klug, A. (1962) Physical principles in the construction of regular viruses. *Cold. Spring. Harb Symp. Quant. Biol.* **27**, 1–24.

Cockburn, J.J., Abrescia, N.G., Grimes, J.M., Sutton, G.C., Diprose, J.M., Benevides, J.M., ThomasJr, G.J., Bamford, J.K., Bamford, D.H. and Stuart, D.I. (2004) Membrane structure and interactions with protein and DNA in bacteriophage PRD1. *Nature* **432**, 122–5.

Demaula, C.D., Bonneau, K.R. and MacLachlan, N.J. (2000) Changes in the outer capsid proteins of bluetongue virus serotype ten that abrogate neutralization by monoclonal antibodies. *Virus Res.* **67**, 59–66.

Diprose, J.M., Burroughs, J.N., Sutton, G.C., Goldsmith, A., Gouet, P., Malby, R., Overton, I., Zientara, S., Mertens, P.P., Stuart, D.I. and Grimes, J.M. (2001) Translocation portals for the substrates and products of a viral transcription complex: The bluetongue virus core. *EMBO J.* **20**, 7229–39.

Diprose, J.M., Grimes, J., Sutton, G., Burroughs, J., Meyer, A., Maan, S., Mertens, P.P.C. and Stuart, D. (2002) The core of bluetongue virus binds double-stranded RNA. *J. Virol.* **76**, 9533–6.

Dormitzer, P.R., Nason, E.B., Prasad, B.V. and Harrison, S.C. (2004) Structural rearrangements in the membrane penetration protein of a non-enveloped virus. *Nature* **430**, 1053–8.

Dormitzer, P.R., Sun, Z.Y., Wagner, G. and Harrison, S.C. (2002) The rhesus rotavirus VP4 sialic acid binding domain has a galectin fold with a novel carbohydrate binding site. *EMBO J.* **21**, 885–97.

Dryden, K.A., Farsetta, D.L., Wang, G., Keegan, J.M., Fields, B.N., Baker, T.S. and Nibert, M.L. (1998) Internal/structures containing transcriptase-related proteins in top component particles of mammalian orthoreovirus. *Virology* **245**, 33–46.

Eaton, B.T., Hyatt, A.D. and Brookes, S.M. (1990) The replication of bluetongue virus. *Curr. Top. Microbiol. Immunol.* **162**, 89–118.

Fillmore, G.C., Lin, H. and Li, J.K. (2002) Localization of the single-stranded RNA-binding domains of bluetongue virus nonstructural protein NS2. *J. Virol.* **76**, 499–506.

Forzan, M., Wirblich, M.C. and Roy, P. (2004) A capsid protein of nonenveloped bluetongue virus exhibits membrane fusion activity. *Proc. Natl. Acad. Sci. USA* **101**, 2100–105.

Funke, L., Dakoji, S. and Bredt, D.S. (2005) Membrane-associated guanylate kinases regulate adhesion and plasticity at cell junctions. *Annu. Rev. Biochem.* **74**, 219–45.

Furuichi, Y. and Miura, K. (1975) A blocked structure at the 5′ terminus of mRNA from cytoplasmic polyhedrosis virus. *Nature* **253**, 374–5.

Gillies, S., Bullivant, S. and Bellamy, A.R. (1971) Viral RNA polymerases: Electron microscopy of reovirus reaction cores. *Science* **174**, 694–6.

Gitlin, L. and Andino, R. (2003) Nucleic acid-based immune system: The antiviral potential of mammalian RNA silencing. *J. Virol.* **77**, 159–65.

Goldbach, R., Bucher, E. and Prins, M. (2003) Resistance mechanisms to plant viruses: An overview. *Virus Res.* **92**, 207–12.

Gouet, P., Diprose, J.M., Grimes, J.M., Malby, R., Burroughs, J.N., Zientara, S., Stuart, D.I. and Mertens, P.P.C. (1999) The highly ordered double-stranded RNA genome of bluetongue virus revealed by crystallography. *Cell* **97**, 481–90.

Grimes, J., Basak, A.K., Roy, P. and Stuart, D. (1995) The crystal structure of bluetongue virus VP7. *Nature* **373**, 167–70.

Grimes, J.M., Burroughs, J.N., Gouet, P., Diprose, J.M., Malby, R., Zientara, S., Mertens, P.P.C. and Stuart, D.I. (1998) The atomic structure of the bluetongue virus core. *Nature* **395**, 470–8.

Handa, N., Nureki, O., Kurimoto, K., Kim, I., Sakamoto, H., Shimura, Y., Muto, Y. and Yokoyama, S. (1999) Structural basis for recognition of the tram RNA precursor by the sex-lethal protein. *Nature* **398**, 579–85.

Hassan, S.H., Wirblich, C., Forzan, M. and Roy, P. (2001) Expression and functional characterization of bluetongue virus VP5 protein: Role in cellular permeabilization. *J. Virol.* **75**, 8356–67.

Hassan, S.S. and Roy, P. (1999) Expression and functional characterization of bluetongue virus VP2 protein: Role in cell entry. *J. Virol.* **73**, 9832–42.

Hewat, E.A., Booth, T.F. and Roy, P. (1992) Structure of bluetongue virus particles by cryoelectron microscopy. *J. Struct. Biol.* **109**, 61–9.

Huismans, H., van der Walt, N.T., Cloete, M. and Erasmus, B.J. (1987a) Isolation of a capsid protein of bluetongue virus that induces a protective immune response in sheep. *Virology* **157**, 172–9.

Huismans, H., van Dijk, A.A. and Els, H.J. (1987b) Uncoating of parental bluetongue virus to core and subcore particles in infected L cells. *Virology* **157**, 180–8.

Huismans, H. and Verwoerd, D.W. (1973) Control of transcription during the expression of the bluetongue virus genome. *Virology* **161**, 421–8.

Hutchinson, I.R. (1999) The vole of VP7(T13) in initiation of infection by Bluetongue virus. PhD thesis.

Jacobs, B.L. and Langland, J.O. (1996) When two strands are better than one: The mediators and modulators of the cellular responses to double-stranded RNA. *Virology* **219**, 339–49.

Jacobs, B.L. and Langland, J.O. (1998) Reovirus sigma 3 protein: dsRNA binding and inhibition of RNA-activated protein kinase. *Curr. Top. Microbiol. Immunol.* **233**, 185–96.

Lawton, J.A., Estes, M.K. and Prasad, B.V.V. (1997a) Three-dimensional visualisation of mRNA release from actively transcribing rotavirus particles. *Nat. Struct. Biol.* **4**, 118–21.

Lawton, J.A., Zeng, C.Q., Mukherjee, S.K., Cohen, J., Estes, M.K. and Prasad, B.V. (1997b) Three-dimensional structural analysis of recombinant rotavirus-like particles with intact and amino-terminal-deleted VP2: implications for the architecture of the VP2 capsid layer. *J. Virol.* **71**, 7353–60.

Le Blois, H., Mertens, P.P.C., French, T., Burroughs, J.N. and Roy, P. (1992) The expressed VP4 protein of bluetongue virus is the guanylyl transferase. *Virology* **189**, 757–61.

Liemann, S., Chandran, K., Baker, T.S., Nibert, M.L. and Harrison, S.C. (2002) Structure of the reovirus membrane-penetration protein, mu1, in a complex with is protector protein, sigma3. *Cell* **108**, 283–95.

Livolant, F. and Leforestier, A. (1996) Condensed phases of DNA: Structures and phase transitions. *Prog. Polym. Sci.* **21**, 1115–64.

Lymperopoulos, K., Noad, R., Tosi, S., Nethisinghe, S., Brierley, I. and Roy, P. (2006) Specific binding of bluetongue virus NS2 to different viral plus-strand RNAs. *Virology* **353**, 17–26.

Lymperopoulos, K., Wirblich, C., Brierley, I. and Roy, P. (2003) Sequence specificity in the interaction of bluetongue virus non-structural protein 2 (NS2) with viral RNA. *J. Biol. Chem.* **278**, 31722–30.

Maan, S., Maan, N.S., Ross-smith, N., Batten, C.A., Shaw, A.E., Anthony, S.J., Samuel, A.R., Darpel, K.E., Veronesi, E., Oura, C.A., Singh, K.P., Nomikou, K., Potgieter, A.C., Attoui, H., van Rooij, E., van Rijn, P., De Clercq, K., Vandenbussche, F., Zientara, S., Bréard, E., Sailleau, C., Beer, M., Hoffman, B., Mellor, P.S. and Mertens, P.P. (August 1, 2008) Sequence analysis of bluetongue virus serotype 8 from the Netherlands 2006 and comparison to other European strains. *Virology* **377**(2), 308–18.

Maan, S.N.S., Rao, A.R.S., Attoui, H. and Mertens, P.P.C. (2007) Analysis and phylogenetic comparisons of full-length VP2 genes of the twenty-four bluetongue virus serotypes. Journal of general. *Virology* **88**, 621–30.

Markotter, W., Theron, J. and Nel, L.H. (September 15, 2004) Segment specific inverted repeat sequences in bluetongue virus mRNA are required for interaction with the virus non structural protein NS2. *Virus Res.* **105**(1), 1–9.

Martinez-Costas, J., Gonzalez-Lopez, C., Vakharia, V.N. and Benavente, J. (2000) Possible involvement of the double-stranded RNA-binding core protein sigmaA in the resistance of avian reovirus to interferon. *J. Virol.* **74**, 1124–31.

Martinez-Costas, J., Sutton, G., Ramadevi, N. and Roy, P. (1998) Guanylyltransferase and RNA 5′-triphosphatase activities of the purified expressed VP4 protein of bluetongue virus. *J. Mol. Biol.* **280**, 859–66.

Mathieu, M., Petitpas, I., Navaza, J., Lepault, J., Kohli, E., Pothier, P., Prasad, B.V., Cohen, J. and Rey, F.A. (2001) Atomic structure of the major capsid protein of rotavirus: Implications for the architecture of the virion. *EMBO J.* **20**, 1485–97.

Mérai, Z., Kerényi, Z., Kertész, S., Magna, M., Lakatos, L. and Silhavy, D. (June, 2006) Double-stranded RNA binding may be a general plant RNA viral strategy to suppress RNA silencing. *J. Virol.* **80**(12), 5747–56.

Mertens, P.P.C. (April, 2004) The dsRNA viruses. *Virus Res.* **101**(1), 3–13.

Mertens, P.P.C., Attoui, H., Duncan, R. and Dermody, T.S. (2005) Reoviridae. In: M.L Cox, Y. Zheng, C. Tickle, R. Jansson, H. Kehrer-Sawatzki, B. Wood, D. Cooper, G. Melino, P. Delves, J. Battista, I. Levitan, K. Roberts, W.F. Bynum, G. Phillips, and D. Harper (eds.), *Virus Taxonomy. Eighth Report of the International Committee on Taxonomy of Viruses.* London: Elsevier/Academic Press, pp. 447–54.

Mertens, P.P.C., Burroughs, J.N. and Anderson, J. (1987) Purification and properties of virus particles, infectious subviral particles, and cores of bluetongue virus serotypes 1 and 4. *Virology* **157**, 375–86.

Mertens, P.P.C., Burroughs, J.N., Wade-Evans, A.M., Le Blois, H., Oldfield, S., Basak, A., Loudon, P. and Roy, P. (1992) Analysis of guanyltransferase and transmethylase activities associated with bluetongue virus cores and recombinant baculovirus-expressed core-like particles. *Bluetongue, African Horse Sickness, and Related Orbiviruses: Proceedings of the Second International Symposium,* 404–415.

Mertens, P.P.C., Burroughs, J.N., Walton, A., Wellby, M.P., Fu, H., O'Hara, R.S., Brookes, S.M. and Mellor, P.S. (1996) Enhanced infectivity of modified bluetongue virus particles for two insect cell lines and for two culicoides vector species. *Virology* **217**, 582–93.

Mertens, P.P.C. and Diprose, J. (2004) The bluetongue virus core: A nano-scale transcription machine. *Virus Res.* **2101**, 29–43.

Modrof, J., Lymperopoulos, K. and Roy, P. (2005) Phosphorylation of bluetongue virus nonstructural protein 2 is essential for formation of viral inclusion bodies. *J. Virol.* **79**, 10023–31.

Mohd Jaafar, F., Attoui, H., Bahar, M.W., Siebold, C., Sutton, G., Mertens, P.P., De Micco, P., Stuart, D.I., Grimes, J.M. and De Lamballerie, X. (2005) The structure and function of the outer coat protein VP9 of banna virus. *Structure* **13**, 17–28.

Mohd Jaafar, F., Attoui, H., Mertens, P.P.C., de Micco, P. and de Lamballerie, X. (2005) Identification and functional analysis of VP3, the guanylyltransferase of banna virus (genus seadornavirus, family reoviridae). *J. Gen. Virol.* **86**, 1147–57.

Muthukrishnan, S., Both, G.W., Furuichi, Y. and Shatkin, A.J. (1975) 5'-Terminal 7-methylguanosine in eukaryotic mRNA is required for translation. *Nature* **255**, 33–7.

Nason, E.L., Rothagel, R., Mukherjee, S.K., Kar, A.K., Forzan, M., Prasad, B.V. and Roy, P. (2004) Interactions between the inner and outer capsids of bluetongue virus. *J. Virol.* **78**, 8059–67.

Ramadevi, N., Burroughs, N.J., Mertens, P.P.C., Jones, I.M. and Roy, P. (1998) Capping and methylation of mRNA by purified recombinant VP4 protein of bluetongue virus. *Proc. Natl. Acad. Sci. USA* **95**, 13537–42.

Ramadevi, N. and Roy, P. (1998) Bluetongue virus core protein VP4 has nucleoside triphosphate phosphohydrolase activity. *J. Gen. Virol.* **79**, 2475–80.

Reinisch, K.M., Nibert, M.L. and Harrison, S.C. (2000a) Structure of the reovirus core at 3.6 Å resolution. *Nature* **404**, 960–7.

Ross-Smith N. (2008) Structure and function of bluetongue virus non-structural protein 2 (NS2). D.Phil Thesis, University of Oxford.

Roy, P. (1989) Bluetongue virus genetics and genome structure. *Virus Res.* **13**, 179–206.

Roy, P., Fukusho, A., Ritter, G.D. and Lyon, D. (1988) Evidence for genetic relationship between RNA and DNA viruses from the sequence homology of a putative polymerase gene of bluetongue virus with that of vaccinia virus: Conservation of RNA polymerase genes from diverse species. *Nucl. Acids Res.* **16**, 11759–67.

Scheper, G.C. and Proud, C.G. (2002) Does phosphorylation of the cap-binding protein eIF4E play a role in translation initiation? *Eur. J. Biochem.* **269**, 5350–9.

Schoehn, G., Moss, S.R., Nuttall, P.A. and Hewat, E.A. (1997) Structure of Broadhaven virus by cryoelectron microscopy: Correlation of structural and antigenic properties of Broadhaven virus and bluetongue virus outer capsid proteins. *Virology* **235**, 191–200.

Shatkin, A.J. (1976) Capping of eucaryotic mRNAs. *Cell* **9**, 645–53.

Smith, R.E. and Furuichi, Y. (1982) Segmented CPV genome dsRNAs are independently transcribed. *J. Virol* **41**, 326.

Smith, R.E., Zweerink, H.J. and Joklik, W.K. (1969) Polypeptide components of virions, top components and cores of reovirus type 3. *Virology* **39**, 791–810.

Stauber, N., Martinez-Costas, J., Sutton, G., Monastyrskaya, K. and Roy, P. (1997) Bluetongue virus VP6 protein binds ATP and exhibits an RNA-dependent ATPase function and a helicase activity that catalyze the unwinding of double-stranded RNA substrates. *J. Virol.* **71**, 7220–6.

Stuart, D.I., Gouet, P., Grimes, J.M., Malby, R., Diprose, J.M., Zientara, S., Burroughs, J.N. and Mertens, P.P.C. (1998) Structural studies of orbivirus particles. *Arch. Virol.* **14**, 235–50.

Stuart, D.I. and Grimes, J.M. (2006) Structural studies on orbivirus proteins and particles. *Curr. Top. Microbiol. Immunol.* **309**, 221–44.

Sutton, G., Grimes, J.M., Stuart, D.I. and Roy, P. (2007) Bluetongue virus VP4 is an RNA-capping assembly line. *Nat. Struct. Mol. Biol.* **14**, 449–51.

Takagi, T., Moore, C.R., Diehn, F. and Buratowski, S. (1997) An RNA 5'-triphosphatase related to the protein tyrosine phosphatases. *Cell* **89**, 867–73.

Tan, B.H., Nason, E., Staeuber, N., Jiang, W., Monastryrskaya, K. and Roy, P. (2001) RGD tripeptide of bluetongue virus VP7 protein is responsible for core attachment to culicoides cells. *J. Virol.* **75**, 3937–47.

Taraporewala, Z., Chen, D. and Patton, J.T. (1999) Multimers formed by the rotavirus nonstructural protein NSP2 bind to RNA and have nucleoside triphosphatase activity. *J. Virol.* **73**, 9934–43.

Taraporewala, Z.F., Chen, D. and Patton, J.T. (2001) Multimers of the bluetongue virus non-structural protein, NS2, possess nucleotidyl phosphatase activity: Similarities between NS2 and rotavirus NSP2. *Virology* **280**, 221–31.

Theron, J., Huismans, H. and Nel, L.H. (1996) Site-specific mutations in the NS2 protein of epizootic haemorrhagic disease virus markedly affect the formation of cytoplasmic inclusion bodies. *Arch. Virol.* **141**, 1143–51.

Theron, J. and Nel, L.H. (1997) Stable protein-RNA interaction involves the terminal domains of bluetongue virus mRNA, but not the terminally conserved sequences. *Virology* **229**, 134–42.

Thomas, C.P., Booth, T.F. and Roy, P. (1990) Synthesis of bluetongue virus-encoded phosphoprotein and formation of inclusion bodies by recombinant baculovirus in insect cells: It binds the single-stranded RNA species. *J. Gen. Virol.* **71**, 2073–83.

Touris-Otero, F., Cortez-San Martin, M., Martinez-Costas, J. and Benavente, J. (2004) Avian reovirus morphogenesis occurs within viral factories and begins with the selective recruitment of sigmaNS and lambdaA to microNS inclusions. *J. Mol. Biol.* **341**, 361–74.

Uitenweerde, J.M., Theron, J., Stoltz, M.A. and Huismans, H. (1995) The multimeric nonstructural NS2 proteins of bluetongue virus, African horse sickness virus, and epizootic hemorrhagic disease virus differ in their single-stranded RNA-binding ability. *Virology* **209**, 624–32.

Urakawa, T., Ritter, D.G. and Roy, P. (1989) Expression of largest RNA segment and synthesis of VP1 protein of bluetongue virus in insect cells by recombinant baculovirus: Association of VP1 protein with RNA polymerase activity. *Nucl. Acids Res.* **17**, 7395–401.

Van Dijk, A.A. and Huismans, H. (1980) The in vitro activation and further characterization of the bluetongue virus-associated transcriptase. *Virology* **104**, 347–56.

Van Staden, V., Theron, J., Greyling, B.J., Huismans, H. and Nel, L.H. (1991) A comparison of the nucleotide sequences of cognate NS2 genes of three different orbiviruses. *Virology* **185**, 500–4.

Vasquez-Del Carpio, R., Gonzalez-Nilo, F.D., Jayaram, H., Spencer, E., Prasad, B.V., Patton, J.T. and Taraporewala, Z.F. (2004) Role of the histidine triad-like motif in nucleotide hydrolysis by the rotavirus RNA-packaging protein NSP2. *J. Biol. Chem.* **279**, 10624–33.

Vasquez-Del Carpio, R., Gonzalez-Nilo, F.D., Riadi, G., Taraporewala, Z.F. and Patton, J.T. (2006) Histidine triad-like motif of the rotavirus NSP2 octamer mediates both RTPase and NTPase activities. *J. Mol. Biol.* **362**, 539–54.

White, C.K. and Zweerink, H.J. (1976) Studies on the structure of reovirus cores. Selective removal of polypeptide lambda 2. *Virology* **70**, 171–80.

Wikoff, W.R., Liljas, L., Duda, R.L., Hendrix, H., RW and Johnson, J.E. (2000) Topologically linked protein rings in the bacteriophage HK97 capsid. *Science* **289**, 2129–33.

Yazaki, K. and Miura, K. (1980) Relation of the structure of cytoplasmic polyhedrosis virus and the synthesis of its messenger RNA. *Virology* **105**, 467–79.

Zarbl, H. and Millward, S. (1983). The reovirus multiplication cycle. In: W. Joklik (ed.), *The Reoviridae,* London/New York: Plenum Press.

Zhao, Y., Thomas, C., Bremer, C. and Roy, P. (1994) Deletion and mutational analyses of bluetongue virusNS2protein indicate that the amino but not the carboxy terminus of the protein is critical for RNA–protein interactions. *Virology* **68**, 2179–85.

Molecular epidemiology studies of bluetongue virus

7

SUSHILA MAAN, NARENDER S. MAAN, KYRIAKI NOMIKOU, SIMON J. ANTHONY, NATALIE ROSS-SMITH, KARAM P. SINGH, ALAN R. SAMUEL, ANDREW E. SHAW AND PETER P. C. MERTENS

Department of Arbovirology, Institute for Animal Health, Pirbright Laboratory, Ash Road, Pirbright, Woking, Surrey, GU24 0NF, UK

Introduction

Bluetongue virus (BTV) is endemic in the majority of the warmer regions around the world, between approximately 45–53°N and 35°S, including areas of North and South America, Africa, the Indian subcontinent, Australasia, and Asia. The BTV strains that have been isolated from these different regions have significant variations in the nucleotide sequence of their dsRNA genome segments that reflect their geographic origins (Gould, 1987; Gould and Pritchard, 1990; Pritchard *et al.*, 1995, 2004). For the majority of the BTV genome segments, the most significant of these variations clearly divides those viruses that have been analysed into 'eastern' and 'western' groups/topotypes (Maan *et al.*, 2007a; Mertens *et al.*, 2007a). The eastern group includes isolates from the Middle and Far East, the Indian subcontinent, Australia, China and Malaysia, while the western group includes viruses from Africa and the Americas.

The BTV RNA polymerase has an 'error rate' that introduces point mutations during successive rounds of virus replication. As the virus is transmitted to new mammalian hosts by individual vector insects (each of which represents a 'bottleneck'), some of these changes become established within the

ISBN-13: 978-0-12-369368-6

virus population. Geographical separation allows the viruses in different regions to acquire unique point mutations, some of which may make them particularly well suited to transmission and survival in their respective local ecosystems. The processes of genome segment reassortment and selection have gradually led to the appearance of distinctive regional variants in most of the BTV genome segments. The level of divergence that can now be detected between the eastern and the western viruses indicates that they have been physically separated for a relatively long period of time.

There are 24 distinct serotypes of BTV, whose identity is determined by the specificity of reactions between the proteins involved in cell attachment and penetration on the surface of the viral capsid, and the neutralizing antibodies that are generated during infection of the mammalian host. The BTV genes encoding these outer-capsid proteins show nucleotide sequence variations that correlate with both virus serotype and the geographical origin of the virus isolate. The distribution and number of serotypes also varies between geographical regions (see www.reoviridae.org/dsrna_virus_proteins//btv-serotype-distribution.htm). The majority of BTV serotypes have been detected in Africa and the Indian subcontinent (see Chapters 2 and 8), with smaller numbers in other parts of Asia, Australia and the Americas (Chapters 9 and 10).

Since 1998, six serotypes of BTV have invaded Europe which appears to represent a 'cross-roads' between east and west, containing a unique mixture of viruses from both geographic areas, adding this continent to the list of bluetongue (BT)-affected regions (see Chapter 11).

'Molecular epidemiology' studies can compare the RNA sequences of individual or multiple genome segments from novel BTV isolates, with those of existing strains from known locations and dates, identifying both virus serotype and topotype. Sequence variations can also identify individual virus lineages and even the presence of reassortant strains that have exchanged genome segments between different 'parental' strains. Smaller differences in RNA sequence can be detected even within a single BT outbreak, and it is now possible to track the spatial and temporal spread of individual viruses, in a manner that was previously impossible using conventional serological assays alone.

Molecular epidemiology studies of BTV depend on the development of sequence databases for the RNA segments of individual virus isolates from defined locations with well-documented isolation dates and passage histories. Ideally these viruses should be held in long-term reference-collections (e.g. the BTV reference-collection at IAH Pirbright – www.reoviridae.org/dsRNA_virus_proteins/ReoID/BTV-isolates.htm). Such virus collections allow sequence data for specific isolates to be linked directly to information concerning their biological characteristics and epidemiology.

Although the majority of the epidemiological data that are discussed in this chapter are derived from virus isolates held at IAH Pirbright (see above), it is important to recognize that this and other reference-collections represent 'work in progress'. To remain relevant and up to date, these collections are dependent on the continuing efforts of the scientists world-wide, who contribute virus strains to them.

Variable and conserved genome segments

The ten linear dsRNA genome segments of BTV range in size from 3944 to 822 bp (for BTV-8) (Mertens *et al.*, 2005; Maan *et al.*, 2008). They are numbered as segment 1 to segment 10 (Seg-1 to Seg-10) in order of decreasing molecular weight and are packaged as exactly one copy of each segment within the three-layered icosahedral protein capsid of the virus (Colour plates 1, 2 and 15) (see Chapter 6).

The BTV genome codes for a total of 10 distinct viral proteins, one from each dsRNA segment (Mertens *et al.*, 1984), seven of which (VP1–VP7) are structural components of the virus particle (Colour plates 1, 2 and 15) (see Chapter 6), while three (NS1, NS2 and NS3) are non-structural proteins that are synthesized during virus replication in infected cells. Cross-hybridization studies (Huismans and Bremer, 1981; Huismans *et al.*, 1987; Mertens *et al.*, 1987b), as well as more recent nucleotide sequence comparisons (Bonneau *et al.*, 1999, 2000; Wilson *et al.*, 2000; Singh *et al.*, 2004; Potgieter *et al.*, 2005; Maan *et al.*, 2007a, b, 2008; Balasuriya *et al.*, 2008) have shown different levels of variation in the individual BTV genome segments and the proteins that they encode. The capsid proteins that are situated on or near to the surface of the virus particle are more variable than the protein components of the inner virus core, or the non-structural proteins (Mertens, 2004; Maan *et al.*, 2008 – Figure 7.1).

Data concerning the individual BTV proteins and RNAs are available at www.reoviridae.org/dsrna_virus_proteins//BTV.htm

BTV outer-capsid protein genes

The outer capsid of the BTV particle is composed of VP2 and VP5, which are the least conserved of the BTV proteins (encoded by Seg-2 and Seg-6, respectively). Variations in these outer-capsid components are considered likely to reflect adaptive responses by the virus to the host and vector, dictated by the virus's requirements to attach to and penetrate cells from different species/tissues and their responses to antibody-selective pressures caused by the vertebrate hosts' immunological defences.

Studies of BTV strains, which were generated by exchange (reassortment) of genome segments between two different BTV serotypes, have demonstrated that both of the BTV outer-capsid proteins VP2 and VP5 can influence the specificity of reactions between the virus particle and the neutralizing antibodies (Cowley and Gorman 1989; Mertens *et al.*, 1989). VP2 is the outermost of the BTV capsid proteins (see Chapter 6) and represents the primary target antigen for binding of

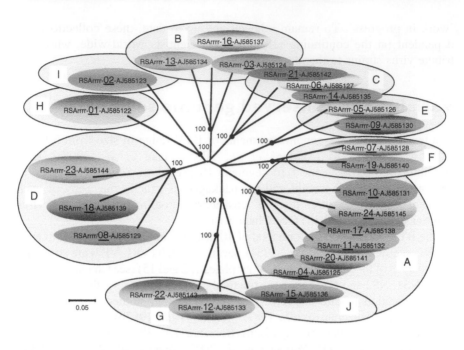

Figure 7.1 Phylogenetic tree for Seg-2 of the 24 BTV serotypes. This neighbour-joining tree was constructed using MEGA2 with the default parameters and the full-length Seg-2 (*VP2* gene) sequences of the reference strains of the 24 BTV serotypes. Nine evolutionary branching points are indicated by black dots (along with their bootstrap values) on the tree, which correlate with the ten Seg-2 'nucleotypes' designated A–J (Maan *et al.*, 2007a). (See colour plate 17).

neutralizing antibodies (Huismans and Van Dijk, 1990; Roy *et al.*, 1990; DeMaula *et al.*, 2000; Maan *et al.*, 2007a, 2008; Mertens *et al.*, 2007a). Sequence analyses of Seg-2 from reference strains of the 24 BTV serotypes have also confirmed that it represents the least conserved region of the BTV genome (Figures 7.1 and 7.2), showing 29% to 59% nucleotide variation between different serotypes (Maan *et al.*, 2007a) (Figures 7.1 and 7.3 and colour plate 17).

Phylogenetic analyses of Seg-2 from over 300 BTV isolates have confirmed that it separates into 24 distinct groups or clades, accurately reflecting virus serotype, with <33% nt sequence variation within each serotype, and 29–59% variation between types (Maan *et al.*, 2007a; Mertens *et al.*, 2007a). However, some BTV serotypes are more closely related than others, allowing them to be grouped into 10 distinct nucleotypes (A–J – Figure 7.1 and colour plate 17), showing 29–33% sequence variation between viruses belonging to different serotypes but the same Seg-2 nucleotype (Figure 7.3). BTV serotypes within the same Seg-2 nucleotype also tend to show closer serological relationships (Figure 7.5) (Maan *et al.*, 2007a).

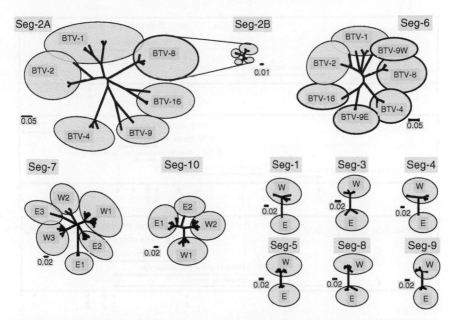

Figure 7.2 Neighbour-joining trees of BTV genome segments 1–10, showing their relative levels of variability. Individual trees for the different BTV genome segments of selected European and other BTV strains are shown on same scale. For the viruses and genome segments compared, Seg-2 and Seg-6 are most variable, followed by Seg-7, then Seg-10, with the remaining 'conserved' segments showing almost equivalent levels of variation and a clear east (E)–west (W) grouping. Tree 2A shows a comparison of Seg-2 from the European BTV serotypes 1, 2, 4, 8, 9 and 16, whereas tree 2B compares only Seg-2 sequences from BTV-8 strains (Maan *et al.*, 2008).

Sequence variation in Seg-2 provides evidence for distinct geographical grouping within individual BTV serotypes, with distinct eastern and western lineages or topotypes (Figure 7.6). BTV strains belonging to the same Seg-2 topotype (within a single serotype) displayed <13% nt sequence variation in Seg-2 (Figure 7.3). The eastern BTV group usually includes viruses from India, Indonesia, China or Australia, whereas the western group includes viruses that are primarily from Africa and North or South America (Figure 7.6) (Maan *et al.*, 2004a, b, 2007a, 2008; Mertens *et al.*, 2007a). However, in a few cases viruses have been detected that do not fit this pattern and appeared to be in the 'wrong' geographical location. In each case this can be linked to the introduction of a virus strain from a different location (as observed in India – see Chapter 8).

As the most variable of the BTV genome segments, sequence variations in Seg-2 provide a basis for molecular epidemiology studies that can be used to identify/confirm virus serotype, the geographical origins of the virus lineage to which an isolate belongs and to distinguish between even closely related virus

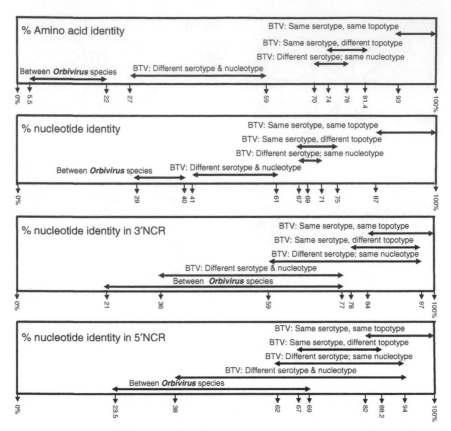

Figure 7.3 Sequence identity in Seg-2 and outer-capsid protein VP2. The range of amino acid and nucleotide sequence identities that were detected within and between BTV serotypes, topotypes and nucleotypes, or between BTV and other *Orbivirus* species are shown schematically. Estimates of the level of similarity within a single serotype are based on multiple datasets for widely distributed isolates of types 1, 2, 4, 9 and 16 (the European serotypes) (Maan *et al.*, 2007a). The comparison of Seg-2 and VP2 from different *Orbivirus* species included available data for EHDV-1 (Ac. no. D10767); EHDV-2 (Ac. no. AB030735); Chuzan virus (Ac. no. AB014725); AHSV-1 (Ac. no. AY163329); AHSV-5 (Ac. no. AY163331); BRDV-2 (Ac. no. M87875) and SCRV (Ac. no. AF133432).

strains within a single epizootic or region. Analyses of Seg-2 from multiple BTV isolates belonging to different serotypes has also supported the design of oligonucleotide primers for use in RT-PCR-based typing assays (Mertens *et al.*, 2007a). These assays have been used to identify the serotype and topotype (origins) of the European field and vaccine strains, and have helped to identify exotic BTV serotypes that have arrived in North America (www. reoviridae.org/dsrna_virus_proteins//ReoID/BTV-S2-Primers-Eurotypes.htm) (see Chapters 9 and 17).

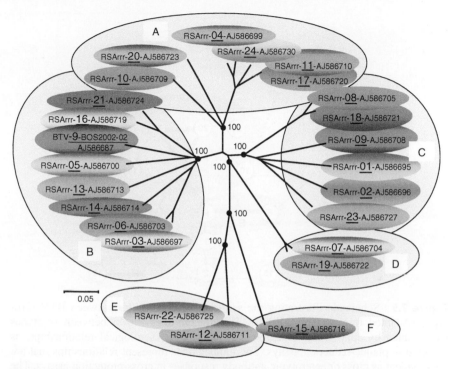

Figure 7.4 Phylogenetic tree for Seg-6 of the 24 BTV serotypes. Unrooted neighbour-joining tree showing relationships between the nucleotide sequences of Seg-6 from reference strains of the 24 BTV serotypes. The tree was constructed using MEGA2 with the default parameters and the full-length genome Seg-6 (*VP5* gene) sequences of the reference strains of the 24 BTV serotypes. Six evolutionary branching points are indicated by black dots (along with their bootstrap values) on the tree, which correlate with the six Seg-6 'nucleotypes', designated A–F. (See colour plate 18).

VP5 (encoded by Seg-6) is the smaller of the BTV outer-capsid proteins (Colour plate 18 – Chapter 6) and is the second most variable of the BTV genome segments (Figure 7.2 – Maan *et al.*, 2008 and Figure 7.4 – Singh *et al.*, 2004). Studies of BTV reassortants have shown that although VP5 can also affect the specificity of reactions with neutralizing antibodies, its influence over BTV serotype is less than that of VP2 (Cowley and Gorman, 1989; Mertens *et al.*, 1989; DeMaula *et al.*, 2000). Variations in Seg-6 also show some correlation with virus serotype, although unlike Seg-2, this is not an absolute relationship (Singh *et al.*, 2004; Maan *et al.*, 2008). In some cases, isolates belonging to different serotypes (e.g. serotypes 7 and 19) can contain an almost identical Seg-6/VP5, while isolates from the same serotype can occasionally contain a Seg-6/VP5 showing large differences that are more

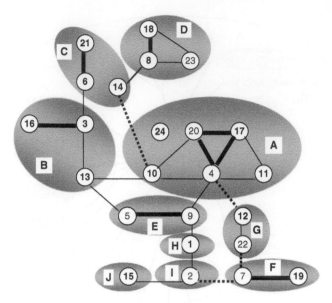

Figure 7.5 Schematic illustrating the serological relationships between BTV serotypes (Adapted from Erasmus, 1990). The thicker lines shown between serotypes
(represented by appropriate numbers) indicate stronger serological relationships, as
detected in plaque-reduction assays. The thinner lines represent relationships that are
only evident as cross or heterotypic antibody responses in cross-protection assays. The
black dotted lines represent interrelationships that are very weak. The different BTV
nucleotypes (A–J) that were identified by phylogenetic analyses are represented by
grey ovals. Serotype 24 is phylogenetically related to serotypes 4, 10, 11, 17 and 20,
but the serological data concerning relationships between this and other types is
limited. We have therefore included it within the oval shown for nucleotype 'A' but
without a line connecting it to other serotypes.

typical of different serotypes (e.g. isolates of serotype 9) (Figure 7.4). Like
Seg-2, Seg-6 also shows variations within a single serotype that correlate with
the geographical origins of individual virus isolates (Seg-6 topotypes – Singh
et al., 2004). In most cases (apart from serotype 9) there are relatively low
levels of variation, with isolates of the same serotype clustering together.

The sequences of Seg-6 also group as a number of distinct clades, shown in
Figure 7.4 as nucleotypes A–F. Most of these groups correlate with the
nucleotypes of Seg-2 (mentioned above). For example, BTV serotypes 4, 10,
11, 17, 20 and 24, are included in nucleotype A of both Seg-2 and Seg-6.
Serotypes 3, 13 and 16 (Seg-2 nucleotype B); serotypes 6, 14 and 21
(Seg-2 nucleotype C); plus serotype 5 and an eastern strain of serotype 9
(Seg-2 nucleotype E) are all included in Seg-6 nucleotype B. However, the
unusually high level of variation detected in Seg-6 of BTV-9 places the
western reference strain of this serotype in Seg-6 nucleotype C, suggesting

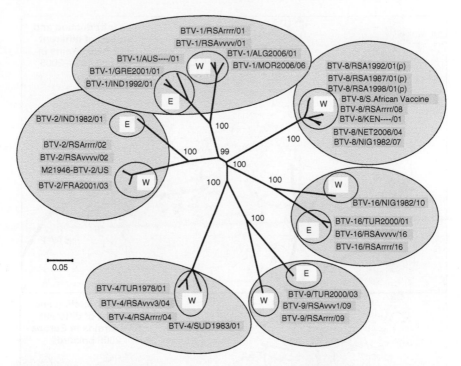

Figure 7.6   Genome segment 2 tree for the European BTV serotypes. Neighbour-Joining tree constructed from full-length BTV genome segment 2 sequences showing the genetic relationships between the six European serotypes of BTV (Mann *et al.*, 2008). The individual virus isolates are numbered according to their designation in the dsRNA virus reference collection at the Institute for animal Health at Pirbright (see www.reoviridae.org/dsRNA_virus_proteins/ReoID/BTV-isolates.htm).

that a reassortment event may have occurred during the evolution of this strain. In this case, the change in Seg-6 between different strains of serotype 9 does not appear to have affected the specificity of the viruses' reactions with neutralizing antibodies.

BTV core-protein genes

The two inner layers of the BTV capsid represent the transcriptionally active 'core' particle that is responsible for BTV mRNA synthesis within the infected host cell (Grimes *et al.*, 1998; Gouet *et al.*, 1999; Diprose *et al.*, 2001; Mertens and Diprose, 2004 – see Chapters 4, 5 and 6). The core-surface layer is composed of 780 copies of the VP7(T13) protein (encoded by Seg-7), which are attached to the outer surface of the 'sub-core' shell that is constructed from 120 copies of the highly conserved VP3(T2) protein (encoded by BTV Seg-3) (Grimes *et al.*, 1998). The BTV subcore surrounds the 10 dsRNA segments of

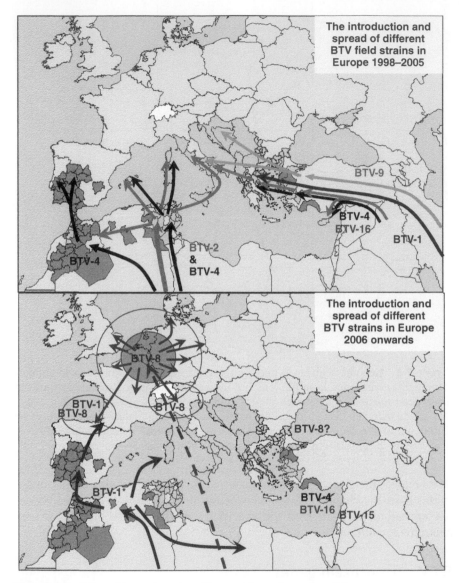

Figure 7.7   Map of incursions into Europe (See colour plate 19).

the virus genome (Verwoerd *et al.*, 1972; Grimes *et al.*, 1995, 1998; Gouet *et al.*, 1999; Mertens *et al.*, 1987a; Mertens 1999), as well as 10–12 transcriptase complexes that are composed of the highly conserved polymerase (VP1), capping enzyme (VP4) and helicase (VP6) (encoded by Seg-1, Seg-4 and Seg-9, respectively) (Colour plates 1 and 2) (Diprose *et al.*, 2001 – Chapter 6).

VP7 is the most immuno-dominant of the virus-species/serogroup-specific BTV antigens, showing high levels of cross-reaction in serological assays between different BTV strains (regardless of serotype or topotype) but little or no cross-reaction with the members of other *Orbivirus* species. Although Seg-7 is more conserved than Seg-2 or Seg-6, it was still the third most variable of the BTV genome segments compared by Maan *et al.* (2008) (Figure 7.2). Seg-7 sequences can be divided into a number of distinct clades that show at least some correlation with the geographical origins of the virus isolates (Seg-7 topotypes) (Wilson *et al.*, 2000; Maan *et al.*, 2008). These variations in VP7 may help to accommodate the more variable outer-capsid proteins VP2 and VP5 that are attached to VP7 on the core surface layer.

VP7 can also mediate initiation of infection by BTV core particles (Mertens *et al.*, 1996), which can be neutralized by antibodies to VP7 (Hutchinson, 1999). Variations observed in Seg-7/VP7 may therefore also reflect host–cell specificity and antibody selection pressure. Despite sequence variations that exist in Seg-7, VP7 represents a target for the majority of serological assays used to identify members of the BTV serogroup (BTV species) (Wade-Evans *et al.*, 1990; Afshar *et al.*, 1992). Oligonucleotide primers have also been designed targeting Seg-7, for use in BTV virus-species/serogroup-specific RT-PCR assays (Anthony *et al.*, 2007).

The BTV sub-core proteins [VP1(Pol), VP3(T2), VP4(CaP) and VP6(Hel)] are more highly conserved than those coding for the outer-core or outer-capsid layers (Mertens, 2004; Maan *et al.*, 2008). These proteins show serological cross-reactions between all BTV strains, although (like VP7) they react poorly or not at all with the members of other *Orbivirus* species (i.e. other serogroups). Although genome segments 1, 3, 4 and 9 are also highly conserved across the BTV species/serogroup, they do show significant levels of variation that correlate with the geographic origins of virus isolates (topotypes) dividing them into eastern and western groups (Figure 7.2).

Variations in Seg-3 and VP3 were originally used to identify sequence variations that correlate with the geographic origins of BTV isolates (Gould, 1987; Pritchard *et al.*, 1995). The high levels of conservation in Seg-3 have made comparisons of Seg-3/VP3 useful for distinguishing and identifying the members of different *Orbivirus* species. This approach provided important evidence leading to the recognition of St Croix River virus (SCRV) as a new species within the genus *Orbivirus* (Attoui *et al.*, 2001).

Seg-1 (coding for the viral polymerase – VP1) is highly conserved not only within the BTV species but also shows a significant level of conservation across the genus *Orbivirus*. Indeed, sequence comparisons of the polymerase gene have even been used to compare distantly related reoviruses making it possible to identify the members of the different genera within the family *Reoviridae* as a whole (see Chapter 3 – Attoui *et al.*, 2000; Mertens, 2004).

The more highly conserved genome segments provide potential targets for the development of real time RT-PCR assays. Two 'monoplex' assays targeting BTV genome segment 1 (Shaw *et al.*, 2007) can be used to identify RNA derived from viruses belonging to either the eastern or western topotypes. The primers and probes that were used can also be combined in a duplex diagnostic assay in order to detect any member of the BTV-serogroup/virus-species (see Chapter 17).

BTV non-structural protein genes

Two of the BTV non-structural proteins, NS1and NS2 (encoded by genome segments 5 and 8 respectively (Figure 7.2) are also highly conserved. However, these segments show clear separation into eastern and western groups, in a manner similar to those coding for the conserved protein components of the virus sub-core (Maan *et al.*, 2008). In contrast the smallest of the BTV non-structural proteins, NS3 (encoded by Seg-10) represents the fourth most variable of the BTV proteins/genes (Maan *et al.*, 2008) with separation into a number of distinct clades that show at least some geographic bias (Figure 7.2 – Bonneau *et al.*, 1999; Nikolakaki *et al.*, 2005; Balasuriya *et al.*, 2008; Maan *et al.*, 2008). However, NS3 of African horse sickness virus (AHSV – an *Orbivirus* species closely related to BTV) shows a significantly higher level of variation. In this case the sequences of Seg-10 group into three distinct clusters – alpha, beta and gamma (Sailleau *et al.*, 1997; Martin *et al.*, 1998; Huismans *et al.*, 2005). There have been relatively few outbreaks of AHS outside Africa and the absence of any clear geographic grouping of AHSV Seg-10 may reflect the general restriction of the virus to the African continent.

NS3 is involved in release of BTV particles from the infected host cell (Hyatt *et al.*, 1993). *Culicoides* cells can become persistently infected with BTV and do not show significant levels of cytopathic effects (cpe) or cell lysis. This protein is therefore thought to be particularly relevant to infection of and transmission by the insect vector, suggesting that variations in Seg-10 may show some correlation with the identity of the major vector species in different geographic regions (see Chapter 14).

BTV incursions into Europe

Prior to 1998, outbreaks of BT were infrequent in southern Europe and western Turkey and were short lived and involved only a single BTV serotype (Mellor and Boorman 1995; Mellor *et al.*, 2000; Mellor and Wittmann 2002 – see Chapter 11). These epizootics have included B TV-10 in the Iberian Peninsula and Morocco (1956–1960) (Manso-Ribeiro *et al.*, 1957), BTV-3 in Cyprus during 1943 (Polydorou, 1978), BTV-4 in Lesbos and Rhodes during 1979 (Vassalos, 1980) and BTV-4 in Turkey during the 1980s (Yonguc *et al.*, 1982). However, BTV-4 has subsequently been detected in Cyprus on several occasions between 1969 and 2004 (Sellers *et al.*, 1979), suggesting that it has persisted and is well adapted to the local ecosystem.

Since 1998, there have been fundamental changes in the incidence of BT in Europe, starting in the Mediterranean region, with outbreaks gradually extending further north into central and northern Europe arriving in the UK during August 2007. The repeated introduction and establishment of new BTV strains in mainland Europe has provided clear evidence of an epidemiological 'step-change' in the region and has already resulted in the deaths of over one million animals (Purse *et al.*, 2005 – Chapter 16). The northern European outbreak, caused by BTV-8 which started in the Netherlands and Belgium during the summer of 2006, is the largest and most economically damaging BT epizootic on record. Its occurrence demonstrates a major increase in the threat posed to European livestock by BTV and potentially by other arboviruses, including orbiviruses such as epizootic haemorrhagic disease virus (EHDV) and AHSV.

Novel sequencing methods recently developed for dsRNA viruses (Shapiro *et al.*, 2005; Maan *et al.*, 2007a, b) have been used in the creation of a sequence database for the genome segments of many different BTV isolates (see www.iah.bbsrc.ac.uk/dsRNA_virus_proteins/btv_sequences.htm). These data, particularly those for Seg-2 and 6 (encoding the variable outer capsid proteins VP2 and VP5) have been used in molecular epidemiology studies of BTV within Europe and have identified or confirmed the serotype of each of the European BTV strains (Maan *et al.*, 2004a, b; Singh *et al.*, 2004; Maan *et al.*, 2007a, 2008; Mertens *et al.*, 2007a). These analyses have also identified the topotype of each BTV incursion into Europe, indicating the origins of the virus strain involved.

Since 1998, there have been at least 11 separate introductions of BTV into Europe, involving nine distinct strains of the virus, belonging to six different serotypes (BTV-1, BTV-2, BTV-4, BTV-8, BTV-9 and BTV-16), with new introductions in almost every year except 2002 (Figure 7.7). There are four distinct routes by which these viruses have arrived in Europe: from the east via Turkey/Cyprus into Greece; from North Africa via Algeria or Tunisia into Italy and the eastern Mediterranean Islands; from Morocco into southern Spain and Portugal and, via an unknown route into northern Europe (the Netherlands/Belgium).

BTV has been recorded on the fringes of Europe for many years with several additional serotypes present in Turkey, Cyprus, Israel and Africa (Taylor and Mellor, 1994). BTV-15 was identified in Israel during 2006, the first time this serotype has been identified in the Mediterranean region, suggesting that it could also represent a future threat to Europe.

Further information concerning BTV outbreaks and viruses is available from www.reoviridae.org/dsrna_virus_proteins/outbreaks.htm#top

BTV-1

Sequencing and phylogenetic analyses of Seg-2 identified three introductions of BTV-1 into the Mediterranean region, in 2001, 2006 and 2007 (see www.reoviridae.org/dsRNA_virus_proteins/EroID/btv-1.htm).

The first European isolates of BTV-1 from Greece (IAH reference collection, isolates: GRE2001/01 to GRE2001/07) were shown to be related to viruses from India and Malaysia. The virus is thought to have entered Europe from the east, possibly via Turkey, although there was no evidence of BTV-1 during a serological survey of Turkey in the early 1980s (Yonguc et al., 1982). The Greek BTV-1 isolates show a close relationship to viruses from India and Malaysia (~95.9% similarity). This raises questions concerning the importance of animal movements via established livestock trade routes (Slingenbergh et al., 2002; Bourn, 2003) which could have allowed this virus to extend westwards from much further east, followed by local wind-borne movement of insects from Turkey to Greece. Interestingly, this strain of BTV-1 only persisted in Greece for a relatively short period and did not spread to other European countries.

In July 2006, BTV-1 was detected in Algeria (isolates ALG2006/01 to ALG2006/06), spreading to Morocco by September (isolates MOR2006/06 to MOR2006/12). The virus was identified by RT-PCR using serotype-specific primers targeting Seg-2 and sequence analysis (Mertens et al., 2007a – www. reoviridae.org/dsrna_virus_proteins//ReoID/BTV-S2-Primers-Eurotypes.htm).   Unlike the earlier Greek strain of BTV-1, Seg-2 of the African isolates is more closely related to western BTV-1 strains from sub-Saharan Africa (95.9% similarity) (www.reoviridae.org/dsRNA_virus_proteins/BTV1-segment2-tree.htm). It was also evident that neither of the European BTV-1 lineages was closely related to the South African live attenuated vaccine strain (IAH virus collection isolate number RSAvvvv/01). During October 2006 BTV-1 was also detected in Sardinia, indicating that it had moved northwards into the western Mediterranean islands, possibly via wind-borne adult *Culicoides*.

In July 2007, BTV-1 belonging to the same western virus lineage was again identified in North Africa, in Tunisia, Libya and Morocco (isolates TUN2007/01; LIB2007/01 to LIB2007/08; and MOR2007/01 to MOR2007/04). In late July, the virus was also identified in the south of the Iberian Peninsula (Spain, Portugal and Gibraltar). Sheep in these areas that had previously been vaccinated against BTV-4 still developed clinical signs of BT as a result of infection with BTV-1 (isolates GIB2007/01 to GIB2007/03). It appears likely that the virus had again moved across the Straits of Gibraltar from Morocco via wind-borne adult *Culicoides*. Within 2007, BTV-1 crossed the Iberian Peninsula (from south to north), arriving in south-west France during November 2007. This provided the first overlap between the northern European outbreak of BTV-8 (see below) and another BTV strain/serotype and is likely to provide opportunities for genome segment exchange between the two virus stains, potentially leading to the generation of novel reassortant viruses.

BTV-2

BT outbreaks caused by BTV-2 were identified in Tunisia during February 2000 (Anon, 2000a). The virus was isolated and is represented in the IAH reference collection by TUN2000/01 and TUN2000/02 (see www.reoviri-dae.org/dsRNA_virus_proteins/ReoID/btv-2.htm).

Further outbreaks, also thought to be caused by BTV-2, were reported in Tunisia and Algeria during 2000 (Anon, 2000b, c), and BTV-specific anti-bodies were detected in northern Morocco. During August 2000, BTV-2 was also identified on the Italian island of Sardinia and by October it had been confirmed in Sicily, Calabria (southern mainland Italy), Corsica and the Spanish islands of Menorca and Mallorca (Anon, 2000c, d). In 2001, outbreaks caused by BTV-2 were again reported in Corsica (IAH isolates FRA2001/01 to FRA2001/06), Sardinia (IAH isolates SAD2001/01 to SAD2001/07) and north of Rome, in Lazio and Tuscany (Anon, 2001a). During 2002, further outbreaks caused by BTV-2 occurred in Sicily (IAH isolates ITL2002/01 to ITL2002/07), Sardinia (IAH isolates SAD2002/01 to SAD2002/03) and mainland Italy (Savini *et al.*, 2003).

Seg-2 of the initial Tunisian isolates of BTV-2 is almost identical (99.8%) to that of isolates made in Corsica and Sardinia, indicating that like the BTV-1 strain from North Africa, BTV-2 had spread northwards from North Africa into Italy and the western Mediterranean islands.

Nucleotide sequence comparisons of Seg-2 show that the BTV-2 isolates from recent outbreaks in the Mediterranean Basin are closely related to viruses from Nigeria, Sudan and South Africa (with >97.3% similarity overall). All of these virus isolates belong to a western group/topotype indicating a sub-Saharan origin for the European outbreak strain (www.iah.bbsrc.ac.uk/dsRNA_virus_proteins/BTV2-segment-2-tree.htm). Comparison with BTV-2 strains isolated in India and China indicate that they form a separate and more distantly related eastern group, with >27.9% variation in Seg-2 when com-pared with western strains of the same serotype. Although the European isolates are related to the live, attenuated South African vaccine strain of BTV-2 (3.7–3.9% nucleotide variation), it has been possible to design oligo-nucleotide primers to distinguish between these field and vaccine strains in RT-PCR assays (Mertens *et al.*, 2007a) (see www.reoviridae.org/dsrna_viru-s_proteins//ReoID/BTV-S2-Primers-Eurotypes.htm). These analyses indicate that the live, attenuated vaccine was not the cause of the initial European BTV-2 outbreaks. However, Seg-2 of some of the subsequent isolates of BTV-2 from non-vaccinated animals in mainland Italy was much more closely related to the BTV-2 vaccine strain used in the region (99.5% similarity) indicating that the vaccine virus had also been transmitted in the field (Ferrari *et al.*, 2005; Batten *et al.*, 2008).

The Sahara represents a major physical barrier to the movement of both insect and mammalian species. However, Zebu cattle that were infected with foot and mouth disease virus (FMDV) were transported across the Sahara of during late 1999 and may have provided an entry route for the virus into North Africa (Samuel and Knowles, 2001). Such cattle could also have provided an entry route for BTV into the region. However, some of the FMDV epidemics that have occurred in North Africa originated in the Middle East, caused by the transport of infected sheep during religious festivals (Samuel *et al.*, 1999). BTV types 2, 4, 6, 9 and 13 were also recorded in Syria and Jordan during the early 1980s with types 2, 4, 9, and 13 in Turkey (Taylor and Mellor, 1994). Although BTV isolates of Middle Eastern origin have not been available for sequencing and comparative studies, it is possible that the North African strain of BTV-2 could have originated from this region.

BTV-4

The level of diversity in Seg-2 between BTV-4 isolates from Africa, South America and the Mediterranean region is relatively low (<10.5%) indicating that they all belong to a western group/topotype, and this may reflect the spread of a single strain originating in southern Africa (see www.reoviridae.org/dsRNA_virus_proteins/ReoID/btv-4.htm). As observed with BTV-1 and BTV-2, comparisons of these strains to a Chinese isolate of BTV-4 showed significantly higher levels of diversity, indicating that it belongs to a distinct eastern Seg-2 topotype (see www.reoviridae.org/dsrna_virus_proteins//btv4-segment-2-tree.htm).

In 1999, the Turkish veterinary authorities vaccinated approximately 60 000 sheep with a live attenuated vaccine strain of BTV-4 (European Commission, 2000), although it was later confirmed that the Turkish outbreaks were caused by BTV-9. BTV-4 was subsequently detected in Greece during 1999 and again in 2000 (IAH isolates GRE1999/01 to GRE2000/07), and it was initially suggested that the Greek BTV-4 outbreaks may have been linked to the Turkish vaccination programme. However, BTV-4 had previously been recorded in a number of countries to the south and east of Europe (e.g. Syria, Jordan, Israel, Cyprus and Turkey – Taylor and Mellor, 1994). Several additional isolates of BTV-4 have also been made recently in Israel (IAH isolates ISR2001/01 to ISR2001/13 and ISR2006/10 to ISR2006/12) that were identified using RT-PCR assays using BTV-4-specific primers targeting Seg-2 (Mertens *et al.*, 2007a) (www.reoviridae.org/dsrna_virus_proteins//ReoID/BTV-S2-Primers-Eurotypes.htm). Phylogenetic analyses have demonstrated that these European and Israeli strains are closely related to earlier isolates from Cyprus (CYP1969/01) and Turkey (TUR1978/01), and to the IAH reference strain of BTV-4 (RSArrrr/04) which also originated from Cyprus (ASOT-1) (Sellers *et al.*, 1979) (Figure 7.6). This suggests the 1999–2000 Greek strain of BTV-4 has been circulating in the Mediterranean region for at least 40 years and was not derived from the Turkish vaccine strain (www.iah.bbsrc.ac.uk/dsRNA_virus_proteins/btv4-segment-2-tree.htm).

In 2003, BTV-4 was detected on several western Mediterranean islands including Menorca (IAH isolates SPA2003/01 to SPA2003/04). Phylogenetic analyses of Seg-2 have shown that this virus is a western strain of BTV-4 and is distinct from the strain circulating in the eastern Mediterranean region (~6%

nucleotide difference). These conclusions were confirmed by Breard *et al.* 2007 for Corsican isolates (IAH isolate FRA2003/01). This strain of BTV-4 is believed to have entered Europe from North Africa, possibly from Tunisia or Algeria, like BTV-1 in 2006 and BTV-2 in 2000, indicating that it may also have a sub-Saharan origin. The same strain of BTV-4 was subsequently isolated from Morocco in 2002 (IAH isolate MOR2004/02) and Spain in 2004 (IAH isolates SPA2004/01 and SPA2004/02) where it persisted into 2005 (IAH isolates SPA2005/02 to SPA2005/05). It was concluded that there had been introductions of BTV-4 into Europe, from North Africa in both 2003 and 2004.

BTV-8

In August 2006, BT was recognized for the first time in northern Europe in Belgium, the Netherlands, Luxemburg, Germany and north-east France (OIE, 2006; Toussaint *et al.*, 2006) (see www.reoviridae.org/dsRNA_virus_proteins/ ReoID/btv-8.htm). Multiple isolates of the virus were made from across the region (IAH isolates NET2006/01 to NET2006/06, GER2006/01 to GER2006/ 04, BEL2006/01, BEL2006/02 and FRA2006/01). Although the outbreak subsided during the winter of 2006/07 it reappeared in the same regions during May–June 2007 (isolates NET2007/01 to NET2007/08) spreading initially across Germany (GER2007/01 to GER2007/22) and central France (FRA2007/01 to FRA2007/05), then to Switzerland (SWI2007/01), the Czech Republic, Denmark (DEN2007/01) and the East Anglian region of the UK during August/September 2007 (UKG2007/01 to UKG2007/64). Although these BTV-8 outbreaks stopped during the winter of 2007/08, export of infected animals into northern Spain and Italy lead to further outbreaks in both these regions (EFSA report, 2007).

Sequence analyses of Seg-2 and Seg-6 from NET2006/04 (representing the first virus isolate from the outbreak) and Seg-2 from multiple other BTV-8 isolates from the northern European outbreak have demonstrated that the virus belongs to a western lineage from sub-Saharan Africa but is distinct from the BTV-8 vaccine strain from South Africa (Maan *et al.*, 2008, see www.reoviridae.org/dsRNA_virus_proteins/BTV-8-Seg-2-tree.htm). It is uncertain how BTV-8 arrived in northern Europe, but the absence of the virus from the Mediterranean region indicates that it did not involve simple incremental extension of earlier outbreaks and is likely to reflect a distinct but unknown entry route and mechanism. However, some serological evidence for BTV-8 infection of sentinel animals was subsequently obtained from Bulgaria (during 2006), although no virus was detected by virus isolation or by RT-PCR assays of blood samples, making it difficult to determine the relative timing of the BTV infections involved or their relevance of these reports to northern Europe.

During January 2008, a group of cows, some of which had previously been infected with BTV-8, were imported from the Netherlands into Northern

Ireland. Although all of these animals were negative using RT-PCR assays of blood samples, three of their calves were born RT-PCR-positive for BTV-8 RNA, providing clear evidence of vertical transmission in utero. Initial sequencing studies indicate that dam and calf were infected by an identical virus strain. Two in-contact adult cows also became infected with the same virus, providing evidence of horizontal transmission despite the apparent absence of adult *Culicoides*. The viruses from these samples are represented by IAH isolates UKG2008/01 to UKG2008/06 and are closely related to other BTV-8 isolates from the Netherlands during 2006–2008 (see SNPS analysis below).

Single nucleotide polymorphism analysis

The availability of large numbers of blood samples and virus isolates from different years and different regions of the northern European BTV-8 outbreak has provided an unprecedented opportunity to examine variations in the virus genome segments as it spread across Europe from a single point of introduction in 2006. Analysis of Seg-2 from different BTV-8 isolates indicated that two very closely related viruses (showing 5 nt changes in Seg-2) both of which were present in the Netherlands, arrived in the UK during September 2007. These viruses continued to spread during 2006 and 2007 gradually acquiring further point mutations in the process. Although these studies are still ongoing, they already indicate a mutation rate within Seg-2 involving as few as 5-point mutations in the 2939 bp between consecutive years. SNPS analysis is providing a way to track the movements of individual viruses from one specific location to the next, as recently described for the foot and mouth outbreak in the UK during 2007 (Cottam *et al.*, 2008).

BTV-9

The first outbreaks of BT in Europe since the 1980s were caused by BTV-9 and occurred on four Greek islands close to the Anatolian coast of Turkey (Rhodes, Leros, Kos and Samos) during October 1998 (IAH isolate GRE1998/01 – see www.iah.bbsrc.ac.uk/dsRNA_virus_proteins/ReoID/btv-9.htm) (Anon, 1998a, b). This was the first time that BTV-9 had ever been recorded in Europe, although there was earlier serological evidence for its presence in Turkey (Taylor and Mellor, 1994). Sequence analysis of Seg-2 demonstrated that this BTV-9 strain belongs to an eastern group of viruses, a finding that is consistent with its arrival in Eastern Europe via Turkey, distinguishing it from the South African BTV-9 vaccine strains that belong to a western group (www.iah.bbsrc.ac.uk/dsRNA_virus_proteins/btv9-segment-2-tree.htm). Using these sequence data, it has been possible to design primers that distinguish the European vaccine and field strains of BTV-9, by simply identifying their different topotypes (Mertens *et al.*, 2007a) (www.reoviridae.org/dsRNA_virus_proteins/ReoID/BTV-S2-Primers-Eurotypes.htm#BTV-9).

BTV-9, from the same eastern lineage, was identified as the cause of outbreaks in mainland Greece during 1999 (IAH isolates GRE199/06 and GRE1999/12) (Anon, 1999b), south-eastern Bulgaria (IAH isolate BUL1999/01) (Anon, 1999a, b) and European Turkey (Anon, 1999b). During 2000, further outbreaks caused by BTV-9 were reported in north-western and central Greece (Anon, 2001e), then in 2001 from Serbia (IAH isolates SER2001/01 and SER2001/02), Montenegro, Kosovo (IAH isolates KOS2001/01-KOS2001/04), Macedonia, Bulgaria, Croatia (Anon, 2001a–f), mainland Italy and Sicily. In 2002, BTV-9 was identified again in Bosnia, Bulgaria, Montenegro, Yugoslavia and Albania, and there was an unconfirmed report of BT in Kosovo (Calistri et. al., 2004).

Although BTV-9 was also isolated in Sicily in 2003 (IAH isolate ITL2003/01), this was from an animal that died within a week after vaccination with the live BTV-9 vaccine, and the virus that was recovered proved to be identical (Seg-2) to the South African vaccine strain. BTV-9 also caused further outbreaks in Italy during 2004.

Sequence comparisons of Seg-2 and Seg-6 showed that the isolates analysed from the recent BTV-9 outbreaks in Europe are almost identical and can be grouped together as a single lineage (<12.9% nt sequence variation in Seg-2). All of these viruses belong to the same eastern geographic group as BTV-9 from Australia, India and Indonesia (www.reoviridae.org/dsRNA_virus_proteins/btv9-segment-2-tree.htm). BTV-9 had previously been reported in Anatolian Turkey, Syria, Jordan and Israel (Taylor and Mellor, 1994). Although these viruses are not available for comparison, some of them could have spread in a westerly direction into Europe, and may have caused the outbreaks of BTV-9 that started in 1998.

Sequencing studies have shown >99% nt identity in Seg-2 between the BTV-9 field isolates from Italy 2001 and the Greek strains of BTV-9 from 1999, indicating that this serotype initially arrived in Italy from an easterly direction (Savini et. al., 2004). Most of the European isolates of BTV-9 that have been characterized are distinct from the South African reference and vaccine strains which belong to a distinct 'western' topotype of BTV-9. No introductions of other distinct BTV-9 field strains have been detected, although it is possible that some of the later cases from Italy, which have not been available for sequencing, may reflect the use and persistence of the live vaccine strain in the region (Savini et al., 2008), e.g. IAH isolate ITL2003/01 (see above).

BTV-15

An isolate that was derived from BT outbreaks that occurred in Israel during 2006 (IAH isolate ISR2006/11) was identified as BTV-15 using RT-PCR assays containing serotype-specific primers targeting Seg-2 (see www.reoviridae.org/dsrna_virus_proteins//ReoID/btv-15.htm). Subsequent phylogenetic analyses confirmed the virus serotype and indicated that it belongs to a western lineage (www.reoviridae.org/dsRNA_virus_proteins/BTV15-Seg2-tree.htm). However, the number of BTV-15 isolates that are available is very limited, and it is not yet possible to be more precise about the origins of this virus strain. Existing data suggest that it is new to the Mediterranean region, and it may therefore represent a future threat to European livestock.

BTV-16

BTV-16 was responsible for some of the outbreaks BT that occurred in Greece during 1999–2000 (IAH isolate GRE1999/13), and in the Turkish province of Izmir during 2000 (IAH isolates TUR2000/01, TUR2000/02, TUR2006 to TUR2000/10) (Anon, 2000e). During 2002 and 2004, there were further outbreaks caused by BTV-16 on Sicily (ITL2004/01), mainland Italy, Sardinia (SAD2004/01 to SAD2004/24) and Corsica (FRA2004/01). BTV-16 was also identified in Cyprus during 2006 (isolates CYP2006/01 and CYP2006/02) (see also www.reoviridae.org/dsrna_virus_proteins//ReoID/btv-16.htm).

Sequence analysis of Seg-2 indicates that all of the Mediterranean/European isolates of BTV-16 belong to an eastern group of viruses and are closely related to the South African BTV-16 vaccine (<0.7% variation). The vaccine strain was derived from a reference strain of BTV-16 that was originally isolated from Hazara in West Pakistan during 1960 (Howell et al., 1970). The BTV-16 isolates from Turkey, Greece and Cyprus all form a closely related group, as might be expected for viruses of the same serotype that are derived from the same region, indicating that they represent components of a single outbreak. These data indicate that BTV-16 first arrived in Europe from the east, possibly from Turkey.

Very close relationships exist between the European field and live attenuated vaccine strains of BTV-16 (Seg-2), suggesting a recent common ancestry (see www.iah.bbsrc.ac.uk/dsRNA_virus_proteins/btv16-segment2-tree.htm). This has made it difficult to design RT-PCR primers that can be used reliably to distinguish between the field and vaccine strains of this serotype (Mertens et al., 2007a, see www.reoviridae.org/dsrna_virus_proteins//ReoID/BTV-S2-Primers-Eurotypes.htm#BTV-16). This close relationship also suggests that the live attenuated vaccine may be involved in the origins of the European incursions of BTV-16. Indeed, a live vaccine strain of BTV-16 has been used repeatedly as part of an annual vaccination campaign in Israel (Shimshony, 2004) and could represent a source for the European viruses. After BTV-16 arrived in Eastern Europe in 1999, it appeared to spread westwards, eventually appearing in mainland Italy during 2002. However, the Italian field strain of BTV-16 appears to have an even closer relationship to the vaccine strain than the viruses from either Greece or Turkey (~99.9% in Seg-2), suggesting that it was not derived directly from the earlier Greek or Turkish outbreaks (Batten et al., 2008; Savini et al., 2008). However, when the Italian BTV-16 strain was isolated, the BTV-16 live vaccine had not yet been used in Italy, and it is possible, therefore, that the Italian outbreak strain was derived more directly from the annual vaccination campaign in Israel. Indeed, the Italian field virus has been shown to be a reassortant containing Seg-2 derived from the BTV-16 vaccine but with Seg-5 derived from the BTV-2 vaccine strain that was also used as part of the multivalent vaccine in Israel (Batten *et al.*, 2008). This suggests that two vaccine strains could have reassorted in Israel before the virus moved to Italy.

The outbreak in Sardinia during 2004 was also caused by BTV-16, although the strain involved (IAH isolates SAD2004/01 to SAD2004/24) proved to be identical to the BTV-16 vaccine that was used in Italy during 2004 (Savini *et al.*, 2008), demonstrating that it was not caused by the eastern European field strain from Greece and Turkey. Indeed, the BTV-16 vaccine strain can cause severe disease in some European breeds of sheep (e.g. Dorset Poll sheep) with viraemia levels ($>10^6$ TCID$_{50}$/ml) that are sufficient to allow infection of feeding *Culicoides* and therefore transmission of the virus (Veronesi *et al.*, 2005). The BTV-16 strain from Sardinia, which was isolated from blood samples taken during 2005, replicated directly in BHK cells, with no requirement for initial passaging in hens' eggs or insect cells as is usually the case with field strains of BTV. This suggests that the virus had previously been adapted to grow in mammalian cell cultures, further supporting the suggestion that it is the vaccine strain.

Phylogenetic analyses of BTV core/ non-structural proteins and identification of reassortants

Since 1998, multiple strains of BTV have arrived in Europe from both eastern (BTV-1, BTV-9, BTV-16) and western (BTV-1, BTV-2, BTV-4, BTV-8) lineages. In attempts to minimize virus circulation, live attenuated monovalent vaccines for BTV-2, BTV-4, BTV-8, BTV-9 (western topotype) and BTV-16 (eastern topotype) have also been used in the Mediterranean region. The South African 'Group B' multivalent live attenuated vaccine (containing types 3, 8, 9, 10 and 11) was also used briefly in Bulgaria during 2000 (Panagiotatos, 2004; Savini *et al.*, 2007). These activities have generated an unprecedented mix of BTV field and vaccine strains within Europe and have provided multiple opportunities for the exchange/reassortment of genome segments (Monaco *et al.*, 2005; Batten *et al.*, 2008).

Comparisons of genome segment 3 from different European strains showed ~99.8% nt similarity between the western BTV-2 from Corsica (Ac No. AY124371) and Italy (IAH isolates ITL2000/01 & ITL2002/07) with the western BTV-4 strains from Morocco and Spain (IAH isolates MOR2004/02 &

SPA2003/01). This indicates that despite belonging to different serotypes and therefore containing Seg-2 from different lineages these strains all share Seg-3 derived from a recent common ancestor, providing direct evidence of genome segment exchange/reassortment in the field (Maan *et al.*, 2008).

Similar comparisons of Seg-5 from the eastern strains of BTV-16 from Greece (GRE1999/13) and BTV-9 from Bulgaria (BUL1999/01) showed 99.9% identity indicating that they have also been involved in reassortment (Maan *et al.*, 2008). The eastern BTV-16 strain from Italy 2002 contains Seg-5 that is identical to that of the western BTV-2 vaccine strain used in the region (Batten *et al.*, 2008) providing further evidence of reassortment, although in this case between two parental vaccine strains that belong to different topotypes.

Conclusions/Discussion

Sequence analyses of the European BTV isolates and comparisons to other selected strains from around the world have shown variations in segments 1, 3, 4, 5, 8 and 9 that correlate primarily with the geographical origins of the virus lineage from which they were derived, dividing the viruses into eastern and western topotypes/groups (Maan *et al.*, 2008). This indicates that the viruses in these different regions have been separated and have acquired point mutations over a long period of time (Gould and Pritchard, 1991, Pritchard *et al.*, 1995, Maan *et al.*, 2008). Lower levels of variation were also detected in these genome segments which indicate further separation into more closely related 'local' topotypes within both the eastern and the western groups (Gould, 1987; Gould and Pritchard, 1990; Pritchard *et al.*, 1995, 2004). Further analyses of additional BTV isolates from around the world will help to define the nature and distribution of these different geographic/topotype groups for each of the BTV genome segments.

Although variations in Seg-2/VP2 and Seg-6/VP5 primarily reflect virus serotype both genome segments also show variations that reflect the geographic origin of the virus isolate (Seg-2 and Seg-6 topotypes). These data suggest that BTV diverged into serotypes before strains of the individual serotypes became geographically separated, allowing them to acquire further point mutations that define the different topotypes within each serotype.

Seg-7 (encoding VP7) and Seg-10 (encoding NS3) also show some evidence of 'east-west' grouping. However, Seg-7 is the third most variable segment of the viruses compared by Maan *et al.* (2008) and separates into at least six distinct groups. Three of these included eastern viruses only, whereas three contained western viruses with the exception of BTV-12 from China which clusters with some of the western stains. Additional groups of Seg-7 were identified by Wilson *et al.* (2000). Seg-10 is the fourth most variable segment of the BTV strains that were compared by Maan *et al.* (2008) and separated the viruses into four distinct groups, two eastern and two western (although one of the eastern groups is represented by only a single strain of BTV-15 from China) (Balasuriya *et al.*, 2008).

The involvement of VP7 in BTV infection of insect cells and of NS3 in the release of virus particles from insect cells suggests that variations in Seg-7/VP7 and Seg-10/NS3 could relate to different groups/populations of insect vectors in the different geographic regions from which these viruses were derived (Balasuriya *et al.*, 2008). It is possible that certain combinations of genome segments 7 and 10 variants may be required for or can enhance BTV transmission within individual ecosystems or by specific insect vector populations. These particular combinations of segments would therefore have a selective advantage in those regions, and as a result of reassortment and selection would tend to dominate the virus population over time.

There is compelling evidence linking local climate change to changes in the distribution of *Culicoides imicola* in southern Europe (Purse *et al.*, 2005). Higher ambient temperatures also appear likely to have increased the vector competence of certain northern Palearctic *Culicoides* species in the region (see Chapters 11 and 14. These changes have coincided with outbreaks of BT in Europe, caused by eight distinct BTV strains from six different serotypes (BTV-1, BTV-2, BTV-4, BTV-8, BTV-9 and BTV-16). Several of these virus lineages (with the exception of the eastern strain of BTV-1 from Greece 2001) have persisted and spread both westwards and northwards across much of southern and central Europe (Figure 7.7), resulting in the deaths of over 2 million animals by 2006. With the arrival of BTV-8 in northern Europe during 2006, there have been incursions of BTV into Europe in 9 of the last 10 years (1998–2007). There have also been disease outbreaks caused by persistence and transmission of BTV-vaccine strains in the field (Ferrari *et al.*, 2005; Monaco *et al.*, 2006; Savini *et al.*, 2008). Molecular epidemiology studies have not only helped to identify the origins and serotypes of the European viruses, they have also confirmed that they arrived through four distinct routes (Figure 7.7). Recently a novel BT-like virus was also isolated from goats in Switzerland (Thür and Hoffman, Personal Communication). Initial molecular epidemiology studies indicate that this virus may belong to a further topotypic group within the BTV species that is distinct from the east/west groups described here. Its serotype had not yet been determined.

Once an arthropod-borne disease, such as BT, becomes established in the mammalian host-population in an area containing competent vector species of *Culicoides*, as in the Mediterranean Basin, outbreaks may occur in successive years. This is particularly true if a mechanism exists for virus survival from one 'vector-season' to the next (BTV-overwintering) (Takamatsu *et al.*, 2003; White *et al.*, 2005). The vertical and horizontal transmission of BTV-8 that was observed in Northern Ireland and several other countries in northern Europe appears likely to provide at least one such overwintering mechanism for the virus. Together with the repeated introductions of BTV in successive years, it is therefore unlikely that the outbreaks of BT in Europe will end in the immediate future. Continuing changes in global climate may also increase the distribution and competence of vector-insect populations in the region allowing the virus to spread still further north. It may therefore already be realistic to regard BT as endemic in Europe.

Earlier serological surveys have shown that many BTV serotypes, including those that have recently caused outbreaks in the Mediterranean region, have been present on the periphery of Europe for several decades, most notably in sub-Saharan Africa (Gibbs and Greiner, 1994), Cyprus, Turkey and the Middle East (Reviewed by Purse *et al.*, 2005). Livestock trade routes, such as the 'Eurasian ruminant street', link Europe and North Africa to western Asia and India (Slingenbergh *et al.*, 2002; Bourn, 2003). Movements of ruminant livestock can also occur across the Sahara into northern Africa and westward from the Middle East across North Africa. Molecular epidemiology studies indicate that individual BTV strains can move large distances (like BTV-1 or BTV-8) possibly from as far away as India or sub-Saharan Africa. The virus has also crossed the Mediterranean Sea and the English Channel by wind-borne movements of infected insects.

The changed environment in Europe suggests that if other strains of BTV are introduced they are also likely to present a significant further threat to animal health in the region. Consequently, an effective control/vaccination programme is required that will reduce disease in affected countries and help to create barrier regions to prevent further spread of these viruses within or into Europe.

The live, attenuated BTV vaccines that are currently available (see Chapter 18) appear to have helped prevent disease in South Africa where many of the vaccine strains were originally developed. They have also been used effectively in some of the recent European vaccination campaigns (cf. a monovalent, live-attenuated BTV-2 vaccine was used successfully in both Corsica and the Balearic islands). However, vaccination with BTV-2, BTV-9 and BTV-16 has not yet eradicated these viruses from Italy and has led to disease caused by the vaccine strains themselves in local host populations (Ferrari *et al.*, 2005; Veronesi *et al.*, 2005; Batten *et al.*, 2008; Maan *et al.*, 2008).

Pentavalent vaccines were also used in Bulgaria but did not stop the spread of BTV-9. Indeed, the only European country that has managed to fully control and even eradicate the recent outbreaks of BT is Greece where the live vaccines were not used.

Further sequencing studies will help to clarify relationships between the European field and vaccine strains of BTV. However, it is clear that the use of live vaccines increases the genetic diversity within the virus population and poses an as yet unquantified risk of genome segment reassortment between vaccine and field viruses potentially generating virus strains with novel biological properties. There is an urgent need for better BTV control and vaccination strategies that are appropriate for use in Europe. Inactivated vaccines based on European BTV strains have recently been developed (e.g. by Merial, Fort Dodge, Intervet, Boxmeer, Holland) which do not depend on virus replication in the vaccinated host and are incapable of causing 'vaccine outbreaks' or of becoming involved in genome segment reassortment.

Members of several other *Orbivirus* species are also transmitted by *Culicoides*, in some cases by the same vector species that transmit BTV, e.g. EHDV and AHSV. The repeated movements of BTV into Europe suggest that other orbiviruses could be introduced via the same routes.

AHSV-2 and AHSV-7 were isolated or detected in Senegal and Mauritania respectively during 2007 (IAH isolates SEN2007/01 to SEN2007/07), and AHSV-9 was detected in Kenya (IAH isolates KEN2007/01 and KEN2007/02). AHSV can cause mortality levels of greater than 90% in susceptible and serologically naive horse populations (Mellor and Hamblin, 2004), and it is considered to represent a significant risk to equid populations in Europe. A previous European outbreak of AHS (1987–1991) caused by the introduction of an infected zebra was limited to the southwest of the Iberian Peninsula probably because of the distribution limits of the insect vector, *C. imicola*, at that time.

Although regarded with less concern, EHDV is also transmitted by the same *Culicoides* species as BTV, and outbreaks of EHD causing significant levels of disease but low levels of mortality were recorded in cattle in Morocco during 2004 and 2006 (IAH isolates MOR2004/03, MOR2006/01 to MOR2006/05 and MOR2006/13 to MOR2006/21) and in Israel in 2006 (IAH isolates ISR2006/01 to ISR2006/09).

Recent studies of orbivirus isolates from Alabama, Missouri and Florida since 1999, using Seg-2 specific RT-PCR assays, have identified eight BTV serotypes that were previously exotic to the USA (BTV types 1. 3, 5, 6, 14, 19, 22 and 24) (Johnson *et al.*, 2007; Mertens *et al.*, 2007b – see Chapter 9). These observations indicate that the effects of climate change are not restricted to Europe. They are also unlikely to be restricted to *Culicoides* species or the viruses that they transmit. These changes may therefore affect other vector groups (e.g. mosquitoes), potentially changing the distribution and incidence of other arthropod-transmitted diseases, both in Europe and elsewhere.

Initial studies of European BTV strains have demonstrated several incidences of genome segment reassortment. Further sequence analyses of the full genomes from these and other virus strains may provide information concerning the identity of the genome segments that are most frequently exchanged. The emergence of a single reassortant strain to dominate the viruses causing an outbreak of disease (see above) suggests that it may have a selective advantage over the other reassortants that would have been generated at the same time as well as over the parental strains from which it was derived. The unique combination of BTV strains that currently exist in Europe, together with the unique characteristics of the European ecosystem including novel vector species of *Culicoides*, could lead to the emergence of reassortant virus strains that are particularly suited to the region. Collectively these processes of reassortment and selection may play a part in the evolution of different BTV topotypes in different parts of the world, and could eventually even lead to the emergence of a distinct European topotype.

Sources of Information

BTV isolates by country: www.reoviridae.org/dsRNA_virus_proteins/ReoID/virus-nos-by-country.htm

BTV isolates by serotype: www.reoviridae.org/dsRNA_virus_proteins/ReoID/BTV-isolates.htm

BTV-Seg-2 Trees: www.reoviridae.org/dsRNA_virus_proteins/orbivirus-phylogenetic-trees.htm

Recent BTV outbreaks: www.reoviridae.org/dsRNA_virus_proteins/outbreaks.htm

BTV accession numbers: www.reoviridae.org/dsRNA_virus_proteins/orbivirus-accession-numbers.htm

European outbreaks: www.reoviridae.org/dsRNA_virus_proteins/ReoID/BTV-mol-epidem.htm

BTV RNA and proteins: www.reoviridae.org/dsRNA_virus_proteins/BTV.htm

BTV serotype distribution: www.reoviridae.org/dsRNA_virus_proteins/btv-serotype-distribution.htm

References

Afshar, A., Eaton, B.T., Wright, P.F., Pearson, J.E., Anderson, J., Jeggo, M. and Trotter, H.C. (1992) Competitive ELISA for serodiagnosis of bluetongue: evaluation of group-specific monoclonal antibodies and expressed VP7 antigen. *J. Vet. Diagn. Invest.* **4**, 231–7.

Anon (1998a) Bluetongue in Greece: confirmation of diagnosis. *Dis. Info.* **11**, 166–7.

Anon (1998b) Disease outbreaks reported to the OIE in November–December 1998. *Bull. Off. Int. Épizoot.* **110**, 506.

Anon (1999a) Bluetongue in Bulgaria: additional information. *Dis. Info.* **12**, 122.

Anon (1999b) Disease outbreaks reported to the OIE July–August 1999. *Bull. Off. Int. Épizoot.* **111**, 379.

Anon (2000a) Disease outbreaks reported to the OIE January–February 2000. *Bull. Off. Int. Épizoot.* **112**, 16.

Anon (2000b) Disease outbreaks reported to the OIE July–August 2000. *Bull. Off. Int. Épizoot.* **112**, 687–8.

Anon (2000c) Disease outbreaks reported to the OIE September–October 2000. *Bull. Off. Int. Épizoot.* **112**, 552.

Anon (2000d) Bluetongue in France: in the island of Corsica. *Dis. Info.* **13**, 195–7.

Anon (2001a) Bluetongue in Greece and Italy. *Dis. Info.* **14**, 215–18.

Anon (2001b) Bluetongue in Croatia: suspected outbreak. *Dis. Info.* **14**, 262–5.

Anon (2001c) Bluetongue in Kosovo (FRY), territory under United Nations interim administration. *Dis. Info.* **14**, 242–3.

Anon (2001d) Bluetongue in the Former Yugoslav Republic of Macedonia: follow-up 1. *Dis. Info.* **14**, 265–8.

Anon (2001e) Bluetongue in Yugoslavia. *Dis. Info.* **14**, 280–1.

Anon (2001f) Bluetongue in Greece: follow-up report no. 2. Dis. Info. 14, 262–5.

Anthony, S., Jones, H., Darpel, K.E., Elliott, H., Maan, S., Samuel, A., Mellor, P.S. and Mertens, P.P.C. (2007) A duplex RT-PCR assay for detection of genome segment 7 (VP7 gene) from 24 BTV serotypes. *J. Virol. Methods* **141**, 188–97.

Attoui, H., Billoir, F., Biagini, P., de Micco, P. and de Lamballerie, X. (2000) Complete sequence determination and genetic analysis of Banna virus and Kadipiro virus: proposal for assignment to a new genus (*Seadornavirus*) within the family *Reoviridae*. *J. Gen. Virol.* **81**, 1507–15.

Attoui, H., Stirling, J.M., Munderloh, U.G., Billoir, F., Brookes, S.M., Burroughs, J.N., de Micco, P., Mertens, P.P.C. and de Lamballerie, X. (2001) Complete sequence characterization of the genome of the St Croix River virus, a new orbivirus isolated from cells of *Ixodes scapularis*. *J. Gen. Virol.* **82**, 795–804.

Balasuriya, U.B., Nadler, S.A., Wilson, W.C., Pritchard, L.I., Smythe, A.B., Savini, G., Monaco, F., De Santis, P., Zhang, N., Tabachnick, W.J. and Maclachlan, N.J. (2008) The NS3 proteins of global strains of bluetongue virus evolve into regional topotypes through negative (purifying) selection. *Vet. Microbiol.* **126**, 91–100.

Batten, C.A., Maan, S., Shaw, A.E., Maan, N.S and Mertens, P.P.C. (2008) A European field strain of bluetongue virus derived from two parental vaccine strains by genome segment reassotment. *Virus Res.* Jul 28, [Epub ahead of print].

Bonneau, K.R., Zhang, N.Z., Wilson, W.C., Zhu, J.B., Zhang, F.Q., Li, Z.H., Zhang, K.L., Xiao, L., Xiang, W.B. and MacLachlan, N.J. (2000) Phylogenetic analysis of the S7 gene does not segregate Chinese strains of bluetongue virus into a single topotype. *Arch. Virol.* **145**, 1163–71.

Bonneau, K.R., Zhang, N., Zhu, J., Zhang, F., Li, Z., Zhang, K., Xiao, L., Xiang, W. and MacLachlan, N.J. (1999) Sequence comparison of the L2 and S10 genes of bluetongue viruses from the United States and the People's Republic of China. *Virus Res.* **61**, 153–60.

Bourn, D. (2003) Livestock dynamics in the Arabian Peninsula: A regional review of national livestock resources and international livestock trade. [on line] <http://ergodd. zoo.ox.ac.uk/download/reports/Livestock%20Dynamics%20in%20the%20Arabian%20 Peninsula.pdf>(Animal Production and Health Division of the Food and Agricultural Organization of the United Nations.

Breard, E., Sailleau, C., Nomikou, K., Hamblin, C., Mertens, P.P.C., Mellor, P.S., El Harrak, M. and Zientara, S. (2007) Molecular epidemiology of bluetongue virus serotype 4 isolated in the Mediterranean Basin between 1979 and 2004. *Virus Res.* **125**, 191–7.

Calistri, P., Giovannini, A., Conte, A., Nannini, D., Santucci, U., Patta, C., Rolesu, S. and Caporale, V. (2004) *Vet. Ital.* **40**(3), 243–51.

Cottam, E.M., Wadsworth, J., Shaw, A.E., Rowlands, R.J., Goatley, L., Maan, S., Maan, N.S., Mertens, P.P.C., Ebert, K., Li, Y., Ryan, E.D., Juleff, N., Ferris, N.P., Wilesmith, J.W., Haydon, D.T., King, D.P., Paton, D.J. and Knowles, N.J. (2008) Transmission pathways of foot-and-mouth disease virus in the United Kingdom in 2007. *PLoS Pathog.* **18**, 4.

Cowley, J.A. and Gorman, B.M. (1989) Cross-neutralization of genetic reassortants of bluetongue virus serotypes 20 and 21. *Vet. Microbiol.* **19**, 37–51.

DeMaula, C.D., Bonneau, K.R. and MacLachlan, N.J. (2000) Changes in the outer capsid proteins of bluetongue virus serotype ten that abrogate neutralization by monoclonal antibodies. *Virus Res.* **67**, 59–66.

Diprose, J.M., Burroughs, J.N., Sutton, G.C., Goldsmith, A., Gouet, P., Malby, R., Overton, I., Zientara, S., Mertens, P.P.C., Stuart, D.I. and Grimes, J.M. (2001) Translocation portals for the substrates and products of a viral transcriptase complex: the Bluetongue virus core. *EMBO J.* **20**, 7229–39.

EFSA report (2007) Scientific Opinion of the Scientific Panel on Animal Health and Welfare on the EFSA Selfmandate on bluetongue origin and occurrence. *EFSA J.* **480**, 1–20.

Erasmus, B. J. (1990). Bluetongue virus.. In: Z. Dinter and B. Morein, (eds), *Virus infections of Ruminants.* New York: Elsevier Science Publishers, pp. 227–37.

European Commission (2000) Possible use of vaccination against bluetongue in Europe: report of the Scientific Committee on animal health and animal welfare (V. Moennig, chairman) (Sanco/C3/AH/R19/2000). European Commission, Brussels, 27pp.

Ferrari, G., De Liberato, C., Scavia, G., Lorenzetti, R., Zini, M., Farina, F., Magliano, A., Cardeti, G., Scholl, F., Guidoni, M., Scicluna, M.T., Amaddeo, D., Scaramozzino, P. and Autorino, G.L. (2005) Active circulation of bluetongue vaccine virus serotype-2 among unvaccinated cattle in central Italy. *Prev. Vet. Med.* **68**, 103–13.

Gibbs, P.J. and Greiner, E.C. (1994) The epidemiology of Bluetongue. *Comp. Immunol. Microbiol. Infect. Dis.* **17**, 207–20.

Gouet, P., Diprose, J. M., Grimes, J. M., Malby, R., Burroughs, J.N., Zientara, S., Stuart, D.I. and Mertens, P.P.C. (1999) The Highly ordered double stranded RNA genome of bluetongue virus revealed by crystallography. *Cell* **97**, 481–90.

Gould, A.R. (1987) The complete nucleotide sequence of bluetongue virus serotype 1 RNA3 and a comparison with other geographic serotypes from Australia, South Africa and the United States of America, and with other orbivirus isolates. *Virus Res.* **7**, 169–83.

Gould, A.R. and Pritchard, L.I. (1990) Relationships amongst bluetongue viruses revealed by comparisons of capsid and outer coat protein nucleotide sequences. *Virus Res.* **17**, 31–52.

Gould, A.R. and Pritchard, L.I. (1991) Phylogenetic analyses of complete nucleotide sequence of the capsid protein (VP3) of Australian epizootic hemorrhagic disease of deer virus (serotype 2) and cognate genes from other orbiviruses. *Virus Res.* **21**, 1–18.

Grimes, J., Basak, A.K., Roy, P. and Stuart, D. (1995) The Crystal-Structure of Bluetongue Virus VP7. *Nature* **373**, 67–170.

Grimes, J.M., Burroughs, J.N., Gouet, P., Diprose, J.M., Malby, R., Zientara, S., Mertens, P.P.C. and Stuart, D. I., (1998) The atomic structure of the bluetongue virus core. *Nature* **395**, 470–8.

Howell, P.G., Kumm, N.A. and Botha, M.J. (1970) The application of improved techniques to the identification of strains of bluetongue virus. *Onderstepoort J. Vet. Res.* **37**, 59–66.

Huismans, H. and Bremer, C.W. (1981) A comparison of an Australian bluetongue virus isolate (CSIRO 19) with other bluetongue virus serotypes by cross-hybridization and cross-immune precipitation. *Onderstepoort J. Vet. Res.* **48**, 59–67.

Huismans, H. and Van Dijk, A.A. (1990) Bluetongue virus structural components. In: P. Roy and B.M. Gorman (eds), *Current Topics in Microbiology and Immunology: Bluetongue Viruses*, vol. 162. Springer-Verlag, Germany, pp. 21–41.

Huismans, H., Cloete, M. and le Roux, A. (1987) The genetic relatedness of a number of individual cognate genes of viruses in the bluetongue and closely related serogroups. *Virology* **161**, 421–8.

Huismans, H., van Staden, V., Fick, W.C., van Niekerk, M. and Meiring, T.L. (2005) A comparison of Different Orbivirus proteins that could affect virulence and pathogenesis. *Vet. Ital.* **40**, 417–25.

Hutchinson, I. R. (1999) The role of VP7(T13) in initiation of infection of bluetongue virus. PhD thesis, University of Hertfordshire.

Hyatt, A.D., Zhao, Y. and Roy, P. (1993) Release of bluetongue virus-like particles from insect cells is mediated by BTV nonstructural protein NS3/NS3a. *Virology* **193**, 592–603.

Johnson, D.J., Mertens, P.P.C., Maan, S. and Ostlund, E.N. (2007) Exotic bluetongue viruses identified from ruminants in the southeastern USA from 1999–2006. Proceedings of Annual Conference of American Association of Veterinary Laboratory Diagnosticians (AAVLD). Reno, NV, October 2007, p. 118.

Maan, S., Maan, N.S., Samuel, A.R., O'Hara, R., Meyer, A.J., Rao, S. and Mertens, P.P.C. (2004a) Completion of the sequence analysis and comparisons of genome segment 2 (encoding outer capsid protein VP2) from representative isolates of the 24 bluetongue virus serotypes. *Vet. Ital.* **40**, 484–8.

Maan, S., Maan, N.S., Samuel, A.R., Rao, S., Attoui, H. and Mertens, P.P.C. (2007a) Analysis and phylogenetic comparisons of full-length VP2 genes of the 24 bluetongue virus serotypes. *J. Gen. Virol.* **88**, 621–30.

Maan, S., Maan, N. S., Ross-smith, N., Batten, C. A., Shaw, A. E., Anthony, S. J., Samuel, A. R., Darpel, K. E., Veronesi, E., Oura, C. A. L., Singh, K. P., Nomikou, K., Potgieter, A. C., Attoui, H., van Rooij, E., van Rijn,P., Clercq, K. D., Vandenbussche, F., Zientara, S., Bréard, E., Sailleau, C., Beer, M., Hoffman, B., Mellor, P. S. and Mertens, P. P. C. (2008) Sequence analysis of bluetongue virus serotype 8 from the Netherlands 2006 and comparison to other European strains. *Virology* **377**, 308–18.

Maan, S., Rao, S., Maan, N.S., Anthony, S.J., Attoui, H., Samuel, A.R. and Mertens, P.P.C. (2007b) Rapid cDNA synthesis and sequencing techniques for the genetic study of bluetongue and other dsRNA viruses. *J. Virol. Methods* **143**, 132–9.

Maan, S., Samuel, A.R., Maan, N.S., Attoui, H., Rao, S. and Mertens, P.P.C. (2004b) Molecular epidemiology of bluetongue viruses from disease outbreaks in the Mediterranean Basin. *Vet. Ital.* **40**, 489–96.

Manso-Ribeiro, J., Rosa-Azevedo, J.A., Noronha, F.O., Braco-Forte, M.C., Grave-Periera, C. and Vasco-Fernande, M. (1957) Fievre catarrhale du mouton (bluetongue). *Bull.Off. Int.Epizoot.* **48**, 350–67.

Martin, L.A., Meyer, A.J., O'Hara, R.S., Fu, H., Mellor, P.S., Knowles, N.J., and Mertens, P.P.C. (1998) Phylogenetic analysis of African horse sickness virus segment 10: sequence variation, virulence characteristics and cell exit. *Arch. Virol. Suppl.* **14**, 281–93.

Mellor, P.S. and Boorman, J. (1995) The transmission and geographical spread of African Horse Sickness and Bluetongue viruses. *Ann. Trop. Med. Parasitol.* **89**, 1–15.

Mellor, P.S. and Hamblin, C. (2004) African horse sickness. *Vet Res.* **35**, 445–66.

Mellor, P.S. and Wittmann, E.J. (2002) Bluetongue virus in the Mediterranean Basin, 1998–2001. *Vet. J.* **164**, 20–37.

Mellor, P.S., Boorman, J. and Baylis, M. (2000) Culicoides biting midges: Their role as arbovirus vectors. *Annu. Rev. Entomol.* **45**, 307–40.

Mertens, P.P.C. (1999) Orbivirus and Coltivirus. In: A. Granoff and R.G. Webster, (eds), *Encyclopedia of Virology*, 2nd edn. vol. 2. Academic Press, London, pp. 1043–74.

Mertens, P.P.C. (2004) dsRNA viruses. *Virus Res.* **101**, 3–13.

Mertens, P.P.C. and Diprose, J. (2004) The Bluetongue virus core: a nano-scale transcription machine. *Virus Res.* **101**, 29–4.

Mertens, P.P.C., Brown, F. and Sangar, D.V. (1984) Assignment of the genome segments of bluetongue virus type 1 to the proteins which they encode. *Virology* **135**, 207–17.

Mertens, P.P.C., Burroughs, J.N. and Anderson, J. (1987a) Purification and properties of virus particles, infectious sub-viral particles and cores of bluetongue virus serotypes 1 and 4. *Virology* **157**, 375–86.

Mertens, P.P.C., Pedley, S., Cowley, J. and Burroughs, J.N. (1987b) A comparison of six different bluetongue virus isolates by cross-hybridization of the dsRNA genome segments. *Virology* **161**, 438–47.

Mertens, P.P.C., Pedley, S., Cowley, J., Burroughs, J.N., Corteyn, A.H., Jeggo, M.H., Jennings, D.M. and Gorman, B.M. (1989) Analysis of the roles of bluetongue virus outer capsid proteins VP2 and VP5 in determination of virus serotype. *Virology* **170**, 561–5.

Mertens, P.P.C., Burroughs, J.N., Walton, A., Wellby, M.P., Fu, H., O'Hara, R.S., Brookes, S.M. and Mellor, P.S. (1996) Enhanced infectivity of modified bluetongue virus particles for two insect cell lines and for two *Culicoides* vector species. *Virology* **217**, 582–93.

Mertens, P.P.C., Maan, N.S., Johnson, D.J., Ostlund, E.N. and Maan, S. (2007b). Sequencing and RT-PCR assays for genome segment 2 of the 24 bluetongue virus serotypes: identification of exotic serotypes in the southeastern USA (1999–2006). World Association of Veterinary Laboratory Diagnosticians (WAVLD) – 13th International Symposium Proceeding. Melbourne, Australia, 12–14 November 2007, p. 55.

Mertens, P.P.C., Maan, N.S., Prasad, G., Samuel, A.R., Shaw, A.E., Potgieter, A.C., Anthony, S.J. and Maan, S. (2007a) The design of primers and use of RT-PCR assays for typing European BTV isolates: differentiation of field and vaccine strains. *J. Gen. Virol.* **88**, 2811–23.

Mertens, P.P.C., Maan, S., Samuel, A. and Attoui, H. (2005). Orbivirus, Reoviridae. In: C.M. Fauquet, M.A. Mayo, J. Maniloff, U. Desselberger, and L.A. Ball, (eds), *Virus Taxonomy, VIIIth Report of the ICTV*. London: Elsevier/Academic Press, 466–83.

Monaco, F., Cammà, C., Serini, S. and Savini, G. (2005) Differentiation between field and vaccine strain of bluetongue virus serotype 16. *Vet. Microbiol.* **116**, 45–52.

Monaco, F., Cammà, C., Serini, S. and Savini, G. (2006) Differentiation between field and vaccine strain of bluetongue virus serotype 16. *Vet. Microbiol.* **116**, 45–52.

Nikolakaki, S.V., Nomikou, K., Koumbati, M., Mangana, O., Papanastassopoulou, M., Mertens, P.P.C. and Papadopoulos, O. (2005) Molecular analysis of the NS3/NS3A gene of Bluetongue virus isolates from the 1979 and 1998–2001 epizootics in Greece and their segregation into two distinct groups. *Virus Res.* **114**, 6–14.

OIE (2006) Bluetongue in Bulgaria, Office International des Epizooties (OIE) Disease Information, 19 October2006.19 (42). www.oie.int/eng/info/hebdo/AIS_74.HTM#Sec8.

Panagiotatos, D.E. (2004) Regional overview of bluetongue viruses, vectors, surveillance and unique features in Eastern Europe between 1998 and 2003. *Vet. Ital.* **40**, 61–72.

Polydorou, K. (1978) The 1977 outbreak of bluetongue in Cyprus. *Trop. Anim. Health Prod.* **10**, 229–32.

Potgieter, A.C., Monaco, F., Mangana, O., Nomikou, K., Yadin, H. and Savini, G. (2005) VP2 segment sequence analysis of some isolates of bluetongue virus recovered in the Mediterranean Basin during the 1998–2003 outbreak. *J. Vet. Med. B. Infect. Dis. Vet. Public Health* **52**, 372–9.

Pritchard, L.I., Gould, A.R., Wilson, W.C., Thompson, L., Mertens, P.P.C. and Wade-Evans, A.M. (1995) Complete nucleotide sequence of RNA segment 3 of bluetongue virus serotype 2 (Ona-A). Phylogenetic analyses reveal the probable origin and relationship with other orbiviruses. *Virus Res.* **35**, 247–61.

Pritchard, L.I., Sendow, I., Lunt, R., Hassan, S.H., Kattenbelt, J., Gould, A.R., Daniels, P.W. and Eaton, B.T. (2004) Genetic diversity of bluetongue viruses in south east Asia. *Virus Res.* **101**, 193–201.

Purse, B.V., Mellor, P.S., Rogers, D.J., Samuel, A.R., Mertens, P.P.C. and Baylis, M. (2005) Climate change and the recent emergence of bluetongue in Europe. *Nat. Rev. Microbiol.* **3**, 171–81.

Roy, P., Marshall, J.J.A. and French, T.J. (1990) Structure of the bluetongue virus genome and its encoded proteins. In: P. Roy and B.M. Gorman, (eds.), *Current Topics in Microbiology and Immunology: Bluetongue Viruses*, vol. 162. Springer-Verlag, Germany, pp. 43–87.

Sailleau, C., Moulay, S. and Zientara, S. (1997) Nucleotide sequence comparison of the segments S10 of the nine African horsesickness virus serotypes. *Arch. Virol.* **142**, 965–78.

Samuel, A.R. and Knowles, N.J. (2001) Foot-and-mouth disease type O viruses exhibit genetically and geographically distinct evolutionary lineages (topotypes). *J. Gen. Virol.* **82**, 609–21.

Samuel, A.R., Knowles, N.J. and Mackay, D.K.J. (1999) Genetic analysis of type O viruses responsible for epidemics of foot-and-mouth disease in North Africa. *Epidemiol. Infect.* **122**, 529–38.

Savini, G., Goffredo, M., Monaco, F., de Santis, P. and Meiswinkel, R. (2003) Transmission of bluetongue virus in Italy. *Vet. Rec.* **152**, 119.

Savini, G., MacLachlan, N.J., Sanchez-Vizcaino, J.M. and Zientara, S. (2008) Vaccines against bluetongue in Europe. *Comp. Immunol. Microbiol. Infect. Dis.* **31**, 101–20.

Savini, G., Potgieter, A.C., Monaco, F., Mangana-Vougiouka, O., Nomikou, K., Yadin, H. and Caporale, V. (2004) VP2 gene sequence analysis of some isolates of bluetongue virus recovered in the Mediterranean Basin during the 1998–2002 outbreak. *Vet Ital.* **40**, 473–8.

Savini, G., Ronchi, G.F., Leone, A., Ciarelli, A., Migliaccio, P., Franchi, P., Mercante, M.T. and Pini, A. (2007) An inactivated vaccine for the control of bluetongue virus serotype 16 infection in sheep in Italy. *Vet. Microbiol.* **124**, 140–6.

Sellers, R.F., Gibbs, E.P., Herniman, K.A., Pedgley, D.E. and Tucker, M.R. (1979) Possible origin of the bluetongue epidemic in Cyprus, August 1977. *J. Hyg. (Lond.)* **83**, 547–55.

Shapiro, A., Green, T., Rao, S., White, S., Carner, G., Mertens, P.P.C. and Becnel, J. (2005) Morphological and molecular characterization of a cypovirus (*Reoviridae*) from the mosquito *Uranotaenia sapphirina* (Diptera: Culicidae). *J. Virol.* **79**, 9430–8.

Shaw, A.E., Monaghan, P., Alpar, H.O., Anthony, S., Darpel, K.E., Batten, C.A., Guercio, A., Alimena, G., Vitale, M., Bankowska, K., Carpenter, S., Jones, H., Oura, C.A., King, D.P., Elliott, H., Mellor, P.S. and Mertens, P.P.C. (2007) Development and validation of a real time RT-PCR assay to detect genome bluetongue virus segment 1. *J. Virol. Methods* **145**, 115–26.

Shimshony, A. (2004) Bluetongue in Israel – a brief historical overview. *Vet. Ital.* **40(3)**, 116–8.

Singh, K.P., Maan, S., Samuel, A.R., Rao, S., Meyer, A., and Mertens, P.P.C. (2004) Phylogenetic analysis of bluetongue virus genome segment 6 (encoding VP5) from different serotypes. *Vet. Ital.* **40**, 479–83.

Slingenbergh, J.H.W., Hendrickx, G., Wint, G.R.W. (2002) Will the new livestock revolution succeed? *Agri World Vision* **2**, 31–3.

Takamatsu, H., Mellor, P.S., Mertens, P.P.C., Kirkham, P.A., Burroughs, J.N. and Park-house, R.M.E. (2003) A possible overwintering mechanism for bluetongue virus in the absence of the insect vector. *J. Gen. Virol.* **84**, 227–35.

Taylor, W.P. and Mellor, P.S. (1994) Distribution of bluetongue virus in Turkey, 1978-81. *Epidemiol. Infect.* **112**, 623–3.

Toussaint, J.F., Vandenbussche, F., Mast, J., Demeestere, L., Goris, N., Van Dessel, W., Vanopdenbosch, E., Kerkhofs, P., Zientara, S., Sailleau, C., Czaplicki, G., Depoorter, G. and Dochy, J.M. (2006) Bluetongue in Northern Europe. *Vet. Rec.* **159**, 327.

Vassalos, M. (1980) Cas de Fièvre catarrhale du muton dans l'ille de Lesbos (Grèce). *Bull. Off. Int.Epizoot.* **92**, 547–55.

Veronesi, E., Hamblin, C. and Mellor, P.S. (2005) Live attenuated bluetongue vaccine viruses in Dorset Poll sheep, before and after passage in vector midges (Diptera: Ceratopogonidae). *Vaccine* **23**, 5509–16.

Verwoerd, D.W., Els, H.J., De Villiers, E. and Huismans, H. (1972) Structure of the bluetongue virus capsid. *J. Virol.* **10**, 783–94.

Wade-Evans, A.M., Mertens, P.P.C. and Bostock, C.J. (1990) Development of the polymerase chain reaction for the detection of bluetongue virus in tissue samples. *J. Virol. Methods* **30**, 15–24.

White, D.M., Wilson, W.C., Blair, C.D. and Beaty, B.J. (2005) Studies on overwintering of bluetongue viruses in insects. *J. Gen. Virol.* **86**, 453–62.

Wilson, W.C., Ma, H.C., Venter, E.H., van Djik, A.A., Seal, B.S. and Mecham, J.O. (2000) Phylogenetic relationships of bluetongue viruses based on gene S7. *Virus Res.* **67**, 141–51. (Erratum in: *Virus Res* (2001) **73**, 201–202).

Yonguc, A.D., Taylor, W.P., Csontos, L. and Worrall, E. (1982) Bluetongue in western Turkey. *Vet Rec.* **111**, 144–6.

Bluetongue in the Indian subcontinent

8

GAYA PRASAD,* D. SREENIVASULU,† KARAM P. SINGH,‡ PETER P.C. MERTENS¶ AND SUSHILA MAAN§,¶

*Department of Animal Biotechnology, College of Veterinary Sciences, CCS HAU, Hisar 125004, Haryana, India.

†Department of Microbiology, College of Veterinary Science, Tirupati 517502, A.P., India

‡Centre for Animal Disease Research and Diagnosis, Indian Veterinary Research Institute, Izatnagar, Bareilly 243122, UP, India

§Previously: Department of Animal Biotechnology, College of Veterinary Sciences, CCS HAU, Hisar 125004, Haryana, India

¶Arbovirus Research Group, Department of Arbovirology, Institute for Animal Health, Ash Road Pirbright Woking Surrey, GU24 0NF, UK

Introduction

The Indian subcontinent is a geographically vast region located between 0–40°N and 60–100°E. It includes all of India, as well as Bangladesh, Bhutan, Nepal, Sri Lanka and Pakistan. The subcontinent also includes several different climate zones, supporting a large and genetically diverse population of domestic and wild ruminants. These animals play an important role in sustaining the human population by providing food security for the entire subcontinent. In the past four decades, several exotic breeds of sheep and cattle have been introduced into the subcontinent, in attempts to improve the native breeds through cross-breeding programmes. Consequently, exotic breeds, cross-breeds and native breeds of sheep and cattle are all farmed in the region. Bluetongue virus (BTV) is known to infect a wide variety of animal species, including most ruminants, with outcomes ranging from inapparent to frank or even fatal clinical disease, depending on the species and breed of host animal and the serotype/strain of BTV.

ISBN-13: 978-0-12-369368-6

Outbreaks of clinical bluetongue (BT) have been reported from both India and Pakistan. Although an exact estimate of the economic losses on the subcontinent has not been calculated, BT appears to have great impact on the livestock sector, with exotic breeds of sheep, including Rambouillet, Russian Merino, Southdown, Corriedale and Suffolk initially thought to be more susceptible to disease. However, from 1981 onwards, native breeds of sheep also developed severe clinical signs of the disease.

Bluetongue was first reported in Pakistan, as early as 1958, (Sarwar, 1962) and subsequently in the Maharashtra state of India (Sapre, 1964). Since then outbreaks of BT have been reported from several other regions of the subcontinent (Prasad *et al.*, 1992; Sreenivasulu *et al.*, 1996; Prasad, 2000), and the virus is now considered to be endemic in both India and Pakistan. However, there have been few reports concerning the incidence and prevalence of BT from other countries in the region, including Afghanistan, Bangladesh, Nepal, Bhutan and Sri Lanka (Anon, 2003).

Information on BT in Pakistan is very limited, and most of the information in this paper is therefore derived from investigations that have been conducted in India. Owing to the immense economic importance of BT in Indian ruminant livestock, the Indian Council of Agricultural Research (ICAR) has launched a National Network Project on BT with 11 centres located in different regions of the country. The mandate of the Network Project is to gain an understanding of BT epidemiology, develop suitable vaccines based on local strains/serotypes and suggest strategies for control. Serological data concerning BTV have been reported for several native breeds of sheep, goats, cattle, buffalo and camels within India. The present status of BT in the subcontinent is described briefly in this chapter.

Geographical distribution and seasonality

India comprises a peninsula that lies between 8.4°–37.6°N and 68.7°–97.3°E. (Figure 8.1) and is divided into five zones, which include the northern mountains, northern plains, great plateau, coastal plains and Indian desert. The occurrence of BT varies depending on region, season and intensity of rainfall. Broadly the climate of the subcontinent has been classified into tropical, subtropical and temperate. Most of India has a tropical monsoon type of climate, although the whole country is influenced by the monsoons. These usually begin in the south, during late May or early June, reaching the north about 6 weeks later. In some years, rains are torrential but in others they may be only light. The Himalayas separate the subcontinent from the rest of Asia. Rain-fed areas, including the costal regions of Andhra Pradesh, Karnataka, Kerala and Tamil Nadu, where land is used for cultivation of rice, experience relatively more outbreaks of BT than do the drier regions of

Figure 8.1   Map showing the states of India and the locations of the regional research centres described within the text and listed in Table 8.7. Numbers refer to the numbers in the table.

the country. Variations in temperature and rainfall occur not only from season to season but also from place to place. Some regions have hot summers and cold winters, while others regions have almost the same climate throughout the year.

Bluetongue has been reported in several states of India, including northern (Haryana, Himachal Pradesh, Jammu and Kashmir, Punjab, Rajasthan and Uttar Pradesh), central (Madhya Pradesh), western (Gujarat and Maharashtra) and southern areas (Andhra Pradesh, Karnataka, Kerala and Tamil Nadu). However, there are no reports of BT in the north-eastern states of India (Prasad *et al.*, 1992; Srivastava *et al.*, 1995; Sreenivasulu *et al.*, 1996; Prasad, 2000).

The presence of BTV has also been reported from Jaba Sheep Farm in the Mansehra District, North West Frontier Province of Pakistan (Sarwar, 1962; Akhtar *et al.*, 1997). With the exception of Pakistan, there is no other published evidence for the prevalence of BTV in other countries of the Indian subcontinent.

Culicoides vectors are significantly affected by the climate and weather, and annual variations frequently influence the incidence and overall severity of the diseases transmitted by these insects. Most of the annual BT outbreaks recorded on the subcontinent have occurred in areas affected by the north-east monsoon (during October to December – the monsoon period), followed by areas affected by the south-west monsoon (during June to September). The number of adult vectors in the south of India reaches a peak during the monsoon season (June to December). Outbreaks of BT in Karnataka, Tamil Nadu and Andhra Pradesh have also been associated with these periods of peak *Culicoides* activity (Sreenivasulu *et al.*, 2004).

Subba Reddy (2004) reported the presence of several species of *Culicoides* in the Chittor and Prakasam districts of Andhra Pradesh, with *Culicoides actoni, Culicoides anophelis, Culicoides inoxius, Culicoides majorinus, Culicoides peregrinus* and *Culicoides oxystoma* being the predominant species. In contrast, the predominant species in Tamil Nadu were *Culicoides imicola* and *C. peregrinus*, while in the Marathwada region and Kolkata, *Culicoides schultzei* was the predominant species (Dasgupta, 1995). Ganesh Udupa (2001) reported that the abundance of *C. imicola* in Tamil Nadu was related to the occurrence of seroconversions in domestic animals.

Clinical bluetongue in sheep

In the Indian subcontinent, BT is almost exclusively a disease of sheep, although antibodies to BTV (indicating infection) have been detected in various animal species, and there are some unconfirmed reports of BT in goats. However, diseases such as foot and mouth disease (FMD), peste des petits ruminants (PPR), sheep pox, goat pox and contagious ecthyma are also endemic in the subcontinent, making differential diagnosis of BT, on the basis of clinical signs alone, difficult and potentially subject to errors. It is therefore quite likely that some of the 'suspected' outbreaks of BT in sheep and goats may be due to other diseases. Outbreaks typically occur either when susceptible animals (breeds) are introduced to endemic areas or when infected midges carry the virus to adjacent areas containing populations of susceptible and immunologically naive sheep. However, many Indian strains of BTV appear to be incapable of causing disease following experimental infection of sheep from breeds that are known to be susceptible to disease in the field.

Experimental reproduction of disease has therefore been very difficult in native Indian breeds of sheep (Channakeshava, 2006; Rudragouda, 2007; G. Prasad, unpublished observations).

The first outbreak of BT in the subcontinent was recorded in sheep in Pakistan during 1958 (Sarwar, 1962). Subsequently, an outbreak of BT was recorded in sheep in the Maharashtra state of India in 1964, which was initially diagnosed on the basis of clinical signs. This was later confirmed by the Onderstepoort OIE International Bluetongue Reference Laboratory, South Africa, by detection of BTV-specific antibodies in the sera of recovered sheep (Sapre, 1964). In 1973, an outbreak of BT was recorded in Russian Merino sheep in Kothipura Farm in the northern Indian state of Himachal Pradesh (Uppal and Vasudevan, 1980). Later, a further outbreak of BT was reported in 1975 in exotic 'Corriedale' sheep in the Central Sheep Breeding Farm (CSBF), Hisar, Haryana (Vasudevan, 1982).

Srinivas et al. (1982) reported an outbreak of BT in sheep in Bidar, Gulbarga and nine other districts of Karnataka, where morbidity was ~50%. During 1982, a very severe outbreak of BT occurred in eastern Maharashtra adjoining the Telangana region of Andhra Pradesh (southern India), which later spread to western Marathwada, affecting sheep in all districts of the region with up to 80% morbidity in some village flocks (Singh et al., 1982). Subsequently, Harbola et al. (1982) recorded an outbreak of BT in Maharashtra state and suggested that crosses of the native Chokla breed with Merinos were more susceptible than the native breed itself, although Deccani sheep (another native breed) also showed clinical signs during the outbreak. Of 8980 exotic, cross-bred and native sheep, 868 were clinically affected, 100 of which died, giving an overall case fatality rate of 11.5% (Harbola et al., 1982).

In 1983, an outbreak of BT was recorded in Rajasthan, a desert state in northern India, in which Rambouillet and Merino breeds were found to be more susceptible than native breeds. The morbidity rate in Merino and Rambouillet was 33 and 24%, respectively (Lonkar et al., 1983). Subsequently, Sharma et al. (1985) conducted a systematic epidemiological study in the Central Sheep and Wool Research Institute (CSWRI), Rajasthan, and reported a fatality rate of 15%. In a retrospective study at the same farm between 1980 and 1987, Srivastava et al. (1989) reported a positive correlation between the occurrence of congenital defects in lambs and BTV infection of the mother. The rare deformities recorded in various breeds of sheep included a foetus with a small snout (suicephaly), the presence of two lateral oral openings, bifid tongue and the absence of various parts of the body, such as the pelvic girdle, rectum, reproductive organs, hind legs, abdominal muscles or skin. Since the incidence of BTV in the flocks at CSWRI coincides with the breeding cycle, a possible role of BTV in causing congenital defects in lambs cannot be ruled out.

A very severe outbreak of BT was observed in 1985 at the CSBF Hisar, Haryana, in Rambouillet sheep imported from Texas, USA (Jain *et al.*, 1986). In the initial stages, affected animals developed a high temperature ranging from 40.5 to 42.2 °C. Severely affected animals showed arched back (torticollis) and were disinclined to move. In acute cases, a dramatic increase in respiration rate was recorded. A nasal discharge, at first watery, then mucopurulent, and in some cases blood stained, appeared and eventually dried to form crusts. The nose was often occluded by a greyish-brown scab composed of desquamated epithelium along with excoriation of the lower lip. The nasal septum was congested and excoriations were also present on the muzzle. Oedema of the face, muzzle, nostrils, upper and lower lips and the tongue, as well as hyperaemia of upper and lower lips, were observed. Excoriation of the epithelium of the gums, inner side of lips and cheeks, bars of the hard palate and the dental pad occurred. A careful examination of the feet of affected sheep revealed inflammation and reddening of the coronary band (which was more frequent in hind feet) leading to lameness. There was no correlation between oral and foot lesions. In some BT-affected sheep, cracking of the skin was also observed. Blood collected from Rambouillet sheep that were pyrexic (41.6–42.2 °C) and exhibiting other typical BT clinical signs was inoculated intradermally and subcutaneously into cross-bred sheep (50% Corriedale, 25% Merino and 25% native Nali breed) but produced only mild BT, evidenced by hyperaemia of the upper and lower lips and pyrexia. Bluetongue virus was also isolated from the blood of the experimentally inoculated sheep (Jain *et al.*, 1986). Subsequently, an epidemiological study carried out in CSBF, Hisar, revealed mortality and case fatality rates of 13.3 and 31%, respectively, in 1985, which increased to 16.7 and 43%, respectively, in 1986. In 1987 and 1988, the mortality rate fell back to 7% (Mahajan *et al.*, 1991).

Clinical BT in India was initially confined to the exotic breeds of sheep, including Southdown, Rambouillet, Russian Merino and Corriedale. However, from 1981 onwards, BT also became established in native Indian breeds of sheep, resulting in severe outbreaks in the region (Table 8.1). This change in the incidence of disease suggests that an exotic strain(s) of the virus may have been introduced to the region at around this time. Initial reports in Karnataka state and in adjoining regions of Maharashtra and Andhra Pradesh recorded mortality rates ranging from 2 to 50%, involving native breeds of sheep. Morbidity rates of up to 80% were also recorded. From 1983 onwards, outbreaks have occurred every year during the monsoon season, particularly in southern India. Over a period of 12 years (1983–2005), the BT outbreaks in Andhra Pradesh have shown periodic and cyclical variations in both their incidence and case fatality rates (2–38%), with intervals of 2–3 years (Figure 8.2a–b).

A similar incidence of disease was also observed in other south Indian states, including Tamil Nadu and Karnataka (Sreenivasulu, unpublished data). The disease appears to have been more severe in small rural flocks

Table 8.1   Breeds of sheep reported to be susceptible to bluetongue virus (BTV) infection in India

Species	Breeds (State)	Clinical BT reported	Inapparent infection detected by serology[a]
Exotic sheep	Corriedale (Haryana, Himachal Pradesh, Rajasthan)	+	+
	Dorset (Rajasthan, Himachal Pradesh)	+	+
	Rambouillet (Haryana, Rajasthan, Himachal Pradesh, Maharashtra)	+	+
	Russian Merino (Rajasthan, Maharashtra. Himachal Pradesh, Jammu and Kashmir)	+	+
	Southdown (Maharashtra, Rajasthan, Himachal Pradesh)	+	+
	Suffolk (Rajasthan)	+	+
	Karakul (Rajasthan)	−	−
Cross-bred Sheep	Southdown × Bannur (Maharashtra, Rajasthan)	+	+
	Rambouillet × Nali (Rajasthan, Haryana)	?	+
	Rambouillet × Sunali (Haryana, Rajasthan)	?	+
	Deccani × Merino (Maharashtra, Andhra Pradesh)	+	+
	Chokla × Merino (Rajasthan, Maharashtra, Gujarat, Andhra Pradesh)	+	+
	Rambouillet × Pole Dorset (Rajasthan, Maharashtra, Gujarat)	+	+
	Hisardale × Nali (Haryana)	?	+
Native sheep	Nali (Rajasthan, Haryana and Gujarat)	−	+
	Sonali (Rajasthan, Haryana and Gujarat)	−	+
	Chokla (Maharashtra, Karnataka, Rajasthan and Andhra Pradesh)	−	+
	Deccani (Maharashtra, Andhra Pradesh, Karnataka)	+	+
	Malpura (Rajasthan)	−	+
	Ramanathapuram (Andhra Pradesh, Karnataka and Tamil Nadu)	?	+

(Continued)

Table 8.1 (*Continued*)

Species	Breeds (State)	Clinical BT reported	Inapparent infection detected by serology[a]
	Vellai (Andhra Pradesh, Karnataka and Tamil Nadu)	?	+
	Kilakarashal (Andhra Pradesh, Karnataka and Tamil Nadu)	?	+
	Mandia (Andhra Pradesh, Karnataka and Tamil Nadu)	?	+
	Mecheri (Andhra Pradesh, Karnataka and Tamil Nadu)	?	+
	Nellore Palla (Andhra Pradesh)	+	+

[a] BTV group-specific antibodies were detected by AGID.

(a)

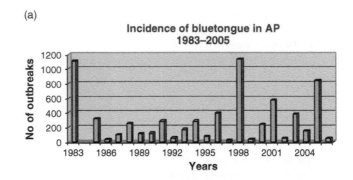

(b)

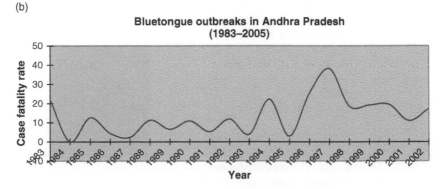

Figure 8.2 Bluetongue (BT) outbreaks (1983–2005) in Andhra Pradesh showing periodic and cyclical variations in both their (a) incidence and (b) case fatality rates, with an interval of 2–3 years.

than in organized farms. Investigations revealed morbidity, mortality and case fatality rates among rural or organized farms as 9, 3 and 29% or 6, 0.5 and 8%, respectively. The higher morbidity and mortality in rural areas may be due to stress factors induced by poor nutrition, high parasitic burden, reduced insecticide use leading to increased insect biting, fatigue due to long walks and the lack of veterinary aid (Sreenivasulu *et al.*, 2004).

Mehrotra *et al.* (1991) reported a widespread outbreak of BT in the native sheep of Tamil Nadu during November and mid-December 1989. The disease was present in two districts, where more than 5000 animals were affected. The clinical signs reported included pyrexia, depression, anorexia, oedema of face, cyanosis of the tongue, conjunctivitis and nasal discharge. The oral mucus membranes exhibited haemorrhagic oedema and small numbers of ulcers on the dental pad. Cyanosis of the tongue and facial oedema were pronounced in some animals. The owners disposed of the majority of affected animals due to the poor carcass quality.

Kulkarni *et al.* (1992) also reported severe outbreaks of BT in native sheep in rural areas of Maharashtra, with overall morbidity, mortality and case fatality rates of 32, 8 and 25%, respectively. All of the southern Indian breeds of sheep were found to be susceptible to disease and developed clinical BT, although Saravanabava (1992) reported variations in susceptibility between breeds. Trichy black and Ramnad white sheep were more susceptible to disease than the Vambur and Mecheri sheep in Tamil Nadu. Clinical disease has not been recorded in sheep in Kerala state although BTV antibodies have been found in their sera (Ravishankar, 2003).

Clinical BT is now reported annually in local breeds of sheep in Andhra Pradesh, Karnataka, Tamil Nadu and Maharashtra, although it has rarely been reported in native sheep breeds in the northern Indian states. The clinical picture of BT in native sheep in north India also appears to be slightly different from that in the south; the major difference being that swelling of lips and face is less conspicuous, though the mucocutaneous borders are very sensitive to touch and bleed easily on handling. The classical signs of cyanosis of the tongue and reddening of the coronary band are uncommon in both northern and southern Indian sheep breeds.

Two of the native breeds of sheep that are severely affected with BT (Nellore Palla and Deccani) are shown in Figure 8.3a–d (see colour plate 20). The presence of BTV was confirmed in these animals by virus isolation. It is uncertain whether the differences observed in clinical disease between native sheep breeds from the south and north of India are due to the strains of virus in each area, the genetic make-up of the sheep breeds, different vector species or a combination of these factors. Further investigation is required to resolve these questions.

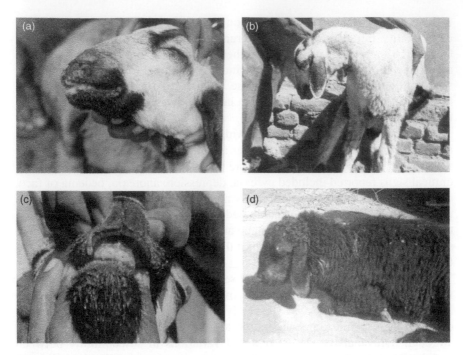

Figure 8.3 (a) Bluetongue (BT)-affected Nellore Palla sheep showing thick muco-purulent nasal discharge; (b) BT-affected Nellore Palla sheep showing muscle stiffness and torticollis; (c) BT-affected Nellore Palla sheep showing ulceration and erosions in the mucocutaneous borders and bleeding lips. (d) BT-affected Deccani sheep showing oedema of head and recumbency (See colour plate 20).

Serological prevalence of bluetongue in other animal species

Serological and clinical evidence concerning BTV infection of goats, cattle, buffalo and camels in the Indian subcontinent has also been reported.

Goats

Sapre (1964) reported the first occurrence of BT in goats in the subcontinent, in Maharashtra state. Subsequently, antibodies to BTV were detected in a number of other states, in several goat breeds and crosses (Table 8.2), although there is no information concerning the prevalence of different BTV serotypes. Sodhi *et al.* (1981) reported 1.4% incidence of BTV antibodies in goats in Punjab State, while Bandyopadhyay and Mullick (1983) reported a 3.0% prevalence of BTV antibodies in goats in Uttar Pradesh. A serological survey carried out between 1990 and 1992 at the Western Regional Station of CSWRI, Avikanagar, showed that 33.3% of

Table 8.2 Breeds of goats reported to be susceptible to bluetongue virus infection in India

Species	Breeds (State)	Clinical BT reported	Inapparent infection detected by serology[a]
Goat	Goat (breed not known) (Maharashtra)	+	+
	Gaddigoat (Himachal Pradesh, Jammu & Kashmir)	−	+
	Alpine goat (Himachal Pradesh, Jammu & Kashmir)	−	+
	Sirohi (Uttar Pradesh, Rajasthan)	−	+
	Nagauri (Rajasthan)	−	+
Cross-bred goat	Alpine × Sirohi (Rajasthan)	−	+
	Tokanburg × Sirohi (Rajasthan and Uttar Pradesh)	−	+

[a] BTV group-specific antibodies were detected by AGID.

goats in Rajasthan, and 35% of goats in the adjoining state of Haryana, were seropositive for BTV antibodies by agar gel immunodiffusion (AGID) tests (N.C. Jain and G. Prasad, unpublished data). Bluetongue virus antibodies have also been detected in goats from the southern Indian states of Andhra Pradesh, Karnataka and Kerala with prevalences of 43.6, 2.63 and 5.3%, respectively (Sreenivasulu *et al.*, 2004; Ravishankar *et al.*, 2005a; Doddamani and Hari Babu, 2006). Recently, Vengadabady *et al.* (2006) reported the occurrence of BT in goats in the Coimbatore district of Tamil Nadu, with 13% morbidity and 2.3% mortality; however, these data are based mainly on historical information and clinical diagnosis.

Cattle and buffalo

Serological evidence of BTV infection in local and exotic cattle breeds and buffalo has been reported from several states of India (Table 8.3), although there are no reports of clinical BT in these animals, and attempts to isolate virus have also been unsuccessful. In a study conducted by Tongaonkar *et al.* (1983) in Gujarat State, 13.4% of buffalo and 15.6% of cattle were seropositive for BTV antibodies. The positive sera revealed the presence of antibodies against serotypes 1, 15 and 17. Bandyopadhyay and Mullick (1983) reported 3.7% of cattle sera positive for BTV antibodies in Andhra Pradesh, Haryana, Rajasthan and Uttar Pradesh. Subsequently, Mehrotra and Shukla (1984) tested 154 cattle serum samples from seven states (Andhra Pradesh, Karnataka, Gujarat, Punjab, Orissa, Himachal Pradesh and West Bengal) and

Table 8.3 Breeds of cattle and buffalo reported to be susceptible to bluetongue virus (BTV) infection in India

Species	Breed (State)	Clinical BT reported	Inapparent infection detected by serology[a]
Native cattle	Haryana (Haryana, Punjab and Rajasthan)	−	+
	Sahiwal (Haryana, Punjab and Rajasthan)	−	+
	Rathi (Haryana, Punjab and Rajasthan)	−	+
	Red Sindhi (Haryana, Punjab and Rajasthan, Gujarat)	−	+
	Tharparker (Haryana, Punjab and Rajasthan, Gujarat)	−	+
	Local cattle (Andhra Pradesh, Karnataka, Tamil Nadu and Kerala)	−	+
Exotic cattle	Jersey (Haryana, Punjab and Himachal Pradesh, Uttar Pradesh and Uttarakhand)	−	+
	Holstein Friesian (Haryana, Punjab, Himachal Pradesh, Uttar Pradesh and Uttarakhand)	−	+
	Reddane (Haryana, Punjab, Himachal Pradesh, Uttar Pradesh and Uttarakhand)	−	+
Cross-bred cattle	Karanswiss × Friesian (Haryana)	−	+
	Rathi × Jersey (Haryana and Punjab)	−	+
	Rathi × Reddane (Haryana and Punjab, Rajasthan, Uttar Pradesh)	−	+
	Jersey × Red Sindhi (Haryana and Punjab, Rajasthan)	−	+
	Friesian × Sahiwal (Haryana and Punjab, Rajasthan)	−	+
	Reddane × Sahiwal (Haryana and Punjab, Rajasthan)	−	+
	Sahiwal × Jersey (Haryana and Punjab, Rajasthan)	−	+
	Tharparker × Jersey (Haryana and Punjab, Rajasthan)	−	+
	Tharparker × Friesian (Haryana and Punjab, Rajasthan)	−	+
	Haryana × Friesian (Haryana and Punjab, Rajasthan)	−	+
Native buffalo	Murrah (Haryana, Punjab and Gujarat)	−	+

[a] BTV group-specific antibodies were detected by AGID.

reported 18% samples positive for BTV antibodies. Sharma *et al.* (1981) conducted a serological survey in cattle and buffalo in Punjab State and reported that 6.8% of cattle sera were positive for BTV antibodies. Subsequently, Oberoi *et al.* (1988) demonstrated the presence of BTV-specific antibodies in 38.5% of buffalo and 70% of cattle sera in Punjab state.

In 1990, cattle and buffalo in Haryana State were surveyed for antibodies to BTV. Serum samples were collected from a total of 549 cattle and 498 buffalo, of which 4.2 and 10.6%, respectively, were positive (Jain *et al.*, 1992). Similar levels of seropositive animals were also detected in cattle and buffalo from southern India (Mehrotra and Shukla, 1984; Sreenivasulu *et al.*, 2004; Ravishankar *et al.*, 2005b). A more recent and more extensive serological survey carried out in northern India (Haryana, Himachal Pradesh, Punjab, Jammu & Kashmir and Rajasthan) detected BTV-specific antibodies in 23.4% of the cattle tested (Kakker *et al.*, 2002). These reports have established that BTV infection is widespread in both cattle and buffalo in India.

Camels

Clinical BT has not been observed in camels in India, and their role in the epidemiology of this disease is unknown. According to the 17th Livestock census conducted by the Department of Animal Husbandry, Dairying and Fisheries (DAHDF) of the Indian Government in 2003, there were 498 000 camels in Rajasthan, 53 000 in Gujarat and 50 000 in Haryana state. There are few reports concerning the prevalence of BTV antibodies in camel sera from these three states. However, Chandel and Kher (1999) reported that 9.3% of sera from Kutchhi and Marwari breeds of camels from Gujarat State were positive for BTV-specific antibodies using AGID tests. Malik *et al.* (2002) also conducted serosurveys (using the same test) of Bikaneri and Jaisalmeri breeds in Rajasthan, showing that 9.9% of the animals tested were seropositive. In a subsequent survey of camels in Gujarat State, Chandel *et al.* (2003) reported 12.5 and 19.3% seroprevalence by the AGID test and cELISA, respectively. Although the BTV serotypes infecting camels have not been widely surveyed, in Gujarat state, Chauhan *et al.* (2004) reported the presence of type-specific antibodies against BTV serotypes 1, 2, 3, 4, 10, 12, 14, 15, 16, 17, 18, 20, 21 and 24.

Wild ruminants

Although systematic studies have not yet been conducted to assess the status of BTV infection in Indian wildlife, the subcontinent does have a large and very diverse wild ruminant population. A limited serological survey conducted in a tiger reserve in Rajasthan state indicated the presence of BTV antibodies in serum samples from two sambar deer (*Cervus unicolor*). Clinical BT has not been reported in wildlife from any part of the subcontinent. However, there is a possibility of two-way transmission of BTV between wildlife species and the large domestic ruminant populations in and around the wildlife reserves in India (Prasad *et al.*, 1998a).

Other species

There have been reports of BTV and African horse sickness virus (AHSV) infections in carnivores in Africa and elsewhere due to ingestion of infected meat or contaminated vaccines (Akita *et al.*, 1994; Alexander *et al.*, 1994, 1995). However, there is no information on the incidence of BTV infections or seroprevalence in carnivores within India. Should it occurs, BTV infection could be a concern in the maintenance of susceptible and 'endangered' wildlife species (both ruminants and carnivores).

Vectors

Several different species of *Culicoides* can act as 'competent vectors' of BTV around the world (see Chapter 14), although the role of Indian *Culicoides* species in the biological transmission of the virus has not yet been fully elucidated. Dasgupta and collaborators have studied the taxonomy, biology and ecology of *Culicoides* species in India, identifying and describing several new midge species but did not cover the entire subcontinent and did not determine their role in pathogen transmission (Sen and Dasgupta, 1958, 1959; Dasgupta, 1961, 1995) (Table 8.4). However, seven *Culicoides* species that were detected are known (or suspected) as BTV vectors elsewhere in the world, including *C. actoni, C. brevitarsis, C. fulvus, C. imicola, C. orientalis, C. oxystoma* and *C. peregrinus* (Mellor *et al.*, 2000; Chapter 14).

Table 8.4 Species of *Culicoides* reported from India

S. No.	Species	Indian record
1	*Culicoides actoni* (Smith)[a]	West Bengal and Bihar
2	*Culicoides anophelis* (Edwards)	West Bengal, Assam, Bihar
3	*Culicoides arakawae* (Arakawa)	West Bengal
4	*Culicoides autumnalis* (Sen and Dasgupta)	West Bengal
5	*Culicoides brevitarsis* (Kieffer)[a]	West Bengal
6	*Culicoides candidus* (Sen and Dasgupta)	West Bengal
7	*Culicoides certus* (Dasgupta)	West Bengal
8	*Culicoides circumscriptus* (Kieffer)	West Bengal, Bihar and Orissa
9	*Culicoides clavipalpis* (Mukerji)	West Bengal, Bihar
10	*Culicoides definitus* (Sen and Dasgupta)	West Bengal
11	*Culicoides distinctus* (Sen and Dasgupta)	West Bengal
12	*Culicoides drydaeus* (Wirth and Hubert)	West Bengal

Table 8.4 (*Continued*)

S. No.	Species	Indian record
13	*Culicoides dumdumi* (Sen and Dasgupta)	West Bengal
14	*Culicoides flaviscutatus* (Wirth and Hubert)	Maharashtra, West Bengal and Assam
15	*Culicoides fortis* (Sen and Dasgupta)	West Bengal
16	*Culicoides fulvus* (Sen and Dasgupta)[a]	West Bengal and Tamil Nadu
17	*Culicoides hegneri* (Causey)	West Bengal
18	*Culicoides homotomus* (Kieffer)	West Bengal
19	*Culicoides huffi* (Causey)	West Bengal and Tamil Nadu
20	*Culicoides imicola* (Kieffer)[a]	West Bengal
21	*Culicoides indianus* (Macfie)	Karnataka
22	*Culicoides inesploratus* (Sen and Dasgupta)	West Bengal
23	*Culicoides innoxius* (Sen and Dasgupta)	West Bengal and Bihar
24	*Culicoides inornatithorax* (Dasgupta)	West Bengal
25	*Culicoides insolens* (Chaudhuri and Dasgupta)	West Bengal
26	*Culicoides kamrupi* (Sen and Dasgupta)	Assam
27	*Culicoides macfiei* (Causey)	West Bengal
28	*Culicoides majorinus* (Chu)	West Bengal
29	*Culicoides orientalis* (Macfie)[a]	West Bengal
30	*Culicoides oxystoma* (Kieffer)[a]	Assam, West Bengal, Bihar, Rajasthan, Tamil Nadu, Punjab, Haryana, Himachal Pradesh
31	*Culicoides palpifer* (Dasgupta and Ghosh)	West Bengal and Bihar
32	*Culicoides paraliui* (Dasgupta)	West Bengal
33	*Culicoides. parararipalpis* (Dasgupta)	West Bengal
34	*Culicoides peliliouensis* (Tokunaga)	West Bengal
35	*Culicoides peregrinus* (Kieffer)[a]	Assam, Bihar, West Bengal, Andhra Pradesh, Orissa
36	*Culicoides raripalpis* (Smith)	West Bengal
37	*Culicoides rariradialis* (Dasgupta)	West Bengal
38	*Culicoides rarus* (Dasgupta)	West Bengal
39	*Culicoides shortti*	Assam and West Bengal
40	*Culicoides sikkimmensis* (Dasgupta)	West Bengal

[a] Species reported as being BTV vectors in other countries.

A survey of *C. oxystoma* (a midge species that is most active between 19.00 and 22.00 h) was carried out at Presidency College, Kolkata (West Bengal), by Sen and Dasgupta (1958) and showed a major peak in abundance during August, with a minor peak in March. The rapid increase in the number of adult *C. oxystoma* during the monsoon season (July to September) coincides with the spread of BTV in northern India. Thirty-six percent of the wild-caught adult female *C. oxystoma* also contained blood meals that reacted with anti-cattle serum (Sen and Dasgupta, 1958). The muddy substrate collected from margins of stock ponds around Kolkata (Calcutta) were found to harbour immature *C. oxystoma*, indicating that such areas are natural breeding sites for this species. These authors also reported that swarms of *C. oxystoma* were often seen early in the morning close to cattle sheds. Each of the gravid females collected from these swarms contained 115–120 mature eggs. The life cycle of *C. oxystoma* is completed in 28–32 days under favourable environmental conditions. *Culicoides oxystoma* has been reported to feed preferentially on the lower, ventral areas of cattle (Sen and Dasgupta, 1958). *Culicoides imicola*, *C. actoni*, *C. fulvus* and *C. brevitarsis* have also been identified in India (Dasgupta, 1995), although there is no information on their biology in the region.

The involvement of the *Culicoides* species in the transmission of BTV in India was indicated by virus isolation from adult insects in Haryana (Jain *et al.*, 1988), although the species was not identified. However, the presence of *C. schultzei*, *C. peregrinus* and *C. actoni* (known BTV vectors elsewhere in the world) was subsequently reported in the Marathwada region of Maharashtra state (Narladakar *et al.*, 1993). *Culicoides oxystoma* was also identified in the animal farms where BTV seroconversions had occurred (Bhatnagar *et al.*, 1997). On this basis, it was suggested that *C. oxystoma* could be a vector for BTV in India. Moreover, *C. oxystoma* was repeatedly trapped near sentinel herds/flocks in Punjab, Haryana, Himachal Pradesh and Rajasthan, indicating a close association with sheep and cattle. There were also continuing seroconversions in these herds, further supporting the hypothesis that *C. oxystoma* is a BTV vector in India (Kakker *et al.*, 1996; Bhatnagar *et al.*, 1997). However, final confirmation would require virus isolation from *C. oxystoma* and experimental transmission of virus to susceptible animals. The Indian subcontinent has several diverse ecological zones, and it is therefore considered likely that different *Culicoides* species are involved in BTV transmission in different areas (see Table 8.4).

Bluetongue virus serotypes reported from Indian subcontinent

To date, 24 serotypes of BTV have been recorded worldwide. Twenty-one serotypes have been reported from India, ten of these on the basis of virus isolation and eleven on the basis of neutralizing antibodies (Table 8.5) (Prasad, 2000: see www.iah.bbsrc.ac.uk/dsRNA_virus_proteins/btv-serotype-distribution.htm). The statewise prevalence of different serotypes of BTV in India is presented in Table 8.6. The largest number of BTV serotypes has been reported from Tamil Nadu, followed by Maharashtra and Gujarat.

Table 8.5   Bluetongue virus (BTV) serotypes reported in India

Basis	Serotypes	Total
Virus isolation	1, 2, 3, 4, 8, 9, 16, 17, 18, 23	10
Neutralizing antibodies (additional)	5, 6, 7, 10, 11, 12, 13, 14, 15, 19, 20	11
Total		21

Table 8.6   Statewise prevalence of different serotypes of bluetongue virus (BTV) in India

State	BTV serotypes on the basis of neutralizing antibodies	BTV serotypes on the basis of virus isolation	Total
Jammu and Kashmir	–	23	1
Himachal Pradesh	4	3, 4, 9, 16, 17	5
Haryana	2, 8, 12, 16	1, 4	6
Rajasthan	–	1	1
Uttar Pradesh	23	9, 18, 23	3
Uttarakhand		23	1
MP	18, 23	23	2
Gujarat	1, 2, 3, 5, 8, 9, 10, 11, 12, 13, 15, 16, 17, 20	6	15
Maharashtra	2, 3, 4, 5, 6, 7, 8, 10, 12, 13, 16	1, 9, 16, 18, 4, 17	15
Andhra Pradesh	4, 6, 12, 13, 14, 17, 18, 19	2, 9	10
Karnataka	1, 2, 4, 12, 16, 17, 20	18, 23	9
Tamil Nadu	1, 3, 4, 5, 6, 7, 11, 12, 14, 15, 16 17, 19, 20	1, 2, 3, 16, 23	16

Although the first BT outbreak was reported as early as 1964, the virus was not isolated (Sapre, 1964). Sera and clinical specimens collected from BT-affected animals during 1970s, 1980s and 1990s were subsequently sent to the OIE BTV Reference Laboratory, Onderstepoort, South Africa (Sapre, 1964). Therefore, most of the reports concerning BTV serotypes in the Indian subcontinent are based on virus isolation from such clinical specimens or on the detection of virus-neutralizing antibodies in the sera of recovered animals. However, antibodies that will neutralize a wide range of BTV types can be generated after sequential infection with different BTV strains (Jeggo et al., 1983a, b, 1984, 1986). This can make the serological identification of BTV serotype very difficult, particularly in endemic areas such as India where multiple serotypes are circulating.

Clinical material and sera from recovered sheep from Tathawada Exotic Sheep Breed Farm, Poona, Maharashtra, were sent to the BTV Reference

Laboratory, Onderstepoort, and yielded isolates of BTV serotypes 1 and 16 in BHK-21 cells, while serotypes 2, 7, 9 and 10 were identified on the basis of serum-neutralizing antibodies only (unpublished findings). Uppal and Vasudevan (1980) reported the presence of BTV serotypes 3, 4, 9, 16 and 17 in Russian Merino sheep affected in two outbreaks at Kothipura farm in Himachal Pradesh, while serotypes 1 and 4 were incriminated in an outbreak in Australian Corriedale Sheep at CSBF, Hisar (Uppal and Vasudevan, 1980). Serum samples subsequently sent from Maharashtra state to the Onderstepoort Reference Laboratory, by Sriguppi (1982) and Choudhary (1982), indicated the presence of antibodies to BTV serotypes 1, 2, 3, 4, 7, 10, 16 and 17.

Kulkarni and Kulkarni (1984) reported the isolation of BTV serotypes 8 and 18 in developing chicken embryos from BT outbreaks in Maharashtra during 1981 and 1983. In 1985, BTV serotype 1 was isolated from Rambouillet sheep affected with BTV, using BHK-21 cells, at CSBF, Hisar (Jain *et al.*, 1986). Subsequently, BTV-1 was also isolated from sheep in CSWRI, Avikanagar (Prasad *et al.*, 1994), and Mehrotra *et al.* (1989) isolated BTV-18 from Maharashtra. Serological surveys conducted in Tamil Nadu showed neutralizing antibodies against BTV serotypes 1, 4, 5, 6, 7, 11, 12, 13, 14, 15, 16, 17, 19 and 20 (Janakiraman *et al.*, 1991; Nachimuthu *et al.*, 1992). Mehrotra *et al.* (1995) reported concurrent outbreaks of BT and PPR in Dehradun, Uttarakhand, and isolated BTV-23. Deshmukh and Gujar (1999) isolated BTV-1 from Maharashtra, while Sreenivasulu *et al.* (1999) recently isolated BTV-2 from an outbreak in native sheep of Andhra Pradesh. Bluetongue virus isolations were also made from outbreaks occurring in Andhra Pradesh, two of these being identified as BTV-9 (Prasad unpublished data, 2005).

There are very few reports on the BTV serotypes present in Pakistan. However, Akhtar *et al.* (1995) identified BTV serotypes 3, 9, 15, 16 and 18 on the basis of virus-neutralizing antibodies in the sera of recovered animals in the North West Province.

Diagnosis

The diagnosis of BT and identification of BTV is carried out in India using both serological- and nucleic acid-based diagnostic tests. These studies are funded by ICAR, under the All India Network Project (AINP), involving 11 regional centres located in different states that represent the different geographical regions of the country (Table 8.7).

Virus isolation

Embyonated chicken eggs (ECE), BHK-21 cells and mosquito cells (C6/36) have been used for the isolation of BTV in India (Kulkarni and Kulkarni, 1984; Jain *et al.*, 1986; Mehrotra *et al.*, 1995, 1996; Bhat *et al.*, 1996; Sreenivasulu

Table 8.7 List of research centres involved in the All India National Network Projects (AINP) on bluetongue (BT)

S. No.	Region	State	Address
1	Northern	Uttar Pradesh	Centre for Animal Disease Research and Diagnosis, Indian Veterinary Research Institute, Izatnagar, Barriely, UP
2	Northern	Uttarakhand	Indian Veterinary Research Institute, Mukteshwar, Uttarakhand,
3	Northern	Haryana	Department of Animal Biotechnology, CCS Haryana Agricultural University, Hisar, Haryana
4	Northern	Rajasthan	Division of Animal Health, Central Sheep and Wool Research Institute, Malpura, Tonk, Rajasthan
5	Central India	Madhya Pradesh	Department of Veterinary Microbiology, College of Veterinary Sciences and Animal Husbandry, Jabalpur, MP
6.	Eastern India	West Bengal	Department of Preventive Medicine, College of Veterinary Sciences, West Bengal University of Veterinary and Fishery Sciences, Kolkata, West Bengal
7	Western India	Gujarat	Department of Veterinary Microbiology, College of Veterinary Sciences, Gujarat; Agriculture University, Dantiwara, Gujarat
8	Western India	Maharashtra	Department of Veterinary Microbiology, College of Veterinary Sciences, Maharashtra; Veterinary and Animal Sciences University, Parbhani, Maharashtra, India
9	Southern India	Andhra Pradesh	Department of Veterinary Microbiology, College of Veterinary Sciences, Andhra Pradesh; Veterinary and Animal Science University, Rajendra Nagar, Hyderabad, Andhra Pradesh
10	Southern India	Tamil Nadu	Department of Veterinary Microbiology, Madras Veterinary College, Tamil Nadu University, Veterinary and Animal Science, Chennai, Tamil Nadu
11	Southern India	Karnataka	Institute of Animal Health and Veterinary Biologicals, Hebbal, Bangalore, Karnataka

et al., 1999). Bluetongue virus has also been isolated from *Culicoides* species using larvae of the mosquito *Toxorhynchites splendens*, which were inoculated with suspensions of pools of the *Culicoides*, and then after 7 days of incubation, the virus was detected by immunoperoxidase and fluorescent antibody tests (FAT) (Rahman and Manickam, 1997). The authors described this as the 'Toxorhynchites fluorescent antibody test' (TFA).

Immunological detection of virus

AGID tests, immunofluorescence, immunoperoxidase, enzyme immunoassays (ELISAs) and dot immunobinding assays (DIA) have been widely used in India for the detection of BTV-specific antibodies and antigens. The samples tested include infected chicken embryos, cell cultures and mononuclear cells (Gupta *et al.*, 1990; Chander *et al.*, 1991; Garg and Prasad, 1994, 1995). Bluetongue virus-specific monoclonal antibodies and various recombinant antigens (such as yeast or baculovirus-expressed VP7) were used in the development of BTV serogroup-specific blocking ELISAs, competitive ELISAs and DIA (Naresh and Prasad, 1995; Naresh *et al.*, 1996).

Nucleic acid-based diagnosis

RNA polyacrylamide gel electrophoresis (PAGE) has been used as a diagnostic tool for the identification BTV in India (Prasad and Minakshi, 1999). RNA-PAGE has also been used to identify different genotypes of the same serotype, as well as to indicate different serotypes of BTV (Prasad *et al.*, 1998b). Group-specific non-radio-labelled probes, based on the NS1 and VP3 genes, have been developed in India for detection of BTV in clinical specimens or infected cell cultures (Kumar, 2000; Malik *et al.*, 2001).

Serogroup-specific RT-PCR, sequencing, restriction enzyme profile analysis (REPA) and phylogenetic analyses (targeting conserved genome segments) are now available in an increasing number of laboratories throughout India for the identification of BTV (Bandyopadhyay *et al.*, 1998; Prasad *et al.*, 1999; Tiwari *et al.*, 2000a, b; Malik *et al.*, 2001; Dahiya *et al.*, 2004, 2005; Kovi *et al.*, 2006). Serotype-specific RT-PCR assays (targeting genome segments 2 or 6) have also been used to identify different BTV serotypes from India (Maan, 2004; Singh *et al.*, 2004; Maan *et al.*, 2004a, b, c; 2007a, b; Mertens *et al.*, 2007).

Molecular epidemiology studies

Sequence analyses of Seg-7 indicate that Indian BTV isolates have comparatively low sequence homology with American, Chinese or French isolates of the virus but have higher homology with at least some Australian isolates

(Kovi *et al.*, 2006). It has been concluded that Seg-7 of BTV-18 and -23 strains from India forms a single monophyletic group with Australian isolates, while the Indian isolates of BTV-1 form a separate group (Kovi *et al.*, 2006; Punia, 2005). These data agree with the separation of Seg-7 nt sequences into several distinct eastern and western clades (Bonneau *et. al.*, 2000; Maan *et. al.*, 2008).

A study of partial Seg-2 sequences (1240–1844 bp region) showed 79–89% identity between Indian and Australian isolates of BTV-1, but only 73–75% identity with South African BTV-1 isolates. Restriction enzyme profile analysis of VP2 gene sequences also indicated a closer relationship between Indian and Australian strains of BTV-1 (Dahiya *et al.*, 2004). These data agree with wider studies of full-length Seg-2 sequences that have shown a relatively close relationship between Seg-2 of Indian and Australian isolates of BTV-1, -9 and -16, all of which belong to 'eastern' phylogenetic group (Maan, 2004; Maan *et al.*, 2004a, b, c, 2007a, 2008; see Chapter 7 – Molecular Epidemiology – Figure 7.6, p. 143) (www.iah.bbsrc.ac.uk/dsRNA_virus_proteins/orbivirus-phylogenetic-trees.htm).

Phylogenetic comparisons of the nucleotide sequence of Seg-2 from two BTV-2 isolates from India (IAH reference strain number IND1982/01[accession number AJ585152] and isolate 'Mahboobnagar (AP)' [accession number DQ462580]) show that they both belong to an eastern topotype. However, later Indian strains of BTV-2 isolated after the importation of sheep from the United States clearly belong to a 'western' topotype (IAH reference strain number IND2003/01 and IND2003/02), with much higher similarities to viruses from North America, Europe and Africa (99.5–99.9% nt identity). Similarly, Seg-2 of Indian strains of BTV-10 isolated in 2004 and 2005 (IND2004/01 and IND2005/04 and IND2005/05) show a high level of similarity (89.5–99.9%) to BTV-10 from the United States. Further details of Indian virus isolates held in the reference collection at IAH Pirbright are available at www.iah.bbsrc.ac.uk/dsRNA_virus_proteins/ReoID/virus-nos-by-country.htm#India.

These data suggest that BTV strains of both serotype 2 and serotype 10 could have been imported to India from the United States. Although it is not clear how this happened, it is possible that some of the animals imported to India from the United States, as part of the breeding programme, were infected with these viruses. The possibility that exotic strains of BTV were imported to India may also help to explain the increased incidence and severity of the disease that has been observed in local breeds of Indian sheep.

Economic impact of bluetongue and control

Recurring outbreaks of clinical BT in sheep and production losses in other domestic ruminants have caused great economic loss in southern India. Native breeds of sheep that are particularly susceptible to clinical BT are reared mainly on a small scale, and therefore the disease is one of the major causes of economic loss to less-affluent farmers. These losses can be attributed to mortality in sheep, overall loss of productivity, weight loss and wool break. The larger and more organized farming sector also experiences losses as the result of a ban on the export of live animals, their germplasms and other animal products. It is difficult to assess the effect of BTV infection on the overall productivity of cattle and buffalo, as they are largely asymptomatic, but previous studies in the United States have suggested that infection can result in a significant overall drop in milk yield and productivity (Tabachnick *et al.*, 1996).

Once established in a particular area, it is very difficult to eradicate BTV. Interruption of the BTV transmission cycle is considered to be the most effective method of prevention and control of the disease. Most of the vector control approaches used in India involve systematic and repeated application of insecticide and are expensive to implement and so may be of use only in large, well-organized farms. However, most of the ruminant animals in India are kept by the less well-organized farming sector, owned by people who cannot afford these methods of vector control. This may be another contributing factor to the higher incidence of disease and losses in this sector.

Several approaches have been used to control *Culicoides* vectors in India, including the use of chemical insecticide sprays in livestock housing and 'pour-on' insecticides applied to animals during the vector-abundant, monsoon season. The treatment of sheep in insecticide dips is an effective method for the control of ectoparasites and has been used against *Culicoides* on several of the more 'organized' sheep farms. Farm managers have informally reported a significant decline in the incidence of BT after the use of Ektomin and Butox (as dips) during the rainy season. Similarly, parenteral administration of certain insecticides (e.g. Ivermectin) has been reported to reduce the abundance of ectoparasites and the incidence of BT. In the authors' experience, administration of Ivermectin to sheep greatly reduced the prevalence of BTV in an 'organized' sheep-breeding farm near Hisar, Haryana.

Although synthetic chemical insecticides are very effective in destroying midges and their larvae, these also pose a significant risk to human and animal health and to the environment. Consequently, efforts are being made to develop alternative pest management strategies that are non-polluting and environmentally benign. One such strategy that has attracted much interest involves the use of naturally occurring bioinsecticides such as products from the Neem tree (*Azadirachta indica*) (Attri and Ravi Prasad, 1980; Blackwell *et al.*, 2004).

The existence of 24 BTV serotypes, which largely fail to cross-protect (most of which are circulating in India), has made the goal of protective immunization against the disease particularly difficult to achieve in the region. Live attenuated or inactivated vaccines, based on local Indian strains of the virus, are not available, and there is no current BTV vaccination programme within India. However, recent developments of inactivated or sub-unit vaccines (including research programmes within India) may in the future help to control the disease in the subcontinent (see Chapter 18).

Acknowledgments

The authors are very grateful to Prof. Mathew Baylis for his help with the Indian map.

Web sites

BTV type-specific primers: www.iah.bbsrc.ac.uk/dsRNA_virus_proteins/ReoID/rt-pcr-primers. htm

BTV reference collection: www.iah.bbsrc.ac.uk/dsRNA_virus_proteins/ReoID/BTV-isolates.htm

Indian BTV isolates: www.iah.bbsrc.ac.uk/dsRNA_virus_proteins/ReoID/virus-nos-by-country.htm#India

BTV serotype-distribution: www.iah.bbsrc.ac.uk/dsRNA_virus_proteins/btv-serotype-distribution.htm

Data on BTV RNAs and proteins: www.iah.bbsrc.ac.uk/dsRNA_virus_proteins/BTV.htm

BTV Seg-2 phylogenetic trees: www.iah.bbsrc.ac.uk/dsRNA_virus_proteins/orbivirus-phylogenetic-trees.htm

References

Akhtar, S., Djallem, N., Shad, G. and Thieme, O. (1997) Bluetongue virus seropositivity in sheep flocks in North West Frontier Province, Pakistan. *Prev. Vet. Med.* **29**, 293–8.

Akhtar, S., Howe, R.R., JadoonNaqvi, J.K. and Naqvi, M.A. (1995) Prevalence of five serotypes of bluetongue virus in a Rambouillet sheep flock in Pakistan. *Vet. Rec.* **136**, 495.

Akita, G.Y., Ianconescu, M., MacLachlan, N.J. and Osburn, B.I. (1994) Bluetongue disease in dogs associated with contaminated vaccine. *Vet. Rec.* **134**, 283–4.

Alexander, K.A., Kat, P.W., House, J., House, C., O'Brien, S.J., Laurenson, M.K., McNutt, J.W. and Osburn, B.I. (1995) African horse sickness and African carnivores. *Vet. Microbiol.* **47**, 133–40.

Alexander, K.A., MacLachlan, N.J., Kat, P.W., House, C., O'Brien, S.J., Lerche, N.W., Sawyer, M., Frank, L.G., Holekamp, K. and Smale, L. (1994) Evidence of natural blue-tongue virus infection among African carnivores. *Am. J. Trop. Med. Hyg.* **51**, 568–76.

Anon (2003) Office international des epizooties (OIE), Animal Health Data., http://www.oie.int/eng/info/en_infold.htm.

Attri, B.S. and Ravi Prasad, G. (1980) Neem oil extractive – an effective larvicide. *Indian J. Entomol.* **42**, 371.

Bandyopadhyay, S.K. and Mullick, B.B. (1983) Serological prevalence of bluetongue antibodies in India. *Indian J. Anim. Sci.* **53**, 1355–6.

Bandyopadhyay, S.K., Kataria, R.S. and Tiwari, A.K. (1998) Detection of bluetongue virus genome segment 6 sequences by RT-PCR. *Indian J. Exp. Biol.* **36**, 1034–7.

Bhat, R.M., Srivastava, R.N. and Prasad, G. (1996) Adaptation and growth pattern studies of bluetongue virus serotype 1 in and insect cell line. *Indian J. Anim. Sci.* **66**, 113–19.

Bhatnagar, P., Prasad, G., Kakker, N.K., Dasgupta, S.K., Rajpurohit, B.S. and Srivastava, R.N. (1997) A potential vector of bluetongue virus in north-western India. *Indian J. Anim. Sci.* **67**, 486–8.

Blackwell, A., Evans, K.A., Strang, R.H. and Cole, C.M. (2004) Toward development of neem-based repellents against the Scottish highland biting midge, *Culicoides impunctatus. Med. Vet. Entomol.* **18**, 449–52.

Bonneau, K.R., Zhang, N.Z., Wilson, C., Zhu, J.B., Zhang, F.Q., Li, Z.H., Zhang, K.L., Xiao, L., Xiang, W.B. and MacLachlan, N.J. (2000) Phylogenetic analysis of the S7 gene does not segregate Chinese strains of bluetongue virus into a single topotype. *Arch. Virol.* **145**, 1163–71.

Chandel, B.S. and Kher, H. (1999) Seroprevalence of bluetongue in dromedary camels in Gujarat, India. India. *J. Camel Pract. Res.* **6**, 83–5.

Chandel, B.S., Chauhan, H.C. and Kher, H.N. (2003) Comparison of the standard AGID test and competitive ELISA for detecting bluetongue virus antibodies in camels in Gujarat, India. *Trop. Anim. Health. Prod.* **35**, 99–104.

Chander, S., Prasad, G. and Jain, N.C. (1991) Dot immunobinding assay for detection of bluetongue virus antibodies in sheep experimentally inoculated with bluetongue virus type. *Vet. Microbiol.* **1**, 289–94.

Channakeshava, S.U. (2006) Molecular pathology and immune response to bluetongue virus in experimentally infected sheep. M.V.Sc thesis submitted to Deemed University, Indian Veterinary Research Institute, Izatnagar, U.P. India.

Chauhan, H.C., Chandel, B.S., Gerdes, T., Vasava, K.A., Patel, A.R., Kher, H.N., Singh, V. and Dongre, R.A. (2004) Seroepidemiology of bluetongue in dromedary camels in Gujarat, India. *J. Camel Pract. Res.* **11**, 141–5.

Choudhary, P.G. (1982) Incidence of bluetongue in State of Maharashtra. In: M.L. Mehrotra (ed.), *Proceedings of National Workshop on Bluetongue*. Izatnagar: IVRI, pp. 153–8.

Dahiya, S., Prasad, G., Minakshi and Kovi, R.C. (2004) VP2 gene based phylogenetic relationship of Indian isolates of bluetongue virus serotype 1 and other serotypes from different parts of the world. *DNA Seq.* **15**, 351–61.

Dahiya, S., Prasad, G. and Minakshi Kovi, R.C. (2005) Restriction analyses of conserved and variable regions of VP2 gene of Indian isolates of bluetongue virus serotype 1. *Indian J. Exp. Biol.* **43**, 272–6.

Dasgupta, S.K. (1961) Breeding habitats of indian *Culicoides. Curr. Sci.* **31**, 465–6.

Dasgupta, S.K. (1995). Morphotaxonomic features and species of Indian *Culicoides* (Diptera: Ceratopogonidae). In: G. Prasad and R.N. Srivastava (eds.), *Bluetongue: Indian Perspectives*. Hisar: HAU Press, pp. 115–88.

Deshmukh, V.V. and Gujar, M.B. (1999) Isolation and adaptation of bluetongue virus in cell culture system. *Indian J. Com. Microbiol. Immunol. Infect. Dis.* **20**, 14–16.

Doddamani, R.S. and Hari Babu, Y. (2006) Study of prevalence of bluetongue in sheep and goats in north Karnataka. *Tamil Nadu J. Vet. Anim. Sci.* **2**, 229–33.

Ganesh Udupa, K. (2001) Culicoides spp. (Diptera: Ceratopogonidae) associated with livestock and their relevance to bluetongue infection in Tamil Nadu. Ph.D. thesis submitted to Tamil Nadu Veterinary and Animal Sciences University. Chennai, India.

Garg, A.K. and Prasad, G. (1994) Susceptibility of cattle lymphocytes and monocytes to bluetongue virus. *Indian J. Virol.* **10**, 97–103.

Garg, A.K. and Prasad, G. (1995) Replication of bluetongue virus in mononuclear cells of sheep. *Indian J. Anim. Sci.* **65**, 1–7.

Gupta, Y., Chand, P., Singh, A. and Jain, N.C. (1990) Dot immunobinding assay in comparison with enzyme linked immunosorbent assay. *Vet. Microbiol.* **22**, 365–71.

Harbola, P.C., Choudhary, P.G., Krishna, L., Sirruguppi, B.S. and Kole, R.S. (1982) Incidence of bluetongue in sheep in Maharashtra. *Indian J. Comp. Microbiol. Immunol. Infect. Dis.* **3**, 121–7.

Janakiraman, D., Venugopalan, A.T., Ramaswamy, V. and Venkatesan, R.A. (1991) Serodiagnostic evidence of prevalence of bluetongue virus serotypes among sheep and goats in Tamil Nadu. *Indian J. Anim. Sci.* **61**, 497–8.

Jain, N.C., Gupta, Y. and Prasad, G. (1992) Bluetongue virus antibodies in buffaloes and cattle in Haryana state of India. In: T.E., Walton and B.I., Osburn (eds.), *African Horse Sickness and Other Related Orbiviruses*. Boca Raton: CRC Press, pp. 188–92.

Jain, N.C., Prasad, G., Gupta, Y. and Mahajan, N.K. (1988) Isolation of bluetongue virus from *Culicoides* in India. *Rev. Sci. Tech. Off. Int. Epizoot.* **7**, 375–8.

Jain, N.C., Sharma, R. and Prasad, G. (1986) Isolation of bluetongue virus from sheep in India. *Vet. Rec.* **119**, 17–18.

Jeggo, M.H., Gumm, I.D. and Taylor, W.P. (1983b) Clinical and serological response of sheep to serial challenge with different bluetongue virus types. *Res. Vet. Sci.* **34**, 205–11.

Jeggo, M.H., Wardley, R.C., Brownlie, J. and Corteyn, A.H. (1986) Serial inoculation of sheep with two bluetongue virus types. *Res. Vet. Sci.* **40**, 386–92.

Jeggo, M.H., Wardley, R.C. and Taylor, W.P. (1983a) Host response to bluetongue virus. *Double Stranded RNA Viruses*, 353–9. eds.

Jeggo, M.H., Wardley, R.C. and Taylor, W.P. (1984) Role of neutralizing antibody in passive immunity to bluetongue infection. *Res.Vet. Sci.* **36**, 81–5.

Kakker, N.K., Prasad, G., Bhatnagar, P. and Srivastava, R.N. (2002) Sero-prevalence of bluetongue virus infection in cattle in Haryana, Himachal Pradesh, Punjab and Rajasthan. *Indian J. Comp. Microbiol. Infect. Dis.* **23**, 147–51.

Kakker, N.K., Prasad, G., Srivastava, R.N. and Bhatnagar, P. (1996) Sentinel herd as tool for studying bluetongue virus infection four northern states of India. *Indian J. Anim. Sci.* **66**, 1254–7.

Kovi, R.C., Dahiya, S., Minakshi and Prasad, G. (2006) Nucleotide sequence analysis of VP7 gene of indian isolates of bluetongue virus vis-à-vis other serotypes from other parts of the world. *DNA Seq.* **17**, 187–98.

Kulkarni, D.D. and Kulkarni, M.N. (1984) Isolation of bluetongue virus from sheep. *Indian J. Comp. Microbiol. Imunol. Infect. Dis.* **5**, 125.

Kulkarni, D.D., Bannilikar, A.S., Karpe, A.S., Gujar, M.B. and Kulkarni, M.N. (1992) Epidemiological observations on bluetongue in sheep in Marathwada region of Maharashtra state of India. In: T.E. Walton and B.I. Osburn (eds.), *Bluetongue, African Horse Sickness and Related Orbiviruses*. Boca Raton, Florida, USA: CRC Press, pp. 193–6.

Kumar, S.V. (2000) VP3 gene based detection of bluetongue virus using polymerase chain reaction and DNA probe. Ph.D. thesis submitted to CCS Haryana Agricultural University, Hisar, India.

Lonkar, P.S., Uppal, P.K., Belwal, L.M. and Mathur, P.B. (1983) Bluetongue in sheep in India. *Trop. Anim. Hlth. Prod.* **15**, 86–7.

Maan, S. (2004) Complete nucleotide sequence analyses of genome segment 2 from the twenty-four serotypes of bluetongue virus: Development of nucleic acid based typing methods and molecular epidemiology. PhD thesis, University of London.

Maan, S., Maan, N.S., Ross-smith, N., Batten, C.A., Shaw, A.E., Anthony, S.J., Samuel, A.R., Darpel, K.E., Veronesi, E., Oura, C.A.L., Singh, K.P., Nomikou, K., Potgieter,

A.C., Attoui, H., van Rooij, E., van Rijn, P., Clercq, K.D., Vandenbussche, F., Zientara, S., Bréard, E., Sailleau, C., Beer, M., Hoffman, B., Mellor, P.S., and Mertens, P.P.C. (2008) Sequence analysis of bluetongue virus serotype 8 from the Netherlands 2006 and comparison to other European strains. *Virology* **377**, 308–18.

Maan, S., Maan, N.S., Samuel, A.R., O'Hara, R., Meyer, A.J., Rao, S. and Mertens, P.P.C. (2004a) Completion of the sequence analysis and comparisons of genome segment 2 (encoding outer capsid protein VP2 from representative isolates of the 24 bluetongue virus serotypes. *Vet. Ital.* **40**, 484–8.

Maan, S., Maan, N.S., Samuel, A.R., Rao, S., Attoui, H. and Mertens, P.P.C. (2007a) Analysis and phylogenetic comparisons of full-length VP2 genes of the 24 bluetongue virus serotypes. *J. Gen. Virol.* **88**, 621–30.

Maan, S., Maan, N.S., Singh, K.P., Samuel, A.R. and Mertens, P.P.C. (2004b) The development of RT-PCR based assays and sequencing for typing European strains of bluetongue virus and differential diagnosis of field and vaccine strains. *Vet. Ital.* **40**, 552–61.

Maan, S., Rao, S., Maan, N.S., Anthony, S.J., Attoui, H., Samuel, A.R. and Mertens, P.P.C. (2007b) Rapid cDNA synthesis and sequencing techniques for the genetic study of bluetongue and other dsRNA viruses. *J. Virol. Methods* **143**, 132–9.

Maan, S., Samuel, A.R., Maan, N.S., Attoui, H., Rao, S. and Mertens, P.P.C. (2004c) Molecular epidemiology of bluetongue viruses from disease outbreaks in the Mediterranean Basin. *Vet Ital.* **40**, 489–95.

Mahajan, N.K., Prasad, G., Jain, N.C., Dhanoea, J.S. and Gupta, Y. (1991) Epizootiological studies on bluetongue at an organized sheep breeding farm near Hissar, Haryana. *Indian J. Anim. Sci.* **61**, 1–5.

Malik, Y., Minakshi, Dahiya, S., Ramesh, K., Pawan, K. and Prasad, G. (2002) Bluetongue virus antibodies in domestic camels(*Camelus dromediarius*)in northern region of Rajasthan, India. *Indian J. Anim. Sci.* **72**, 551–2.

Malik, Y., Prasad, G., Minakshi and Maan, S. (2001) Polymerase chain reaction and non-radiolabeled DNA probe for detection of bluetongue viruses. *Indian J. Anim. Sci.* **71**, 501–3.

Mehrotra, M.L. and Shukla, D.C. (1984) Seroprevalence of bluetongue virus in sheep, cattle, buffaloes in India. *Indian J. Anim. Sci.* **54**, 718.

Mehrotra, M.L., Shukla, D.C. and Khanna, P.N. (1996) Studies on bluetongue disease in India – isolation and serotyping of field isolate. *Indian J Com Microbiol. Immunol. Infect Dis.* **17**, 8–13.

Mehrotra, M.L., Shukla, D.C. and Kataria, J.M. (1989) Isolation and identification of viral agents resembling bluetongue virus in an outbreak in sheep. *Indian J. Anim. Sci.* **59**, 1072–4.

Mehrotra, M.L., Shukla, D.C., Singh, K.P., Khanna, P.N. and Saikumar, G. (1995) Isolation of bluetongue virus from an outbreak of mobilli virus infection in sheep and goat. *Indian J. Comp. Microbiol. Imunol. Infect. Dis.* **16**, 135–6.

Mehrotra, M.L., Singh, R. and Shukla, D.C. (1991) Seroepidemiology and isolation of virus from and outbreak of bluetongue in Tamil Nadu. *Indian J. Anim. Sci.* **61**, 1282–3.

Mellor, P.S., Boorman, J. and Baylis, M. (2000) *Culicoides* biting midges: Their role as arbovirus vectors. *Ann. Rev. Entomol.* **45**, 307–40.

Mertens, P.P.C., Maan, N.S., Prasad, G., Samuel, A.R., Shaw, A.E., Potgieter, A.C., Anthony, S.J. and Maan, S. (2007) The design of primers and use of RT-PCR assays for typing European BTV isolates: Differentiation of field and vaccine strains. *J. Gen. Virol.* **88**, 2811–23.

Nachimuthu, K., Thangavelu, A., Dhinakarraj, G. and Venkatesan, R.A. (1992) Isolation of bluetongue virus from sheep. *Indian J. Anim. Sci.* **62**, 112–14.

Naresh, A. and Prasad, G. (1995) Detection of bluetongue virus antibodies by monoclonal antibody and recombinant antigen based c-ELISA. *Indian J. Exp. Biol.* **33**, 880–2.

Naresh, A., Roy, P. and Prasad, G. (1996) Evaluation of recombinant bluetongue virus antigens using dot immunobinding assay. *Am. J. Vet. Res.* **57**, 1556–8.

Narladakar, B.W., Shastri, U.V. and Shivpuje, P.R. (1993) Studies on culicoides spp (Dipetra: Ceratopogonidae) prevalent in Marathwada region (Maharashtra) and their host preference. *Indian Vet. J.* **70**, 116–18.

Oberoi, M.S., Singh, G. and Kwatra, M.S. (1988) Serological evidences of bluetongue virus activity in cattle and buffalo population. *Indian J. Virol.* **4**, 50–1.

Prasad, G. (2000) Whither bluetongue research in India. *Indian J. Microbiol.* **40**, 163–75.

Prasad, G. and Minakshi (1999) Comparative evaluation of sensitivity of RNA-polyacrylamide gel electrophoresis and dot immunobinding assay for detection of bluetongue virus in cell culture. *India J. Exp. Biol.* **37**, 157–60.

Prasad, G., Garg, A.K., Minakshi, Kakker, N.K. and Srivastava, R.N. (1994) Isolation of bluctongue virus from sheep in Rajasthan. *India. Rev. Sci. Tech. Off. Int. Epiz.* **13**, 935–7.

Prasad, G., Jain, N.C. and Gupta, Y. (1992) Bluetongue virus infection in India: A review. *Rev. Sci. Tech. Off. Int. Epiz.* **11**, 699–711.

Prasad, G., Malik, P., Malik, P.K. and Minakshi (1998a) Serological survey of bluetongue virus antibody in domestic and wild ruminants in and around Sariska tiger reserve, Rajasthan. *Indian J. Virol.* **14**, 51–3.

Prasad, G., Minakshi, Maan, S. and Malik, Y. (1998b) Genomic diversity in indian isolates of bluetongue virus serotype 1. *Indian J. Microbiol.* **38**, 161–3.

Prasad, G., Minakshi, Malik, Y. and Maan, S. (1999) RT-PCR and its detection limit for cell culture grown bluetongue virus serotype 1 using NS1 gene group specific primers. *India J. Exp. Biol.* **70**, 103–9.

Punia, S. (2005) Cloning and expression of genome segment 7 of Indian isolates of Blue tongue virus. PhD thesis, CCS HAU Hisar.

Rahman, A.H. and Manickam, R. (1997) *Toxorhynchites* -fluorescent antibody system for the detection of bluetongue virus from *Culicoides* midges (Diptera: Ceratopogonidae). *Onderstepoort J. Vet. Res.* **64**, 301–7.

Ravishankar, C. (2003) Seroprevalence of bluetongue in sheep and goats in Kerala, M.V.Sc Thesis submitted to Kerala Agricultural University, Kerala, India.

Ravishankar, C., Nair, G.K., Mini, M. and Jayaprakasan, V. (2005a) Seroprevalence of bluetongue virus antibodies in sheep and goats in Kerala State, India. *Rev. Sci. Tech. Off. Int. Epiz.* **24**, 953–8.

Ravishankar, C., Nair, G.K., Mini, M., Jayaprakasan, V. and Sunilkumar, N.S. (2005b) Seroprevalence of bluetongue among cattle in Kerala. *Indian Vet. J.* **82**, 568–9.

Rudragouda, C. (2007) Immunopathology of bluetongue virus serotype-1 in experimentally infected and vaccinated sheep. M.V.Sc. Thesis submitted to Deemed University, Indian Veterinary Research Institute, Izatnagar, U.P. India.

Sapre, S.N. (1964) An outbreak of bluetongue in goats and sheep. *Indian Vet. Rev. (M&B)* **15**, 69–71.

Saravanabava, K. (1992) Studies on bluetongue virus in sheep, Ph.D. thesis, Tamil Nadu Veterinary and Animal Sciences University, Madras, India.

Sarwar, M.M. (1962) A note on bluetongue in sheep in West Pakistan. *Pakistan J. Anim. Sci.* **1**, 1–5.

Sen, P. and Dasgupta, S.K. (1958) Determination of blood meal of Indian *Culicoides*. *Curr. Sci.* **27**, 122.

Sen, P. and Dasgupta, S.K. (1959) Studies on indian *Culicoides*. *Annals Entomol Soc. Am.* **52**, 617–43.

Sharma, M.M., Lonkar, P.S., Srivastva, C.P., Dubey, S.C., Maru, A. and Kalra, D.B. (1985) Epidemiology of bluetongue in sheep at an organised farm in semi arid part of Rajasthan, India. *Indian J. Comp. Microbiol., Immunol. Infect. Dis.* **6**, 188–92.

Sharma, S.N., Oberoi, M.S., Sodhi, S.S. and Baxi, K.K. (1981) Bluetongue virus precipitating antibodies in dairy animals in the Punjab. *Trop. Anim. Health. Prod.* **13**, 193.

Singh, B., Anantwar, L.G. and Samod, A. (1982) A severe outbreak of bluetongue in sheep in Maharashtra. In: S.C. Dube (ed.), *National Seminar on Sheep and Goat Diseases.* Avikanagar: Central Sheep Wool Research Institute, p. 16.

Singh, K.P., Maan, S., Samuel, A.R., Rao, S., Meyer, A. and Mertens, P.P.C. (2004) Phylogenetic analysis of bluetongue virus genome segment 6 (encoding VP5) from different serotypes. *Vet. Ital.* **40**, 479–83.

Sodhi, S.S., Oberoi, M.S., Sharma, S.N. and Baxi, K.K. (1981) Prevalence of bluetongue virus precipitating antibodies in sheep and goats in Punjab, Indian. *Zentbl. Vet. Med.* **28**, 421–3.

Sreenivasulu, D., Subba Rao, M.V. and Gard, G.P. (1996) Bluetongue virus in India: A review. In: T.D. St George and P. Kegao (eds.), *Bluetongue Disease in Southeast Asia and Pacific.* Canberra, Australia: ACIAR, pp. 15–19.

Sreenivasulu, D., Subba Rao, M.V. and Gard, G.P. (1999) Isolation of bluetongue virus serotype 2 from native sheep in India. *Vet. Rec.* **144**, 452–3.

Sreenivasulu, D., Subba Rao, M.V., Reddy, Y.N. and Gard, G.P. (2004) Overview of bluetongue disease, viruses, vectors, surveillance and unique features: The indian subcontinent and adjacent regions. *Vet. Ital.* **40**, 73–7.

Sriguppi, B.S. (1982). A note on seroepidemiology of bluetongue. In: M.L. Mehrotra (ed.), *Proceedings of National Workshop on Bluetongue.* Izatnagar: IVRI, pp. 114–22.

Srinivas, R.P., Moogi, V.M., Ananth, M., Gopal, T. and Keshavamurthy, B.S. (1982) Bluetongue disease outbreaks among sheep and goats of Karnataka. In: S.C. Dube (ed.), *Proceedings of National Seminar on Sheep and Goat Diseases.* Avikanagar: Central Sheep Wool Research Institute, pp. 14–20.

Srivastava, C.P., Maru, A. and Dubey, S.C. (1989) Epidemiology of congenital defects in foetal lambs in bluetongue endemic flocks. *Indian J. Anim. Sci.* **9**, 14–20.

Srivastava, R.N., Prasad, G., Kakker, N.K. and Bhatnagar, P. (1995) Epizootiology of bluetongue virus in particular reference to India. In: G. Prasad and R.N. Srivastava (eds.), *Bluetongue: Indian Perspective.* Hisar: 1CCS HAU Press, pp. 1–3.

Subba Reddy, C.V. (2004) Studies on certain aspects of Culicoides species. M.V.Sc thesis submitted to Acharya, N.G. Ranga Agricultural University, Tirupati. India.

Tabachnick, W.J., Robertson, M.A. and Murphy, K.E. (1996) *Culicoides variipennis and bluetongue disease.* Ann. NY Acad. Sci. **791**, 219–26.

Tiwari, A.K., Kataria, R.S., Butchaiah, G. and Das, S.K. (2000a) One step RT-PCR for detection of bluetongue virus RNA. *Indian J. Anim. Sci.* **70**, 582–3.

Tiwari, A.K., Kataria, R.S., Desai, G., Butchaiah, G. and Bandyopadhyay, S.K. (2000b) Characterization of India bluetongue virus isolate by RT-PCR and restriction enzyme analysis of the VP7 gene sequence. *Vet. Res. Commun.* **24**, 401–9.

Tongaonkar, S.S., Ayangar, S.K., Singh, B.K. and Kant, R. (1983) Seroprevalence of bluetongue in indian buffalo(*Bubalus bubalis*). *Vet. Rec.* **112**, 326.

Uppal, P.K. and Vasudevan, B. (1980) Occurrence of bluetongue in India. *Indian. J. Comp. Microbiol. Immunol. Infect. Dis.* **1**, 18–20.

Vasudevan, B. (1982) A note on bluetongue infection among sheep in India. In. In: M.L. Mehrotra (ed.) *Proceedings of National Workshop on Bluetongue*. Izatnagar: IVRI, pp. 129–44.

Vengadabady, N., Selvaraj, R. and Ramaswami, A.M. (2006) A report on the outbreak of bluetongue in sheep and goats in Coimbatore district. *Tamil Nadu J. Vet. Anim. Sci.* **2**, 246–7.

Current status of bluetongue virus in the Americas

9

WILLIAM C. WILSON,* JAMES O. MECHAM,* EDWARD SCHMIDTMANN,* CARLOS JIMENEZ SANCHEZ,† MARCO HERRERO† AND IRENE LAGER‡

*Arthropod-Borne Animal Diseases Research Laboratory, USDA, ARS, Laramie, WY, 82071, United States of America
†Escuela de Medicina Veterinaria, Heredia, 86-3000, Costa Rica
‡Instituto de Virologia, INTA-Castelar Hurlingham, Buenos Aires, Argentina

Introduction

The presence of bluetongue virus (BTV) in the Americas is important in part owing to the direct impact of clinical disease on the livestock industry, but more importantly due to the economic impact of restrictions on international trade in animals and animal germplasm. Although, North, Central, and South America are geographically distinct regions and epidemiologically diverse, there is some overlap in terms of the serotypes of BTV and the biting midge vectors that transmit them. Distinctive groups of BTV serotypes have been found in North America (1, 2, 3, 5, 6, 10, 11, 13, 14, 17, 19 and 22) (Barber, 1979; Johnson et al., 2005, 2007; Mertens et al., 2007b, 2007c), Central America, the Caribbean Basin (1, 3, 4, 6, 8, 12, and 17) (Mo et al., 1994), and South America (4, 6, 12, 14, 17, 19, and 20) (Lager, 2004), with only BTV serotypes 6 and 17 being common to all three areas. Although there are recent reports of introductions of serotypes not previously found in the United States, only serotypes 2, 10, 11, 13, and 17 have been shown to be circulating (Johnson et al., 2007). The distribution of *Culicoides* vector species also varies between the three regions. In North America, the primary proven vector of BTV is *Culicoides sonorensis* (Jones, 1965; Jones et al., 1981; Jones, 1985). A second species that has been implicated as a vector of BTV in the United States is *Culicoides insignis*, which is found in southern Florida but may extend as far north as Alabama (Greiner et al., 1985; Kline and Greiner, 1985;

ISBN-13: 978-0-12-369368-6

Kramer *et al.*, 1985; Tanya *et al.*, 1993). In Central and South America, available evidence indicates that *C. insignis* is the primary vector of BTV (Ronderos *et al.*, 2003); although, *Culicoides pusillus* may also be a vector. Although Central America shares a common vector and several serotypes of BTV with North and South America, BTV serotype 8 (BTV-8) is unique to this region of the Americas. This chapter will discuss the disease prevalence in domestic and wild ruminants, current status of BTV types in the Americas, status of the biting midge species that transmit BTV, economic impact, and control strategies for each continent. The relevance of virus–vector interactions on the epidemiology of bluetongue (BT) in the Americas will also be discussed.

North America

Disease status in domestic and wild ruminants

The first documentation of BT-like disease in North America was in sheep in Texas in 1948 (Hardy and Price, 1952). Severe signs and lesions included nasal discharge, inflammation in the nasal and muzzle area with some bleeding ('soremuzzle'), ulcers around the teeth and on the tongue, rapid weight loss, depression, diarrhea, and inflammation in the coronary band and laminar corium of the hoof resulting in lameness. Morbidity was estimated at 10–30%. Although the causative agent was not determined, similarities of disease signs to those described for BT in South Africa (Spreull, 1905; Theiler, 1906) strongly suggest that this outbreak of 'soremuzzle' was caused by BTV.

The first comprehensive serological survey for BTV in the United States. was conducted during the winter months of 1977/1978 on blood serum samples from slaughter cattle (Metcalf *et al.*, 1981). Serum BTV antibody prevalence varied from 0 to 79%, with the highest prevalence in the southwestern states and the lowest prevalence in the northeastern states. Serological surveys conducted on a regular basis between 1979/1980 and 1990/1991 by the USDA, APHIS confirmed previous findings of geographically distinct regions in terms of low and high incidence of BTV activity. Compilation of data for virus isolations on a state-by-state basis between 1976 and 1990 paralleled the serological data (Pearson *et al.*, 1991). Serological and virus isolation data collected since then have further verified the geographic stratification of BTV activity in the United States (Table 9.1) (Ostlund, 1996–2004). There is also serological evidence for BTV infection of sheep and cattle in Mexico (Suzan *et al.*, 1983; Stott *et al.*, 1989; Barajas Rojas *et al.*, 1993). There are occasional incursions of BTV, causing mild clinical disease in cattle, into the Canadian Okanagan Valley, British Columbia, but the disease does not appear to be endemic (Dulac *et al.*, 1992).

As noted above, sheep infected with BTV often manifest clinical signs of disease (Luedke *et al.*, 1964, 1969). Mortality as high as 50% and morbidity as high as 70% have been reported in naïve sheep populations in the United States

Table 9.1 Compiled United States Department of Agriculture, Animal Plant Health Inspection Service, National Veterinary Service Laboratory's annual reports of bluetongue virus (BTV) isolations or polymerase chain reaction (PCR)-positive samples from diagnostic and surveillance submissions

Year	Submissions	Positives	Serotypes	Species	States
2004	233	109[a]	1, 17	Cattle, deer, and sheep	AL, CO, FL, LA, MT, NE, TX
2003	216	20	2, N/A	Alpaca, Bighorn sheep, cattle, deer, and sheep	CA, CO, FL, IL, OK, TX
2002	503	50	10, 17	Cattle, deer, and sheep	AZ, CA, FL, KS, NM, SC, TX
2001	386	12	13, 17	Bighorn sheep, cattle, deer, elk, goats, and sheep	AZ, CA, FL, KS, MO, NM, SD, TX
2000	197	11	11, 13, 17	Cattle and sheep	CA, KS, MD, MT, NE, NM, OR
1999	184	18	2, 10, 11, 13, 17	Bighorn sheep, cattle, and sheep	CA, FL, ID, LA, OK, WA
1998	159	26	10, 11, 17	Bighorn sheep, cattle, deer, gerenuk, and sheep	AZ, CA, FL, ID, KS, NV, NM, WA
1997	150	15	11, 13	Bighorn sheep, cattle, deer, gerenuk, and sheep	AZ, CA, FL, NM
1996	139	51[b]	11, 13, 17[c]	Bighorn sheep, bongo, cattle, deer, and sheep	AZ, CA, FL, LA, NM, OR, TX
1995	170	15	11	Cattle	ID

Source: Summarized from annual reports by E. Ostlund and colleagues to the Bluetongue and Bovine Retrovirus Committee Meetings published in the Proceeding of the United States Animal Health Association (Ostlund, 1996–2004).

[a]Not including one isolate from Costa Rica.

[b]AL, Alabama; AZ, Arizona; CA, California; FL, Florida; IL, Illinois; LA, Louisiana; KS, Kansas; ID, Idaho; NE, Nebraska; NM, New Mexico; NV, Nevada; MO, Missouri; MT, Montana; OK, Oklahoma; OR, Oregon; SC, South Carolina; SD, South Dakota; TX, Texas; WA, Washington.

[c]Not including seven isolates (serotypes. 1, 2, 10, 11, 18, and 19) from cattle from the Dominican Republic.

(Hourrigan and Klingsporn, 1975a). Cattle usually develop minimal clinical responses to BTV infection, even though they develop viremia and good antibody responses (Luedke *et al.*, 1969; Hourrigan and Klingsporn, 1975b; MacLachlan, 1994). It had been reported that infection in utero can result in abortion or fetal and newborn abnormalities in both sheep and cattle (Luedke, 1985). However, additional work has suggested that BTV does not typically cross the placental barrier unless it is modified by passage in cell culture (MacLachlan *et al.*, 2000).

The involvement of wildlife in the epidemiology of BT in the United States has been shown by both serology and virus isolation (Trainer and Jochim, 1969; Couvillion *et al.*, 1980; Couvillion *et al.*, 1985; Jessup, 1985; Drolet *et al.*, 1990; Stallknecht *et al.*, 1991a; Stallknecht *et al.*, 1991b; Stallknecht and Davidson, 1992; Dunbar *et al.*, 1999). The affected species include white-tailed deer, black-tailed deer, mule deer, elk, pronghorn, and bighorn sheep. In addition to these ruminant wildlife species, there is one report of antibodies to BTV detected in Florida black bears (Dunbar *et al.*, 1998). Serological surveys in Mexico also have demonstrated antibodies to BTV in white-tailed deer and mule deer (Aguirre *et al.*, 1992; Martinez *et al.*, 1999). The only published study on the pathology of BTV in US wildlife was done in white-tailed deer by experimental inoculation with the virus. Clinical signs and lesions were similar to those that have been described in domestic ruminants suffering from BT (Howerth *et al.*, 1988).

Status of virus types

The first isolation of BTV from a sheep in the United States was reported from California in 1953 (McKercher *et al.*, 1953). Plaque reduction neutralization tests identified four distinct BTV serotypes circulating in North America (Barber and Jochim, 1973). By 1975, BTV had been reported in 13 states (Hourrigan, 1975). Until 1982, only four serotypes of BTV (10, 11, 13, and 17) had been identified in the United States (Barber, 1979). Annual reports from the National Veterinary Services Laboratories (NVSL) to the United States Animal Health Association indicate that these serotypes still predominate in the United States (see Table 9.1). These data provide a rough estimate of BTV activity based on samples submitted to the NVSL for testing. Canada is free of BTV, except for the occasional incursion of serotype 11 into the Okanagan Valley, British Columbia (Thomas *et al.*, 1982; Dulac *et al.*, 1992). Investigation of an outbreak of BT in Ona, Florida, in 1982 resulted in the isolation of BTV-2 (Gibbs *et al.*, 1983; Barber and Collisson, 1985). It was later shown that the original isolate of BTV-2 had changed genetically by either mutation or reassortment with an indigenous serotype to form a new electropherotype (Collisson *et al.*, 1985). Recent evidence indicates that this serotype has persisted in the Southeastern United States and has reassorted with other BTV serotypes cocirculating in the region (Mecham and Johnson, 2005). Interestingly, there are no reports of BTV-2 isolations in other parts of the Western Hemisphere.

In 2004, BTV-1 was isolated from a deer in Louisiana (Johnson *et al.*, 2005). Serological evidence suggests that BTV-1 may have over-wintered in the region (Emery, 2005). It has been suggested that BTV-1 was introduced to Louisiana from Central America as a result of 2004 hurricanes. However, sequence studies to determine this are inconclusive (Mecham *et al.*, 2005). Further investigations are needed to elucidate the origin of BTV-1 and to determine whether it will persist and disseminate from the Louisiana coastal region. The possible persistence of BTV-1 in the Louisiana coastal region has not been found by entomological and serological surveys (Reeves *et al.*, 2007).

Several other BTV isolates have recently been made in North America from samples taken in the period 1999–2005, mainly from Florida but including one sample from Mississippi. These viruses failed to 'type' in conventional serological assays using antisera to the BTV serotypes that had already been identified within the United States. However, subsequent reverse transcription-polymerase chain reaction (RT-PCR) assays using serotype-specific primers targeting genome segment 2 (Mertens *et al.*, 2007a, 2007c), followed by sequence analyses and comparisons to genome segment 2 from reference strains of the 24 BTV serotypes (Maan *et al.*, 2007), demonstrated that these included BTV serotypes 3, 5, 6, 14, 19, and 22 (Johnson *et al.*, 2007; Mertens *et al.*, 2007b, 2007c). These six serotypes had previously been regarded as exotic to the United States and there is no evidence that these viruses have become established. There is insufficient molecular epidemiological data to determine the origin of these isolates, even though BTV-3, -6, -14, and -19 have been identified in Central and South America.

Status of insect vectors

Serological studies in 1981 demonstrated the importance of the presence of the insect vector for virus transmission (Metcalf *et al.*, 1981; Osburn *et al.*, 1981). *Culicoides* (Diptera: Ceratopogonidae) were first implicated as a vector of BTV in North America by the infection of sheep after injection of triturated *Culicoides* captured during an outbreak of BT in Texas (Price and Hardy, 1954). In 1963, it was demonstrated in the laboratory that adult *Culicoides* could transmit BTV from sheep to sheep (Foster *et al.*, 1963). The North American *Culicoides variipennis* complex (Wirth and Morris, 1985) was first described as five subspecies *C. v. variipennis, C. v. sonorensis, Culicoides v. occidentalis, Culicoides v. australis*, and *Culicoides v. albertensis* (Wirth and Jones, 1957). Downes (1978) recognized two species *C. variipennis* and *C. occidentalis*, with *C. albertensis, C. australis*, and *C. sonorensis* regarded as subspecies of *C. occidentalis*. Studies demonstrating that *C. v. variipennis, C. v. occidentalis*, and *C. v. sonorensis* were genetically distinct taxa (Tabachnick, 1992; Holbrook and Tabachnick, 1995) led to a comprehensive analysis of the population genetics of the *C. variipennis* complex (Holbrook *et al.*, 2000). This study determined that larval populations of *C. sonorensis*, sympatric in aquatic habitats with the larvae of *C. variipennis* or *C. occiden-talis*, were genetically distinct, with no evidence of genetic introgression and thus deserving of separate species status.

The incrimination of *C. sonorensis* as the primary vector of BTV in North America is based on the evidence of (1) blood feeding on susceptible livestock species during BTV outbreaks (Jones, 1965; Jones *et al.*, 1981), (2) oral susceptibility, with replication and dissemination of BTV (Foster and Jones, 1979), (3) transmission of BTV to vertebrate animals under experimental conditions (Foster *et al.*, 1968), and (4) isolation of BTV from *C. sonorensis* field populations (Nelson and Scrivani, 1972). These attributes fulfill the criteria required for incrimination of a vector species (Barnett, 1960). Transovarial transmission of BTV has not been demonstrated with colonized *C. sonorensis* in the laboratory (Jones and Foster, 1971). Bluetongue virus virogenesis studies supported this finding by failing to detect viral antigen in ovarian follicles and eggs (Ballinger *et al.*, 1987). However, later studies did demonstrate BTV antigen in protein yolk bodies and in the vitelline membrane of developing oocytes from colonized *C. sonorensis* fed a BTV-containing blood meal but, failed to demonstrate virus in progeny (Nunamaker *et al.*, 1990). It is well documented that the developmental stages of *C. sonorensis* overwinter in temperate winter environments (Jones, 1967; Barnard and Jones, 1980). Thus, transovarial transmission of BTV could be a mechanism of regional virus persistence. A recent study demonstrated the presence of viral RNA in larvae, and pupae of *C. sonorensis* collected from the field in a BTV endemic area suggests that transovarial transmission may occur (White *et al.*, 2005, see also Chapter 16, this volume).

Despite the presence of at least 17 species of North American *Culicoides* that blood feed on livestock, only *C. sonorensis* is recognized as a proven vector of BTV (see review: Tabachnick, 1996). Other North American *Culicoides* species, including members of the potential-vector subgenus *Avaritia* and *Culicoides stellifer*, have been investigated as vectors of BTV in regional studies (Jones *et al.*, 1983; Mullen *et al.*, 1985a; Mullen *et al.*, 1985b; Mullens and Dada, 1992; Weiser-Schimpf *et al.*, 1993). However, their status as vectors of BTV is yet to be established. Bluetongue virus was isolated from *Culicoides cockerellii* in Colorado (Kramer *et al.*, 1990), but further understanding of the role of this western species as a potential vector of BTV has not been pursued.

Immature populations of *C. sonorensis* are common and widespread in littoral sediments of standing and slow-moving aquatic habitats that are often contaminated with cattle manure (Wirth and Morris, 1985) and exposed to sunlight (Mullens and Rodriguez, 1988). Aquatic habitats with larval *C. sonorensis* commonly have higher levels of dissolved salts, because of contamination with manure and surface water evaporation, than habitats with larval *C. variipennis* in the Eastern United States (Schmidtmann *et al.*, 2000). Further, a test of the relationship between dissolved salts and the *C. variipennis* complex examined widespread aquatic habitats that were classified by the species present and soil chemistry (Schmidtmann, 2006). The results showed that conspecific populations occurred in habitats that had similar levels of dissolved salts, which differed from habitats with sister species, regardless of geographic location. Reverse stepwise discriminant analysis indicated that 'electrical conductivity' and 'chloride' best differentiated between habitats with *C. sonorensis* and *C. variipennis* and the combination of 'electrical conductivity' and 'calcium' between *C. sonorensis* and *C. occidentalis*. More recent studies demonstrated on the basis of light/suction trap sampling that extensive areas east of the Missouri River in the upper Midwestern United States are devoid of *C. sonorensis* (Schmidtmann, unpublished data). Because precipitation in this region exceeds evaporation, salts are leached from surface soils that were also modified by repeated Pleistocene continental glaciations. In contrast, evaporation exceeds precipitation west of the Missouri River, where aquatic habitats with *C. sonorensis* span millions of square miles of nonglaciated soils with moderate to high levels of salinity. Dissolved salts are known to have strong influence on aquatic invertebrate communities, and thus the available evidence indicates that they play a role in regulating the distribution of the *C. variipennis* complex. As such, differing levels of soil salinity represent a plausible explanation for the presence and absence of *C. sonorensis*, hence transmission of BTV, on a geographic scale.

The most prevalent North American BTV serotypes (10, 11, 13, and 17) (Barber, 1979) and the vector *C. sonorensis* comprise the North American BTV 'episystem' – a distinct grouping of viruses and their unique vector that differ from the BTV-vector episystems that exist on other continents (Tabachnick, 2004). Bluetongue virus serotype 2 was first reported in the United States in 1982 (Gibbs *et al.*, 1983; Barber and Collisson, 1985). This virus is currently restricted to the Southeastern United States (Ostlund *et al.*, 2004), where it may be associated with a geographically limited vector species. Initial laboratory studies indicated that *C. insignis* in Florida were competent for BTV-2 (Tanya *et al.*, 1993), and it was subsequently shown that this serotype does not replicate in colonized *C. sonorensis* as well as other US serotypes do (Mecham and Nunamaker, 1994). When colonized *C. sonorensis* were fed a blood meal containing a high titer of BTV-2, held for a standard extrinsic incubation period, then allowed to feed on naïve sheep, it required a minimum of 500 bites per animal for virus transmission (Wilson, unpublished data). This suggests that *C. sonorensis* may not be a very capable vector for BTV-2. *Culicoides insignis* is present in southern Florida (Greiner *et al.*, 1985; Kline and Greiner, 1985), where its presence may be necessary for the persistence of BTV-2.

The detection in 2004 of a Neotropical virus, BTV serotype-1 (BTV-1), in Louisiana and of the exotic serotypes 3, 5, 6, 14, 19, and 22 in Florida and BTV-3 in Mississippi raises questions about the origin of these viruses, their vectors, vertebrate host range, and potential for economic impact in North America. The Neotropical species *C. insignis* is found in peninsular Florida and may occur elsewhere in the Southeastern United States (Kramer *et al.*, 1985). It is possible that US populations of *C. insignis* could transmit BTV-1 and other serotypes as it apparently does in Central America (Lager, 2004). Nevertheless, both C. *sonorensis* and C. *variipennis* are widespread across the Southeastern United States (Holbrook *et al.*, 2000) and should also be considered potential vectors of BTV-1, -2, -3, -5, -6, -14, -19, and -22 in the region until proven otherwise. An assessment of the role of C. *variipennis,* C. *sonorensis,* and C. *insignis* in the transmission of these previously exotic BTV types could be used as a test of the integrity of the North American BTV episystem. The distribution of these *Culicoides* species in relation to BTV epidemiology was reviewed previously (Tabachnick, 1996).

Understanding the distribution and vector biology of *Culicoides* is important because the absence of C. *sonorensis* and the low seroprevalence to BTV in the Northeastern United States form the basis for recognition of the region as BTV-free (Walton *et al.*, 1992). The sister species C. *variipennis* is common in dairy farms across the Northeastern United States, where larval populations are found in various manure-contaminated aquatic habitats (Schmidtmann *et al.*, 1983; Mullens and Rutz, 1984; Holbrook and Tabachnick, 1995). It is widely accepted that C. *variipennis* is an inefficient vector of BTV, although only unpublished vector competence data from field-collected populations support this statement (Tabachnick and Holbrook, 1992; Tabachnick, 1996). The other sister species, C. *occidentalis,* also appears to have a low vector competence for BTV (Tabachnick, 1996). Considerable variation exists in the infectivity rates of field-collected C. *sonorensis* populations in the United States (Tabachnick and Holbrook, 1992; Tabachnick, 1996), which is not surprising, since variable BTV susceptibility among field populations has been reported previously (Jones and Foster, 1978). In addition to differences in susceptibility among *Culicoides* species and populations, environmental conditions such as temperature also affect virogenesis in C. *sonorensis* (Mullens *et al.*, 2004). Thus, in Southwestern Canada, climatic conditions affect the life history components of C. *sonorensis* populations associated with prairie and feedlot habitats (Lysyk, 2006), such that C. *sonorensis* in the Southern US climatic conditions have ~1.8–2.6-fold greater vectorial capacity for BTV due to the effects of temperature (Lysyk and Danyk, 2007).

Repeated serological surveys have shown a low (<2%) prevalence of antibodies to BTV cattle in the Northeastern United States (Ostlund *et al.*, 2004). The low BTV seroprevalence in Northeastern US cattle, where C. *variipennis* predominates, is likely due to the movement of cattle into this area because C. *variipennis* is an inefficient vector species. The higher prevalence to BTV in Southern and Western US cattle is associated with the presence of C. *sonorensis,* a competent vector. The difference in species vector competence is likely to be due to genetic factors that are influenced by the environment (Campbell *et al.*, 2004). The absence of C. *sonorensis* and BTV antibodies in Northeastern US cattle has led to the relaxation of restrictions on export of cattle from these areas to BTV-free countries (Walton, 2004).

Economic impact

Bluetongue has a considerable economic impact in North America, with losses attributed to effects on animal health and productivity as well as nontariff trade restrictions that affect the sale and movement of animals. In 1998 alone, 66 countries required animal testing to demonstrate the lack of recent BTV exposure before importation of US ruminants and their products (Kahrs, 1998). Annual losses to US livestock industries attributed to the inability to trade with BTV-free counties have been estimated at $144 million (Hoar *et al.*, 2003). The low BTV seroprevalence in the Northeastern United States has led to the concept of regionalization (Ahl, 1998). Risk model analysis suggests that the United States could be divided into more than two regions with regard to BTV export risk (Hoar *et al.*, 2003). This concept is further supported by the low seroprevalence to BTV in Montana beef cattle, suggesting that there is little risk from movement of these animals to Canada (Van Donkersgoed *et al.*, 2004). Cooperative studies between Canadian and US researchers may help reduce economic losses due to unreasonable restrictions on animal movements between Canada and Montana.

Control strategies

Currently, there is one nationally licensed vaccine for BT in the United States for use in sheep only (Kemeny and Drehle, 1961). This vaccine uses attenuated BTV-10 and is effective only against that serotype (Luedke and Jochim, 1968). Bluetongue virus serotype 10, 11, and 17 attenuated vaccines are also available from the California Wool Growers Association for use only in California (www.woolgrowers.org). Vaccination of pregnant ewes with attenuated vaccines may cause abortions (Schultz and Delay, 1955). Vaccination with inactivated vaccines has shown variable results (Berry *et al.*, 1981; Campbell *et al.*, 1985; Mahrt and Osburn, 1986). An attenuated vaccine containing serotypes 10, 11, 13, and 17 has been used to vaccinate wildlife species in California (McConnell *et al.*, 1985). Neither adverse affects nor abortions were noted in this study, and the authors suggest that this approach could be used in a wildlife-restocking effort. Homologous killed vaccines are used mostly for captive wildlife breeding farms; however, the primary recommendation for control of BTV in the United States is insect vector control. Recommendations to control the insect vector include removal of potential breeding sites and use of insecticides and repellents.

Central America

Disease status in domestic and wild ruminants

Two epidemiological studies were conducted by an inter-American BTV team of investigators in the late 1980s–early 1990s (Homan *et al.*, 1990; Mo *et al.*,

1994). In both studies, clinical signs of BT-like disease were not observed in sentinel animals, but there was high incidence of infection as determined by serology. The authors suggested several explanations for this observation. The Central American and the Caribbean Basin BTV strains might be less virulent than those encountered in temperate zones. Most animals in the region are infected at a young age and so are protected against disease by the time they become adults. Also, because of many differential diagnoses, BT may be overlooked because it mimics other, better-recognized diseases.

Status of virus types

Prior to 1990s, it was known that over 60% of livestock in some countries of the Caribbean region had antibodies to BTV (Gibbs *et al.*, 1983) and that BTV serotypes 10, 11, 13, and 17 had been identified in Mexico (Stott *et al.*, 1989). Serological surveys of ruminants in the Caribbean region indicated the presence of antibody to BTV serotypes 6, 14, and 17 (Gumm *et al.*, 1984). The first inter-American team of investigators isolated BTV serotypes 1, 3, 6, and 12 from sentinel animals in the nine participating countries in Central America and the Caribbean Basin (Homan *et al.*, 1990). A subsequent study from 1987 to 1990, which included 11 countries in this region, showed that serotypes 4, 8, and 17 were also present (Mo *et al.*, 1994). These studies suggested that new disease outbreaks appear to involve a serotype that was absent in previous years. For example, BTV-1 predominated in Central America in 1991 but it was not detected in sentinel animals in the previous year (Mo *et al.*, 1994). Several key investigators in Central America and the Caribbean Basin have reported that there have not been any new research activities on BTV in Central America since the last paper from the inter-American BTV Team in the 1990s. Thus the current disease status and the economic impact of BTV infections in domestic and wild ruminants in the countries of this region are not known. Finally, because the disease status is unknown, there are no official control strategies for BTV infections in the region.

Status of vector species

Investigations of potential biting midge vectors in the Virgin Islands identified *Culicoides furens* and *C. insignis* as the predominant phototactic species; whereas *C. pusillus*, *Culicoides trilineatus*, *Culicoides jamaicensis*, and *Culicoides phlebotomus* were collected less frequently in light traps (Greiner *et al.*, 1990). The predominant species found feeding on cattle were *C. furens, C. insignis, C. pusillus*, and *C. trilineatus* (Greiner *et al.*, 1990). The primary species found in association with Barbados sheep were *C. insignis, C. pusillus, C. phlebotomus, C. furens, C. jamaicensis*, and *C. trilineatus* (Greiner *et al.*, 1990). Overall, *C. insignis* was the dominant species followed by *C. pusillus* (Greiner *et al.*, 1990). The first proven biological vector of BTV in southern

Florida, Central America and the Caribbean Basin was *C. insignis* (Tanya *et al.*, 1992). Although 44 species of *Culicoides* have been identified in this region, *C. insignis* accounts for 90% of collections, further implicating it as the primary vector species in the area (Greiner *et al.*, 1993).

South America

Disease status in domestic and wild ruminants

The first published report of BTV infection in South America was from Brazil in 1978 (Cunha *et al.*, 1982). Serological evidence of livestock infection was found in the states of São Paulo and Rio de Janeiro. Since then several serological surveys have determined that BTV infection is widespread in South America but generally without overt disease (Gibbs *et al.*, 1983; Gumm *et al.*, 1984; Rosadio *et al.*, 1984; Homan *et al.*, 1985; Lopez *et al.*, 1985; Tamayo *et al.*, 1985; Cunha *et al.*, 1987; Cunha *et al.*, 1988; Da Silva *et al.*, 1988; Brown *et al.*, 1989; Cunha, 1990; Castro *et al.*, 1992; Lage *et al.*, 1996; Melo *et al.*, 1999; Gonzalez *et al.*, 2000; Melo *et al.*, 2000). Most of these serological surveys were based on the results of a BTV agar gel immunodiffusion technique. Because the antigen used in this test can result in cross-reactivity, it is possible that some animals infected with related orbiviruses such as epizootic hemorrhagic disease virus may have been incorrectly recorded as seropositive for BTV.

The domestic species involved in these surveys were cattle (Castro *et al.*, 1982; Cunha *et al.*, 1982; Gibbs *et al.*, 1983; Viegas de Abreu, 1983; Gumm *et al.*, 1984; Homan *et al.*, 1985; Lopez *et al.*, 1985; Tamayo *et al.*, 1985; Cunha *et al.*, 1987; Cunha, 1990; Gonzalez *et al.*, 2000; Melo *et al.*, 2000; Gorch *et al.*, 2002), sheep (Gibbs *et al.*, 1983; Gumm *et al.*, 1984; Rosadio *et al.*, 1984; Cunha *et al.*, 1988; Cunha, 1990), and goats (Gibbs *et al.*, 1983; Cunha *et al.*, 1988; Da Silva, *et al.*, 1988; Brown *et al.*, 1989; Cunha, 1990).

There was a wide range of antibody prevalence reported in the South American surveys even within the same country (Table 9.2). This could be due in part to differences in climatic and environmental factors that affect the distribution of the vector(s) and/or the susceptible host(s) or the timescale over which the samples were collected, but it is suggestive of different levels of BTV transmission (Gibbs *et al.*, 1983; Cunha *et al.*, 1987; Gibbs and Greiner, 1994).

Other ruminant species have been analyzed as possible hosts or reservoirs for BTV by looking for the presence of antibodies. In Argentina, free-ranging llamas, guanacos, vicuñas and Pampean deer lacked BTV antibodies (Karesh *et al.*, 1998; Leoni *et al.*, 2001), but investigators in Peru found that alpaca can be infected with BTV (Rivera and Ameghino, 1987). In Brazil, a hemorrhagic disease affected cervids in the Rio de Janeiro

Table 9.2 Bluetongue virus (BTV) serological surveys in South American countries

Country	Location	No. of samples/species/techniques[a]	Seroprevalence (%)
Argentina	Misiones	1325/bov/AGID	40.7
		248/bov/AGID	35.88
		20/ov/AGID	95
	Corrientes	1528/bov/AGID	0.7
		295/bov/AGID	2.7
		415/ov/AGID	3.13
		93/bov/c-ELISA	21.50
	Chaco, Formosa	2021/bov/AGID	0
	Entre Ríos	956/bov/AGID	0
	Santa Fé	311/ov/AGID	0
Brazil	Paraiba	137/bov/AGID	4.82
	Sergipe	97/bov/AGID	89.69
	Mina Gerais	410/bov/AGID	76.3
		340/cap/AGID	5.9
		329/buf/AGID	54.4
	Rio de	553/bov/AGID	40.86
	Janeiro	626/cap/AGID	44.08
		66/ov/AGID	24.24
	Sao Pablo	214/bov/AGID	53.73
	Paraná	106/bov/AGID	19.81
	Sta Catarina	174/bov/AGID	37.75
	Rio Grande do Sul	409/bov/AGID	1.22
Chile	X° Region	1752/bov/AGID	19.6
		434/bov/ns	0
	Not specified (ns)	1139/bov/ns	0
		1224/ov/ns	0
Colombia	Antioquia, Cordoba	635/bov/AGID	51.8
	Valle de Aburra	86/bov/AGID	56
Ecuador	El Oro	87/bov/AGID	10
Guyana	Geographical diverse parts	719/bov/AGID	56
		387/bov/AGID	50
		255/cap/AGID	40
	Rupununi	50/bov/AGID	8
		25/ov/AGID	0
		25/cap/AGID	4
Peru	North	8/ov/AGID	87.5
	Central	17/ov/AGID	41
	South	9/ov/AGID	55.5
Suriname	Geographical diverse parts	451/bov/AGID	82
		77/ov/AGID	88
		68/cap/AGID	91
Venezuela	Aragua	151/bov/AGID	74.8
		151/bov/c-ELISA	94.7

Source: Lager (2004).

[a] bov = bovine, AGID = Agar gel immunodiffusion assay, ov = ovine, c-ELISA = competitive enzyme immunosorbent assay, cap = caprine, ns = not specified.

Zoo in January–February 1992 and was described in a herd of brown brocket deer (*Mazama gouazoubira*) in the campus of the University of São Paulo State. One out of four brockets died. At this same institution, six more brown brockets died in 1993. Serological studies based on agar gel immunodiffusion tests recorded the presence of antibodies against BTV or related orbiviruses (Cubas, 1996). In July 1992, BT was documented in one specimen of marsh deer (*Blastocerus dichotomus*) in the Ilha Solteira Zoo in Brazil (Cubas, 1996). In Bolivia, free-ranging gray brocket deer lacked antibodies to BTV (Deem *et al.*, 2004).

Status of virus types

The first BTV isolation from naturally infected animals in South America was BTV-4 from zebu cattle that were imported from Brazil to the United States (Groocock and Campbell, 1982). Brazil and Argentina are the only South American countries where BTV has been isolated. The Brazilian isolation was made at the Pan American Foot-and-Mouth Disease Centre from an April 2001 outbreak of clinically affected sheep and goats. The isolation was confirmed by RT-PCR and determined to be BTV-12 by virus neutralization (VN) (Clavijo *et al.*, 2002). In Argentina, BTV was isolated at INTA-Castelar from the blood of clinically normal sentinel animals and identified as serotype 4 by VN and RT-PCR using primers to genome segment 2 (Gorch *et al.*, 2002; Mertens and Bamford, 2005).

By serology, the serotypes that may be present in South America are 4, 6, 14, 17, 19, and 20 in Brazil (Groocock and Campbell, 1982; Cunha *et al.*, 1987); 12, 14, and 17 in Colombia (Homan *et al.*, 1985); 14 and 17 in Guyana; and 6, 14, and 17 in Suriname (Gumm *et al.*, 1984). These results should be considered as suggestive because of the potential for cross-reactions among virus serotypes. No other reports are available from the remaining South American countries.

Status of vector species

Very little information is available about the vector/s involved in the transmission of BTV in South America. *Culicoides insignis* is the predominant presumptive vector detected in the area (Homan *et al.*, 1985; Cunha *et al.*, 1987; Gorch *et al.*, 2002; Ronderos *et al.*, 2003). Bluetongue virus has been isolated from *C. insignis* in Central America and the Caribbean (Mo *et al.*, 1994), and *C. insignis* has also been shown competent for the transmission of BTV in southern Florida (Tanya *et al.*, 1992). However, as BTV was isolated from *C. pusillus* in Central America and the Caribbean, and this species is also present in South America, it is possible that *C. pusillus* could be a BTV vector in the area (Mo *et al.*, 1994; Ronderos *et al.*, 2003). Information available in South America is insufficient to exclude the possibility that additional species of *Culicoides* may also transmit BTV. Bluetongue virus has not yet been isolated from any *Culicoides* species in South America.

Economic impact

Bluetongue virus infection is endemic in most South American countries, but only four outbreaks of the disease have been reported. However, South America concentrates its control and eradication campaigns on other diseases such as foot-and-mouth disease, so it is possible that BT goes undetected. Nevertheless, BT is slowly coming into consideration by the livestock sanitary authorities of the region.

Bluetongue diagnosis has improved in South America in recent years. This is extremely important for control of the infection and to mitigate trade barriers, and it will also improve the ability to make a differential diagnosis from other vesicular diseases, which are the main problems for this continent. All of the South American countries and the Southern Common Market (Brazil, Argentina, Uruguay, and Paraguay) have import–export regulations concerning livestock, animal products, and germplasm, which are based on the OIE International Zoosanitary Code.

Control strategies

The control measures applied in respect of BTV by 8 of the 13 South American countries and reported to OIE are 'notifiable disease' and 'precautions at the border'. However, only two countries have screening, surveillance, and/or monitoring measures in place.

Virus–vector Interactions

Because BTV serotypes and vector species vary between geographic regions, it has been postulated that these orbiviruses have evolved in response to selective pressures from the invertebrate hosts. Phylogenetic studies based on sequence analysis have defined geographically distinct, genetic virus groupings (topotypes) (Pritchard *et al.*, 1995; Bonneau *et al.*, 1999; Pritchard *et al.*, 2004) that could be related to insect vector distribution. The S3 gene, which encodes the inner core protein VP3, was the first BTV gene shown to display distinct topotypes (Pritchard *et al.*, 1995). The NS3 protein has been shown to be involved in virus budding and is highly expressed in insect cells (Beaton *et al.*, 2002). Phylogenetic analysis based on this RNA segment segregated virus serotypes and strains into distinct topotypes; however, there is no evidence that this was the result of positive selection (Bonneau *et al.*, 1999; Balasuriya *et al.*, 2008). The phylogenetic groupings could be attributed to founder effects, as previously described (Bonneau *et al.*, 2001), rather than to positive selection by the insect vector. Founder effects may also account for the regionalization of segment 2 genotypes from US strains of the epizootic hemorrhagic disease virus complex (Cheney *et al.*, 1996) and also suggest that orbiviruses do not readily move across continents. Stronger evidence of positive selection of virus topotypes by the insect vector comes from sequence analysis of RNA genome segment 7, which encodes VP7. This protein is involved in virus binding and virogenesis in insect cells (Mertens *et al.*, 1996; Xu *et al.*, 1997; Tan *et al.*, 2001). Phylogenetic analysis based on the segment 7 sequence of BTV isolates from around the world showed a partial correlation between virus topotypes and insect vector distributions (Wilson *et al.*, 2000).

The relationship between the BTVs and their insect vectors is a complex issue and cannot be addressed solely by studies of virus phylogeny. Verification of these relationships will require controlled studies between paired viruses and insect vectors. It has been shown that a mesenteron infection barrier is of primary importance in determining vector competence of *Culicoides* species (Fu *et al.*, 1999). Studies on the innate immune response of *Culicoides* to BTV infection are just beginning (Campbell and Wilson, 2002; Campbell *et al.*, 2003) and will shed light on infection, replication, and transmission of viruses by an insect vector. Databases of genes expressed in the tissues that affect vector competence (salivary gland, midgut) have been established. Understanding the genetics of both the virus and the insect vector, as well as the interaction of the two, is necessary to fully understand the epidemiology of BT. In addition to genetic variation, there are environmental factors that may dramatically affect the distribution or competence of insect vectors. Salinity in larval habitats appears to have a dramatic effect on the distribution of vector species (Linley, 1986; Schmidtmann *et al.*, 2000). The composition of insect midgut microbiota differs between competent and incompetent vector species (Campbell *et al.*, 2004). Whether salinity can affect midgut microbiota composition or whether cach factor independently effects vector distribution and/or competence has not been shown. Studies on the effects of temperature on the ontogeny of the insect, on virus replication in the insect, and transmission of the virus by the vector to a mammalian host (Mullens and Holbrook, 1991; Mullens *et al.*, 1995; Wittmann *et al.*, 2002) are beginning to provide predictors of BTV outbreaks (Gerry *et al.*, 2001). For example, Canada has *C. sonorensis*, but it does not have endemic BTV, and recent evidence suggests that Canadian *C. sonorensis* females appear to be refractory to BTV infection (Mullens *et al.*, 2004). Whether this apparent susceptibility difference is due to genetic or environmental differences is as yet unclear. The objective evaluation of this complex list of genetic and environmental factors affecting vector competence and vectorial capacity is needed to fully understand the epidemiology of BT and to formulate reasonable animal regulatory statutes to reduce its economic impact.

Summary

The status of BTV and insect vector distribution in the Americas is fairly well-defined. With the exception of serotypes 6 and 17, which are in common, North, Central, and South America have unique groups of viruses. The predominant vector species in North America is *C. sonorensis*, while the predominant vector species in Central and South America is *C. insignis*, with, perhaps, some involvement of *C. pusillus* in South America. In North America, infection is enzootic, with occasional epizootics; in Central and South America, the disease is primarily enzootic, with few overt cases of disease. The economic impact of BTV infection is primarily from a trade perspective, with regulations and restrictions controlling the movement of animals and animal products from countries with BTV activity to countries that are BTV-free. Control strategies include both vaccination and insect control measures. Virus–vector interactions are complex and are influenced by both genetic and environmental factors. The biodiversity in the Americas provides opportunities to investigate this interaction. Understanding vector competence will require an integrated approach to elucidate these interactions.

Acknowledgments

The authors thank Drs K. Bennett, B.S. Drolet, G.J. Letchworth, and T.E. Walton for critical reviews of this manuscript.

References

Aguirre, A.A., McLean, R.G., Cook, R.S. and Quan, T.J. (1992) Serologic survey for selected arboviruses and other potential pathogens in wildlife from Mexico. *J. Wildl. Dis.* **28**, 435–42.

Ahl, R. (1998) Regionalization, risk analysis and exotic agents. *Proc. U.S. Anim. Health Assoc.* **98**, 127–8.

Balasuriya, U.B., Nadler, S.A., Wilson, W.C., Pritchard, L.I., Smythe, A.B., Savini, G., Monaco, F., De Santis, P., Zhang, N., Tabachnick, W.J. and Maclachlan, N.J. (2008) The NS3 proteins of global strains of bluetongue virus evolve into regional topotypes through negative (purifying) selection. *Vet. Microbiol.* **126**, 91–100.

Ballinger, M.E., Jones, R.H. and Beaty, B.J. (1987) The comparative virogenesis of three serotypes of bluetongue virus in *Culicoides variipennis* (Diptera: Ceratopogonidae). *J. Med. Entomol.* **24**, 61–5.

Barajas Rojas, J.A., Riemann, H.P. and Franti, C.E. (1993) Application of enzyme-linked immunosorbent assay for epidemiological studies of diseases of livestock in the tropics of Mexico. *Rev. Sci. Tech.* **12**, 717–32.

Barber, T.L. (1979) Temporal appearance, geographic distribution, and species of origin of bluetongue virus serotypes in the United States. *Am. J. Vet. Res.* **40**, 1654–6.

Barber, T.L. and Collisson, E.W. (1985) Epidemiologic implications of the genetic variations of bluetongue virus. *Prog. Clin. Biol. Res.* **178**, 597–606.

Barber, T.L. and Jochim, M.M. (1973) Serologic characterization of selected bluetongue virus strains from the United States. *Proc. U.S Anim. Health Assoc.* **77**, 352–9.

Barnard, D.R. and Jones, R.H. (1980) *Culicoides variipennis*: Seasonal abundance, over-wintering and voltinism in north-eastern Colorado. *Environ. Entomol.* **9**, 709–12.

Barnett, H.C. (1960) The incrimination of arthropods as vectors of disease. *Proc. 11th Int. Congr. Entomol.* **2**, 341–5.

Beaton, A.R., Rodriguez, J., Reddy, Y.K. and Roy, P. (2002) The membrane trafficking protein calpactin forms a complex with bluetongue virus protein NS3 and mediates virus release. *Proc. Natl. Acad. Sci. USA* **99**, 13154–159.

Berry, L.J., Osburn, B.I., Stott, J., Farver, T., Heron, B. and Patton, W. (1981) Inactivated bluetongue virus vaccine in lambs: Differential serological responses related to breed. *Vet. Res. Commun.* **5**, 289–93.

Bonneau, K.R., Mullens, B.A. and MacLachlan, N.J. (2001) Occurrence of genetic drift and founder effect during quasispecies evolution of the VP2 and NS3/NS3A genes of bluetongue virus upon passage between sheep, cattle, and *Culicoides sonorensis*. *J. Virol.* **75**, 8298–305.

Bonneau, K.R., Zhang, N., Zhu, J., Zhang, F., Li, Z., Zhang, K., Xiao, L., Xiang, W. and MacLachlan, N.J. (1999) Sequence comparison of the L2 and S10 genes of bluetongue viruses from the United States and the People's Republic of China. *Virus Res.* **61**, 153–60.

Brown, C.C., Olander, H.J., Castro, A.E. and Behymer, D.E. (1989) Prevalence of anti-bodies in goats in north-eastern Brazil to selected viral and bacterial agents. *Trop. Anim. Health Prod.* **21**, 167–9.

Campbell, C.H., Barber, T.L., Knudsen, R.C. and Swaney, L.M. (1985) Immune response of mice and sheep to bluetongue virus inactivated by gamma irradiation. *Prog. Clin. Biol. Res.* **178**, 639–47.

Campbell, C.L., McNulty, M., Letchworth, G.J. and Wilson, W.C. (2003) Molecular investigations of virus/vector interactions. *Vet. Ital.* **40**, 390–5.

Campbell, C.L., Mummey, D.L., Schmidtmann, E.T. and Wilson, W.C. (2004) Culture-independent analysis of midgut microbiota in the arbovirus vector *Culicoides sonorensis* (Diptera: Ceratopogonidae). *J. Med. Entomol.* **41**, 340–8.

Campbell, C.L. and Wilson, W.C. (2002) Differentially expressed midgut transcripts in *Culicoides sonorensis* (Diptera: Ceratopogonidae) following Orbivirus (Reoviridae) oral feeding. *Insect Mol. Biol.* **11**, 595–604.

Castro, R.S., Leite, R.C., Abreu, J.J., Lage, A.P., Lobato, Z. and Balsamao, S.L. (1982) Prevalence of antibodies to selected viruses in bovine embryo donors and recipients from Brazil, and its implications in international embryo trade. *Trop. Anim. Health Prod.* **24**, 173–9.

Castro, R.S., Leite, R.C., Abreu, J.J., Lage, A.P., Ferraz, I.B., Lobato, Z.I. and Balsamao, S.L. (1992) Prevalence of antibodies to selected viruses in bovine embryo donors and recipients from Brazil, and its implications in international embryo trade. *Trop. Anim. Health Prod.* **24**, 173–6.

Cheney, I.W., Yamakawa, M., Roy, P., Mecham, J.O. and Wilson, W.C. (1996) Molecular characterization of the segment 2 gene of epizootic hemorrhagic disease virus serotype 2: Gene sequence and genetic diversity. *Virology* **224**, 555–60.

Clavijo, A., Sepulveda, L., Riva, J., Pessoa-Silva, M., Tailor-Ruthes, A. and Lopez, J.W. (2002) Isolation of bluetongue virus serotype 12 from an outbreak of the disease in South America. *Vet. Rec.* **151**, 301–2.

Collisson, E.W., Barber, T.L., Gibbs, E.P. and Greiner, E.C. (1985) Two electropherotypes of bluetongue virus serotype 2 from naturally infected calves. *J. Gen. Virol.* **66**, 1279–86.

Couvillion, C.E., Jenney, E.W., Pearson, J.E. and Coker, M.E. (1980) Survey for antibodies to viruses of bovine virus diarrhea, bluetongue, and epizootic hemorrhagic disease in hunter-killed mule deer in New Mexico. *J. Am. Vet. Med. Assoc.* **177**, 790–1.

Couvillion, C.E., Nettles, V.F., Davidson, W.R., Pearson, J.E. and Gustafson, G.A. (1985) Hemorrhagic disease among white-tailed deer in the southeast from 1971 through 1980. *Proc. U.S. Anim. Health Assoc.* **855**, 522–37.

Cubas, Z.S. (1996) Special challenges of maintaining wild animals in captivity in South America. *Rev. Sci. Tech.* **15**, 267–87.

Cunha, R.G. (1990) Neutralizing antibodies for different serotypes of bluetongue virus in sera of domestic ruminants for Brazil. *Rev. Bras. Med. Vet.* **12**, 3–7.

Cunha, R.G., de Souza, D.M. and da Silva Passos, W. (1987) Antibodies to bluetongue virus in sera from cattle of Sao Paulo's state and Brazil's south region. *Rev. Bras. Med. Vet.* **9**, 121–4.

Cunha, R.G., de Souza, D.M. and Texeira, A.C. (1982) Anticorpos precipitantes para o virus da lengua azul em soro de bovinos do estado do Rio de Janeiro. *Biologico, Sao Paulo* **48**, 99–103.

Cunha, R.G., de Souza, D.M. and Texeira, A.C., (1988) Incidencia de anticorpos para o virus da lingua azul em soros de caprinos e ovinos do estado do Rio de Janeiro. *Arq. Flum. Med. Vet.* **3**, 53–6.

Da Silva, J., Modena, C., Moreira, E., Machado, T., Viana, F. and Abreu, V. (1988) Frequency of foot and mouth disease, bluetongue and enzootic bovine leucosis in goats, in the state of Mina Gerais, Brazil. *Arq. Bras. Med. Vet. Zoot.* **40**, 393–403.

Deem, S.L., Noss, A.J., Villarroel, R., Uhart, M.M. and Karesh, W.B. (2004) Disease survey of free-ranging grey brocket deer (Mazama gouazoubira) in the Gran Chaco, Bolivia. *J. Wildl. Dis.* **40**, 92–8.

Downes, J.A. (1978) *Culicoides variipennis* complex: A necessary re-alignment of nomenclature (Diptera: Ceratopogonidae). *J. Med. Ent.* **12**, 63–9.

Drolet, B.S., Mills, K.W., Belden, E.L. and Mecham, J.O., (1990) Enzyme-linked immunosorbent assay for efficient detection of antibody to bluetongue virus in pronghorn (*Antilocapra americana*). *J. Wildl. Dis.* **26**, 34–40.

Dulac, G.C., Sterritt, W.G., Dubuc, C., Afshar, A., Myers, D.J., Taylor, E.A., Jamieson, B.R. and Martin, M.W. (1992) Incursions of Orbivirues in Canada and their serologic monitoring in the native animal population between 1962–1991. In: T.E. Walton and B.I. Osburn, (eds.), *Bluetongue, African Horse Sickness and Related Orbiviruses*. Boca Raton: CRC Press, pp. 120–7.

Dunbar, M.R., Cunningham, M.W. and Roof, J.C. (1998) Seroprevalence of selected disease agents from free-ranging black bears in Florida. *J. Wildl. Dis.* **34**, 612–19.

Dunbar, M.R., Velarde, R., Gregg, M.A. and Bray, M. (1999) Health evaluation of a pronghorn antelope population in Oregon. *J. Wildl. Dis.* **35**, 496–510.

Emery, M.P., DeJean, S.K., Palmer, T.J. and Ostlund, E.N. (2005) Serological survey of domestic ruminants in Louisiana for antibodies to BTV-1 (Ostlund presenter) in Report of the Committee on Bluetongue and Bovine Retroviruses. *Proc. U.S. Anim. Health Assoc.* **109**, 236–41.

Foster, N.M. and Jones, R.H. (1979) Multiplication rate of bluetongue virus in the vector Culicoides variipennis (Diptera: Ceratopogonidae) infected orally. *J. Med. Entomol.* **15**, 302–3.

Foster, N.M., Jones, R.H. and Luedke, A.J. (1968) Transmission of attenuated and virulent bluetongue virus with *Culicoides variipennis* infected orally via sheep. *Am. J. Vet. Res.* **29**, 275–9.

Foster, N.M., Jones, R.H. and McCrory, B.R. (1963) Preliminary investigations on insect transmission of bluetongue virus in sheep. *Am. J. Vet. Res.* **24**, 1195–200.

Fu, H., Leake, C.J., Mertens, P.P. and Mellor, P.S. (1999) The barriers to bluetongue virus infection, dissemination and transmission in the vector, *Culicoides variipennis* (Diptera: Ceratopogonidae). *Arch. Virol.* **144**, 747–61.

Gerry, A.C., Mullens, B.A., Maclachlan, N.J. and Mecham, J.O. (2001) Seasonal transmission of bluetongue virus by *Culicoides sonorensis* (Diptera: Ceratopogonidae) at a southern California dairy and evaluation of vectorial capacity as a predictor of bluetongue virus transmission. *J. Med. Entomol.* **38**, 197–209.

Gibbs, E.P. and Greiner, E.C. (1994) The epidemiology of bluetongue. *Comp. Immunol. Microbiol. Infect. Dis.* **17**, 207–20.

Gibbs, E.P., Greiner, E.C., Alexander, F.C., King, T.H. and Roach, C.J. (1983) Serological survey of ruminant livestock in some countries of the Caribbean region and South America for antibody to bluetongue virus. *Vet. Rec.* **113**, 446–8.

Gibbs, E.P., Greiner, E.C., Taylor, W.P., Barber, T.L., House, J.A. and Pearson, J.E. (1983) Isolation of bluetongue virus serotype 2 from cattle in Florida: Serotype of bluetongue virus hitherto unrecognized in the Western Hemisphere. *Am. J. Vet. Res.* **44**, 2226–8.

Gonzalez, M.C., Perez, N. and Siger, J. (2000) Serologic evidence of bluetongue virus in bovine from Aragua state, Venezuela. *Rev. Fac. Cs. Vets. UCV.* **41**, 3–12.

Gorch, C., Vagnozzi, A., Duffy, S., Miquet, J., Pacheco, J., Bolondi, A., Draghi, G., Cetra, B., Soni, C., Ronderos, M., Russo, S., Ramirez, V. and Lager, I. (2002) Bluetongue: Isolation and characterization of the virus and identification of vectors in North-eastern Argentina. *Rev. Argent. Microbiol.* **34**, 150–6.

Greiner, E.C., Alexander, F.C., Roach, J., St. John, V.S., King, T.H., Taylor, W.P. and Gibbs, E.P. (1990) Bluetongue epidemiology in the Caribbean region: Serological and entomological evidence from a pilot study in Barbados. *Med. Vet. Entomol.* **4**, 289–95.

Greiner, E.C., Barber, T.L., Pearson, J.E., Kramer, W.L. and Gibbs, E.P. (1985) Orbiviruses from *Culicoides* in Florida. *Prog. Clin. Biol. Res.* **178**, 195–200.

Greiner, E.C., Knausenberger, W.I., Messersmith, M., Kramer, W.L. and Gibbs, E.P. (1990) *Culicoides* spp. (Diptera: Ceratopogonidae) associated with cattle in St. Croix, Virgin Islands, and their relevance to bluetongue virus. *J. Med. Entomol.* **27**, 1071–3.

Greiner, E.C., Mo, C.L., Homan, E.J., Gonzalez, J., Oviedo, M.T., Thompson, L.H. and Gibbs, E.P. (1993) Epidemiology of bluetongue in Central America and the Caribbean: Initial entomological findings. Regional Bluetongue Team. *Med. Vet. Entomol.* **7**, 309–15.

Groocock, C.M. and Campbell, C.H. (1982) Isolation of an exotic serotype of bluetongue virus from imported cattle in quarantine. *Can. J. Comp. Med.* **46**, 160–4.

Gumm, I.D., Taylor, W.P., Roach, C.J., Alexander, F.C., Greiner, E.C. and Gibbs, E.P. (1984) Serological survey of ruminants in some Caribbean and South American countries for type-specific antibody to bluetongue and epizootic haemorrhagic disease viruses. *Vet. Rec.* **114**, 635–8.

Hardy, W.T. and Price, D.A. (1952) Soremuzzle of sheep. *J. Amer. Vet. Med. Assoc.* **120**, 23.

Hoar, B.R., Carpenter, T.E., Singer, R. and Gardner, I. (2003) Regional risk of exporting cattle seropositive for bluetongue virus from the United States. *Am. J. Vet. Res.* **64**, 520–9.

Holbrook, F.R. and Tabachnick, W.J. (1995) *Culicoides variipennis* (Diptera: Ceratopogonidae) complex in California. *J. Med. Entomol.* **32**, 413–19.

Holbrook, F.R., Tabachnick, W.J., Schmidtmann, E.T., McKinnon, C.N., Bobian, R.J. and Grogan, W.L. (2000) Sympatry in the *Culicoides variipennis* complex (Diptera: Ceratopogonidae): A Taxonomic Reassessment. *J. Med. Entomol.* **37**, 65–76.

Homan, E.J., Mo, C.L., Thompson, L.H., Barreto, C.H., Oviedo, M.T., Gibbs, E.P. and Greiner, E.C. (1990) Epidemiologic study of bluetongue viruses in Central America and the Caribbean: 1986–1988. *Am. J. Vet. Res.* **51**, 1089–94.

Homan, E.J., Taylor, W.P., de Ruiz, H.L. and Yuill, T.M. (1985) Bluetongue virus and epizootic haemorrhagic disease of deer virus serotypes in northern Colombian cattle. *J. Hyg. Lond.* **95**, 165–72.

Hourrigan, J.L. and Klingsporn, A.L. (1975a) Epizootiology of bluetongue: The situation in the United States of America. *Aust. Vet. J.* **51**, 203–8.

Hourrigan, J.L. and Klingsporn, A.L. (1975b) Bluetongue: The disease in cattle. *Aust. Vet. J.* **51**, 170–4.

Howerth, E.W., Greene, C.E. and Prestwood, A.K. (1988) Experimentally induced bluetongue virus infection in white-tailed deer: Coagulation, clinical pathologic, and gross pathologic changes. *Am. J. Vet. Res.* **49**, 1906–13.

Jessup, D.A. (1985) Epidemiology of two orbiviruses in California's native wild ruminants: Preliminary report. *Prog. Clin. Biol. Res.* **178**, 53–65.

Johnson, D.J., Mertens, P.P.C., Maan, S. and Ostlund, E.N. (2007) Exotic bluetongue viruses identified from ruminants in the south-eastern U.S. from 1999–2006. In: *Proceedings of the Annual AAVLD Conference. Reno, NV.* p. 118.

Johnson, D.J., Stallknecht, D.E., et al. (2005) First report of bluetongue virus type 1 in the United States. *Proceedings of U.S. Animal Health Association. Hershey, PA.*

Jones, R.H. (1967) An overwintering population of *Culicoides* in Colorado. *J. Med. Entomol.* **4**, 461–3.

Jones, R.H., Schmidtmann, E. and Foster, N. (1983) Vector-competence studies for bluetongue and epizootic hemorrhagic disease viruses with *Culicoides venustus* (Ceratopogonidae). *Mosq. News* **43**, 184–6.

Jones, R.H. (1985) Vector research with the orbiviruses. *Prog. Clin. Biol. Res.* **178**, 147–9.

Jones, R.H. and Foster, N.M. (1971) Transovarian transmission of bluetongue virus unlikely for *Culicoides variipennis*. *Mosq. News* **31**, 434–7.

Jones, R.H. and Foster, N.M. (1978) Heterogeneity of *Culicoides variipennis* field populations to oral infection with bluetongue virus. *Am. J. Trop. Med. Hyg.* **27**, 178–83.

Jones, R.H., Luedke, A.J., Walton, T.E. and Metcalf, H.E. (1981) An entomological perspective toward control. *World Anim. Rev.* **38**, 2–8.

Jones, W.W. (1965) Epidemiological notes: Incidence of *Culicoides variipennis* in an outbreak of bluetongue disease. *Mosq. News* **25**, 217–18.

Kahrs, R.F. (1998) The impact of bluetongue on international trade. *Proc. U.S. Anim. Health Assoc.* **98**, 125–7.

Karesh, W.B., Uhart, M.M., Dierenfeld, E.S., Braselton, W.E., Torres, A., House, C., Puche, H. and Cook, R.A. (1998) Health evaluation of free-ranging guanaco (*Lama guanicoe*). *J. Zoo Wildl. Med.* **29**, 134–41.

Kemeny, L. and Drehle, L.E. (1961) The use of tissue culture-propagated bluetongue virus for vaccine preparation. *Am. J. Vet. Res.* **22**, 921–5.

Kline, D.L. and Greiner, E.C. (1985) Observations on larval habitats of suspected Culicoides vectors of bluetongue virus in Florida. *Prog. Clin. Biol. Res.* **178**, 221–7.

Kramer, W.L., Greiner, E.C. and Gibbs, E.P.J. (1985) Seasonal variations in population size, fecundity, and parity rates of *Culicoides insignis* (Diptera: Ceratopogonidae) in Florida, USA. *J. Med. Entomol.* **22**, 163–9.

Kramer, W.L., Jones, R.H., Holbrook, F.R., Walton, T.E. and Calisher, C.H. (1990) Isolation of arboviruses from *Culicoides* midges (Diptera: Ceratopogonidae) in Colorado during an epizootic of vesicular stomatitis New Jersey. *J. Med. Entomol.* **27**, 487–93.

Lage, A.P., Castro, R.S., Melo, M., Aquiar, P., Barreto Filho, J.B. and Leite, R.C. (1996) Prevalence of antibodies to bluetongue, bovine herpesvirus 1 and bovine viral diarrhoea/ mucosal disease viruses in water buffaloes in Mina Gerais State, Brazil. *Rev. Elev. Med. Vet. Pays Trop.* **49**, 195–7.

Lager, I.A. (2004) Bluetongue virus in South America: Overview of viruses, vectors, surveillance and unique features. *Vet. Ital.* **40**, 89–93.

Leoni, L., Cheetham, S., Lager, I., Parreño, V., Fondevila, N., Rutter, B., Martinez Vivot, M., Fernádez, F. and Schudel, A. (2001). Prevalencia de anticuerpos contra efermedades virales del ganado en llama (*Lama glama*), guanaco (*Lama guanicoe*) y vicuña (*Vicugna vicugna*) en Argentina. II Congreso Latinoamericano de Expecialidad en Pequenos Rumantes y Camelidos Sudamcricanos y XI congreso Nacional de Ovenocultura., Mérida, Yucatán, Mexico.

Linley, J.R. (1986) The effect of salinity on oviposition and egg hatching in *Culicoides variipennis sonorensis* (Diptera: Ceratopogonidae). *J. Am. Mosq. Control Assoc.* **2**, 79–82.

Lopez, W.A., Nicoletti, P. and Gibbs, E.P. (1985) Antibody to bluetongue virus in cattle in Ecuador. *Trop. Anim. Health Prod.* **17**, 82.

Luedke, A.J. (1985) Effect of bluetongue virus on reproduction in sheep and cattle. *Prog. Clin. Bio. Res.* **178**, 71–8.

Luedke, A.J., Bowne, J.G., Jochim, M.M. and Doyle, C. (1964) Clinical and pathologic features of bluetongue in sheep. *Am. J. Vet. Res.* **25**, 963–70.

Luedke, A.J. and Jochim, M.M. (1968) Clinical and serologic responses in vaccinated sheep given challenge inoculation with isolates of bluetongue virus. *Am. J. Vet. Res.* **29**, 841–51.

Luedke, A.J., Jochim, M.M. and Jones, R.H. (1969) Bluetongue in cattle: Viremia. *Am. J. Vet. Res.* **30**, 511–6.

Lysyk, T.J. (2006) Abundance and species composition of *Culicoides* (Diptera: Ceratopo gonidae) at cattle facilities in Southern Alberta, Canada. *J. Med. Entomol.* **43**, 840–9.

Lysyk, T.J. and Danyk, T. (2007) Effect of temperature on life history parameters of adult *Culicoides sonorensis* (Diptera: Ceratopogonidae) in relation to geographic origin and vectorial capacity for bluetongue virus. *J. Med. Entomol.* **44**, 741–51.

Maan, S., Maan, N.S., Samuel, A.R., Rao, S., Attoui, H. and Mertens, P.P.C. (2007) Analysis and phylogenetic comparisons of full-length VP2 genes of the 24 bluetongue virus serotypes. *J. Gen. Virol.* **88**, 621–30.

MacLachlan, N.J. (1994) The pathogenesis and immunology of bluetongue virus infection of ruminants. *Comp. Immunol. Microbiol. Infect. Dis.* **17**, 197–206.

MacLachlan, N.J., Conley, A.J. and Kennedy, P.C. (2000) Bluetongue and equine viral arteritis viruses as models of virus-induced fetal injury and abortion. *Anim. Reprod. Sci.* **60–61**, 643–51.

Mahrt, C.R. and Osburn, B.I. (1986) Experimental bluetongue virus infection of sheep; effect of vaccination: Pathologic, immunofluorescent, and ultrastructural studies. *Am. J. Vet. Res.* **47**, 1198–203.

Martinez, A., Salinas, A., Martinez, F., Cantu, A. and Miller, D.K. (1999) Serosurvey for selected disease agents in white-tailed deer from Mexico. *J. Wildl. Dis.* **35**, 799–803.

McConnell, S., Morrill, J.C. and Livingston, C.W.J. (1985) Use of a quadrivalent modified-live bluetongue virus vaccine in wildlife species. *Prog. Clin. Biol. Res.* **178**, 631–8.

McKercher, D., McGowan, B., Howarth, J. and Saito, J. (1953) A preliminary report on the isolation and identification of the bluetongue virus from sheep in California. *J. Am. Vet. Med. Assoc.* **122**, 300–1.

Mecham, J.O. and Johnson, D.J. (2005) Persistence of bluetongue virus serotype 2 (BTV-2) in the southeast United States. *Virus Res.* **113**, 116–22.

Mecham, J.O., Johnson, D.J. and Wilson, W.C. (2005). Sequence analysis of BTV-1 isolated from a deer in Louisiana. *Proceedings of U.S. Animal Health Association,* Hershey, PA.

Mecham, J.O. and Nunamaker, R.A. (1994) Complex interactions between vectors and pathogens: *Culicoides variipennis sonorensis* (Diptera: Ceratopogonidae) infection rates with bluetongue viruses. *J. Med. Ent.* **31**, 903–7.

Melo, C.B., Oliveira, A.M., Azevedo, E.O., Lobato, Z. and Leite, R.C. (2000) Antibodies to bluetongue virus in bovines of Praiba state, Brazil. *Bras. Med. Vet. Zoot.* **52**, 19–20.

Melo, C.B., Oliveira, A.M., Castro, R.S., Lobato, Z. and Leite, R.C. (1999) Precipitating antibodies against bluetongue virus in bovine from Sergipe, Brazil. *Cienc. Vet. Trop. Recife* **2**, 125–7.

Mertens, P. and Bamford, D. The RNAs and proteins of dsRNA viruses http://www.iah.bbsrc.ac.uk/dsRNA_virus_proteins, 2005.

Mertens, P.P.C, Burroughs, J.N., Walton, A., Wellby, M.P., Fu, H., O'Hara, R.S., Brookes, S.M. and Mellor, P.S. (1996) Enhanced infectivity of modified bluetongue virus particles for two insect cell lines and for two *Culicoides* vector species. *Virology* **217**, 582–93.

Mertens, P.P.C, Maan, N., Johnson, D., Ostlund, E. and Maan, S. (2007a) Sequencing and RT-PCR assays for genome segment 2 of the 24 bluetongue virus serotypes: Identification of exotic serotypes in the South eastern USA (1999–2006) World, Association of Veterinary Laboratory Diagnosticians, Symposium Abstracts, Melbourne, Australia, November 2007,

Mertens, P.P.C., Maan, N.S., Prasad, G., Samuel, A.R., Shaw, A.E., Potgieter, A.C., Anthony, S.J. and Maan, S. (2007b) Design of primers and use of RT-PCR assays for typing European bluetongue virus isolates: Differentiation of field and vaccine strain. *J. Gen. Virol.* **88**, 2811–23.

Mertens, P.P., Maan, N.S., Prasad, G., Samuel, A.R., Shaw, A.E., Potgieter, A.C., Anthony, S.J., and Maan, S. (2007c) Design of primers and use of RT-PCR assays for typing European bluetongue virus isolates: Differentiation of field and vaccine strains. *J. Gen. Virol.* **88**, 2811–23.

Metcalf, H.E., Pearson, J.E. and Klingsporn, A.L. (1981) Bluetongue in cattle: A serologic survey of slaughter cattle in the United States. *Am. J. Vet. Res.* **42**, 1057–61.

Mo, C.L., Thompson, L.H., Homan, E.J., Oviedo, M.T., Greiner, E.C., Gonzalez, J. and Saenz, M.R. (1994) Bluetongue virus isolations from vectors and ruminants in Central America and the Caribbean. *Am. J. Vet. Res.* **55**, 211–5.

Mullen, G.R., Hayes, M.E. and Nusbaum, K.E. (1985a) Potential vectors of bluetongue and epizootic hemorrhagic disease viruses of cattle and white-tailed deer in Alabam. *Prog. Clin. Bio. Res.* **178**, 201–6.

Mullen, G.R., Jones, R.H., Braverman, Y. and Nusbaum, K.E. (1985b) Laboratory infections of *Culicoides debilipalpis* and *C. stellifer* (Diptera: Ceratopogonidae) with bluetongue virus. *Prog. Clin. Biol. Res.* **178**, 239–43.

Mullens, B.A. and Dada, C.E. (1992) Insects feeding on desert bighorn sheep, domestic rabbits, and Japanese quail in the Santa Rosa mountains of southern California. *J. Wildl. Dis.* **28**, 476–80.

Mullens, B.A., Gerry, A.C., Lysyk, T.J. and Schmidtmann, E.T. (2004) Environmental effects on vector competence and virogenesis of bluetongue virus in *Culicoides*. *Vet. Ital.* **40**, 160–6.

Mullens, B.A. and Holbrook, F.R. (1991) Temperature effects on the gonotrophic cycle of *Culicoides variipennis* (Diptera: Ceratopogonidae). *J. Am. Mosq. Control Assoc.* **7**, 588–91.

Mullens, B.A. and Rodriguez, J.L. (1988) Colonization and response of *Culicoides variipennis* (Diptera: Ceratopogonidae) to pollution levels in experimental dairy wastewater ponds. *J. Med. Entomol.* **25**, 441–51.

Mullens, B.A. and Rutz, D.A. (1984) Age structure and survivorship of *Culicoides variipennis* (Diptera: Ceratopogonidae) in central New York State, USA. *J. Med. Entomol.* **21**, 194–203.

Mullens, B.A., Tabachnick, W.J., Holbrook, F.R. and Thompson, L.H. (1995) Effects of temperature on virogenesis of bluetongue virus serotype 11 in *Culicoides variipennis sonorensis*. *Med. Vet. Entomol.* **9**, 71–6.

Nelson, R.L. and Scrivani, R.P. (1972) Isolations of arboviruses from parous midges of the *Culicoides variipennis* complex, and parous rates in biting populations. *J. Med. Entomol.* **9**, 277–81.

Nunamaker, R.A., Sieburth, P.J., Dean, V.C., Wigington, J.G., Nunamaker, C.E. and Mecham, J.O. (1990) Absence of transovarial transmission of bluetongue virus in *Culicoides variipennis*: Immunogold labelling of bluetongue virus antigen in developing oocytes from *Culicoides variipennis* (Coquillett). *Comp. Biochem. Physiol.* **A96**, 19–31.

Osburn, B.I., McGowan, B., Heron, B., Loomis, E., Bushnell, R., Stott, J.L. and Utterback, W. (1981) Epizootiologic study of bluetongue: Virologic and serologic results. *Am. J. Vet. Res.* **42**, 884–7.

Ostlund, E., Moser, K.M., Johnson, D.J., Pearson, J.E. and Schmitt, B.J. (2004) Distribution of bluetongue in the United States of America, 1991–2002. *Vet. Ital.* **40**, 83–8.

Ostlund, E.N. (1996–2004) Report of the committee on bluetongue and bovine retroviruses. *Proc. U.S. Anim. Health Assoc.*

Pearson, J.E., Gustafson, G.A., Shafer, A.L. and Alstad, A.D. (1991) Distribution of bluetongue in the United States. In: T.E. Walton and B.I. Osburn, (eds.), *Bluetongue, African Horse Sickness and Related Orbiviruses*. Boca, Raton: CRC Press, pp. 128–39.

Price, D.A. and Hardy, W.T. (1954) Isolation of the bluetongue virus from Texas sheep; *Culicoides* shown to be a vector. *J. Am. Vet. Med. Assoc.* **12**, 255–258.

Pritchard, L.I., Gould, A.R., Wilson, W.C., Thompson, L., Mertens, P.P. and Wade-Evans, A.M. (1995) Complete nucleotide sequence of RNA segment 3 of bluetongue virus serotype 2 (Ona-A). Phylogenetic analyses reveal the probable origin and relationship with other orbiviruses. *Virus Res.* **3**, 247–61.

Pritchard, L.I., Sendow, I., Lunt, R., Hassan, S.H., Kattenbelt, J., Gould, A.R., Daniels, P.W. and Eaton, B.T. (2004) Genetic diversity of bluetongue viruses in south east Asia. *Virus Res.* **101**, 193–201.

Reeves, W.K., Kato, C.Y., Mayer, R.T. and Foil, L. (2007) in Report of the Committee on Bluetongue and Bovine Retroviruses. *Proc. U.S. Anim. Health Assoc.* (in press).

Rivera, H.M., Madewell, B.R. and Ameghino, E. (1987) Serologic survey of viral antibodies in the Peruvian alpaca (*Lama pacos*). *Am. J. Vet. Res.* **48**, 189–91.

Ronderos, M.M., Greco, N.M. and Spinelli, G.R. (2003) Diversity of biting midges of the genus *Culicoides* Latreille (Diptera: Ceratopogonidae) in the area of the Yacyreta Dam Lake between Argentina and Paraguay. *Mem. Inst. Oswaldo Cruz* **98**, 19–24.

Rosadio, R.H., Evermann, J.F. and DeMartini, J.C. (1984) A preliminary serological survey of viral antibodies in Peruvian sheep. *Vet. Microbiol.* **10**, 91–6.

Schmidtmann, E.T., Bobian, R.J. and Belden, R.P. (2000) Soil chemistries define aquatic habitats with immature populations of the *Culicoides variipennis* complex (Diptera: Ceratopogonidae). *J. Med. Entomol.* **37**, 58–64.

Schmidtmann, E.T., Mullens, B.A., Schwager, S.J. and Spear, S. (1983) Distribution, abundance, and a probability model for *Culicoides variipennis* on dairy farms in New York State. *Environ. Entomol.* **12**, 768–73.

Schmidtmann, E.T. (2006) Testing the relationship between dissolved salts and immature populations of the *Culicoides variipennis* complex (Diptera: Ceratopogonida. *Environ. Entomol.* 35, 1154–160.

Schultz, G. and Delay, P.D. (1955) Losses in newborn lambs associated with bluetongue vaccination of pregnancy ewes. *J. Am. Vet. Med. Assoc.* **127**, 224–6.

Spreull, J. (1905) Malarial catarrhal fever (bluetongue) of sheep in South Africa. *J. Comp. Pathol.* **18**, 321–37.

Stallknecht, D., Kellogg, M., Blue, J. and Pearson, J. (1991a) Antibodies to bluetongue and epizootic hemorrhagic disease viruses in a barrier island white tailed deer populatio. *J. Wildl. Dis.* **27**, 668–74.

Stallknecht, D.E., Blue, J.L., Rollor, E.A., Nettles, V.F., Davidson, W.R. and Pearson, J.E. (1991b) Precipitating antibodies to epizootic hemorrhagic disease and bluetongue viruses in white-tailed deer in the South-eastern United States. *J. Wildl. Dis.* **27**, 238–47.

Stallknecht, D.E. and Davidson, W.R. (1992) Antibodies to bluetongue and epizootic hemorrhagic disease viruses from white-tailed deer blood samples dried on paper strips. *J. Wildl. Dis.* **28**, 306–10.

Stott, J.L., Blanchard Channell, M., Osburn, B.I., Riemann, H.P., and Obeso, R.C. (1989) Serologic and virologic evidence of bluetongue virus infection in cattle and sheep in Mexico. *Am. J. Vet. Res.* **50**, 335–40.

Suzan, V.M., Onuma, M., Aguilar, R.E. and Murakami, Y. (1983) Prevalence of bovine herpes-virus-1, parainfluenza-3, bovine rotavirus, bovine viral diarrhea, bovine adenovirus-7, bovine leukemia virus and bluetongue virus antibodies in cattle in Mexico. *Jpn J. Vet. Res.* **31**, 125–32.

Tabachnick, W. (2004) *Culicoides* and the global epidemiology of bluetongue virus infection. *Vet. Ital.* **40**, 145–50.

Tabachnick, W.J. (1992) Genetic differentiation among populations of *Culicoides variipennis* (Diptera: Ceratopogonidae), the North American vector of bluetongue virus. *Ann. Entomol. Soc. Am.* **85**, 140–7.

Tabachnick, W.J. (1996) *Culicoides variipennis* and bluetongue-virus epidemiology in the United States. *Ann. Rev. Ent.* **41**, 23–43.

Tabachnick, W.J. and Holbrook, F.R. (1992) The *Culicoides variipennis* complex and the distribution of bluetongue viruses in the United States. *Proc. U.S. Anim. Health Assoc.* **96**, 207–12.

Tamayo, R., Schoebitz, R., Alonso, O. and Wenzel, J. (1985) First report of bluetongue antibody in Chile. *Prog. Clin. Biol. Res.* **178**, 555–8.

Tan, B.H., Nason, E., Staeuber, N., Jiang, W., Monastryrskaya, K. and Roy, P. (2001) RGD tripeptide of bluetongue virus VP7 protein is responsible for core attachment to *Culicoides* cells. *J. Virol.* **75**(8), 3937–47.

Tanya, V.N., Greiner, E.C. and Gibbs, E.P. (1992) Evaluation of *Culicoides insignis* (Diptera: Ceratopogonidae) as a vector of bluetongue virus. *Vet. Microbiol.* **32**, 1–14.

Tanya, V.N., Greiner, E.C., Shroyer, D.A. and Gibbs, E.P. (1993) Vector competence parameters of *Culicoides variipennis* (Diptera: Ceratopogonidae) for bluetongue virus serotype 2. *J. Med. Entomol.* **30**, 204–8.

Theiler, A. (1906) Bluetongue in sheep. *Ann. Rept. Dir. Agric. Transvaal.* **1904–1905**, 110–21.

Thomas, F.C., Skinner, D.J. and Samagh, B.S. (1982) Evidence for bluetongue virus in Canada: 1976–1979. *Can. J. Comp. Med.* **46**, 350–3.

Trainer, D.O. and Jochim, M.M. (1969) Serologic evidence of bluetongue in wild ruminants of North America. *Am. J. Vet. Res.* **30**, 2007–11.

Van Donkersgoed, J., Gertonson, A., Bridges, M., Raths, D., Dargatz, D., Wagner, B., Boughton, A., Knoop, D. and Walton, T.E. (2004) Prevalence of antibodies to bluetongue virus and *Anaplasma marginale* in Montana yearling cattle entering Alberta feedlots: Fall 2001. *Can. Vet. J.* **45**, 486–92.

Viegas de Abreu, V.L. (1983) Prevalence of reactions to the immunodiffusion test for bluetongue antibodies among cattle and buffaloes in northern Brazil. *Arquivo Brasileiro de Medicina Veterinaria e Zootecnia* **5**, 759–760.

Walton, T. (2004) The history of bluetongue and a current global overview. *Vet. Ital.* **40**, 31–8.

Walton, T.E., Tabachnick, W.J., Thompson, L.H. and Holbrook, F.R. (1992) An entomologic and epidemiologic perspective for bluetongue regulatory changes for livestock movement from the United Sates and observations on bluetongue in the Caribbean Basin. In: T.E. Walton and B.I. Osburn, (eds.), *Bluetongue, African Horse Sickness, and Related Orbiviruses*. Boca Raton: CRC Press, pp. 952–60.

Weiser-Schimpf, L., Wilson, W.C., French, D.D., Baham, A. and Foil, L.D. (1993) Bluetongue in sheep and cattle and *Culicoides variipennis* and *C. stellifer* (Diptera: Ceratopogonidae) in Louisiana. *J. Med. Entomol.* **30**, 719–24.

White, D.M., Wilson, W.C., Blair, C.D. and Beaty, B.J. (2005) Studies on overwintering of bluetongue viruses in insects. *J. Gen. Virol.* **86**, 453–62.

Wilson, W.C., Ma, H.C., Venter, E.H., van Djik, A.A., Seal, B.S. and Mecham, J.O. (2000) Phylogenetic relationships of bluetongue viruses based on gene S7. *Virus Res.* **67**, 141–51 (*Erratum* **73**, 201–2)

Wirth, W.W. and Jones, R.H. (1957) The North American subspecies of *Culicoides variipennis* (Diptera: Heleidae). *U.S. Dep. Agric. Tech. Bull.* 1170.

Wirth, W.W. and Morris, C. (1985) The taxonomic complex, *Culicoides variipennis. Prog. Clin. Biol. Res.* **178**, 165–75.

Wittmann, E.J., Mello, P.S. and Baylis, M. (2002) Effect of temperature on the transmission of orbiviruses by the biting midge, *Culicoides sonorensis. Med. Vet. Entomol.* **16**, 147–56.

Xu, G., Wilson, W.C., Mecham, J.O., Murphy, K., Zhou, E.M. and Tabachnick, W.J. (1997) VP7: An attachment protein of bluetongue virus for cellular receptors in *Culicoides variipennis. J. Gen. Virol.* **78**, 1617–23.

Bluetongue viruses in Australasia and East Asia

10

PETER DANIELS, ROSS LUNT AND IAN PRICHARD

Australian Animal Health Laboratory, CSIRO Livestock Industries, PMB 24, Geelong, Vic 3220, Australia

Introduction

The major bluetongue virus (BTV) episystem encompassing East Asia, Southeast Asia and Australia is incompletely, even poorly, described. Although comprehensive surveillance and detailed studies have been conducted over several decades in Australia, the information from Asian areas is mainly the result of studies conducted over relatively short periods of time, mostly 10–15 years ago.

Bluetongue (BT) disease is not a major animal health problem in the countries to the east and southeast of the Indian subcontinent. Bluetongue outbreaks have been reported rarely, involving only a relatively few animals and usually affecting imported European breeds of sheep. Hence there has been little need for most countries in the region to give priority to studies to isolate BTVs or to describe their ecology. Several countries in the region have not even recognized the presence of BTV, as indicated by data submitted to the Office International des Epizooties (OIE).

This chapter will summarize the main aspects of the knowledge of BTV in this extensive geographical region, including reports of disease, the BTV serotypes identified either through outbreak investigations or surveillance, the genetic relationships among isolates and the *Culicoides* species implicated or suspected as vectors. It is of interest to develop an understanding of whether the region essentially comprises a single episystem or a number of discrete ecological foci of BTV activity. Identification of trends in the parameters just outlined is important to help anticipate the possible emergence of new BT issues in the region.

ISBN-13: 978-0-12-369368-6

Major studies of bluetongue virus in the region

Much of the knowledge of BTV in east and Southeast Asia come from structured, prospective surveillance programmes conducted over a limited number of years rather than from passive surveillance based on routine laboratory testing over long periods. In many cases, a laboratory diagnostic capability was created in the study areas specifically for the purpose of the surveillance projects.

An early study was conducted in Indonesia in collaboration with Australia, based on the establishment of sentinel cattle herds at a number of geographical locations across the length of the country (Daniels *et al.*, 1995). Numerous BTVs were isolated over a number of years, serotyped (Sendow *et al.*, 1991b, 1992, 1993a, 1993b) and later partially characterized genetically (Sendow *et al.*, 1997; Pritchard *et al.*, 2004). Serological studies showed that BTV infections were widespread in ruminants, that large ruminants had a higher seroprevalence than small ruminants and confirmed the wider distribution across the country of serotypes that had been previously identified by virus isolation from the discrete groups of monitored sentinel cattle (Sendow *et al.*, 1991a). The studies included a large programme of trapping and identification of *Culicoides* spp (Sukarsih *et al.*, 1993, 1996).

A major study in Malaysia followed in response to disease observed in sheep imported to that country from Australia (Sharifah *et al.*, 1995). Again the study was based mainly on the systematic sampling of sentinel groups of animals. Isolates from the study were serotyped and investigated for virulence (Hassan *et al.*, 1996) and have subsequently become available for genetic analyses (Pritchard *et al.*, 2004).

In Thailand, a serological study for BTV antibodies was carried out in 1991 and demonstrated seroprevalences in the range 60–75% in indigenous cattle, sheep and goats (Apiwatnakorn *et al.*, 1996).

In China, BT has been recognized since 1979 (Zhang and Kirkland, 1998; Zhang *et al.*, 2004), the details being summarized by Zhang *et al.* (1996). A number of BTV strains were isolated during the investigations and a large epidemiological study initiated in collaboration with Australia (Zhang and Kirkland, 1998; Kirkland *et al.*, 2002). The sentinel cattle programme in this study yielded further isolates, and these isolates from both sheep and cattle together form the basis of much of our understanding of the BTV in the east Asian region.

In Japan, seroepidemiological studies have been conducted since 1974 (Goto *et al.*, 2004). Cases have been reported, particularly in imported animals, and BTV has been isolated. Bluetongue virus has also been isolated from asymptomatic, locally bred cattle in structured sentinel herd-based studies. Bluetongue virus infections are considered endemic in the more southerly areas of the country.

Bluetongue disease in Australasia and East Asia

Bluetongue is not reported in Australia. Although the geographical regions where sheep husbandry is important are largely separate from the zone where BTV is transmitted, there is overlap, and serological evidence indicates that sheep are occasionally infected asymptomatically. Experimental infections, using non-cell culture-adapted strains of BTV, confirm that many of the BTVs in Australia are of limited virulence or non-virulent. However, some isolates of Australian BTV from the Northern Territory have shown moderate virulence (Johnson et al., 1992; Hooper et al., 1996).

On two documented occasions in Southeast Asia, imported European breeds of sheep have shown clinical BT with some mortality, in situations where clinical BT is unknown in indigenous ruminants. In the Indonesian outbreak in 1981, morbidity was estimated at 50% with a case fatality rate of 3.3% (Sudana and Malole, 1982). In 1987, a similar incident was reported in Malaysia (Chiang, 1989) in which the morbidity was less, with 159 of 2249 imported animals being affected, but with a case fatality rate of 52% (Sharifah et al., 1995).

To investigate the inherent virulence of BTV in Southeast Asia, pathogenesis studies were conducted in both Malaysia and Indonesia under conditions that have allowed demonstration of disease in other countries. The viruses used in the trials had been maintained as wild-type virus in blood of the field-infected animal or had been maintained only by ruminant to ruminant passage. The challenged sheep were European breeds that had been shown previously to be susceptible to BT, and the animals in the studies were shown initially to be serologically naïve to BTV. In Malaysia, five serotypes were tested without eliciting clinical BT (Hassan et al., 1996). In the Indonesian studies, the tested strains were also apparently non-virulent or non-pathogenic under the conditions of the study (Sendow et al., 1997a).

Apiwatnakorn et al. (1996) reported a serological study for BTV antibodies in 1991, which demonstrated seroprevalences of 75% in indigenous sheep in Thailand without any reports of disease.

Hence across Southeast Asia, there is a pattern of only moderate pathogenicity of the local BTV fauna, with indigenous sheep being resistant to disease. European sheep showed moderate disease susceptibility under some field conditions, but without virulence being readily demonstrable in controlled experimental studies.

Bluetongue virus studies were initiated in China because of BT being reported in sheep in at least six provinces (Zhang et al., 2004). Case fatality rates of up to 30% were reported, but the data do not indicate a significant ongoing morbidity problem (Zhang et al., 1996). The extent of the BT disease problem in Japan is also uncertain, with BT having been reported in relatively small numbers in both cattle and sheep (Goto et al., 2004). Bluetongue virus was isolated from some affected sheep but not from the cattle.

This BT situation throughout Southeast Asia and Australia, and to a certain extent in east Asia, contrasts with the situation in south Asia as described in India. There, pathogenic BTV infections have been reported in both Merino and local breeds of sheep (Prasad *et al.*, 1992; Sreenivasulu *et al.*, 2004). The severity of the disease with characteristic pathological changes indicates that highly virulent BTVs are circulating in that country. There may also be BTV of moderate and low virulence, but published data are limited. India experiences BT disease of far greater severity than has been observed in countries to the east and southeast, as reviewed above, and BT has been nominated as one of the most important sheep diseases in India (Sreenivasulu *et al.*, 2004). Although case fatality rates are of an order reported elsewhere, up to 30%, it is the susceptibility of indigenous breeds of sheep that creates a notable difference in the severity of the disease in India.

Bluetongue virus serotypes isolated in the region

The BTV serotypes known to occur in the countries of Australasia and east Asia are shown in Figure 10.1, based on published information (Prasad *et al.*, 1992; Daniels *et al.*, 1995, 2004; Sharifah *et al.*, 1995; Goto *et al.*, 2004; Sreenivasulu *et al.*, 2004; Zhang *et al.*, 2004). Although the BTV situation in India is not the subject of this chapter, it is necessary to note the BTV serotypes present in that country to develop a full understanding of the ecology of BTV throughout Asia to the east and southeast.

It is interesting to observe how widely serotypes BTV-1, -3, -9, -16 and -23 are distributed, having been isolated in all the countries where structured surveillance programmes have been conducted as well as in India. There are seven serotypes not yet reported either in India or the countries further east; BTV-8 and -18 have not been reported east of India, and a single report of BTV-17 from China has been listed as unconfirmed (Kirkland *et al.*, 2002). The commonality of serotypes present as demonstrated in Table 10.1, and conversely the considerable number of serotypes not present in the region, confirms that this broad geographical area has elements of a common BTV episystem. India at this time appears to have a richer BTV fauna than the other countries, except China, so a potential role of this region as a possible source of BTV genetic material within the broader episystem may be interesting to explore.

Within the dominant pattern, smaller groupings of serotypes can be noted. BTV-2 is present in India, Malaysia and China but not in Indonesia or Australia further to the south and east. Conversely, serotypes 20 and 21 have not been isolated in India but are present in various countries further east, in Southeast Asia, Australia, China and Japan (see Table 10.1). These findings suggest that much more study is needed to identify the full range of virus

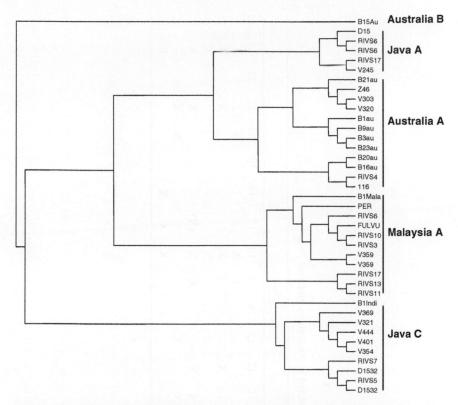

Figure 10.1   Genetic relationships among bluetongue virus (BTV) isolates from Australia and Southeast Asia, showing discrete genetic groupings and a relationship to an isolate from India.

serotypes actually present across the region to enable a comprehensive interpretation to be made.

It is beyond the scope of this chapter to consider all the within-country patterns of BTV distribution, but details are available in the cited references. Local factors will be important in determining specific BTV ecological niches within the broader episystem. An example is the distribution of BTV serotypes in Australia, which leads to an interpretation that there are at least two basic BTV ecological systems in that country, a somewhat restricted focus in the wet tropics of the north of the Northern Territory where all known Australian serotypes have been isolated and a broad zone of BTV distribution throughout the northern and eastern Australian pastoral areas in which only two of the eight recorded serotypes in Australia are found, BTV-1 and BTV-21 (Daniels *et al.*, 1997, 2004). Annual sequence analysis of BTV isolates from around the Australian BTV-infected zone (Anon, 2007) shows that this situation remains stable until the present time.

Table 10.1 Bluetongue virus (BTV) serotypes reported as isolated in Australasia and East Asia

Country	Serotype																							
	1	2	3	4	5	6	7	8	9	10	11	12	13	14	15	16	17	18	19	20	21	22	23	24
Australia	X		X						X						X	X				X	X		X	X
Indonesia	X		X				X		X			X				X					X		X	
Malaysia	X	X	X						X							X							X	
China	X	X	X	X					X		X	X			X	X	(X)			X	X		X	
Japan				X									X					X		X	X			
(India)	X	X	X	X				X	X		X	X			X	X	X						X	

The bluetongue virus genotypes in the region

Pioneering studies of sequence analysis of BTV isolates from different parts of the world established that viruses in separate geographical locations had consistent similarities among themselves and differences from BTV from other locations (Gould, 1987). Hence BTV from South Africa, Australia and North America were shown to have identifiable 'topotypes'.

Early sequence analyses of BTV from the study sites in Southeast Asia and Australia showed that the viruses from both Malaysia and Indonesia were in the same broad topotype, described by Gould (1987), as the Australian topotype but could also be sub-grouped into a number of constant genotypes, some of which were apparently confined to local areas and others were more widespread in the region (Melville *et al.*, 1997; Sendow *et al.*, 1997b). A close genetic relationship between the Southeast Asian isolates and some isolates from India was also reported, and the likelihood that the south Asian–Southeast Asian geographical region hosted a BTV episystem with mobility of viruses in the direction of the prevailing climatic forces was raised (Daniels *et al.*, 1997). A summary of these analyses is presented in Figure 10.1, which shows that some Indian BTVs group with Indonesia in the Java C genotype, Indonesian and Malaysian isolates group together in the Malaysia A genotype and Australian and Indonesian isolates group together in the Australia A genotype.

It is enormously valuable to have numerous BTV isolates collected over significant periods of time for analysis. This has been demonstrated by the evidence of incursions of viruses from Southeast Asia into the northern Australian ecosystem (Daniels *et al.*, 1997; Melville *et al.*, 1997). Such observations are not possible where the BTV fauna in adjacent regions is not adequately described. Five separate observations of BTV genes consistent with a Southeast Asian origin have been made in isolates of BTV in the Northern Territory of Australia since 1992. These observations are interpreted as representing five separate incursions of BTV into northern Australia, indicating international movements of BTV within the broader episystem due to natural influences such as wind-borne dispersal of infected vectors.

Pritchard *et al.* (2004) extended the significance of these observations by documenting evidence for a reassortment between an introduced BTV and the pre-existing strains, with a hybrid, naturally genetically recombinant BTV strain established as an ongoing part of this BTV ecological niche. Since the reassortant virus persists in this northern ecological niche until the present time, a focus of current surveillance programmes in Australia is to detect any movement of this reassortant virus beyond the area to which it is currently restricted.

An analysis of the S10 gene of isolates from China confirmed these BTV strains to be of the Australasian topotype and clearly genetically different from the BTVs present in North America (Zhang *et al.*, 1999).

The known and suspected *Culicoides* spp vectors

Bluetongue viruses are transmitted among ruminant hosts by *Culicoides* spp insect vectors. The species of *Culicoides* that are competent to act as vectors differ in some cases from continent to continent but are basically few in number compared with the number of species in the genus. It is hypothesized that different vector species are crucial in defining the various BTV episystems and may substantially determine the expression of certain phenotypic characteristics of BTV strains such as virulence (Daniels *et al.*, 1997, 2004; Tabachnick, 2004).

In Australia, the competent vector species have been defined experimentally as *Culicoides fulvus*, *Culicoides wadai*, *Culicoides actoni* and *Culicoides brevitarsis* (Standfast *et al.*, 1985). These four species are also widely distributed in Indonesia as defined by BTV surveillance programmes (Sukarsih *et al.*, 1996). More recently, it has been recognized that some of the insects identified as *C. fulvus* in northern Australia are actually *Culicoides dumdumi*, so Australia has both these species confirmed as competent vectors (Bellis and Dyce, 2005). Throughout south, east and Southeast Asia, specific vector competence studies have not yet been attempted. However, the geographical distribution of these five vector species in Asia has been reviewed (Wirth and Hubert, 1989).

Most of the species of vector present in Southeast Asia and Australia have also been reported in India (Wirth and Hubert, 1989). *Culicoides imicola*, the predominant vector in South Africa where virulent BTV can cause severe disease problems, is present in India but not in the countries of Southeast Asia where BTV has been studied in detail. *Culicoides imicola* has been reported from Thailand and Vietnam (Wirth and Hubert, 1989) but sheep husbandry is not a significant industry in these countries and BT has not been reported there. Li *et al.* (1996) reported that *C. imicola* is found in parts of China where BT has also been observed, as are other vector species that are also widely distributed throughout this episystem, namely *C. actoni*, *C. fulvus and C. brevitarsis*. In Japan, BTV has been isolated from *C. brevitarsis* on a number of occasions (Goto *et al.*, 2004).

To fully describe the episystem(s) of the region, it is important to establish whether other *Culicoides* in the *Orientalis* taxonomic group, which are also widely distributed from India eastward and to the southeast (Wirth and Hubert, 1989), such as *Culicoides orientalis*, *Culicoides maculatus* and *Culicoides nudipalpis* may also be competent as vector species. However, the necessary investigations have not yet been attempted.

Conclusions

All molecular analyses to date confirm that the BTVs of south, east and Southeast Asia and Australia are of the one topotype, consistent with that initially described by Gould (1987), and so it is tempting to consider this large geographical area as a single episystem. However, similar analyses would group the BTV of Africa and the America into just one other, second, topotype. Dividing the globe into these broad groupings has been useful in developing an understanding of the likely origins of BTV in the Middle East and Europe in recent years, but in general it is too broad to assist with the development of a more detailed understanding of BTV ecology and BT pathogenesis.

Tabachnick (2004) described the concept of an episystem as 'consisting of the species and environmental aspects of an epidemiological system in a particular ecosystem which affects the distribution and dynamics of a pathogen and disease'. Tabachnick (2004) was particularly interested in the concept of episystems based on vector species, and Daniels *et al.* (1997, 2004) particularly noted variability in BTV isolates from India through to eastern Australia in aspects such as apparent virulence and genotype that were associated spatially with differences in the species of known vectors.

The distribution of BTV serotypes in Australia has been mentioned above, with only BTV-1 and BTV-21 being found in the broad pastoral areas. Here the predominant vector species is *C. brevitarsis*. Although *C. brevitarsis* is also present in the far northern area where multiple BTV serotypes have been detected, species such as *C. actoni* and *C. fulvus* are much more abundant, and as they have much greater infection and susceptibility rates, experimentally, than *C. brevitarsis* (Standfast *et al.*, 1985), they are assumed to play a much greater role in BTV transmission in this northern area. The BTVs from this region show moderate virulence similar to that reported in the field outbreaks in Indonesia and Malaysia, but the incursions of BTV with Southeast Asian genotypes have not spread beyond this ecological niche in the north where *C. actoni* and *C. fulvus* are believed to be more important as vectors. Although a clear cause-and-effect relationship between the vectors and the genetic and phenotypic features of the BTV has not been established, their association with this geographical location could be said to describe a small episystem.

The main vector species that are present in northern Australia are present throughout Southeast Asia and have also been described in China. The discussion of serotypes earlier in this chapter noted the substantial similarity in serotypes present in and absent from this region, with some local variations (see Table 10.1). Differences in genotypes independent of serotype have been detected (see Figure 10.1), suggesting that across the Asian region, there is substantial stability in the strains of BTV present in certain geographical areas. More sites are needed in countries not yet studied, and new collections of BTV are needed in those countries studied previously, to allow a more comprehensive description of the BTVs of the whole region and their relationships to each other. Such analyses might be expected to provide evidence for patterns of movement within the region, such as the demonstration of the likely incursions of BTV from Southeast Asia into northern Australia over the last 15 years (Pritchard *et al.*, 2004).

The presence of the vector *C. imicola* in India has been discussed above, in conjunction with its presence in China and the countries in between and its absence from Southeast Asia and Australia. India has some other apparent differences from east and Southeast Asia; a greater range of serotypes and an apparently more severe disease problem, suggestive of BTV of greater virulence (Daniels *et al.*, 1997, 2004). Hence it may be appropriate to consider the Indian subcontinent as a discrete episystem, but whether *C. imicola* could be considered to be the dominant determining factor of that episystem is open to question, given its occurrence in countries to the east. It can only be reiterated that a useful understanding of the complex relationships in putative episystems can develop only from the availability of a wide range of BTV isolates for study from the regions concerned.

It has been hypothesized that expression of virulence by BTV strains may be influenced by the vector species in which they are transmitted (Daniels *et al.*, 1997, 2004). Such a relationship would strongly link the important features of a BT episystem to the dominant vector species.

References

Anon. (2007) *The National Arbovirus Monitoring Program*, http://www.animalhealthaustralia.-com.au/programs/adsp/namp/namp_home.cfm. accessed 2 May 2007.

Apiwatnakorn, B., Bura, P. and Pasavorakul, O. (1996) Serological study for bluetongue in Thailand. In: T.D. St George and P. Kegao (eds), *Bluetongue Disease in Southeast Asia and the Pacific*. Canberra: ACIAR. pp. 20–2.

Bellis, G.A. and Dyce, A.L. (2005) Intraspecific variation in three species of culicoides in the orientalis complex of the subgenus avaritia within the Australasian zoogeographic region. *Arbovirus Res. Aust.* **9**, 33–6.

Chiang, B.L. (1989) An outbreak of bluetongue in imported sheep. In: Proceedings, 1st Congress, Veterinary Association of Malaysia, pp. 49–52.

Daniels, P.W., Melville, L., Pritchard, I., Sendow, I., Sreenivasulu, D. and Eaton, B. (1997) A review of bluetongue viral variability along a South Asia – eastern Australia transect. *Arbovirus Res. Aust.* **7**, 66–71.

Daniels, P.W., Sendow, I., Pritchard, I., Sukarsih, and Eaton, B. (2004) Regional overview of bluetongue viruses in Southeast Asia: viruses, vectors and surveillance. In: Proceedings, 3rd International Bluetongue Symposium, OIE, Paris. *Vet. Ital.* **40**, 94–100.

Daniels, P.W., Sendow, I., Soleha, E., Sukarsih, Hunt, N.T. and Bahri, S. (1995) Australian-Indonesian collaboration in veterinary arbovirology. A review. *Vet. Microbiol.* **46**, 151–74.

Goto, Y., Yamaguchi, O. and Kubo, M. (2004) Epidemiological observations on bluetongue in sheep and cattle in Japan. *Vet. Ital.* **40**, 78–82.

Gould, A.R. (1987) The nucleotide sequence of bluetongue virus serotype 1 RNA 3 and a comparison with other geographic serotypes from Australia, South Africa, the United States of America and other orbivirus isolates. *Virus Res.* **7**, 169–83.

Hassan, S.S., Gard, G.P., Polkinghorne, I., Abdul Rahman, M.A., Hussin, A.A., Zainal, M. and Aminah Kadariah, L. (1996) Recent studies on bluetongue in peninsular Malaysia.

In: T.D. St George and P. Kegao (eds), *Bluetongue Disease in Southeast Asia and the Pacific*. Canberra: ACIAR, pp. 23–7.

Hooper, P.T., Lunt, R.A. and Stanislawek, W.L. (1996) A trial comparing the pathogenicity of some South African and Australian bluetonguc viruses. *Aust. Vet. J.* **73**, 36–7.

Johnson, S.J., Hoffmann, D., Flanagan, M., Polkinghorne, I.G. and Bellis, G.A. (1992) Clinico-pathology of Australian bluetongue virus serotypes for sheep. In: E. Walton and B.I. Osburn (eds), *Bluetongue, African Horse Sickness and Related Orbiviruses*. Boca Raton: CRC Press, pp. 737–43.

Kirkland, P.D., Zhang, N., Hawkes, R.A., Li, Z., Zhang, F., Davis, R.J., Sanders, D.A., Li, H., Ben, J., He, G.F., Hornitzky, C.L. and Hunt, N.T. (2002) Studies on the epidemiology of bluetongue virus in China. *Epidemiol. Infect.* **128**, 257–63.

Li, H., Li, Z., Zhang, K., Liu, G., Zhou, F. and Shun, Y. (1996) Bluetongue vector surveillance in Yunnan and Sichuan provinces of China. In: T.D. St George and P. Kegao (eds), *Bluetongue Disease in Southeast Asia and the Pacific*. Canberra: ACIAR, pp. 148.

Melville, L.F., Pritchard, L.I., Hunt, N.T., Daniels, P.W., and Eaton, B. (1997) Genotypic evidence of incursions of new strains of bluetongue viruses in the northern territory. *Arbovirus Res. Aust.* **7**, 181–6.

Prasad, G., Jain, N.C. and Gupta, Y. (1992) Bluetongue virus infection in India: a review. *Rev. Sci. Tech. Off. Int. Epizoot.* **11**, 699–711.

Pritchard, L.I., Daniels, P.W., Melville, L.F., Kirkland, P.D., Johnson, S.J., Lunt, R. and Eaton, B.T. (2004) Genetic diversity of bluetongue viruses in Australasia. *Vet. Ital.* **40**, 438–45.

Sendow, I., Daniels, P.W., Cybinski, D.H., Young, P.L. and Ronohardjo, P. (1991a) Antibodies against certain bluetongue and epizootic haemorrhagic disease viral serotypes in Indonesian ruminants. *Vet. Microbiol.* **28**, 111–8.

Sendow, I., Daniels, P.W., Soleha, E., Erasmus, B.J., Sukarsih and Ronohardjo, P (1993a) Isolation of bluetongue virus serotypes new to Indonesia from sentinel cattle in West Java. *Vet. Rec.* **133**, 166–8.

Sendow, I., Daniels, P.W., Soleha, E., Hunt, N. and Ronohardjo, P. (1991b) Isolation of bluetongue viral serotypes 7 and 9 from healthy sentinel cattle in West Java, Indonesia. *Aust. Vet. J.* **68**, 405.

Sendow, I., Daniels, P.W., Soleha, E. and Sukarsih (1992) Epidemiological studies of bluetongue viral infections in Indonesian livestock. In: T.D. St George and P. Kegao (eds), *Bluetongue, African Horse Sickness and Related Orbiviruses*. Boca Raton: CRC Press, pp. 147–54.

Sendow, I., Hamid, H., Sukarsih, Soleh, E., Bahri, S., Pearce, M. and Daniels, P.W. (1997a) Clinical and pathological changes associated with the propagation of Indonesian bluetongue viral isolates in Merino sheep. *Hemera Zoa* **79**, 124–34.

Sendow, I., Pritchard, L.I., Eaton, B.T. and Daniels, P.W. (1997b) Genotypic relationships of bluetongue viruses from Indonesia. *Arbovirus Res. Aust.* **7**, 263–6.

Sendow, I., Soleha, E., Daniels, P.W., Sebayang, D., Achdiyati, J., Karma, K. and Erasmus, B.J. (1993b) Isolation of bluetongue virus serotypes 1, 21 and 23 from healthy cattle in Irian Jaya, Indonesia. *Aust. Vet. J.* **70**, 229–30.

Sharifah, S.H., Ali, M.A., Gard, G.P. and Polkinghorne, I.G. (1995) Isolation of multiple serotypes of bluetongue virus from sentinel livestock in Malaysia. *Trop. Anim. Health Prod.* **27**, 37–42.

Sreenivasulu, D., Subba Rao, M.V., Reddt, Y.N. and Gard, G.P. (2004) Overview of bluetongue disease, viruses, vectors, surveillance and unique features: the Indian subcontinent and adjacent regions. *Vet. Ital.* **40**, 73–7.

Standfast, H.A., Dyce, A.L. and Muller, M.J. (1985) Vectors of bluetongue virus in Australia. In: T.L. Barber, M.M. Jochim, and B.I. Osburn (eds), *Bluetongue and Related Orbiviruses.*. New York: Alan R. Liss Inc, pp. 177–86.

Sudana, I.G. and Malole, M. (1982) Penyidikan penyakit hewan bluetongue di desa Caringin, Kabupaten Bogor. In: Annual Report of Disease Investigation in Indonesia during the Period of 1975–1981, Jakarta: Direktorat Kesehatan Hewan, Direktorat Jenderal Peternakan, pp. 110–21.

Sukarsih, Daniels, P.W., Sendow, I. and Soleha, E. (1993) Longitudinal studies of culicoides spp associated with livestock in Indonesia. In: M.F. Uren and B.H. Kay (eds), *Arbovirus Research in Australia – Proceedings of 6th Symposium*. Brisbane: CSIRO/QIMR, 203–209.

Sukarsih, Sendow, Bahri, S., Pearce, M and Daniels, P.W. (1996) *Culicoides* surveys in Indonesia. In: T.D. St George and P. Kegao (eds), *Bluetongue Disease in Southeast Asia and the Pacific*. Canberra: ACIAR, pp. 123–8.

Tabachnick, W.J. (2004) *Culicoides* and the global epidemiology of bluetongue virus infection. *Vet. Ital.* **40**, 145–50.

Wirth, W.W. and Hubert, A.A. (1989) *The* Culicoides *of Southeast Asia (Diptera: Ceratopogonidae)*. Gainesville, Florida, USA: The American Entomological Institute.

Zhang, N. and Kirkland, P.D. (1998) Bluetongue research in china. *Vet. Rec.* **143**, 231–2.

Zhang, N., Li, Z., Zhang, K., Hu, Y., Li, G., Peng, K., Li, H., Zhang, F., Ben, J., Li, X., Zhou, F. and Liu, G. (1996) Bluetongue history, serology and virus isolation in china. In: T.D. St George and P. Kegao (eds), *Bluetongue Disease in Southeast Asia and the Pacific*. Canberra: ACIAR, pp. 43–50.

Zhang, N., Li, Z., Zhang, F. and Zhu, J. (2004) Studies on bluetongue disease in the people's Republic of China. *Vet. Ital.* **40**, 51–6.

Zhang, N., MacLachlan, N.J., Bonneau, K.R., Zhu, J., Zhang, K., Zhang, F., Xia, L. and Xiang, W. (1999) Identification of seven serotypes of bluetongue virus from the People's Republic of China. *Vet. Rec.* **145**, 427–9.

Bluetongue in Europe and the Mediterranean Basin

11

PHILIP S. MELLOR[*], SIMON CARPENTER[*],
LARA HARRUP[*], MATTHEW BAYLIS[†],
ANTHONY WILSON[*] AND PETER P.C. MERTENS[*]

[*]Institute for Animal Health, Pirbright Laboratory, Pirbright, Woking, Surrey, UK
[†]University of Liverpool, Veterinary Clinical Science, Leahurst, Cheshire, UK

Introduction

Bluetongue virus (BTV) is traditionally understood as occurring around the world in a broad band stretching from about 35°S to 40°N although in certain areas (i.e. western North America and northern China), it may extend up to around 50°N (Mellor et al., 2000; Mellor and Wittmann, 2002). As BTV is transmitted between its vertebrate hosts almost entirely via the bites of *Culicoides* biting midges, its world distribution is limited to geographical areas where competent vector species of *Culicoides* are present and its transmission to those times of the year when climatic conditions are (1) favourable for adult vector activity and (2) sufficiently warm for long enough to allow virus replication in, and transmission by, the vectors.

In Europe, until recently, the above conditions were apparently fulfilled only in parts of Portugal, SW Spain, certain Greek Islands adjacent to the Anatolian Turkish coast and Cyprus since, prior to the 1990s, these were the only areas to have experienced BTV incursions. During the period 1956–1960, BTV-10 entered Portugal and SW Spain from North Africa and caused the deaths of almost 180 000 sheep (Manso-Ribeiro et al., 1957; Campano Lopez and Sanchez Botija, 1958; Manso-Ribeiro and Noronha, 1958; Gorman, 1990), while during the years 1979–1980, BTV-4 entered the Greek Islands of Rhodes and Lesbos (Vassalos, 1980; Dragonas, 1981). Cyprus seemed to be the only area of Europe where BTV occurred regularly, and virus activity had been detected there in at least 21 of the 50 years between 1924 and 1973 (Sellers, 1975).

In the late 1990s, however, Europe has witnessed a dramatic change in the epidemiology of bluetongue (BT).

Bluetongue virus in the Mediterranean Basin, 1998–2005

Between 1998 and 2005, incursions of BTV occurred into many countries around the Mediterranean Basin that are usually free from the virus. Indeed, most of the countries involved had no previous record of BTV on their territory. At least five serotypes of virus were involved, originating from several sources to the east and south of the Mediterranean Basin. In all, virus activity was detected in at least 12 European countries, three countries in N. Africa and Israel, in the Near East. The total number of sheep dying from disease or culled for control and welfare purposes has been estimated at well over one million, which makes this by far the most severe outbreak of BT on record (Purse *et al.*, 2005).

Bluetongue virus in Greece, Turkey, the Balkans, Cyprus and Israel

Greece
In October 1998, after an absence of almost 20 years from Europe, BTV was confirmed in four Greek Islands (Rhodes, Leros, Kos and Samos) close to the Anatolian Turkish coast (Figures 11.1 and 11.2; Anon, 1998a, b; Anon,

Figure 11.1 Bluetongue virus (BTV) activity in the Eastern Mediterranean Basin, 1998–2005, overall area affected (base maps: NASA and Earthsat, 2004).

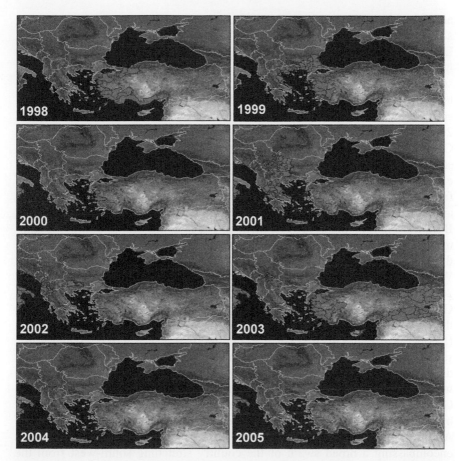

Figure 11.2 Bluetongue virus (BTV) activity in the Eastern Mediterranean Basin by year, from 1998 to 2005 (base maps: NASA and Earthsat, 2004).

1999a). The virus was identified as being BTV serotype 9 (Anon, 2000h), and this was the first occasion that this serotype had been recorded anywhere in Europe, though it had been identified serologically in western Anatolian Turkey during the period 1979–1980 (Taylor and Mellor, 1994a).

Outbreaks in the four Greek islands were recorded until late December 1998 (Anon, 1999a, 2000d), and then ended, probably because activity of the major vector in the area, *Culicoides imicola* Kieffer, is at a minimum during this period and for the remainder of the winter (Anon, 2000f; M. Patakakis, Personal Communication). No further outbreaks of BT anywhere in eastern Europe or Asia Minor were reported over the next 5 months.

In August 1999, however, the Greek authorities reported the presence of BTV in mainland Greece for the first time, initially in the north-eastern Prefecture adjacent to the Turkish and Bulgarian borders (see Figures 11.1

and 11.2; Anon, 1999b) and just subsequent to the announcement of BTV in adjacent areas of Bulgaria (in June) and European Turkey (in July) (see below). During September and October, the virus spread across northern Greece travelling in a westerly direction as far west as Thessaloniki and Larissa and as far south as Magnisia and Evia, thus rendering approximately one-third of Greek territory 'infected' (see Figures 11.1 and 11.2; Anon, 1999d). New cases of BT continued to be reported in some areas of Greece into December 1999 (Anon, 2000a). Also, during September 1999, a separate incursion of BTV into the Greek island of Lesbos was reported, and further outbreaks were reported in several Dodecanese islands (Anon, 1999c). By the end of 1999, serological evidence of the virus and/or clinical signs of BT in Greece had been reported from 13 Aegean islands and also from 10 mainland prefectures (see Figures 11.1 and 11.2; Anon, 2000f; Mellor and Wittmann, 2002). During 2000, low-level BTV transmission in Greece continued to occur, sporadically, during the summer months and BTV-4 was detected in the Prefecture of Arta in the NW of the country, a previously unaffected area (Anon, 2000b, 2001a; Mellor and Wittmann, 2002; Nomikou, 2006, Personal Communication), but in late August 2001, evidence of a rather more active BTV focus was once more detected in Greek territory, initially in the NW of the country. The virus was first detected in the Prefectures of Grevena, Ioannina and Kastoria close to the Albanian border (Anon, 2001c), and the disease was reported to advance into the country in a south/south-westerly direction, along the routes of rivers and canyons, in the direction of the prevailing winds (see Figures 11.1 and 11.2). This information was interpreted as suggesting that the outbreak represented a new incursion into Greece (presumably from areas to the north) rather than resurgence. Eventually by November 2001, the incursion had extended to include the Prefectures of Aitoloakarnania, Evritania, Larissa, Preveza, Thesprotia and Trikala also, thus encompassing a broad swath of the country stretching south-west from the Albanian border to Larissa on the Aegean coast (see Figures 11.1 and 11.2). The island of Lesbos close to the Anatolian coast of Turkey also experienced 19 new outbreaks in what, because of the distances involved, was presumably a separate incursion. Since autumn 2001, however, the Greek territories have been free of reported BTV activity.

A complicating factor that emerged during this series of Greek BTV incursions was that in addition to BTV-9, three further serotypes were identified: 1, 4 and 16 (Anon, 2000h). Both BTV-4 and -9 had been reported previously from regions immediately to the east of Europe, i.e. BTV-4 and -9 from Lesbos, Anatolian Turkey, Syria and Jordan (Taylor and Mellor 1994a; Taylor, 1987; Mellor and Pitzolis, 1979; Vassalos, 1980;) and BTV-4 from Cyprus and Israel (Mellor and Pitzolis, 1979; Shimshony, 1987) at various times during the period 1978–1980. BTV-16 occurs regularly in Israel (Shimshony, 1987; Taylor and Mellor, 1994a), and during the present series of outbreaks, it was also reported, in August 2000, from Anatolian Turkey (see below). BTV-1, however, had not been previously reported from any country in or adjacent to the Mediterranean Basin.

During the course of the Greek outbreaks, control was implemented through surveillance, animal movement restrictions, slaughter and insecticide treatment of infected and 'at-risk' premises; at no time was vaccination practiced anywhere in Greek territory.

Turkey

In July 1999, the Turkish authorities reported a BTV incursion into European Turkey involving animals in two provinces bordering Bulgaria and mainland Greece (see Figures 11.1 and 11.2; Anon, 1999b). Then in October and November 1999, outbreaks of BT were reported, for the first time in the current epizootic, from western Anatolian Turkey involving animals in four provinces (see Figures 11.1 and 11.2; Anon, 1999d). Retrospectively, unofficial sources from Turkey also suggested that BTV had been active throughout western and southern Anatolian Turkey, from early 1998 (see Figures 11.1 and 11.2; A. Ozkul, Personal Communication). In August 2000, BT was again reported (from Izmir Province) but this time serotype 16 of the virus was isolated (see Figures 11.1 and 11.2; Anon, 2000c; C. Hamblin, Personal Communication). In 2003, serological surveys in cattle were carried out in five provinces in European Turkey and 23 provinces in Anatolian Turkey (six in the west, eight in the NE and nine in the SE) (see Figures 11.1 and 11.2; Final Report EU project QLK2-CT-2000-00611; A. Ozkul, Personal Communication). Seroprevalence rates of up to 90% were detected in some regions. Serum neutralization tests against BTV-4, -9 and -16 showed that all three serotypes were active during 2003 in Anatolian Turkey and that BTV-9 and -16 were also active in European Turkey (Final Report EU project QLK2-CT-2000-00611; A. Ozkul, Personal Communication). Control has been implemented in Turkey via clinical surveillance, animal movement restrictions and also by the use of a locally produced live attenuated virus vaccine against BTV-4 (Mellor and Wittmann, 2002).

Bulgaria

In June 1999, BTV invaded Bulgaria for the first time ever and serotype 9 was confirmed in south-east part of the country (see Figures 11.1 and 11.2; Anon, 1999e). The virus spread rapidly in a westerly direction across southern Bulgaria, eventually involving ruminant animals in four districts: Bourgas, Yambol, Haskovo and Kardjaly (see Figures 11.1 and 11.2; Anon, 1999b). In total, 85 villages were affected, and clinical signs were recorded in every flock examined in these villages, albeit at a low prevalence rate of 1.7% within flocks (Purse et al., 2006). At the same time, nine out of 19 sentinel herds of mixed ovine and bovines situated in Yambol, Haskovo and Kardjaly districts also seroconverted confirming BTV transmission at least 30 km from disease outbreak areas and up to 70 km from the Bulgarian border, which suggests that the areas of BTV transmission were considerably more widespread than the reported outbreaks of disease (Purse et al., 2006). Disease outbreaks persisted until late October 1999, at which time daily temperatures fell sharply and transmission apparently ceased. Subsequent to this time and for almost two years, no serological or clinical evidence of BTV transmission was recorded; however, in September 2001, a second incursion, also of BTV-9, was detected in the extreme western part of the country involving 76 villages in seven districts (Vidin, Montana, Vratsa, Sofia, Pernik, Kiustendil and Blagoevgrad) on or close to the borders with the FYR of Macedonia, Yugoslavia and Romania (see Figures 11.1 and 11.2; Purse *et al.*, 2006). Unlike the first incursions, a characteristic of these new outbreaks was that clinical signs were rarely observed (i.e. in only 27 of 25 929 susceptible sheep in affected areas) and almost all of the outbreaks were within 40 km of the international borders. Operation of a sentinel herd system in the region during 2001 indicated that, just as in 1999, transmission ceased in October with the advent of cooler weather. The sentinel system continued to operate in western and southern Bulgaria during 2002 when seroconversion to BTV, but no clinical disease was recorded in the south of the Smolyan and Blagoevgrad Districts in SW Bulgaria, close to the Greek border (see Figures 11.1 and 11.2). Subsequent to 2002 and up to mid-2006, BTV activity has not been detected anywhere within the territory of Bulgaria.

Control has been implemented in Bulgaria via clinical and serological surveillance, animal movement restrictions, insecticide treatment and, in 1999–2000, restricted use of an imported, live attenuated, pentavalent vaccine against BTV serotypes 3, 8, 9, 10 and 11.

Other Balkan countries (Macedonia, Albania, Yugoslavia, Kosovo, Croatia and Bosnia)

Following on from the reports of BTV recrudescence in Greece and Bulgaria during 2001, the Macedonian veterinary authorities introduced compulsory clinical surveillance for BT. In September 2001, typical signs of BT were identified in sheep in the Kriva Palanka Region in the north of the country close to the border with Yugoslavia (see Figures 11.1 and 11.2; Anon, 2001h). The affected animals were slaughtered, zoosanitary control measures implemented and clinical surveillance extended throughout the whole country with serological surveillance in some areas. During early October, further areas of sporadic BTV activity were identified in an additional seven regions scattered across the country from east to west, with most cases being in the regions of Kriva Palanka, Berovo and Krusevo (see Figures 11.1 and 11.2; Anon, 2001j). The number of animals affected was reported to be low and the clinical signs mild. The last clinical case was identified on 16 October 2001, subsequent to which time the ambient temperature decreased. No further BTV activity has been reported from the country.

At more or less the same time as in the FYR of Macedonia, BT was also identified for the first time in 2001 in the Federal Republic of Yugoslavia, in the south, south-east and south-west of Serbia, the south of Montenegro, and also in the autonomous province of Kosovo (see Figures 11.1 and 11.2; Anon,

2001k; Anon, 2000l). Thirty-seven outbreaks in 16 municipalities were recorded in Serbia proper, and the virus was identified as BTV-9 (Anon, 2001f). In 2002, serological surveys indicated that the virus had extended into more northerly areas of Serbia and Montenegro reaching the River Sava, which at about 44°50′N is the furthest north that BTV has ever reached in Europe (see Figures 11.1 and 11.2; Djuricic *et al.*, 2004). In September 2003, BTV activity was again reported from Yugoslavia, from four municipalities in Serbia. In Kosovo, subsequent to 2001, BTV has not been reported officially, but serological evidence from cattle and sheep in the province clearly show that the virus has continued to circulate during 2002, 2003 and 2004 (see Figures 11.1 and 11.2; Osmani *et al.*, 2006).

The presence of BTV in Croatia was first reported in late 2001 on the basis of serological and clinical findings in sheep in the Dubrovnik Region (see Figures 11.1 and 11.2; Anon, 2001m). In late 2002, a sentinel system of bovines was operated in parts of southern Croatia and seroconversions to BTV-9 were detected during October–November of that year, with the areas affected being rather more northerly than in 2001 (see Figures 11.1 and 11.2; Labrovic *et al.*, 2004; Listes *et al.*, 2004). In 2004, serological evidence of BTV transmission in Croatia was again reported from sentinel bovines also in the Dubrovnik Region and the presence of BTV-9 and -16 identified – this was and is the only report of BTV-16 from any part of the Balkans. Subsequent to October 2004, BTV activity has not been reported in any part of Croatia.

In Bosnia and Herzegovina, BTV activity was detected for the first time ever in August 2002 in 15 municipalities in the south and south-east of the country close to the borders with Montenegro and Serbia (see Figures 11.1 and 11.2), and the virus was identified as BTV-9 (Anon, 2002). Further information on BTV in the country has not been forthcoming.

Albania also reported the presence of BTV for the first time in 2002, and in a serological survey of cattle, sheep and goats carried out between October and November of that year, seroconversions to BTV-9 were detected in all of 15 districts surveyed, cattle showing the highest prevalence of infection (see Figures 11.1 and 11.2; Di Ventura *et al.*, 2004). The areas surveyed were arrayed mainly along the borders with Greece, Macedonia, Kosovo and Montenegro, although two districts on the Adriatic coast were also covered and Tirane District in the centre of the country exhibited the highest prevalence of BTV infection (i.e. 61%). No further information of BTV infection in Albania, subsequent to 2002, has been forthcoming.

Cyprus

Cyprus is the one country in Europe that could be historically considered to fall within the BTV endemic zone. However, since 1998 and up to mid-2006, BTV activity has been detected only during October–December 2003 in sentinel herds of cattle in the Larnaca District (see Figures 11.1 and 11.2). The virus was identified as BTV-16.

Israel

Israel is situated in a region where BTV occurs and has been reported regularly since 1924 (Sellers, 1975; Taylor *et al.*, 1985; Hassan, 1992; Braverman *et al.*, 2004). Control involves vaccination with a pentavalent, live attenuated virus vaccine, which includes types 2, 4, 6, 10 and 16 (Shimshony, 2004). During the course of the current series of outbreaks in Europe and up to mid-2006, activity only by BTV type 16 has been reported from Israel and only in the month of September 2003 (Anon, 2003a, b).

Bluetongue virus in North Africa, Italy, France, Spain and Portugal

North Africa – Tunisia

In January 2000, BT was reported for the first time in Tunisia, in areas in the NE of the country (Figures 11.3 and 11.4). The time of incursion was estimated as early December 1999 and clinical cases continued to be detected in January 2000 (Anon, 2000a; Hammami, 2004). The virus was typed as BTV-2 in late January 2000 (Anon, 2000g). The origin of this incursion is uncertain but is likely to be separate from that involving Turkey, Greece and the Balkan countries in eastern Europe. Since foot and mouth disease virus had also entered Tunisia (and Algeria) during 1999, probably via cattle smuggled from Côte d'Ivoire and Guinea into Algeria (Knowles and Davies, 2000), it is possible that BTV could have followed a similar route. Cattle in Africa often experience subclinical infections with BTV, and BTV-2 is common in several areas of sub-Saharan West Africa (Herniman *et al.*, 1980, 1983).

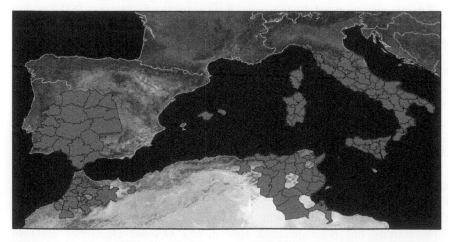

Figure 11.3 Bluetongue virus (BTV) activity in the Central and Western Mediterranean Basin, 1999–2005, overall area affected (base maps: NASA and Earthsat, 2004).

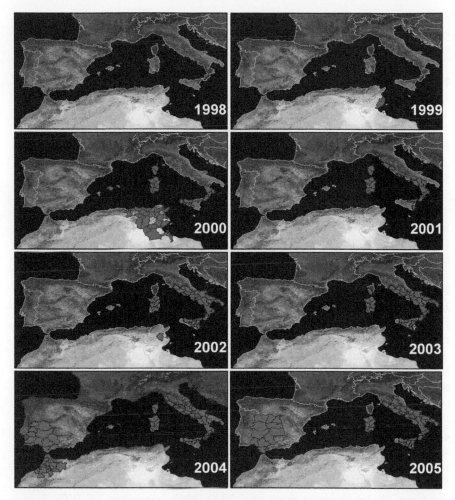

Figure 11.4 Bluetongue virus (BTV) activity in the Central and Western Mediterranean Basin by year, from 1998 to 2005 (base maps: NASA and Earthsat, 2004).

New cases of BT were not detected in Tunisia during February–May 2000, but from June, further outbreaks of BTV-2 were reported from 10 districts mostly distributed in the eastern and central parts of the country (see Figures 11.3 and 11.4; Anon, 2000d, i; Hammami, 2004). These persisted in some areas until October 2000 (Anon, 2000d). Control measures implemented in 2000 included the isolation of infected flocks, spraying of animal holdings with insecticide and mass vaccination (in 2000, 2001 and 2002) with a mono-valent, attenuated BTV-2 vaccine (Hammami, 2004).

In December 2002, a limited number of new outbreaks caused by BTV-2 occurred in the central part of Tunisia in unvaccinated flocks, and subsequent to that time, no evidence of virus circulation has been reported.

North Africa – Algeria
In July 2000, BTV-2 was reported from eastern Algeria (see Figures 11.3 and 11.4) and outbreaks continued into September (Anon, 2000b, e; Hammami, 2004). All of the areas initially affected were close to the Tunisian border. However, in September 2000, BTV was also confirmed in areas ranging up to 250 km to the west of Tunisia (see Figures 11.3 and 11.4; Anon, 2000c). Samples taken from animals near Algiers, over 400 km to the west of Tunisia, were also positive for BTV-specific antibodies (Mellor and Wittmann, 2002). The control strategy adopted by the Algerian veterinary authorities was based upon the use of insecticides to control the vectors and clinical surveillance (Hammami, 2004). Further, BTV activity in the country has not been reported till mid-2006.

North Africa – Morocco
In late 2000, BTV-specific antibodies were apparently recorded in animals from a number of provinces stretching across northern Morocco, but there was no evidence of clinical disease, and details of the locations of the affected areas are not available (M. Abaddi, Personal Communication; Y. Lhor, unpublished observations). Despite serological monitoring, no further evidence of BTV activity in Morocco was detected in the succeeding three years (Hammami, 2004).

However, in 2004, the situation changed and starting in August and continuing to the end of the year, a further incursion of BTV into Morocco was recorded, this time clinically affecting sheep in 14 provinces in the north-west of the country (see Figures 11.3 and 11.4; Anon, 2004a–g). Surprisingly, this incursion was caused by BTV-4 (Anon, 2004b) and is therefore totally separate from the 2000–2002 incursions of BTV-2 into other areas of North Africa. So far the origin of this incursion of BTV-4 into Morocco is unknown.

Italy
In August 2000, BTV (serotype 2) was confirmed for the first time ever in Italian territory. The island of Sardinia was affected first (Anon, 2000c) but by October, the virus had spread to Sicily and southern mainland Italy (see Figures 11.3 and 11.4; Calabria; Anon, 2000k). The incursion into Sardinia was particularly severe with more than 6000 flocks affected and over 260 000 sheep dying, either directly from BT or in the course of implementing control measures (i.e. slaughter of animals suspected of being infected) (Anon, 2000k; Calistri et al., 2004). Outbreaks continued in the three regions through the summer and autumn, and also at a low level through the winter, until May

2001 by which time over 260 000 animals in 6869 flocks had been affected (Calistri *et al.*, 2004). There then followed what is termed the second epidemic, which lasted until April 2002. Once again Sardinia, Sicily and Calabria were affected but the virus spread into Basilicata, Campania, Latium and Tuscany also causing over 6800 outbreaks and the loss of approximately 250 000 sheep and goats (see Figures 11.3 and 11.4; Calistri *et al.*, 2004).

During the first epidemic and for much of the second, vaccination was not employed, but in January 2002, vaccination of all susceptible domestic species with a monovalent, attenuated BTV-2 vaccine commenced in affected regions (Calistri *et al.*, 2004).

In Italy, the third epidemic began in April 2002 and extended until April 2003. During this period, BTV-9 and BTV-16, in addition to BTV-2, were detected for the first time in Italy and clinical disease occurred in eight provinces (Basilicata, Calabria, Campania, Lazio, Molise, Puglia, Sardinia and Sicily) (see Figures 11.3 and 11.4). Both BTV-9 and -16 had previously been reported from the eastern Mediterranean area (Turkey, Greece and the Balkans), but the route of incursion into Italy is uncertain, though illegal trade in animals is suspected (Giovannini *et al.*, 2004a). In those areas where BTV-2 and -9 occurred together (southern mainland Italy and Sicily), an attenuated, bivalent vaccine was employed in place of the monovalent BTV-2 vaccine (Patta *et al.*, 2004; Giovannini *et al.*, 2004b). The results of the vaccination campaigns suggest that in regions where vaccine cover reached approximately 80%, clinical disease either disappeared or was much reduced. In such regions, the spread of infection was also significantly reduced. However, in regions with lower levels of vaccine cover, no such significant reduction in the spread of disease and infection occurred (Patta *et al.*, 2004).

During the summer of 2003, the fourth epidemic commenced. Seroconversions were detected in many provinces in south and central Italy but clinical disease was apparently seen only in Sardinia, where, for the first time in Italy, BTV-4 was isolated. Sadly, this was not the end of the Italian outbreaks and by the end of 2004, circulation of various combinations of BTV-2, -4, -9 and -16 had been recorded in 12 regions extending from Sicily in the south to Tuscany in the north and from Puglia in the east to Sardinia in the west (see Figures 11.3 and 11.4; see http://www.te.izs.it/). Vaccination was carried out in all of the affected regions using the appropriate combinations of BTV-2, -4, -9 and -16 live virus vaccines.

During 2005 for the seventh consecutive year, BTV circulation continued in Italy, in at least 10 provinces: Lazio (BTV-2 and -16), Liguria (BTV-16), Marche (BTV-16), Molise (BTV-2 and -9), Campania (BTV-2 and -9), Puglia (BTV-2, -9 and -16), Basilicata (BTV-2, -4 and -16), Calabria (BTV-2 and -16), Sicily (BTV-2, -4 and -16) and Sardinia (BTV-2, -4 and -16), and at least until November (see Figures 11.3 and 11.4; see http://www.te.izs.it/). Again vaccination was carried out using the appropriate combinations of serotypes in each of the affected areas.

France (Corsica)

In October 2000, BTV was recorded for the first time ever in the French island of Corsica, the virus being identified as BTV-2 (see Figures 11.3 and 11.4; Anon, 2000j). The French authorities speedily authorised a vaccination campaign, using a monovalent, attenuated virus vaccine, which commenced during the winter of 2000–2001. However, apparently full protection was not attained as a second epidemic of BTV-2 commenced in July 2001, affecting animals, mainly in unvaccinated flocks, in both the north and south of the island. In all, in 2001, 335 flocks were affected, a seven-fold increase in 2000 (Breard *et al.*, 2004). Accordingly, a second vaccination campaign was mounted during the winter and spring of 2001–2002, which seems to have been successful as no cases of BT were reported during 2002. However, in October 2003, a further BTV incursion into Corsica was recorded affecting vaccinated sheep in the south part of the island (see Figures 11.3 and 11.4). The virus was identified as BTV-4 on the 31st of October and a third vaccination campaign was mounted from early November using a bivalent, live attenuated vaccine against BTV-2 and -4. In August 2004 and for the fourth year, BT was reported in Corsica, this time two virus serotypes being identified: BTV-4 in the north and west of the island and BTV-16 in the south-west (Anon, 2004h). In response, a further vaccination campaign was mounted using live attenuated vaccines against BTV 2, 4 and 16, and no further evidence of BTV transmission was detected after September 2004.

Spain

The first outbreaks of BT in Spain commenced on two of the Balearic Islands (Menorca and Majorca) in September–October 2000 (see Figures 11.3 and 11.4; Anon, 2000l; Miranda *et al.*, 2003; Gomez-Tejedor, 2004). Three hundred and five outbreaks occurred and the virus was identified as BTV-2. A vaccination campaign in sheep based upon a monovalent, attenuated BTV-2 vaccine was mounted on both islands commencing in October 2000 and was completed before the end of the year. Evidence of BTV transmission, as assessed in sheep and sentinel herds of bovines, did not extend beyond November 2000, but a further vaccination campaign in sheep was mounted in the spring of 2001 and was extended to cover the island of Ibiza, even though no cases of BT were detected there (Gomez-Tejedor, 2004).

However, in October 2003 and extending until the end of the year, a second incursion of BTV into the eastern part of the island of Menorca was reported (see Figures 11.3 and 11.4). The virus was identified as BTV-4, and a vaccination campaign based upon an attenuated monovalent vaccine against that serotype was rapidly mounted (Gomez-Tejedor, 2004).

In October 2004, a further incursion of BTV-4 into Spain was identified by seroconversions in sentinel bovines, this time into mainland Spain (see Figures 11.3 and 11.4; Gomez-Tejedor, 2004). The outbreak apparently commenced in the southern part of mainland Spain in the province of Cadiz in Andalucia and spread rapidly through four provinces in Andalucia and two in Extramadura; the Spanish enclave of Ceuta on the north coast of Morocco was also affected (Gomez-Tejedor, 2004). Once again a vaccination campaign was launched using an attenuated, monovalent BTV-4 vaccine in sheep. However, commencing in July 2005 and running through until the time of writing, the outbreak of BTV-4 in mainland Spain continued to spread, reaching further north than ever before, until by November 2005 it involved provinces in Andalucia (4), Castilla la Mancha (2), Castilla y Leon (2), Extramadura (2) and Madrid (1) (see Figures 11.3 and 11.4; see: http://rasve.mapya.es/). Interestingly, in 2005, all but one of the Spanish outbreaks occurred in bovines, whereas in 2004, 43 were in bovines, 200 in sheep and 77 in goats. Also in 2005, a single isolate of BTV-2 was made in July 2005 from a herd of sentinel bovines in mainland Spain (Elliott, Personal Communication). The origin of the 2004–2005 outbreaks of BTV-4 in mainland Spain was undoubtedly north-west Morocco, where the same serotype of virus was detected just prior to and during the Spanish outbreaks. Molecular studies also confirm the close similarity between the Spanish and Moroccan isolates of BTV-4 (Mertens, unpublished observations). The origin of the single isolate of BTV-2 from mainland Spain in 2005 is more difficult to determine. However, molecular analysis has confirmed that it is virtually indistinguishable from the BTV-2 vaccine virus previously used in several countries and zones in southern Europe including the Spanish Balearic islands.

Portugal
Bluetongue virus was first confirmed in sheep in Portugal in November 2004 in the Evora District of Alentejo, close to the border with Spain, and by December had spread to the Castelo Branco District of Beira Interior (see Figures 11.3 and 11.4; Anon, 2004a, b; Boinas et al., 2005). The virus was identified as BTV-4 (Fevereiro et al., 2005) and a vaccination campaign in sheep was mounted in the affected areas using a live attenuated BTV-4 vaccine (Boinas et al., 2005). At the same time a country-wide serological surveillance system in bovines was initiated followed by an entomological surveillance system. In 2005, seroconversion to BTV but no disease was detected in sentinel herds of bovines until at least October (Boinas, Personal Communication). Further information from Portugal is not yet available. The Portuguese outbreaks, in view of the serotype of virus involved, the areas infected and the times of infection are clearly an extension of the Spanish mainland incursions that originated in Morocco.

Origins of the outbreaks

There are a number of sources of the 1998–2005 BTV incursions into the countries of southern and eastern Europe. The incursions of BTV serotypes 1, 4, 9 and 16 into the eastern part of the Mediterranean Basin (Greece, European

Turkey and the Balkans) some of which then spread further west (4, 9 and 16) clearly originated from the east of Europe. In this context it should be borne in mind that BTV serotypes 2, 4, 6, 9, 10, 13, and 16 have all been reported over a number of years from Anatolian Turkey, Syria, Jordan and/or Israel. Indeed, the westward movement of some of these and of other viruses (e.g. Akabane) through Turkey is well documented (Urman *et al.*, 1980; Vassalos, 1980; Yonguc *et al.*, 1982; Burgu *et al.*, 1992; Taylor and Mellor, 1994a, b). BTV-1, however, has not previously been reported from these areas but the work of Maan *et al.* (2004) has shown that the European isolates of this virus have a genetic lineage closely related to isolates of the serotype from India. It is therefore likely that BTV-1 is also present in Turkey and/or the Middle East but, by chance, has not yet been isolated from these regions, probably because of the limited amount of work on BTV that has been carried out there. These findings strongly suggest that the area to the east of Europe contains a pool of assorted BTV serotypes some of which have previously been identified from Europe and some, like BTV-1, which have not. Under the right environmental conditions, these BTV serotypes, clearly, can extend into and expand within Europe and as we have seen in the period 1998–2005 constitute a significant threat to European livestock (see Figures 11.1 and 11.3). The trick is to understand what these 'right conditions' are so as to be able to predict such incursions and thereby minimise their impact. How this can be achieved is dealt with elsewhere in this book.

The European incursion of BTV-2 has a different origin from the eastern incursions. Prior to entry into Europe this virus was first reported from Tunisia in late 1999 and Algeria in 2000 but the virus (see Figures 11.3 and 11.4), at least until recently, has clearly not been endemic to this region as neither country had previously experienced a BTV incursion of any serotype, and Morocco to the west had not recorded the presence of BTV since 1956 (Placidi, 1957; Hammami, 2004). So from where might this incursion into North Africa and Europe have come? The origin is uncertain but as foot and mouth disease virus had also entered Tunisia (and Algeria) during 1999, probably via cattle smuggled from Cote d'Ivoire and Guinea into Algeria (Knowles and Davies 2000), it is very possible that BTV could have followed a similar route. Cattle in Africa often experience sub-clinical infections with BTV, and BTV-2 is common in several areas of sub-Saharan West Africa (Herniman *et al.*, 1983). It is also possible, though less likely, that BTV-2 could have expanded westwards across North Africa through Egypt and Libya from the Middle East where it also occurs; however, no evidence of such an expansion has entered the public domain. Sequence comparisons of BTV-2 isolates from Tunisia and Algeria, with recent isolates from Israel and other locations in the Middle East, and also from locations in West Africa south of the Sahara might help to resolve this uncertainty. Having reached North Africa, BTV-2 then expanded rapidly northwards during the summer and autumn of 2000 into Sardinia, Corsica, two of the Spanish Balearics, Sicily and southern mainland Italy (Calabria) – in that order (see Figures 11.3 and 11.4). The movement of infected animals and their products has, in many instances, been ruled out as the cause of these incursions (Miranda *et al.*, 2003; Calistri *et al.*, 2004), but the question – how else might it have been achieved? – remains. The most likely mode is via the passive transportation of infected vectors on the wind as aerial plankton. Over the years, many authors have written extensively on this subject providing both circumstantial and real evidence to suggest that *Culicoides* can travel hundreds of kilometres by such means (Sellers *et al.*, 1979a, b; Hayashi *et al.*, 1979; Murray, 1987; Sellers and Maarouf, 1991; Sellers and Mellor, 1993). In the case of the current BTV-2 outbreaks in Sardinia, Calistri *et al.* (2004) have shown the following:

- The first cases of BT occurred in the southernmost part of the island, which, due to its geographic proximity with North Africa, is where landfall of wind-borne *Culicoides* would be expected.
- During the first 15 days after notification of the first outbreak, 262 infected flocks were detected along 200 km of the south-west coast of the island.
- The BTV incubation period is compatible with the hypothesis of a common source for these infections.
- In the same time frame, BTV-2 was also reported in the Balearic Islands, over 200 km to the west.
- The period of the initial outbreak was preceded by and coincided with unusual climatic conditions in which dust storms originating in the infected areas of North Africa moved northwards across southern Italy and Sardinia.

Similarly, an incursion of BTV-4 into the Balearics occurred in October 2003 following outbreaks of the same serotype during the summer and autumn in Sardinia and Corsica, islands situated over 300 km to the east of the Balearics (see Figures 11.3 and 11.4).

In 2004, a further incursion of BTV-4 into the Mediterranean Basin occurred (see Figures 11.3 and 11.4). The outbreak was first detected on 28 August 2004 in NE Morocco but had spread across the Straits of Gibraltar into the southernmost province of mainland Spain (Cadiz) by 30 September (see Figures 11.3 and 11.4). Further spread of the virus both northwards and westwards within Spain and then into Portugal was connected with the movement of infected animals, but the initial incursion was not linked to animal movements and may, therefore, have been due to windborne *Culicoides*. Unpublished analyses have shown that southerly winds blowing over NW Morocco and then the Cadiz area of Spain were prevalent at the time of the incursion and occurred during the time of the diel (evening) when *Culicoides* activity is maximal (Gloster, unpublished observations).

All of this information provides strong circumstantial evidence that infected *Culicoides* can be dispersed on the wind distances of several 100 km, thereby introducing disease into regions remote from the source of infection. It also suggests that movement of *Culicoides* vectors on the wind over such long distances is not a rare event. This is a matter of some concern since elsewhere in this book (Chapter 15) Baylis and his co-authors show that transmission of BTV from infected vectors to susceptible ruminants is an exceptionally efficient process, which suggests that the introduction of BTV via wind-borne vector *Culicoides* as described above may be a regular and very successful survival strategy of the virus.

Vector species of *Culicoides* in Europe and the Mediterranean Basin

Culicoides imicola has long been known as a vector of BTV in those parts of the region affected or suspected of being affected by the virus prior to 1998 (i.e. Israel, Cyprus, Anatolian Turkey, the Greek islands of Rhodes and Lesbos, Morocco, SW Spain and Portugal) (see Mellor and Wittmann, 2002). Indeed, the northernmost locations from which *C. imicola* had been identified in the Mediterranean Basin prior to 1998 (Portugal, SW Spain, Morocco, Algeria, Israel, Anatolian Turkey, Rhodes and Lesbos) are virtually identical to the list of the places affected, at one time or another by BTV prior to 1998. Furthermore, although *C. imicola* had been sought in many rather more northerly but adjacent areas in the region that had not been affected by BTV (northern and eastern Spain, Tunisia, Sicily, parts of mainland Italy, Bulgaria and mainland Greece), it had never been recorded from any of them (Callot *et al.*, 1964; Callot and Kremer, 1969; Kremer *et al.*, 1971; Chaker 1981; Mellor *et al.*, 1984; Gloukova *et al.*, 1991; Boorman *et al.*, 1996; Scaramozzino *et al.*, 1996; Dilovski *et al.*, 1992; Patakakis, Personal Communication; Georgiev and Nedelchev, unpublished observations; Wilkinson, Personal Communication). This suggested that *C. imicola* was not only the most important BTV vector in the region but was also the *only* important vector species, otherwise the virus would, presumably, on some occasions have occurred in its absence. Consequently, the 1998–2005 BTV incursions into many areas of the Mediterranean Basin previously unaffected by the virus, including areas where *C. imicola* had been sought for but not found, was a new and totally unexpected turn of events. The entry of BTV into such areas, therefore, suggested that either the range of *C. imicola* had recently expanded or novel vector species of *Culicoides* were, for the first time, mediating BTV incursions in the region. In the event, the answer seems to include both of these possibilities. Vector surveys carried out in the BTV-affected areas during 2000 and subsequently have recorded the presence of populations of *C. imicola* at many locations in mainland Greece, mainland Italy, Sicily, Sardinia, Corsica, the Balearics, much of eastern Spain and parts of southern mainland France

(Goffredo *et al.*, 2001, 2004; Miranda *et al.*, 2003; Monteys and Saiz-Ardanaz, 2003; Giovannini *et al.*, 2004a; Monteys *et al.*, 2004; Monteys *et al.*, 2005; Purse *et al.*, 2005). Interestingly, and in context with the above findings, Baylis *et al.* (2001) had already used a series of remotely sensed variables to attempt to predict areas of the Mediterranean Basin that were climatically suitable for populations of *C. imicola* and hence at risk to BT, and this early work correctly identified almost all of the above locations as being suitable for *C. imicola*. In essence, therefore, the range of *C. imicola* does seem to have expanded northwards to include much of the northern coast of the Mediterranean Sea and most of the leg of Italy as far north as 44°N (Goffredo *et al.*, 2001). The work of Purse *et al.* (2005) suggests that this sudden expansion is being driven by recent and ongoing changes in the European climate.

However, even though *C. imicola* has extended northwards so dramatically, the range of BTV has extended over 400 km beyond the distribution of this vector in certain parts of Europe, particularly the Balkans (Mellor, 2004), and even within the overall distribution of *C. imicola*, in some areas, BTV transmission has taken place in locations where *C. imicola* is either very rare or absent (e.g. parts of Sicily, Lazio and Tuscany in Italy) (De Liberato *et al.*, 2005; Purse *et al.*, 2004, 2005). This clearly means that in these areas, novel vector species of *Culicoides* are transmitting the virus – but what is the identity of these new vectors and why have they not previously been involved in BTV transmission in the region? Recent studies have shown that two widespread and abundant Palearctic *Culicoides* species complexes (*Culicoides obsoletus* and *Culicoides pulicaris*) comprise very large proportions of the *Culicoides* populations in the non-*C. imicola* BT areas and these species also show fine-scale spatial and temporal correlation with BT outbreaks (De Liberato *et al.*, 2003, 2005; Torina *et al.*, 2004). Also, during the current outbreaks, BTV has been regularly isolated in the field from wild-caught individuals of both species complexes (Caracappa *et al.*, 2003; Savini *et al.*, 2003, 2004, 2005; De Liberato *et al.*, 2005). Earlier vector competence studies with these species indicated that they had a very low oral susceptibility to BTV (Mellor and Jennings, 1988; Jennings and Mellor, 1988), but the recent climate warming in Europe may well have increased their importance as vectors – by increasing their population sizes and survival rates to compensate for their low competence levels and by increasing levels of susceptibility through the temperature-controlled virus developmental effects as described elsewhere in this book (see Chapter 14). So climate warming may be the underlying reason why these novel vectors seem, for the first time, to be playing an important part in BTV transmission. The future role of the *C. obsoletus* and *C. pulicaris* species complexes in the transmission of BTV is a matter of real concern as these species are common and widespread across the whole of central and northern Europe (Purse *et al.*, 2005). What, in a time of climate change, this means in terms of increasing risk from BT to livestock of the more northerly countries in Europe may become apparent in future years.

Bluetongue virus in Europe in 2006 and 2007

During the period from late 2006 to 2007, BTV appears to have fulfilled or exceeded all of the concerns previously raised about its northerly extension in range due to climate change and the involvement of Palearctic vector species of *Culicoides* in the more northerly areas (Mellor and Boorman, 1995). During 2006 a series of incursions of BTV into various parts of the European region occurred, the most significant by far being the introduction of BTV-8, a serotype new to Europe, into locations far beyond the usual range of any BTV and indeed further north than BTV has ever been recorded anywhere in the world (e.g. up to 53°N). This outbreak was first detected in the Maastricht area of the Netherlands on 14 August 2006 (International Society for Infectious Diseases, 2006a; World Organisation for Animal health, 2006) and spread rapidly along a broadly east–west axis to involve most of the country, virtually the whole of Belgium, much of NW Germany, Luxembourg and the northern borders of France. By the end of 2006, over 2000 outbreaks had been declared (EFSA, 2007a). The virus was reported to cause significant disease not only in sheep but also in cattle, which is indicative of a totally naïve population (Howell, 1963). At the time of writing, the origin of the incursion and its mode of entry are unknown, though there has been much speculation (see EFSA, 2007b), but molecular epidemiological investigations conducted at the Community Reference Laboratory (CRL) for bluetongue, based at IAH-Pirbright in the UK, have shown that the strains of BTV-8 isolated in NW Europe are most similar to BTV-8 isolates from sub-Saharan West Africa (International Society for Infectious Diseases, 2006b). However, this preliminary information is not conclusive as very few isolates of this serotype are available for comparison, and work is therefore ongoing. In the locations where BTV-8 has occurred in NW Europe, the traditional vector, *C. imicola*, is totally absent but members of the *Obsoletus* complex are extremely common. Indeed, in early October 2006, BTV nucleic acids were detected in a pool of non-blood–engorged, parous *C. dewulfi* in the Netherlands (Meiswinkel *et al.*, 2007), a species that most workers have traditionally placed in the *Obsoletus* complex because of their close morphological similarity (Campbell and Pelham-Clinton, 1960). Recently, however, this species has been separated into a group on its own (Meiswinkel *et al.*, 2004; Gomulski *et al.*, 2005). This outbreak therefore provides a further example to those cited above in the Balkans and parts of Italy, showing that BTV can be efficiently transmitted by native Palearctic *Culicoides* species in the absence of *C. imicola*. The summer of 2006 was exceptionally warm, and July, when the incursion was first detected, was the warmest in northern Europe since records began in 1850. Temperatures across Belgium, the Netherlands, northern France and NW Germany during July were more than 5°C warmer than the 1961–1990 average. These are precisely the conditions that would be expected to enhance the competence levels of vector *Culicoides* in the region (see Chapter 14, this volume).

Towards the end of 2006 and with the advent of cooler weather, the number of BTV-8 outbreaks in northern Europe decreased and ended. Subsequently, there was no evidence of BTV transmission in northern Europe during the first few months of 2007. However, sadly in late April 2007, a sentinel bovine seroconverted in NW Germany (International Society for Infectious Diseases, 2007). This was the first indication that BTV-8 had successfully overwintered in northern Europe, and the seroconversion occurred close to the earliest date at which vector–host transmission would be predicted to be capable of resuming (Wilson *et al.*, 2007).

During 2007, the outbreak of BTV-8 spread far beyond its greatest extent during 2006, reaching as far north as Denmark (Anon, 2007a) and as far east as the Czech Republic (Anon, 2007b) (Figure 11.5). Towards the end of 2007, the southern extension of the BTV-8 outbreak reached southern France and the first cases of BTV-8 in northern Spain were reported in January 2008 (Anon, 2008) (see Figure 11.5). The first UK case was reported in early August (Defra, 2007), although retrospective climatological analysis suggests that introduction probably took place in early August (Anon, 2007c). By the end of 2007, cases had been reported on 66 holdings in the UK, although over 40 more infected holdings were subsequently identified as a result of pre-movement testing requirements.

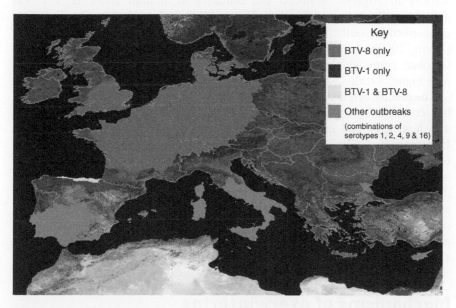

Figure 11.5 Bluetongue virus (BTV) restriction zones in the European Union at the time of going to press, showing the current BTV-8 epizootic in Northern Europe and the recent incursion of BTV-1 into Northern Spain and SW France (Adapted from: http://cc.europa.eu/food/animal/diseases/controlmeasures/bluetongue_en.htm) (See colour plate 21).

The 2007 epizootic of BTV-8 in northern Europe was far more aggressive than in 2006. Between July and October 2007, around 25 000 more sheep died in Belgium than during the same period in 2006, representing approximately one-sixth of the national flock (Wilson and Mellor, 2008). By the end of 2007, well over 30 000 farms in northern Europe had been infected, causing over €150 million damage through direct costs and many times that in lost trade (Hoogendam, 2007). The 2006 present epizootic of BTV-8 in northern Europe is believed to have caused greater economic damage than any previous single-serotype BT outbreak (Wilson and Mellor, 2008).

In addition to the above, fresh incursions of BTV-1 into Algeria and Morocco occurred during summer 2006 (Anon, 2006a, b) and extended into Sardinia (Anon, 2006c). Incursions of BTV-16 into Cyprus and Israel also occurred during October and November of the year and of BTV-4 and -15 into Israel in November and December (Anon, 2006d; http://www.reoviridae.org/dsrna_virus_proteins/ReoID/BTV-isolates.htm). The incursion of BTV-15 into Israel in late 2006 was the first occasion on which this serotype has been reported in the Mediterranean Basin. Serological evidence of BTV-8 was also detected in SE Bulgaria in late 2006 but no virus was isolated (International Society for Infectious Diseases, 2006c). Following the BTV-1 activity in Algeria, Morocco and Sardinia in autumn 2006, BTV-1 was detected in Spain for the first time in July 2007 (Anon, 2007d). This incursion established in southern Spain during August spread to Portugal in September (Anon, 2007e) before jumping to the Basque region in northern Spain in October, from which it spread to SW France in November (Anon, 2007f) (see Figure 11.5).

The precise origins of all of these incursions are uncertain but previous work has confirmed the presence of a number of BTV serotypes in the Middle and Near East and in sub-Saharan West Africa (Taylor and Mellor, 1994a, Herniman et al., 1983), and some of these have now been recorded in Europe. However, the list of serotypes from such areas is at best an out of date snapshot, and it is certain that others remain to be discovered. It is also certain that the recent unprecedented incursions of BTV-8 into NW Europe mean that as a result of changing climate, virtually the whole of Europe must now be considered to be at risk to this disease and probably to other *Culicoides*-transmitted diseases such as epizootic haemorrhagic disease, African horse sickness and Akabane.

Mechanisms of overwintering

In temperate regions, the activity of the *Culicoides* vectors of BTV generally drops to low levels or zero during the winter. Reports of BTV activity in temperate areas may therefore be separated by many months during which no new cases are detected. However, the same virus strains sometimes reappear in the following year, strongly suggesting that the virus is capable of persisting in such regions without detectable activity (Osmani *et al.*, 2006). This phenomenon, termed 'overwintering', has been recognised for many years (Nevill, 1971) and various potential mechanisms for it have been suggested, most of which fall into one of two categories: virus persistence in the host or virus persistence in the vector.

Vertical transmission of BTV in vector midges has long been considered as a possible mechanism for the persistence of the virus over the vector-free winter period, as it is known that in temperate countries and zones, most *Culicoides* survive the winter as fourth instar larvae. Until recently, experiments designed to detect such vertical transmission in *Culicoides* species have consistently reported negative results (Jones and Foster, 1971; Nunamaker *et al.*, 1990; Mellor, 2000) but a recent American study detected fragments of BTV RNA in *Culicoides* larvae and pupae, suggesting that this mechanism may deserve further attention (see Chapter 14, this volume and White *et al.*, 2006). In addition, although midges are short-lived during the summer, several studies have suggested that they may live longer at cold temperatures than previously believed (Gerry and Mullens, 2000; Lysyk, 2007), raising the possibility that a small number of BTV-infected *Culicoides* adults may survive to carry BTV from one season to the next. Such a mechanism would facilitate the overwintering of BTV in regions where the winter period is relatively mild and short.

In the vertebrate hosts of BTV, cattle and sheep are considered to be infectious for up to 60 days following infection (Anon, 2007g), although most infections last for 2–4 weeks (Singer *et al.*, 2001; Gubbins *et al.*, 2008). However, one experimental study has suggested that BTV may be capable of persistently infecting certain inflammatory cells in sheep for substantially longer periods, the virus reappearing in the skin when an inflammatory response is provoked (Takamatsu *et al.*, 2003). The authors suggest that such a response may be triggered by the springtime resumption of biting, by midges, providing an elegant mechanism for virus to persist over the winter period. However, this laboratory-based study has not yet been shown to occur in the field nor has it been supported by similar studies carried out in the United States and Australia (Melville *et al.*, 2004; White and Mecham, 2004).

An alternative mechanism for persistence in the ruminant host is via transplacental infection. Such a mechanism has been described and demonstrated (Gibbs *et al.*, 1979) but is believed to be either a laboratory artefact induced by the use of tissue culture-passaged viruses (Kirkland and Hawkes, 2004) or extremely rare and so not of relevance in the field (MacLachlan *et al.*, 1992; Melville and Gard 1992). However, recent findings in England where PCR-positive and infectious virus-positive calves were born to dams infected in 2007 with BTV-8 during pregnancy suggest that transplacental transmission may be occurring in the field. Such a mechanism, if common, would help explain the widespread overwintering of BTV-8 in NW Europe from 2006 to

2007, which anecdotal information suggests, occurred particularly frequently in areas where transmission had been very intense and so presumably in locations where there were many infected animals. At the time of writing, further evidence supporting the occurrence of overwintering of BTV-8 via transplacental transmission in the field has been provided by a recent report from Northern Ireland where a number of calves born to seropositive cattle during the vector-free period were viraemic (Gildernew, 2008). Clearly, there is an urgent need to confirm that BTV-8 is being transmitted in this way and if it is, to elucidate the basis of the mechanism and determine the duration and titre of infectious viraemia in the newborn animals. It will also be important to determine whether such a mechanism is a characteristic of the particular strain of BTV-8 present in northern Europe or whether it occurs with other field strains and serotypes of BTV also. Confirmation of the existence of transplacental transmission of BTV in the field is likely to have a significant impact upon regulations governing the international movement of animals.

Summary

Bluetongue exists around the world in a broad band covering much of the America, Africa, southern Asia, northern Australia and, occasionally, the southern fringe of Europe. It is considered to be one of the most important diseases of domestic livestock. Recently the virus causing this disease has extended its range northwards into areas of Europe never before affected and has persisted in many of these locations causing the greatest epizootic of the disease on record. Indeed, the most recent outbreaks of BT in Europe are further north than this virus has ever previously occurred anywhere in the world. The reasons for this dramatic change in BT epidemiology are complex but are linked to recent extensions in the distribution of its major vector, *C. imicola*, to the involvement of novel *Culicoides* vector(s) and to ongoing climate change. This paper explores these areas, highlights prospects for the future and discusses recent findings relating to the ability of BTV-8 to overwinter in northern Europe.

References

Anon (1998a) Disease outbreaks reported to the OIE in November–December 1998. *Bull. Off. Int. Epizoot.* **110**, 506.
Anon (1998b) Bluetongue in Greece: confirmation of diagnosis. *Off. Int. Epizoot. Dis. Info* **11**, 166–7.
Anon (1999a) Disease outbreaks reported to the OIE in January–February 1999. *Bull. Off. Int. Epizoot.* **111**, 18.
Anon (1999b) Disease outbreaks reported to the OIE in July–August 1999. *Bull. Off. Int. Epizoot.* **111**, 379.

Anon (1999c) Disease outbreaks reported to the OIE in September–October 1999. *Bull. Off. Int. Epizoot.* **111**, 449.

Anon (1999d) Disease outbreaks reported to the OIE in November–December 1999. *Bull. Off. Int. Epizoot.* **111**, 543.

Anon (1999e) Bluetongue in Bulgaria: additional information. *Off. Int. Epizoot. Dis. Info.* **12**, 122.

Anon (2000a) Disease outbreaks reported to the OIE in January–February 2000. *Bull. Off. Int. Epizoot.* **112**, 16.

Anon (2000b) Disease outbreaks reported to the OIE in July–August 2000. *Bull. Off. Int. Epizoot.* **112**, 449.

Anon (2000c) Disease outbreaks reported to the OIE in September–October 2000. *Bull. Off. Int. Epizoot.* **112**, 552.

Anon (2000d) Disease outbreaks reported to the OIE in November–December 2000. *Bull. Off. Int. Epizoot.* **112**, 687–8.

Anon (2000e) Bluetongue in Algeria. *EMPRES Transboundary Anim. Dis. Bull.* **14**, 20–1.

Anon (2000f) Bluetongue in Greece: follow-up report. *Off. Int. Epizoot. Dis. Info.* **13**, 9–13.

Anon (2000g) Bluetongue in Tunisia: additional information. *Off. Int. Epizoot. Dis. Info.* **13**, 21.

Anon (2000h) Bluetongue in Greece: typing of virus strains isolated in 1998–1999. *Off. Int. Epizoot. Dis. Info.* **13**, 33–4.

Anon (2000i) Bluetongue in Tunisia: follow-up report no. 2. *Off. Int. Epizoot. Dis. Info.* **13**, 176.

Anon (2000j) Bluetongue in France: in the island of Corsica. *Off. Int. Epizoot. Dis. Info.* **13**, 195–7.

Anon (2000k) Bluetongue in Italy: follow-up report no. 1. *Off. Int. Epizoot. Dis. Info.* **13**, 209–10.

Anon (2000l) Bluetongue in Spain: in the Balearic islands. *Off. Int. Epizoot. Dis. Info.* **13**, 224.

Anon (2001a) Disease outbreaks reported to the OIE in January–February 2001. *Bull. Off. Int. Epizoot.* **113**, 17–18.

Anon (2001b) Bluetongue in France: in the island of Corsica. *Off. Int. Epizoot. Dis. Info.* **14**, 178.

Anon (2001c) Bluetongue in Greece and Italy. *Off. Int. Epizoot. Dis. Info.* **14**, 215–18.

Anon (2001d) Bluetongue in Bulgaria: follow-up report no. 1. *Off. Int. Epizoot. Dis. Info.* **14**, 230–1.

Anon (2001e) Bluetongue in Kosovo (FRY), territory under united nations interim administration. *Off. Int. Epizoot. Dis. Info.* **14**, 242–3.

Anon (2001f) Bluetongue in Yugoslavia. *Off. Int. Epizoot. Dis. Info.* **14**, 280–1.

Anon (2001g) Bluetongue in Greece: follow-up report no. 2. *Off. Int. Epizoot. Dis. Info.* **14**, 262–5.

Anon (2001h) Bluetongue in the former Yugoslav republic of Macedonia. *Off. Int. Epizoot. Dis. Info.* **14**, 234.

Anon (2001i) Bluetongue in Croatia: suspected outbreak. *Off. Int. Epizoot. Dis. Info.* **14**, 291–2.

Anon (2001j) Bluetongue in the former Yugoslav republic of Macedonia: follow-up report no. 1. *Off. Int. Epizoot. Dis. Info.* **14**, 265–8.

Anon (2001k) Bluetongue in Yugoslavia. *Off. Int. Epizoot. Dis. Info.* **14**, 252.

Anon (2001l) Bluetongue in Yugoslavia (additional information). *Off. Int. Epizoot. Dis. Info.* **14**, 280.

Anon (2001m) Bluetongue in Croatia: suspected outbreak. *Off. Int. Epizoot. Dis. Info.* **14**, 291.

Anon (2002) Bluetongue in Bosnia and Herzegovina. *Off. Int. Epizoot. Dis. Info.* **15**, 171.

Anon (2003a) Bluetongue in Israel. *Off. Int. Epizoot. Dis. Info.* **16**, 247.

Anon (2003b) Bluetongue in Israel (additional information). *Off. Int. Epizoot. Dis. Info.* **16**, 258.

Anon (2004a) Bluetongue in Morocco. *Off. Int. Epizoot. Dis. Info.* **17**, 273.

Anon (2004b) Bluetongue in Morocco (follow up report no. 1.). *Off. Int. Epizoot. Dis. Info.* **17**, 297.

Anon (2004c) Bluetongue in Morocco (follow-up report no. 2). *Off. Int. Epizoot. Dis. Info.* **17**, 318.

Anon (2004d) Bluetongue in Morocco (follow-up report no. 3). *Off. Int. Epizoot. Dis. Info.* **17**, 329.

Anon (2004e) Bluetongue in Morocco (follow-up report no. 4). *Off. Int. Epizoot. Dis. Info.* **17**, 346.

Anon (2004f) Bluetongue in Morocco (follow-up report no. 5). *Off. Int. Epizoot. Dis. Info.* **17**, 368.

Anon (2004g) Bluetongue in Morocco (follow-up report no. 6). *Off. Int. Epizoot. Dis. Info.* **17**, 383.

Anon (2004h) Bluetongue in France in the island of Corsica. *Off. Int. Epizoot. Dis. Info.* **17**, 266.

Anon (2006a) Bluetongue in Algeria (immediate notification). *Off. Int. Epizoot. Dis. Info.* **19**, 30.

Anon (2006b) Bluetongue in Morocco (immediate notification). *Off. Int. Epizoot. Dis. Info.* **19**, 44.

Anon (2006c) Bluetongue in France (immediate notification). *Off. Int. Epizoot. Dis. Info.* **19**, 35.

Anon (2006d) Bluetongue in Israel (immediate notification). *Off. Int. Epizoot. Dis. Info.* **19**, 48.

Anon (2007a) 13/10/2007: Bluetongue, Denmark (immediate notification). *Off. Int. Epizoot. Dis. Info.* **20**, 42.

Anon (2007b) 29/11/2007: Bluetongue, Czech Republic (immediate notification). *Off. Int. Epizoot. Dis. Info.* **20**, 48.

Anon (2007c) Bluetongue arrived on August 4/5, says DEFRA. *Vet. Rec.* **161**, 602.

Anon (2007d) 27/07/2007: Bluetongue, Spain (immediate notification). *Off. Int. Epizoot. Dis. Info.* **20**, 30.

Anon (2007e) 18/09/2007: Bluetongue, Portugal (immediate notification). *Off. Int. Epizoot. Dis. Info.* **20**, 39.

Anon (2007f) 18/09/2007: Bluetongue, France (immediate notification). *Off. Int. Epizoot. Dis. Info.* **20**, 48.

Anon (2007g) *Chapter 2.2.13 Bluetongue. Terrestrial Animal Health Code.* World Organisation for Animal Health, pp. 131–6.

Anon (2008) 17/01/2008: Bluetongue, Spain, (immediate notification). *Off. Int. Epizoot. Dis. Info.* **21**, 3.

Baylis, M., Mellor, P.S., Wittmann, E.J. and Rogers, D.J. (2001) Prediction of areas around the Mediterranean at risk of bluetongue by modelling the distribution of its vector using satellite imaging. *Vet. Rec.* **149**, 639–43.

Boinas, F., Amador, R., Pereira da Fonseca, I., Nunes, T., Fevereiro, M. and Pinheiro, C. (2005) Perspectives of bluetongue control in Portugal. In: Proc. Third Vet. Sci. Cong. Santarém, 13–15 October 2005, Lisbon, Portugal: Sociedade Portuguesa de Ciências Veterinárias, p. 18.

Boorman, J., Mellor, P.S. and Scaramozzino, P. (1996) A new species of *Culicoides* (Diptera: Ceratopogonidae) from southern Italy. *Parassitologia* **38**, 501–3.

Braverman, Y., Baylis, M., Baylis, M., Tatem, A.J., Rogers, D.J., Mellor, P.S. and Purse, B.V. (2004) What factors determine when epidemics occur in the Mediterranean? Prediction of disease risk through time by climate-driven models of the temporal distribution of outbreaks in Israel. *Vet. Ital.* **40**, 235–42.

Breard, E., Hamblin, C., Hammami, S., Sailleau, C., Dauphin, G. and Zientara, S. (2004) The epidemiology and diagnosis of bluetongue with particular reference to Corsica. *Res. Vet. Sci.* **77**, 1–8.

Burgu, I., Urman, H.K., Akca, Y., Yonguc, A.D., Mellor, P.S. and Hamblin, C. (1992) Serological survey and vector surveillance for bluetongue in southern turkey. In: T.E. Walton and B.I. Osburn (eds), *Proc. Second Int. Symp. Bluetongue, African Horse Sickness & Related Orbiviruses*. Boca Raton, Fl: CRC Press, pp. 168–74.

Calistri, P., Giovannini, A., Conte, A., Nannini, D., Santucci, U., Patta, C., Rolesu, S. and Caporale, V. (2004) Bluetongue in Italy: part 1. *Vet. Ital.* **40**, 243–51.

Callot, J. and Kremer, M. (1969) Description d'un *Culicoides* nouveau: *C. jumineri* (Diptera, Ceratopogonidae) trouve en Tunisie. *Bull. Soc. Pathol. Exot.* **62**, 1112–118.

Callot, J., Kremer, M. and Juminer, B. (1964) Contribution a l'etude des *Culicoides* (Diptera, Ceratopogonidae) de Tunisie. *Arch. Inst. Pasteur Tunis* **41**, 357–64.

Campano Lopez, A. and Sanchez Botija, C. (1958) L'epizootie de fievre catarrhale ovine en Espagne (blue tongue). *Bull. Off. Int. Epizoot.* **50**, 65–93.

Campbell, J.A. and Pelham-Clinton, B.A. (1960) A taxonomic review of the British species of *Culicoides latreille* (Diptera: Ceratopogonidae). *Proc. R. Soc. Edinb, B* **67**, 181–302.

Caracappa, S., Torina, A., Guercio, A., Vitale, F., Calabro, A., Purpari, G., Ferrantelli, V., Vitale, M. and Mellor, P.S. (2003) Identification of a novel bluetongue virus vector species of *Culicoides* in Sicily. *Vet. Rec.* **153**, 71–4.

Chaker, E. *Memoire presente en vue de l'obtention du diplome d'etudes et de recherches en biologie humaine*. Universite Louis Pasteur de Strasbourg, p. 196.

Defra (2007). News release (Ref: 316/07, 22 September 2007): Bluetongue detected in Suffolk. http://www.defra.gov.uk/news/2007/070922a.htm.

De Liberato, C., Purse, B.V., Goffredo, M., Scholl, F. and Scaramozzino, P. (2003) Geographical and seasonal distribution of the bluetongue virus vector, *Culicoides imicola*, in central Italy. *Med. Vet. Entomol.* **17**, 388–94.

De Liberato, C., Scavia, G., Lorenzetti, R., Scaramozzino, P., Amaddeo, D., Cardeti, G., Scicluna, M., Ferrari, G. and Autorino, G.L. (2005) Identification of *Culicoides obsoletus* (Diptera: Ceratopogonidae) as a vector of bluetongue virus in central Italy. *Vet. Rec.* **156**, 301–304.

Di Ventura, M., Tittarelli, M., Semproni, G., Bonfini, B., Savini, G., Conte, A. and Lika, A. (2004) Serological surveillance of bluetongue virus in cattle, sheep and goats in Albania. *Vet. Ital.* **40**, 101–104.

Dilovski, M., Nedelchev, N. and Petkova, K. (1992) Studies on the species composition of *Culicoides* – potential vectors of the virus of the bluetongue in Bulgaria. *Vet. Sci.* **26**, 52–6.

Djuricic, B., Nedic, D., Lausevic, D. and Pavlovic, M. (2004) The epizootiological occurrence of bluetongue in the central Balkans. *Vet. Ital.* **40**, 105–107.

Dragonas, P.N. (1981) Evolution of bluetongue in Greece. *Off. Int. Monthly Epizoot. Circ.* **9**, 10–11.

EFSA (2007a) Bluetongue Serotype 8 Epidemic Bulletin 15. http://www.efsa.europa.eu/etc/medialib/efsa/in_focus/bluetongue.Par.0041.File.dat/bluetongue_update_1_February_2007.pdf.

EFSA (2007b) Report on Epidemiological analysis of the 2006 bluetongue virus serotype 8 epidemic in north-western Europe. http://www.efsa.europa.eu/EFSA/DocumentSet/Report_bluetongue_S8_en,0.pdf.

Fevereiro, M., Luis, T., Vaz, A., Ramos, F., Fagulha, M.T., Barros, S., Duarte, M.M. and Cruz, M.B. (2005) Bluetongue virus serotype 4 in Portugal In: *Proc. Third Vet. Sci. Cong. Santarém, 13–15 October 2005*, Lisbon, Portugal: Sociedade Portuguesa de Ciências Veterinárias, p. 107.

Gerry, A.C. and Mullens, B.A. (2000) Seasonal abundance and survivorship of *Culicoides sonorensis* (Diptera: Ceratopogonidae) at a southern California dairy, with reference to potential bluetongue virus transmission and persistence. *J. Med. Entomol.* **37**, 675–88.

Gibbs, E.P.J., Lawman, M.J.P. and Herniman, K.A.J. (1979) Preliminary observations on transplacental infection of bluetongue virus in sheep – a possible overwintering mechanism. *Res. Vet. Sci.* **27**, 118–20.

Gildernew, M. (2008) Ministerial Statement to Assembly: Update on bluetongue. Official Report, Northern Ireland Assembly, 19 February 2008.

Giovannini, A., Calistri, P., Conte, A., Savini, L., Nannini, D., Patta, C., Santucci, U. and Caporale, V. (2004a) Bluetongue surveillance in a newly infected area. *Vet. Ital.* **40**, 188–97.

Giovannini, A., Calistri, P., Nannini, C., Paladini, C., Santucci, U., Patta, C. and Caporale, V. (2004b) Bluetongue in Italy. *Vet. Ital.* **40**, 252–9.

Gloukova, V.M., Nedelchev, N.K., Rousev, I. and Tanchev, T. (1991) On the fauna of blood-sucking midges of the genus *Culicoides* (Diptera: Ceratopogonidae) in Bulgaria. *Vet. Sci.* **25**, 63–6.

Goffredo, M., Conte, A. and Meiswinkel, R. (2004) Distribution and abundance of *Culicoides imicola*, obsoletus complex and pulicaris complex in Italy. *Vet. Ital.* **40**, 270–4.

Goffredo, M., Satta, G., Torina, A., Federico, G., Scaramozzino, P., Cafiero, M.A., Lelli, R. and Meiswinkel, R. (2001) The 2000 bluetongue virus (BTV) outbreak in Italy: distribution and abundance of the principle vector *Culicoides imicola* Kieffer. In: *Proc. Tenth Int. Symp. Amer. Assoc. Vet. Lab. Diagnosticians (AAVLD)*, Salsomaggiore, Parma, 4–7 July, Ames: AAVLD, pp. 308–9.

Gomez-Tejedor, C. (2004) Brief overview of the bluetongue situation in Mediterranean Europe, 1998–2004. *Vet. Ital.* **40**, 57–60.

Gomulski, L.M., Meiswinkel, R., Delecolle, J.-C., Goffredo, M. and Gasperi, G. (2005) Phylogenetic relationships of the subgenus avaritia fox, 1955 including *Culicoides obsoletus* (Diptera: Ceratopogonidae) in Italy based on internal transcribed spacer 2 ribosomal DNA sequences. *Syst. Entomol.* **30**, 619–31.

Gorman, B.M. (1990) The bluetongue viruses. *Curr. Top. Microbiol. Immunol.* **162**, 1–19.

Gubbins, S., Carpenter, S., Baylis, M., Wood, J.L.N. and Mellor, P.S. (2008) Assessing the risk of bluetongue to UK livestock: uncertainty and sensitivity analysis of a temperature-dependent model for the basic reproductive number. *J. R. Soc. Interface* **5**, 363–71.

Hammami, S. (2004) North Africa: a regional overview of bluetongue virus, vectors, surveillance and unique features. *Vet. Ital.* **40**, 43–6.

Hassan, A. (1992) Status of bluetongue in the Middle East and Africa. In: T.E. Walton and B.I. Osburn (eds), *Bluetongue, African Horse Sickness and Related Orbiviruses. 2nd Int. Symp*, Paris: CRC Press: pp. 38–41.

Hayashi, K., Suzuki, H. and Asahina, S. (1979) Notes on the transoceanic insects-captured on East China sea in 1976, 1977, and 1978. (In Japanese with English summary). *Trop. Med.* **21**, 1–10.

Herniman, K.A.J., Boorman, J.P.T. and Taylor, W.P. (1983) Bluetongue virus in a Nigerian dairy cattle herd. 1. Serological studies and correlation of virus activity to vector population. *J. Hyg. Camb.* **90**, 177–93.

Herniman, K.A.J., Gumm, I.D., Owen, L., Taylor, W.P. and Sellers, R.F. (1980) Distribution of bluetongue virus and antibodies in some countries of the eastern hemisphere. *Bull. Off. Int. Epizoot.* **92**, 581–6.

Hoogendam, K. (2007) *International Study on the Economic Consequences of Outbreaks of Bluetongue Serotype 8 in North-Western Europe.* Leeuwarden: Van Hall Institute.

Howell, P.G. (1963) *Bluetongue. In: Emerging Diseases of Animals.* Rome: FAO, pp. 111–53.

International Society for Infectious Diseases (2006a) Bluetongue, ovine – Netherlands: confirmed. ProMED-mail 2006; 18 August: 20060818.2311. http://www.promedmail.org.

International Society for Infectious Diseases (2006b) Bluetongue – Netherlands, Belgium, Germany (06): BTV-8. ProMED-mail 2006; 28 August: 20060828.2448. http://www.promedmail.org.

International Society for Infectious Diseases (2006c). Bluetongue – Europe (21): Bulgaria, BTV-8 suspected. ProMED-mail 2006; 24 November: 20061124.3347. http://www.promedmail.org.

International Society for Infectious Diseases (2007). Bluetongue – Europe (14): BTV-8, Germany (Nordrhein-Westfalen), confirmed. ProMED-mail 2007; 13 June: 20070613.1928), http://www.promedmail.org.

Jennings, D.M. and Mellor, P.S. (1988) The vector potential of British *Culicoides* species for bluetongue virus. *Vet. Microbiol.* **17**, 1–10.

Jones, R.H. and Foster, N.M. (1971) Transovarial transmission of bluetongue virus unlikely for *Culicoides variipennis. Mosq. News* **31**, 434–7.

Kirkland, P.D. and Hawkes, R.A. (2004) A comparison of laboratory and 'wild' strains of bluetongue virus – is there any difference and does it matter? *Vet. Ital.* **40**, 448–55.

Knowles, N.J.D. and Davies, P.R. (2000) Origin of recent outbreaks of foot-and-mouth disease in North Africa, the Middle East and Europe. Report of the Session of the Research Group of the Standing Technical Committee of the European Commission for the Control of Foot-and-Mouth Disease, 3 edn., Borovets, Bulgaria. FAO Rome, pp. 39-45.

Kremer, M., Leberre, G. and Beaucournu-Saguez, F. (1971) Notes sur le *Culicoides* (Diptera, Ceratopogonidae) de Corse. Description de *C. corsicus* n.sp. *Ann. Parasitol. Hum. Comp.* **46**, 635–60.

Labrovic, A., Poljak, Z., Separovic, S., Jukic, B., Lukman, D., Listes, E. and Bosnic, S. (2004) Spatial distribution of bluetongue in cattle in southern Croatia in the last quarter of 2002. *Vet. Ital.* **40**, 217–20.

Listes, E., Bosnic, S., Benic, M., Lojkic, M., Cac, Z., Cvetnic, Z., Madic, J., Separovic, S., Labrovic, A., Savini, G. and Goffredo, M. (2004) Serological evidence of bluetongue and a preliminary entomological study in southern Croatia. *Vet. Ital.* **40**, 221–5.

Lysyk, T.J. (2007) Seasonal abundance, parity, and survival of adult *Culicoides sonorensis* (Diptera: Ceratopogonidae) in southern Alberta, Canada. *J. Med. Entomol.* **44**, 959–69.

Maan, S., Samuel, A.R., Maan, N.S., Attoui, H., Rao, S. and Mertens, P.P.C. (2004) Molecular epidemiology of bluetongue viruses from disease outbreaks in the Mediterranean Basin. *Vet. Ital.* **40**, 489–96.

Manso-Ribeiro, J., Rosa-Azevedo, J.A., Noronha, F.M.O., Braco-Forte-Junior, M.C., Grave-Pereira, C. and Vasco-Fernandes, M. (1957) Fievre catarrhale du mouton (Bluetongue). *Bull. Off. Int. Epizoot.* **48**, 350–367.

Manso-Ribeiro, J. and Noronha, F.M.O. (1958) Fievre catarrhale du mouton au Portugal (blue tongue). *Bull. Off. Int. Epizoot.* **50**, 46–64.

McLachlan, N.J., Barratt-Boyes, S.M. and Brewer, A.W. (1992) Bluetongue virus infection of cattle. In: T.E. Walton and B.I. Osburn, (eds.), *Bluetongue, African horse sickness and related Orbiviruses,* CRC Press, Boca Raton, pp. 725–736.

Meiswinkel, R., Gomulski, L.M., Delecolle, J.-C., Goffredo, M. and Gasperi, G. (2004) The taxonomy of *Culicoides* vector complexes – unfinished business. *Vet. Ital.* **40**, 151–9.

Meiswinkel, R., van Rijn, P., Leijs, P. and Goffredo, M. (2007) Potential new *Culicoides* vector of bluetongue virus in northern Europe. *Vet. Rec.* **161**, 564–5.

Mellor, P.S. (2000) Replication of arboviruses in insect vectors. *J. Comp. Pathol.* **123**, 231–47.

Mellor, P.S. (2004) Infection of the vectors and bluetongue epidemiology in Europe. *Vet Ital.* **40**, 167–74.

Mellor, P.S. and Jennings, D.M. (1988) The vector potential of British Culicoides species for bluetongue virus. In: P. Roy and B. Osburn (eds.), *Proceedings of a BTV Workshop, Double Stranded RNA Virus Symposium,* Oxford, 9-13th September 1986, pp. 12–21.

Mellor, P.S. and Boorman, J. (1995) The transmission and geographical spread of African horse sickness and bluetongue viruses. *Ann. Trop. Med. Parasitol.* **89**, 1–15.

Mellor, P.S., Boorman, J. and Baylis, M. (2000) *Culicoides* biting midges: their role as arbovirus vectors. *Ann. Rev. Entomol.* **45**, 307–40.

Mellor, P.S., Jennings, M. and Boorman, J.P.T. (1984) *Culicoides* from Greece in relation to the spread of bluetongue virus. *Rev. Elev. Med. Vet. Pays. Trop.* **37**, 286–9.

Mellor, P.S. and Pitzolis, G. (1979) The transmission and geographical spread of African horse sickness and bluetongue viruses. *Ann. Trop. Med. Parasitol.* **89**, 1–15.

Mellor, P.S. and Wittmann, E.J. (2002) Bluetongue virus in the Mediterranean Basin 1998–2001. *Vet. J.* **164**, 20–37.

Melville, L.F. and Gard, G.P. (1992) Investigations of the natural infection with Orbiviruses on reproduction in cattle. In: T.E. Walton and B.I. Osburn, (eds.), *Bluetongue, African horse sickness and related Orbiviruses,* CRC Press, Boca Raton, pp. 744–750.

Melville, L.F., Hunt, N.T., Davis, S.S. and Weir, R.P. (2004) Bluetongue virus does not persist in naturally infected cattle. *Vet. Ital.* **40**, 502–7.

Miranda, M.A., Borras, D., Rincon, C. and Alemany, A. (2003) Presence in the Balearic islands (Spain) of the midges *Culicoides imicola* and *Culicoides obsoletus* group. *Med. Vet. Entomol.* **17**, 52–4.

Monteys, V.S., Aranda, C., Escosa, R., Pages, N. and Ventura, D. (2004) Results of current surveillance of likely bluetongue virus vectors of the genus *Culicoides* in Catalonia, Spain. *Vet. Ital.* **40**, 130–2.

Monteys, V.S.I. and Saiz-Ardanaz, M. (2003) *Culicoides* midges in Catalonia (Spain), with special reference to likely bluetongue virus vectors. *Med. Vet. Entomol.* **17**, 288–93.

Monteys, V.S.I., Ventura, D., Pages, N., Aranda, C. and Escosa, R. (2005) Expansion of *Culicoides imicola,* the main bluetongue virus vector in Europe, into Catalonia, Spain. *Vet. Rec.* **156**, 415–17.

Murray, M.D. and Nix, H.A. (1987) Southern limits of distribution and abundance of the biting-midge *Culicoides brevitarsis* Kieffer (Diptera, Ceratopogonidae) in south-eastern Australia – an application of the growest model. *Aust. J. Zool.* **35**, 575–85.

NASA and Earthsat (2004) World Imagery Base Maps. http:// www.geographynetwork.com.

Nevill, E.M. (1971) Cattle and *Culicoides* biting midges as possible overwintering hosts of bluetongue virus. *Onderstepoort J. Vet. Res.* **38**, 65–72.

Nunamaker, R.A., Sieburth, P.J., Dean, V.C., Wigington, J.G., Nunamaker, C.E. and Mecham, J.O. (1990). Absence of transovarial transmission of bluetongue virus in *Culicoides variipennis*: immunogold labelling of bluetongue virus antigen in developing oocytes from *Culicoides variipennis* (Coquillett). *Comp. Biochem. Physiol. A* **96**, 19–31.

Osmani, A., Murati, B., Kabashi, Q., Goga, I., Berisha, B., Wilsmore, A.J. and Hamblin, C. (2006) Evidence for the presence of bluetongue virus in Kosovo between 2001 and 2004. *Vet. Rec.* **158**, 393–6.

Patta, C., Giovannini, A., Rolesu, S., Nannini, D., Davini, G., Calistri, P., Santucci, U. and Caporale, V. (2004) Bluetongue vaccination in Europe: the Italian experience. *Vet. Ital.* **40**, 601–610.

Placidi, L. (1957) La bluetongue au Maroc. *Bull. Acad. Vét. Fr.* **30**, 79–84.

Purse, B.V., Mellor, P.S., Rogers, D.J., Samuel, A.R., Mertens, P.P.C. and Baylis, M. (2005) Climate change and the recent emergence of bluetongue in Europe. *Nature Rev. Microbiol.* **3**, 171–81.

Purse, B.V., Nedelchev, N., Georgiev, G., Veleva, E., Boorman, J., Veronesi, E., Carpenter, S., Baylis, M. and Mellor, P.S. (2006) Spatial and temporal distribution of bluetongue and its *Culicoides* vectors in Bulgaria. *Med. Vet. Entomol.* **20**, 335–44.

Purse, B.V., Tatem, A.J., Caracappa, S., Rogers, D.J., Mellor, P.S., Baylis, M. and Torina, A. (2004) Modelling the distributions of *Culicoides* bluetongue virus vectors in Sicily in relation to satellite-derived climate variables. *Med. Vet. Entomol.* **18**, 90–101.

Savini, G., Goffredo, M., Monaco, F., Di Gennaro, A., Cafiero, M.A., Baldi, L., De Santis, P., Meiswinkel, R. and Caporale, V. (2005) Bluetongue virus isolations from midges belonging to the *Obsoletus* complex(*Culicoides*, Diptera: Ceratopogonidae) in Italy. *Vet. Rec.* **157**, 133–9.

Savini, G., Goffredo, M., Monaco, F., Di Gennaro, A., de Santis, P., Meiswinkel, R. and Caporale, V. (2004) The isolation of bluetongue virus from field populations of the obsoletus complex in central Italy. *Vet. Ital.* **40**, 286–91.

Savini, G., Goffredo, M., Monaco, F., de Santis, P. and Meiswinkel, R. (2003) Transmission of bluetongue virus in Italy. *Vet. Rec.* **152**, 119.

Scaramozzino, P., Boorman, J.P.T., Vitale, F., Semproni, G. and Mellor, P.S. (1996) Entomological survey on Ceratopogonidae in Central-Southern Italy. *Parassitologia* **38**, 1–2.

Sellers, R.F. (1975) Bluetongue in Cyprus. *Aust. Vet. J.* **51**, 198–203.

Sellers, R.F., Gibbs, E.P., Herniman, K.A., Pedgley, D.E. and Tucker, M.R. (1979b) Possible origin of the bluetongue epidemic in Cyprus, August 1977. *J. Hyg. (Lond.)* **83**, 547–55.

Sellers, R.F. and Maarouf, A.R. (1991) Possible introduction of epizootic haemorrhagic-disease of deer virus (serotype-2) and bluetongue virus (serotype-11) into British-Columbia in 1987 and 1988 by infected *Culicoides* carried on the wind. *Can. J. Vet. Res.* **55**, 367–720.

Sellers, R.F. and Mellor, P.S. (1993) Temperature and the persistence of viruses in *Culicoides* during adverse conditions. *Off. Int. Epizoot. Rev. Sci. Tech.* **12**, 733–55.

Sellers, R.F., Pedgley, D.E. and Tucker, M.R. (1979a) Possible windborne spread of blue-tongue to Portugal, June–July 1956. *J. Hyg. (Lond.)* **81**, 189–96.

Shimshony, A. (1987) Bluetongue activity in Israel, 1950–1985: the disease, virus prevalence, control methods In: W.P. Taylor (ed.), Bluetongue in the Mediterranean region. Proc. Meeting in the Community Programme for Coordination of Agricultural Research. Istituto Zooprofilattico Sperimentale dell' Abruzzo e del Molise, 3–4 October 1985, Teramo, Italy: Commission of the European Communities, Brussels, Luxembourg, pp. 1–22.

Shimshony, A. (2004) Bluetongue in Israel - a brief historical overview. *Vet. Ital.* **40**, 116–118.

Singer, R.S., MacLachlan, N.J. and Carpenter, T.E. (2001) Maximal predicted duration of viraemia in bluetongue virus-infected cattle. *J. Vet. Diagn. Invest.* **13**, 43–9.

Takamatsu, H.-H., Mellor, P.S., Mertens, P.P.C., Kirkham, P.A., Burroughs, J.N. and Parkhouse, R.M.E. (2003) A possible overwintering mechanism for bluetongue virus in the absence of the insect vector. *J. Gen. Virol.* **84**, 227–35.

Taylor, W.P. (1987) Bluetongue in Syria and Jordan. *Proc. Meeting in the Community Programme for Coordination of Agricultural Research. Istituto Zooprofilattico Sperimentale Dell' Abruzzo E Del Molise, 3–4 October 1985*, 39–47, ed.

Taylor, W.P. and Mellor, P.S. (1994a) Bluetongue virus distribution in Turkey 1978–1981. *Epidemiol. Infect.* **112**, 623–33.

Taylor, W.P. and Mellor, P.S. (1994b) The distribution of Akabane virus in the Middle East. *Epidemiol. Infect.* **113**, 175–85.

Taylor, W.P., Sellers, R.F., Gumm, I.D., Herniman, K.A.J. and Owen, L. (1985) Bluetongue epidemiology in the Middle East In: T.L. Barber and M.M. Jochim (eds), *Bluetongue and Related Orbiviruses*. New York: Alan R. Liss Inc, pp. 527–30.

Torina, A., Caracappa, S., Mellor, P.S., Baylis, M. and Purse, B.V. (2004) Spatial distribution of bluetongue virus and its *Culicoides* vectors in Sicily. *Med. Vet. Entomol.* **18**, 81–9.

Urman, H.K., Mille, U.È., Mert, N., Berkin, S.Ë., Kahramen, M.M., Yue, H. and Avvurran, H. (1980) Tuerkiye'de buzagilarda konjenital epizootik arthrogryposis ve hydranencephalie olaylari. Ankara universitesi. *Veteriner Fakgltesi Dergisi* **26**, 287–95.

Vassalos, M. (1980) Cas de fièvre catarrhale du mouton dans l'ile de Lesbos (Grèce). *Bull. Off. Int. Epizoot.* **92**, 547–55.

White, D.M. and Mecham, J.O. (2004) Lack of detectable bluetongue virus in skin of seropositive cattle: implications for vertebrate overwintering of bluetongue virus. *Vet. Ital.* **40**, 513–19.

White, D.M., Blair, C.D. and Beaty, B.J. (2006) Molecular epidemiology of bluetongue virus in northern Colorado. *Virus Res.* **118**, 39–45.

Wilson, A., Carpenter, S., Gloster, J. and Mellor, P.S. (2007) Re-emergence of BTV-8 in northern Europe in 2007. *Vet. Rec.* **161**, 487–9.

Wilson, A. and Mellor, P.S. (2008) Bluetongue in Europe: vectors, epidemiology and climate change. *Parasitol. Res.*, in press.

World Organisation for Animal Health (2006) Bluetongue detected for first time in Northern Europe. http://www.oie.int/eng/press/en_060823.htm.

Yonguc, A.D., Taylor, W.P., Csontos, L. and Worrall, E. (1982) Bluetongue in western Turkey. *Vet. Rec.* **111**, 144–6.

Bluetongue virus in the mammalian host and the induced immune response

12

KARIN E. DARPEL, PAUL MONAGHAN,
SIMON J. ANTHONY, HARU-HISA TAKAMATSU
AND PETER P.C. MERTENS

Institute for Animal Health, Pirbright Laboratory, Ash Rd., Pirbright, Woking, Surrey, GU24 0NF, UK

Introduction

Bluetongue virus (BTV) is an arthropod-borne virus (arbovirus) transmitted between its ruminant hosts via the bites of haematophagous midges of the genus *Culicoides spp.* Like all arboviruses, BTV is able to infect and replicate in a wide range of different cell types and at a range of different temperatures, allowing it to productively infect both insect and mammalians hosts.

Intracellular aspects of BTV replication in mammalian cells are reviewed elsewhere in this book (see Chapters 4 and 5). This chapter reviews current knowledge concerning the BTV replication and transmission cycle within the ruminant host, as well as the immune mechanisms and antiviral responses. This includes considerations of organ tropism and the cellular targets for BTV infection within these organs. Bluetongue virus has evolved to optimise its transmission efficiency within the context of the insect (vector)/ruminant (host) relationship. The BTV replication mechanisms and strategies within

ISBN-13: 978-0-12-369368-6

the ruminant host will therefore also be related to their consequences for transmission to the insect vector.

Bluetongue virus infection of the ruminant host

When adult female *Culicoides* bite their hosts in the field, small amounts of midge saliva are injected into the skin. If the midges are fully infected with BTV (i.e. the infection has disseminated within the insect and has reached the salivary glands), then BTV may also be deposited in the skin with the saliva. Bluetongue virus replication within the insect vector is reviewed by Mellor *et al.* (Chapter 14). Although the amount of virus in insect saliva is very low (Fu, 1995), a single bite from a transmission-competent *Culicoides sonorensis* midge will reliably infect a susceptible sheep, indicating that the BTV transmission process is very efficient (Foster *et al.*, 1963; O'Connell, 2002; Baylis *et al.*, Chapter 15). However, most studies of BTV replication in the ruminant host are based exclusively on needle inoculations rather than on infection via the bites of infected midges (Pini, 1976; Lawman, 1979; Mahrt and Osburn, 1986; Parsonson *et al.*, 1987; MacLachlan *et al.*, 1990; Barratt-Boyes and MacLachlan, 1994). These studies also used different routes of virus administration, ranging from intravenous (i/v), subcutaneous (s/c) to intradermal (i/d). In order to mimic the clinical signs of infected sheep following artificial virus inoculations, it has been usually necessary to infect individual animals with much higher amounts of virus (100–100 000 times) than delivered during a single bite from an infected midge. In view of the fact that the normal route of infection is via midge bite through the mammalian host's skin, the authors of the later studies suggested that subcutaneous or intradermal injections are more appropriate.

Although it is at present unclear why transmission from the vector to the host is so efficient, recent studies have indicated that the arthropod vector itself may play an important role. Saliva proteins of many haematophagous arthropods, including ticks, stable flies, black flies, mosquitoes, Tabanidae and *Culicoides* spp, can have profound immunomodulatory effects on mammalian cells and hosts (Bissonnette *et al.*, 1993; Cross *et al.*, 1993a; Cross *et al.*, 1993b; Cross *et al.*, 1994; Rohousova *et al.*, 2005). Responses to the saliva proteins, or to insect biting, include chemotactic effects, inhibition of lymphocyte proliferation, inhibition of cytokine and chemokine production and changes to the dermal cell populations (Anjili *et al.*, 1995; Costa *et al.*, 2004; Wanasen *et al.*, 2004; Wasserman *et al.*, 2004; Bishop *et al.*, 2006). Such interactions between blood-feeding insects and the mammalian immune system could have important consequences for pathogens that are transmitted by blood-feeding arthropods. Although saliva-enhanced transmission was originally thought to be more important for tick-borne pathogens (Jones *et al.*, 1992), it has now been demonstrated that the saliva of biting insects can also

01 have similar effects. More potent infections, as shown by increased pathogen
02 load and/or higher virulence, have been reported for certain pathogens
03 co-administered with vector saliva, including *Leishmania* and certain arbo-
04 viruses (including West Nile virus, Sindbis virus and vesicular stomatitis
05 virus) (Titus and Ribeiro, 1988; Mbow *et al.*, 1998; Limesand *et al.*, 2000,
06 2003; Norsworthy *et al.*, 2004; Schneider *et al.*, 2004; Schneider *et al.*, 2006).
07 The effect of *Culicoides* saliva on the BTV transmission mechanism is not
08 currently known. However, hypersensitivity and severe allergy towards the
09 biting of adult *Culicoides* has previously been reported in several host species,
10 including sheep (Connan and Lloyd 1988; Yeruham *et al.*, 1993; Yeruham
11 *et al.*, 2000). *Culicoides* biting is also the cause of 'sweet itch' or 'summer
12 eczema' (SE), a hypersensitivity reaction to insect biting, which results in an
13 intense but seasonal pruritic skin condition in horses (Mellor and McCaig
14 1974; Foster and Cunningham, 1998). Exposure to the bites of the BTV vector
15 *C. sonorensis* has been shown to dramatically change the cellular composition
16 of mammalian skin (Foster and Cunningham, 1998; Takamatsu *et al.*, 2003;
17 Darpel, 2007). Consequently, BTV is likely to encounter a different range of
18 mammalian cells in the skin when transmitted by midges, as opposed to needle
19 inoculation. Further studies are necessary to fully investigate the complex
20 nature of BTV (and other arbovirus) transmission processes and relevant
21 host–vector–pathogen interactions.

Bluetongue virus replication in the ruminant host

Cellular targets

29 The pathology of bluetongue (BT) in the mammalian host is characterised largely
30 by its haemorrhagic nature (see MacLachlan and Gard, Chapter 13), which led to
31 early suggestions that microvascular endothelial cells are a major target for BTV
32 infection (Hardy and Price, 1952; Erasmus 1975; Pini 1976). This was later
33 confirmed for several other *Culicoides*-transmitted orbiviruses including epizootic
34 haemorrhagic disease virus (EHDV) and African horse sickness virus (AHSV),
35 using a range of techniques including in-situ hybridisation, in-situ PCR, electron
36 microscopy and immunofluorescence microscopy (Mahrt and Osburn, 1986;
37 MacLachlan *et al.*, 1990; Barratt-Boyes and MacLachlan, 1994; Brown *et al.*,
38 1994; Wohlsein *et al.*, 1997; Brodie *et al.*, 1998a; Brodie *et al.*, 1998b; Wohlsein
39 *et al.*, 1998; Gomez-Villamandos *et al.*, 1999). However, most of these methods
40 cannot distinguish between active virus replication and the simple physical
41 presence of virus in infected animal cells.
42 Recently, confocal microscopy has been used to identify the cellular and
43 tissue tropism of BTV in sheep. By using antibodies that detect and distinguish
44 between the structural proteins of the virus and the non-structural (NS)

proteins, synthesised during virus replication, it has been possible to identify cells in which the virus is actively replicating (Darpel, 2007). These studies showed that BTV infects and replicates in the endothelial cells of capillaries and smaller blood vessels within many organs of the infected host. Viral proteins are able to be detected inside infected endothelial cells, with the presence of NS2 indicating active replication, as NS2 is produced only in cells in which the virus replicates (Mertens *et al.*, 1984; Mertens *et al.*, 1987). On some occasions, the BTV core protein VP7 was detected within the lumen of blood vessels, in close association with erythrocytes, indicating the presence of virus particles (Figure 12.1). Later on in infection [6–9 days post-infection (d.p.i.)], both NS2 and VP7 were detected in the endothelial cells of larger blood vessels, and of vessels resembling lymphatic ducts, suggesting that at the peak of virus replication, less-susceptible endothelial cells may become infected (Darpel, 2007).

Bluetongue virus can infect and replicate in primary endothelial cells and endothelial cell lines causing cytopathic effects (CPE) (Wechsler and McHolland, 1988; Wechsler and Luedke, 1991). The CPE caused by BTV infection of endothelial cells, which results in damage to the endothelial lining of blood vessels together with subsequent immune and repair responses, may represent the major pathological mechanism behind clinical manifestation of the disease (Pini, 1976; Lawman, 1979; Howerth and Tyler, 1988; MacLachlan *et al.*, 1990). However, oedema and haemorrhage have also been observed in the absence of any obvious endothelial cell necrosis. Changes to the tight-junction endothelial cells in blood vessels have also been suggested as an additional mechanism leading to the pathology seen in both BT and African horse sickness (AHS) (Carrasco *et al.*, 1999; Gomez-Villamandos *et al.*, 1999; Chiang *et al.*, 2006).

Several studies have shown that leucocytes can become infected by BTV. Indeed it has been suggested that a wide range of different cell types including neutrophils, monocytes, macrophages and lymphocytes can be involved in BTV dissemination during the early stages of infection in the mammalian host (Erasmus, 1975; Lawman, 1979; Barratt-Boyes and MacLachlan, 1994; MacLachlan and Gard, Chapter 13). However, the identification of individual leucocyte subsets in vivo is difficult to demonstrate (Mahrt and Osburn, 1986; MacLachlan *et al.*, 1990; Darpel, 2007), and there is some disagreement concerning the cell types involved (Lawman, 1979; Morrill and McConnell 1985; MacLachlan *et al.*, 1990). Bluetongue virus replication in agranular leucocytes that are morphologically similar to lymphocytes, monocytes and dendritic cells/macrophages has recently been confirmed using confocal microscopy (Darpel, 2007). Monocytes infected with BTV in vivo showed CPE after a few days of post-infection, while freshly isolated lymphocytes were resistant to infection (Whetter *et al.*, 1989; Barratt-Boyes *et al.*, 1992). However, bovine T lymphocytes that undergo blastogenesis can be infected with BTV (Stott *et al.*, 1990; Barratt-Boyes *et al.*, 1992).

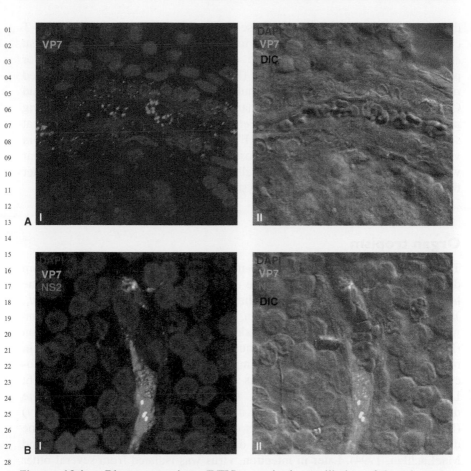

Figure 12.1   Bluetongue virus (BTV) protein in capillaries of lymph nodes. A paraformaldehyde-fixed, microtome cut thick section of lymph nodes from BTV-2 (RSA1971/03) infected sheep labelled with different anti-BTV antibodies and analysed via confocal microscopy. In both panels A and B, cellular nuclei are stained in blue (DAPI) and tissue morphology is shown in II using differential interference contrast (DIC). Viral proteins are stained either red or green as colour coded in the pictures, sometimes using double labelling for different proteins. (Panel A) BTV protein VP7 in the lumen of a small blood vessel in the prescapular lymph node at 3 days post-infection (d.p.i.). (Panel B) BTV proteins NS2 and VP7 in the endothelium of a capillary in the mandibular lymph node at 8 d.p.i. (See colour plate 22)

The fate of BTV-infected lymphocytes is still poorly understood, and there is experimental evidence suggesting that different subsets can be lytically or non-lytically (persistently) infected (Stott *et al.*, 1990; Barratt-Boyes *et al.*, 1992). Lytic infection of CD4$^+$ cultures, chronic infection of CD8$^+$ cells and persistent infection of 'null cell' cultures have all apparently been demonstrated (Stott *et al.*, 1992). These 'null cells' are now thought to be γδ T cells, which have also been shown to be persistently infected (Takamatsu *et al.*, 2003). Several studies have demonstrated that leucocytes and/or peripheral blood mononuclear cells (PBMC) collected from BTV-infected ruminants, then cultured after stimulation with interleukin-2 (IL-2) and Concanavalin A (Con A), can be maintained in an infected state for long periods (Takamatsu *et al.*, 2003; Lunt *et al.*, 2006). Overall, it seems likely that several leucocyte subsets can support BTV replication and may be involved in viral dissemination in the ruminant host. However, currently it is still unclear whether infected leucocytes are lytically or non-lytically infected or whether they are eliminated by the immune system in vivo.

Organ tropism

The organ tropism of BTV has been investigated by isolation of virus from different tissues over the time course of infection in both ovine and bovine hosts (Pini, 1976; Lawman, 1979; Mahrt and Osburn, 1986; MacLachlan *et al.*, 1990; Barratt-Boyes and MacLachlan, 1994). The first round of replication was detected in the regional lymph nodes draining the site of BTV inoculation (Pini, 1976; Lawman, 1979). Subsequent rounds of replication occurred in other lymphatic organs (e.g. other lymph nodes and the spleen), with leuco-cytes as the most likely cellular targets within the different organs (Pini, 1976; Lawman, 1979; MacLachlan *et al.*, 1990). Large amounts of virus were detected in lymphatic tissues (even if the virus was inoculated intravenously) (MacLachlan *et al.*, 1990), and disruption of efferent lymphatic flow from the draining lymph node resulted in a delay of viraemia (Barratt-Boyes and MacLachlan, 1994). Later in infection, virus was isolated from several other organs (including lungs, tongue, muscle and heart) where replication is thought to occur mostly in endothelial cells. Virus replication in these target organs then leads to a generalised viraemia (Pini, 1976; Lawman, 1979). Bluetongue virus was detected only in some organs (e.g. liver and kidney) while the host was viraemic, suggesting that this reflects the presence of virus in the blood within the organ.

A recent study of infection in sheep using confocal microscopy also demon-strated a positive BTV tropism for lymphatic tissues. Early and high levels of virus replication were observed in the mandibular and prescapular (draining) lymph nodes and tonsil. Although the leucocytes in these tissues were infected, a very high level of replication was also detected in endothelial cells of the small blood vessels and capillaries (see Figure 12.1) (Darpel, 2007). Large amounts of BTV NS proteins, indicating a high level of virus replication, were also detected at an early stage of infection in the endothelial cells of capillaries and small blood vessels in the skin (right and left inner thigh) (Figure 12.2). It was concluded from this that the skin can be a major site of BTV replication,

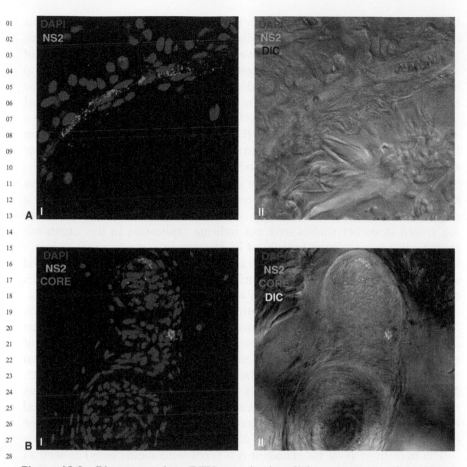

Figure 12.2 Bluetongue virus (BTV) proteins in cellular components of the skin. A paraformaldehyde-fixed microtome cut thick section of lymph nodes from BTV-infected sheep labelled with different anti-BTV antibodies and analysed via confocal microscopy. In both panels A and B, cellular nuclei are stained in blue (DAPI) and tissue morphology is shown in II using differential interference contrast (DIC). Viral proteins are stained either red or green as colour coded in the pictures, sometimes using double labelling for different proteins. (Panel A) BTV protein NS2 in the endothelium of a small blood vessel of the dermis of the skin at 3 days post-infection (d.p.i) with BTV-2 (RSA1971/03). (Panel B) BTV core proteins and NS2 in a glandular structure branching of a hair at 8 d.p.i. with BTV-8 (NET2006/01) (See colour plate 23).

with peak production of virus proteins occurring at ~6–8 d.p.i. At the same time point, BTV was detected in capillaries surrounding skin glands (resembling sebaceous glands). Other studies have also reported the presence of BTV or EHDV in the skin, especially around hair follicles, or in the papillae of the hair (Mahrt and Osburn, 1986; Brodie *et al.*, 1998a; Brodie *et al.*, 1998b).

01 Originally this was not considered to be important, because attempted virus
02 isolation from skin samples had indicated low or inconsistent virus loads
03 (Pini, 1976; Lawman, 1979). However, the fibrous and compact nature of
04 skin tissue makes isolation of infectious virus difficult, which may have led to
05 an underestimation of viral load and consequently an underestimate of the
06 skin's role as a site of BTV replication. The recent detection of high levels of
07 virus replication in the skin (Darpel, 2007) suggests that this organ may have a
08 significant role in the transmission of BTV between insect vector and ruminant
09 host. The skin may be both an important site for early infection and replication of
10 the virus and a source of virus for blood-feeding vector insects (Darpel, 2007).
11 In the confocal microscopic study of Darpel (2007), low levels of virus
12 replication were detected in the spleen, an organ previously thought to be a
13 major site for BTV replication. It is possible that some of the virus detected in
14 the spleen in earlier studies may not indicate replication in this organ but
15 merely virus accumulation because of its attachment to erythrocytes or infec-
16 tion of other blood cells. The spleen is a blood reservoir and recycles red blood
17 cells, which accumulate in the red pulp. In previous studies, BTV particles
18 attached to erythrocytes in the spleen would have been detected using techni-
19 ques such as PCR and or virus isolation, which do not allow differentiation
20 between presence and active replication. However, conversely, the detection
21 of small numbers of BTV particles embedded in the cell membrane of
22 erythrocytes, which can block antigenic sites, can be difficult by confocal
23 microscopy (Barratt-Boyes and MacLachlan, 1995; MacLachlan, 1995;
24 MacLachlan and Gard, Chapter 13). Consequently, techniques like virus
25 isolation may overestimate replication levels in an organ, while confocal
26 microscopy may underestimate the presence of virus particles in the absence
27 of high levels of replication (Darpel, 2007).
28 The tropism of BTV for abundant cell types (endothelial cells) as well as
29 certain mobile cell subsets (leucocytes) is reflected by its widespread distribu-
30 tion within the mammalian host (Darpel, 2007). Although virus was detected
31 in almost every organ investigated at the peak of viraemia, the level of
32 replication, as assessed by the number of infected cells, in most cases appeared
33 to be relatively low. The high level of viraemia often detected during BTV
34 infection may therefore reflect the widespread nature of infection in cells that
35 have direct contact with the blood (i.e. the vascular endothelium). This repli-
36 cation strategy of BTV contrasts with viruses that replicate to high levels in the
37 cells of only a few organs, e.g. foot-and-mouth disease virus, which replicates
38 in the basal layer of the epithelium in tongue, soft palate and coronary band
39 (Monaghan *et al.*, 2005), or viruses like rotavirus that replicate in cells at the
40 apices of small intestinal villi (Ramig, 2004). Differences in the susceptibility
41 of vascular endothelial cells from individual organs may explain the organ
42 tropisms observed during BTV infection. Organs where endothelial cells
43 express higher susceptibility (e.g. head, lymph nodes, skin and lung) are likely
44 to be infected earlier and the virus may replicate to higher levels than in other

organs. It has been reported that endothelial cells taken from different sites in the host animal vary in their morphology and in their response to the same stimulus including infections with different pathogens (Craig *et al.*, 1997; Craig *et al.*, 1998; Gomez-Villamandos *et al.*, 1999). It has also been suggested that differences in the susceptibility of bovine and ovine endothelial cells to BTV infection may explain the differences in the clinical manifestation of BT in cattle and sheep (Coen *et al.*, 1991; DeMaula *et al.*, 2001; DeMaula *et al.*, 2002a; DeMaula *et al.*, 2002b; MacLaclan and Gard, Chapter 13). A clinical study of the northern European strain of BTV-8 from 2006 (Darpel *et al.*, 2007) suggested that differences observed in lung pathology could be explained by varying susceptibilities of lung endothelial cells. This may be one of the factors determining the severity of the clinical signs of BT in individual animals.

Immune response

Humoral immune response

Bluetongue virus-infected ruminants develop various antiviral host immune responses. These include serogroup-specific (BTV species-specific) antibodies that are directed against several of the more conserved virus proteins. VP7 is strongly immunodominant and raises group-specific antibodies that can be used for the diagnosis and identification of any BTV type and strain by ELISA (Huismans and Erasmus 1981; Gumm and Newman 1982; Anderson 1984; Afshar *et al.*, 1987; Afshar *et al.*, 1992). Although these non-neutralising and serogroup-specific antibodies are important for diagnosis, their role, if any, in protective immunity is not clear. Attempts to demonstrate a role for these serogroup-specific antibodies in antibody-dependent cell-mediated cytotoxicity (ADCC) and/or complement-facilitated ADCC studies in sheep have not been successful, although only few such investigations have been undertaken (Jeggo *et al.*, 1983b; Jeggo *et al.*, 1986).

Neutralising antibodies raised against one serotype of BTV generally provide protection only against challenge with BTV strains belonging to the same serotype, not against strains of heterologous serotypes (serotype-specific anti-bodies) (Jeggo *et al.*, 1983a; Jeggo *et al.*, 1984c; Huismans *et al.*, 1987; Stott and Osburn, 1992), but a wider spectrum of cross-reactive neutralising antibody responses and serotype cross-protection has been induced by sequential infec-tion with two or more BTV serotypes (Jeggo *et al.*, 1983a; Jeggo *et al.*, 1984c; Jeggo *et al.*, 1986). However, simultaneous inoculation with three different BTV serotypes not only failed to raise heterotypic antibodies but also resulted in the replication of only two of the three virus serotypes (Jeggo *et al.*, 1984b; Jeggo *et al.*, 1986). These results therefore raise concerns regarding the use of polyvalent, live attenuated vaccines, as such vaccination strategy may result in the failure to develop an immune response against some of the administered serotypes (Jeggo *et al.*, 1984b; Jeggo and Wardley, 1985).

The outer capsid protein VP2 is the main determinant of serotype specificity and is the primary antigen targeted by neutralising antibodies (Huismans and Erasmus, 1981; Mertens *et al.*, 1989; Stott and Osburn, 1992). The second outer capsid protein VP5 appears to have a synergetic effect in the development of a serotype-specific, neutralising antibody response, possibly through physical interactions with VP2 (Villiers, 1974; Cowley and Gorman, 1987; Mertens *et al.*, 1989). In this context, a number of ovine helper T-cell clones that are specific to BTV-1 have been shown to recognise VP2 from the same serotype, when virus core particles and VP5 were not recognised (Takamatsu *et al.*, 1990; Takamatsu *et al.*, 1992). One of the T-cell clones examined produced large amounts of IL-2 and interferon (IFN)-γ when stimulated with VP2 from BTV-1. Because helper T cells are essential for the development of neutralising-antibody-producing B cells, VP2 is likely to contain epitopes recognised by both neutralising antibodies and helper T cells.

Cellular immune response

Although cellular immune responses play an important role in protection against BTV (Jeggo *et al.*, 1983b), the study of these responses, following infection of the ruminant hosts, is surprisingly limited.

Bluetongue virus can efficiently induce IFN *in vitro* and *in vivo*, which can be detected in serum from infected sheep and cattle shortly after infection (MacLachlan and Thomson, 1985). However, the influence of the IFN response on virus clearance and adaptive immune responses is unknown. Indeed, viraemia continues beyond the detection of IFN in serum, suggesting that BTV may be at least partially resistant to this antiviral mechanism (Jeggo, 1983; MacLachlan and Thomson 1985; MacLachlan 1994).

Ruminants infected with BTV or EHDV develop lymphopenia, although the underlying mechanism is unknown. It has been suggested that BTV-specific lymphocyte destruction or lymphocyte sequestering at virus replication sites (e.g. lymph nodes) may be involved (Ellis *et al.*, 1990; Quist *et al.*, 1997).

Cytotoxic T cells (CTLs) are important for protection against intracellular pathogens, and anti-BTV CTLs (which show serotype cross-reactivity) have been demonstrated in BTV-infected mice (Jeggo and Wardley 1982b, 1982a, 1982c) and sheep (Jeggo *et al.*, 1984a, 1985), with maximum CTL activity observed between 7 and 21 d.p.i. (Jeggo and Wardley, 1985; Jeggo *et al.*, 1986). Furthermore, passively transferred T-cell-enriched thoracic duct lymphocytes, taken from a sheep previously infected with BTV, have been shown to partially protect recipient sheep against infection with BTV strains belonging to either the homologous or heterologous serotypes, confirming that T cells (assumed to be CTLs) can mediate cross-protection and that the observed effect is not due to B cells and subsequent antibody production (Jeggo *et al.*, 1984a).

Bluetongue virus-specific ovine T-cell lines (with CTL activity) were able to reduce BTV replication in autologous skin fibroblast cultures through BTV-specific but heterotypic lysis of infected cells (Takamatsu and Jeggo, 1989). The cross-reactivity of these BTV-infected sheep T cells against heterologous BTV strains did not correlate with the cross-reactions shown by neutralising antibodies, in virus neutralisation assays (VNT), between closely related serotypes. This may be due to the recognition of short, linear peptides by T cells via the MHC pathway, while neutralising antibodies primarily recognise conformational epitopes (Takamatsu and Jeggo, 1989).

Bluetongue virus NS1, and to a lesser extent VP2, proteins have been suggested as the major ovine recognition antigens for CTLs, with VP5, VP7 and NS3 being minor antigens (Andrew et al., 1995; Janardhana et al., 1999). Immunisation of mice with recombinant baculovirus-expressed NS2 (from BTV-10) partially protected mice from challenge with vaccinia virus-expressing NS2 (Jones et al., 1997). This protection was shown to be mediated by CTLs (Jones et al., 1997). Although this is an artificial system, it shows the potential of CTL antigens for vaccine development (Alpar et al., Chapter 18).

Overall, relatively few studies of the immune response to BTV have been published, leading to an incomplete understanding of the nature of BTV/ mammalian host interactions, information that is essential for the rational development of effective vaccines (Alpar et al., Chapter 18). The recent spread of multiple BTV serotypes in southern Europe (Purse et al., 2005), the damage and losses caused by the BTV-8 epidemic in northern Europe (2006 onwards) and the recent spread of several exotic BTV types to the United States are all likely to renew interest in the immunobiology of BTV infection in the mammalian host. Most published studies have focussed solely on the clearance of systemic viraemia from the blood. However, the ruminant skin is both the route of transmission between host and vector and an important site for virus replication. More information concerning immune mechanisms operating within the local environment of the skin is therefore required, which includes a specialised/unique immune tissue [skin-associated lymphatic tissue (SALT)].

A potential bluetongue virus overwintering mechanism in the ruminant host

In some regions of the world, severe winter conditions, or dry seasons, kill off adults of BTV vector species, making continuous year-round transmission of the virus unlikely or impossible. Survival of the virus in such areas is therefore dependent on other mechanisms that allow the virus to persist in the absence of active transmission or overt cases of disease (Gibbs et al. 1992). The term 'overwintering' is commonly used to describe the survival of BTV and other arboviruses from one vector season to the next (Sellers and Mellor 1993).

During the outbreak caused by BTV-8 in northern Europe, the virus persisted in the absence of overt disease for many months (overwinter, 2006–2007), reappearing in the following summer when temperatures and insect vector populations were more favourable for BTV transmission. The ability of the virus to survive through vector-free periods is of major epidemiological importance, although the nature of the BTV overwintering mechanism or mechanisms is uncertain and has been the subject of considerable speculation. Several mechanisms have been suggested, which include repeated introduction of infected *Culicoides* from other areas; persistence of the virus in *Culicoides* larvae (transovarial transmission in the insect vector); persistence in the mammalian host and low but undetected levels of the normal insect–host transmission cycle (Gibbs *et al.*, 1992; Brodie *et al.*, 1998b; Takamatsu *et al.*, 2003; White and Mecham, 2004; White *et al.*, 2005; Mellor *et al.*, Chapter 14; MacLachlan and Gard, Chapter 13).

Although viral RNA can be detected for long periods by RT-PCR in cattle and sheep blood (up to 220 days), attempts to recover infectious virus beyond 60 days (rarely 100 days) from PCR-positive bloods were unsuccessful (as outlined in more detail by MacLachlan and Gard, Chapter 13).

Research in the 1960s and 1970s also proposed the possibility that BTV could pass the ruminant placenta to infect the foetus in utero, resulting in viraemic offspring being born, which could act as virus reservoirs (Luedke *et al.*, 1977a–c; Gibbs *et al.*, 1979). However, the demonstrated ability of BTV to cross the placenta from the infected mother to the foetus was later attributed to the fact that these viruses had been passaged in artificial tissue culture systems, which was believed to have altered their characteristics (Flanagan and Johnson, 1995; Kirkland and Hawkes, 2004). Consequently, it was concluded that transplacental transmission could occur only in situations where live attenuated vaccines were in use and that field strains of BTV rarely, if ever, possessed this ability (Parsonson *et al.*, 1987; MacLachlan *et al.*, 2000; MacLachlan and Osburn, 2006). However, in the light of the recent demonstrated ability of the BTV-8 wild-type virus currently circulating in northern Europe to overwinter from 2006 to 2007 in the absence of adult vector *Culicoides*, the ability of this virus to cross the ruminant placental barrier is now being closely investigated.

Only a single study, which investigated the possibility of persistence in the mammalian host, has reported isolation of infectious BTV from post-viraemic animals (Takamatsu *et al.*, 2003). The virus was recovered from cultured skin cells, established from sheep skin samples taken at 63 d.p.i., at a time when no infectious virus could be detected in the blood. A mechanism involving $\gamma\delta$ T cells was subsequently suggested for persistence in the ruminant host (Takamatsu *et al.*, 2003). However, similar investigations in other laboratories have failed to detect infectious virus in the skin of infected sheep or cattle for any significant period of post-viraemia (Melville *et al.*, 2004; White and Mecham, 2004; Lunt *et al.*, 2006). Overwintering in the ruminant host was

01 previously suggested by Luedke *et al.* (1977c), who reported that live virus
02 could be recovered for long periods from a persistently infected bull after
03 repeated exposure to biting *Culicoides* midges. However, this study was later
04 discredited for a number of reasons including the fact that the animal holding
05 facilities were non-insect proof.
06   It has been suggested that some other arboviruses (e.g. West Nile virus)
07 can survive adverse conditions using a combination of overwintering
08 mechanisms in the vector (transovarial transmission) and the host (persistent
09 infection) (Reisen *et al.*, 2006; Reisen and Brault, 2007). Bluetongue virus
10 overwintering could also occur by more than one mechanism. However, the
11 widespread and near simultaneous recrudescence of BTV-8 in Germany, the
12 Netherlands, Belgium and northwest France during 2007 demonstrates that
13 the mechanism(s) involved is/are both effective and of major epidemiological
14 significance.

Conclusive remarks

19 A multidisciplinary approach combining veterinary medicine, virology, ento-
20 mology, immunology and biochemistry is required to fully investigate the
21 complex BTV transmission processes and host–pathogen–vector interactions.
22 The replication of BTV in the ruminant host has inevitably evolved to increase
23 the chances of successful transmission to the insect. It has been proposed that
24 only ruminant hosts that develop a high viraemia represent a risk for virus
25 transmission to the insect vector. However, this is clearly not always the case
26 as the work of Bonneau *et al.* (2002) showed that one *C. sonorensis* feeding on
27 a sheep at 21 days post-infection became infected, despite the fact that no
28 infectious virus was isolated from the sheep blood at this time point (Bonneau
29 *et al.*, 2002). More recently, Darpel (2007) showed that local replication of the
30 virus in the skin may provide another virus source for feeding *Culicoides*
31 (Darpel, 2007).
32   The replication strategy of BTV, as well as its cellular targets for replication
33 in the mammalian host, may provide two distinct mechanisms for virus to be
34 transmitted to the vector insect. These include the hypothesis that development
35 of a high-level viraemia is required for infection of vector insects via ingestion
36 of viraemic blood. Indeed, current control strategies are designed around the
37 concept that a certain minimum 'threshold' viraemia is required for effective
38 virus transmission. However, replication in cells present in the skin, and in
39 mobile cells (leucocytes) that are involved in the inflammatory response to
40 *Culicoides* biting, may provide another route for transmission. Further
41 research will be required to determine whether BTV transmission to the vector
42 insect can occur in the absence of a 'threshold' viraemia and to clarify both
43 BTV replication mechanisms and immune mechanisms within the ruminant
44 host.

References

Afshar, A., Dulac, G.C. and Riva, J. (1992) Comparison of blocking dot Elisa and competitive Elisa, using a monoclonal antibody for detection of bluetongue virus antibodies in cattle. *Vet. Microbiol.* **31**, 33–9.

Afshar, A., Thomas, F.C., Wright, P.F., Shapiro, J.L., Shettigara, P.T. and Anderson, J. (1987) Comparison of competitive and indirect enzyme linked immunosorbent assays for detection of bluetongue virus antibodies in serum and whole blood. *J. Clin. Microbiol.* **25**, 1705–10.

Anderson, J. (1984) Use of monoclonal antibody in a blocking Elisa to detect group specific antibodies to bluetongue virus. *J. Immunol. Methods* **74**, 139–49.

Andrew, M., Whiteley, P., Janardhana, V., Lobato, Z., Gould, A. and Coupar, B. (1995) Antigen specificity of the ovine cytotoxic T lymphocyte response to bluetongue virus. *Vet. Immunol. Immunopathol.* **47**, 311–22.

Anjili, C.O., Mbati, P.A., Mwangi, R.W., Githure, J.I., Olobo, J.O., Robert, L.L. and Koech, D.K. (1995) The chemotactic effect of *Phlebotomus duboscqi* (Diptera, Psychodidae) salivary gland lysates to murine monocytes. *Acta Trop.* **60**, 97–100.

Barratt-Boyes, S.M. and MacLachlan, N.J. (1994) Dynamics of viral spread in bluetongue virus infected calves. *Vet. Microbiol.* **40**, 361–71.

Barratt-Boyes, S.M. and MacLachlan, N.J. (1995) Pathogenesis of bluetongue virus infection of cattle. *J. Am. Vet. Med. Assoc.* **206**, 1322–9.

Barratt-Boyes, S.M., Rossitto, P.V., Stott, J.L. and MacLachlan, N.J. (1992) Flow cytometric analysis of *in vitro* bluetongue virus infection of bovine blood mononuclear cells. *J. Gen. Virol.* **73**, 1953–60.

Bishop, J.V., Mejia, J.S., De Leon, A.A.P., Tabachnick, W.J. and Titus, R.G. (2006) Salivary gland extracts of *Culicoides sonorensis* inhibit murine lymphocyte proliferation and no production by macrophages. *Am. J. Trop. Med. Hyg.* **75**, 532–6.

Bissonnette, E.Y., Rossignol, P.A. and Befus, A.D. (1993) Extracts of mosquito salivary gland inhibit tumor necrosis factor alpha release from mast cells. *Parasite Immunol.* **15**, 27–33.

Bonneau, K.R., DeMaula, C.D., Mullens, B.A. and MacLachlan, N.J. (2002) Duration of viraemia infectious to *Culicoides sonorensis* in bluetongue virus-infected cattle and sheep. *Vet. Microbiol.* **88**, 115–25.

Brodie, S.J., Bardsley, K.D., Diem, K., Mecham, J.O., Norelius, S.E. and Wilson, W.C. (1998a) Epizootic haemorrhagic disease: analysis of tissues by amplification and *in situ* hybridization reveals widespread orbivirus infection at low copy numbers. *J. Virol.* **72**, 3863–71.

Brodie, S.J., Wilson, W.C., O'Hearn, P.M., Muthui, D., Diem, K. and Pearson, L.D. (1998b) Effects of pharmacological and lentivirus-induced immune suppression on orbivirus pathogenesis: assessment of virus burden in blood monocytes and tissues by reverse transcription *in situ* PCR. *J. Virol.* **72**, 5599–609.

Brown, C.C., Meyer, R.F. and Grubman, M.J. (1994) Presence of African horse sickness virus in equine tissues, as determined by *in-situ* hybridization. *Vet. Pathol.* **31**, 689–94.

Carrasco, L., Sanchez, C., Gomez-Villamandos, J.C., Laviada, M.D., Bautista, M.J., Martinez-Torrecuadrada, J., Sanchez-Vizcaino, J.M. and Sierra, M.A. (1999) The role of pulmonary intravascular macrophages in the pathogenesis of African horse sickness. *J. Comp. Pathol.* **121**, 25–38.

Chiang, E.T., Persaud-Sawin, D.A., Kulkarni, S., Garcia, J.G.N. and Imani, F. (2006) Bluetongue virus and double-stranded RNA increase human vascular permeability: role of p38 MAPK. *J. Clin. Immunol.* **26**, 406–16.

Coen, M.L., Ellis, J.A., O'Toole, D.T. and Wilson, W.C. (1991) Cytokine modulation of the interaction between bluetongue virus and endothelial-cells *in vitro*. *Vet. Pathol.* **28**, 524–32.

Connan, R.M. and Lloyd, S. (1988) Seasonal allergic dermatitis in sheep. *Vet. Rec.* **123**, 335–7.

Costa, D.J., Favali, C., Clarencio, J., Afonso, L., Conceicao, V., Miranda, J.C., Titus, R.G., Valenzuela, J., Barral-Netto, M., Barral, A. and Brodskyn, C.I. (2004) *Lutzomyia longipalpis* salivary gland homogenate impairs cytokine production and co-stimulatory molecule expression on human monocytes and dendritic cells. *Infect. Immun.* **72**, 1298–305.

Cowley, J.A. and Gorman, B.M. (1987) Genetic reassortants for identification of the genome segment coding for the bluetongue virus haemagglutinin. *J. Virol.* **61**, 2304–6.

Craig, L.E., Nealen, M.L., Strandberg, J.D. and Zink, M.C. (1997) Differential replication of ovine lentivirus in endothelial cells cultured from different tissues. *Virology* **238**, 316–26.

Craig, L.E., Spelman, J.P., Strandberg, J.D. and Zink, M.C. (1998) Endothelial cells from diverse tissues exhibit differences in growth and morphology. *Microvasc. Res.* **55**, 65–76.

Cross, M.L., Cupp, M.S., Cupp, E.W., Galloway, A.L. and Enriquez, F.J. (1993a) Modulation of murine immunological responses by salivary-gland extract of *Simulium vittatum* (Diptera, Simuliidae). *J. Med. Entomol.* **30**, 928–35.

Cross, M.L., Cupp, M.S., Cupp, E.W., Ramberg, F.B. and Enriquez, F.J. (1993b) Antibody-responses of balb/C mice to salivary antigens of haematophagous black flies (Diptera, Simuliidae). *J. Med. Entomol.* **30**, 725–34.

Cross, M.L., Cupp, E.W. and Enriquez, F.J. (1994) Modulation of murine cellular immune-responses and cytokines by salivary-gland extract of the black fly *Simulium vittatum*. *Trop. Med. Parasitol.* **45**, 119–24.

Darpel, K.E. (2007) The bluetongue virus 'ruminant host - insect vector' transmission cycle; the role of Culicoides saliva proteins in infection. A thesis submitted in partial fulfilment of the requirements of the University of London for the degree of Doctor of Philosophy.

Darpel, K.E., Batten, C.A., Veronesi, E., Shaw, A.E., Anthony, S., Bachanek-Bankowska, K., Kgosana, L., Bin-Tarif, A., Carpenter, S., Mueller-Doblies, U.U., Takamatsu, H.-H., Mellor, P.S., Mertens, P.P.C. and Oura, C.A.L. (2007) Clinical signs and pathology shown by British sheep and cattle infected with bluetongue virus serotype 8 derived from the 2006 outbreak in Northern Europe. *Vet. Rec.* **2007**, 253–61.

DeMaula, C.D., Jutila, M.A., Wilson, D.W. and MacLachlan, N.J. (2001) Infection kinetics, prostacyclin release and cytokine-mediated modulation of the mechanism of cell death during bluetongue virus infection of cultured ovine and bovine pulmonary artery and lung microvascular endothelial cells. *J. Gen. Virol.* **82**, 787–94.

DeMaula, C.D., Leutenegger, C.M., Bonneau, K.R. and MacLachlan, N.J. (2002a) The role of endothelial cell-derived inflammatory and vasoactive mediators in the pathogenesis of bluetongue. *Virology* **296**, 330–7.

DeMaula, C.D., Leutenegger, C.M., Jutila, M.A. and MacLachlan, N.J. (2002b) Bluetongue virus-induced activation of primary bovine lung microvascular endothelial cells. *Vet. Immunol. Immunopathol.* **86**, 147–57.

Ellis, J.A., Luedke, A.J., Davis, W.C., Wechsler, S.O., Mecham, J.O., Pratt, D.L. and Elliot, J.D. (1990) T lymphocyte subset alterations following bluetongue virus infection in sheep and cattle. *Vet. Immunol. Immunopathol.* **24**, 49–67.

Erasmus, B.J. (1975) Bluetongue in sheep and goats. *Aust. Vet. J.* **51**, 165–70.

Flanagan, M. and Johnson, S.J. (1995) The effects of vaccination of merino ewes with an attenuated Australian bluetongue virus serotype 23 at different stages of gestation. *Aust. Vet. J.* **72**, 455–7.

Foster, A.P. and Cunningham, F.M. (1998) The pathogenesis and immunopharmacology of equine insect hypersensitivity. *Adv. Vet. Dermatol.* **3**, 177–89.

Foster, N.M., Jones, R.H. and McCrory, B.R. (1963) Preliminary investigations on insect transmission of bluetongue virus in sheep. *Am. J. Vet. Res.* **24**, 1195–9.

Fu, H. (1995). Mechanisms controlling the infection of Culicoides biting midges with bluetongue virus: a thesis submitted in partial fulfilment of the requirements of the University of Hertfordshire for the degree of Doctor of Philosophy.

Gibbs, E.P.J., Homan, E.J., Mo, C.L., Greiner, E.C., Gonzalez, J., Thompson, L.H., Oveido, M.T., Walton, T.E. and Yuill, T.M. (1992) Epidemiology of bluetongue viruses in the American tropics. *Ann. N.Y. Acad. Sci.* **653**, 243–50.

Gibbs, E.P.J., Lawman, M.J.P. and Herniman, K.A.J. (1979) Preliminary observations on transplacental infection of bluetongue virus in sheep – possible overwintering mechanism. *Res. Vet. Sci.* **27**, 118–20.

Gomez-Villamandos, J.C., Sanchez, C., Carrasco, L., Laviada, M.D., Bautista, M.J., Martinez-Torrecuadrada, J., Sanchez-Vizcaino, J.M. and Sierra, M.A. (1999) Pathogenesis of African horse sickness: ultrastructural study of the capillaries in experimental infection. *J. Comp. Pathol.* **121**, 101–106.

Gumm, I.D. and Newman, J.F.E. (1982) The preparation of purified bluetongue virus group antigen for use as a diagnostic reagent. *Arch. Virol.* **72**, 83–93.

Hardy, W.T. and Price, D.A. (1952) Soremuzzle in sheep. *J. Am. Vet. Med. Assoc.* **120**, 23–5.

Howerth, E.W. and Tyler, D.E. (1988) Experimentally induced bluetongue virus-infection in white-tailed deer – ultrastructural findings. *Am. J. Vet. Res.* **49**, 1914–22.

Huismans, H. and Erasmus, B.J. (1981) Identification of the serotype-specific and group-specific antigens of bluetongue virus. *Onderstepoort J. Vet. Res.* **48**, 51–8.

Huismans, H., Vanderwalt, N.T., Cloete, M. and Erasmus, B.J. (1987) Isolation of a capsid protein of bluetongue virus that induces a protective immune-response in sheep. *Virology* **157**, 172–9.

Janardhana, V., Andrew, M.E., Lobato, Z.I.P. and Coupar, B.E.H. (1999) The ovine cytotoxic T lymphocyte responses to bluetongue virus. *Res. Vet. Sci.* **67**, 213–21.

Jeggo, M.H. (1983) The immune response to bluetongue virus infection. Thesis submitted to the University of Surrey in partial fulfilment of the requirements for the degree of Doctor of Philosophy.

Jeggo, M.H., Gumm, I.D. and Taylor, W.P. (1983a) Clinical and serological response of sheep to serial challenge with different bluetongue virus types. *Res. Vet. Sci.* **34**, 205–11.

Jeggo, M.H. and Wardley, R.C. (1982a) Generation of cross-reactive cytotoxic T lymphocytes following immunization of mice with various bluetongue virus types. *Immunology* **45**, 629–35.

Jeggo, M.H. and Wardley, R.C. (1982b) The induction of murine cytotoxic T lymphocytes by bluetongue virus. *Arch. Virol.* **71**, 197–206.

Jeggo, M.H. and Wardley, R.C. (1982c) Production of murine cytotoxic T lymphocyte by bluetongue virus following various immunization procedures. *Res. Vet. Sci.* **33**, 212–15.

Jeggo, M.H. and Wardley, R.C. (1985) Bluetongue vaccine – cells and/or antibodies. *Vaccine* **3**, 57–8.

Jeggo, M.H., Wardley, R.C. and Brownlie, J. (1984a) A study of the role of cell-mediated-immunity in bluetongue virus-infection in sheep, using cellular adoptive transfer techniques. *Immunology* **52**, 403–10.

Jeggo, M.H., Wardley, R.C. and Brownlie, J. (1985) Importance of ovine cytotoxic T cells in protection against bluetongue virus infection. *Prog. Clin. Biol. Res.* **178**, 477–87.

Jeggo, M.H., Wardley, R.C., Brownlie, J. and Corteyn, A.H. (1986) Serial inoculation of sheep with two bluetongue virus types. *Res. Vet. Sci.* **40**, 386–92.

Jeggo, M.H., Wardley, R.C. and Taylor, W.P. (1983b) Host response to bluetongue virus. In: R.W. Compans and D.H.L. Bishop (eds), *Double-stranded RNA viruses*. Amsterdam: Elsevier, pp. 353.

Jeggo, M.H., Wardley, R.C. and Taylor, W.P. (1984b) Clinical and serological outcome following the simultaneous inoculation of three bluetongue virus types into sheep. *Res. Vet. Sci.* **37**, 368–70.

Jeggo, M.H., Wardley, R.C. and Taylor, W.P. (1984c) Role of neutralizing antibody in passive immunity to bluetongue infection. *Res. Vet. Sci.* **36**, 81–6.

Jones, L.D., Hodgson, E., Williams, T., Higgs, S. and Nuttall, P.A. (1992) Saliva activated transmission (Sat) of thogoto virus: relationship with vector potential of different haematophagous arthropods. *Med. Vet. Entomol.* **6**, 361–5.

Jones, L.D., Williams, T., Bishop, D. and Roy, P. (1997) Baculovirus expressed nonstructural protein NS2 of bluetongue virus induces a cytotoxic T-cell response in mice which affords partial protection. *Clin. Diagn. Lab. Immunol.* **4**, 297–301.

Kirkland, P.D. and Hawkes, R.A. (2004) A comparison of laboratory and 'wild' strains of bluetongue virus – is there any difference and does it matter? *Vet. Ital.* **40**, 448–55.

Lawman, M.J.P. (1979) Observations on the pathogenesis of BTV infections in sheep: a thesis submitted in part fulfilment for the Degree of Doctor of Philosophy, University of Surrey.

Limesand, K.H., Higgs, S., Pearson, L.D. and Beaty, B.J. (2000) Potentiation of vesicular stomatitis New Jersey virus infection in mice by mosquito saliva. *Parasite Immunol.* **22**, 461–7.

Limesand, K.H., Higgs, S., Pearson, L.D. and Beaty, B.J. (2003) Effect of mosquito salivary gland treatment on vesicular stomatitis New Jersey virus replication and interferon alpha/beta expression in vitro. *J. Med. Entomol.* **40**, 199–205.

Luedke, A.J., Joachim, M.M. and Jones, R.H. (1977a) Bluetongue in cattle: effects of *Culicoides variipennis* – transmitted bluetongue virus on pregnant heifers and their calves. *Am. J. Vet. Res.* **38**, 1687–95.

Luedke, A.J., Joachim, M.M. and Jones, R.H. (1977b) Bluetongue in cattle: effects of vector-transmitted bluetongue on calves previously infected in utero. *Am. J. Vet. Res.* **38**, 1697–700.

Luedke, A.J., Jones, R.H. and Walton, T.E. (1977c) Overwintering mechanism for bluetongue virus: biological recovery of latent virus from a bovine by bites of *Culicoides variipennis*. *Am. J. Trop. Med. Hyg.* **26**, 313–25.

Lunt, R.A., Melville, L., Hunt, N., Davis, S., Rootes, C.L., Newberry, K.M., Pritchard, L.I., Middleton, D., Bingham, J., Daniels, P.W. and Eaton, B.T. (2006) Cultured skin fibroblast cells derived from bluetongue virus-inoculated sheep and field-infected cattle are not a source of late and protracted recoverable virus. *J. Gen. Virol.* **87**, 3661–6.

MacLachlan, N.J. (1994) The pathogenesis and immunology of bluetongue virus infection of ruminants. *Comp. Immunol. Microbiol. Infect. Dis.* **17**, 197–206.

MacLachlan, N.J. (1995) The pathogenesis of bluetongue virus infection of cattle: a novel mechanism of prolonged infection in which the erythrocyte functions as Trojan horse. In: T.D. St George and P. Kegao (eds), *Bluetongue Disease in Southeast Asia and the Pacific: Proceeding of the First Regional Bluetongue Symposium, Kunming, China.* Canberra, Australia: ACIAR, p. 151.

MacLachlan, N.J., Conley, A.J. and Kennedy, P.C. (2000) Bluetongue and equine viral arteritis viruses as models of virus-induced foetal injury and abortion. *Anim. Reprod. Sci.* **60**, 643–51.

MacLachlan, N.J., Jagels, G., Rossitto, P.V., Moore, P.F. and Heidner, H.W. (1990) The pathogenesis of experimental bluetongue virus infection of calves. *Vet. Pathol.* **27**, 223–229.

MacLachlan, N.J. and Osburn, B.I. (2006) Impact of bluetongue virus infection on the international movement and trade of ruminants. *J. Am. Vet. Med. Assoc.* **228**, 1346–9.

MacLachlan, N.J. and Thomson, J. (1985) Bluetongue virus – induced interferon in cattle. *Am. J. Vet. Res.* **46**, 1238–41.

Mahrt, C.R. and Osburn, B.I. (1986) Experimental bluetongue virus infection of sheep; effect of vaccination: pathologic, immunofluorescent, and ultrastructural studies. *Am. J. Vet. Res.* **47**, 1198–203.

Mbow, M.L., Bleyenberg, J.A., Hall, L.R. and Titus, R.G. (1998) Phlebotomus papatasi sand fly salivary gland lysate down-regulates a Th1, but up-regulates a Th2, response in mice infected with Leishmania major. *J. Immunol.* **161**, 5571–7.

Mellor, P.S. and McCaig, J. (1974) Probable cause of sweet-itch in England. *Vet. Rec.* **95**, 411–15.

Melville, L.F., Hunt, N.T., Davis, S.S. and Weir, R.P. (2004) Bluetongue virus does not persist in naturally infected cattle. *Vet. Ital.* **40**, 502–7.

Mertens, P.P.C., Brown, F. and Sangar, D.V. (1984) Assignment of the genome segments of bluetongue virus type 1 to the proteins which they encode. *Virology* **135**, 207–17.

Mertens, P.P.C., Burroughs, J.N., and Anderson, J. (1987) Purification and properties of virus-particles, infectious subviral particles, and cores of bluetongue virus serotype-1 and serotype-4. *Virology* **157**, 375–86.

Mertens, P.P.C., Pedley, S., Cowley, J., Burroughs, J.N., Corteyn, M., Jeggo, M.H., Jennings, D.M. and Gorman, B.M. (1989) Analysis of the roles of bluetongue virus outer capsid proteins VP2 and VP5 in determination of virus serotype. *Virology* **170**, 561–5.

Monaghan, P., Simpson, J., Murphy, C., Durand, S., Quan, M. and Alexandersen, S. (2005) Use of confocal immunofluorescence microscopy to localize viral nonstructural proteins and potential sites of replication in pigs experimentally infected with foot-and-mouth disease virus. *J. Virol.* **79**, 6410–8.

Morrill, J.C. and McConnell, S. (1985) An electron microscopic study of blood cells from calves experimentally infected with bluetongue virus. In: T.L. Barber (ed.), *Bluetongue and Related Orbiviruses.* New York: Alan R. Liss, Inc, pp. 279–88.

Norsworthy, N.B., Sun, J.R., Elnaiem, D., Lanzaro, G. and Soong, L. (2004) Sand fly saliva enhances Leishmania amazonensis infection by modulation interleukin-10 production. *Infect. Immun.* **72**, 1240–7.

O'Connell, L. (2002). Entomological aspects of the transmission of arboviral diseases by Culicoides biting midges. A dissertation submitted to the University of Bristol in accordance with the requirements of the degree of Doctor of Philosophy in the Faculty of Science.

Parsonson, I.M., Della-Porta, A.J., McPhee, D.A., Cybinski, D.H., Squire, K.R.E. and Uren, M.F. (1987) Bluetongue virus serotype 20: experimental infection of pregnant heifers. *Aust. Vet. J.* **64**, 14–17.

Pini, A. (1976) A study on pathogenesis of bluetongue: replication of virus in organs of infected sheep. *Onderstepoort J. Vet. Res.* **43**, 159–64.

Purse, B.V., Mellor, P.S., Rogers, D.J., Samuel, A.R., Mertens, P.P.C. and Baylis, M. (2005) Climate change and the recent emergence of bluetongue in Europe. *Nat. Rev. Microbiol.* **3**, 171–81.

Quist, C.F., Howerth, E.W., Bounous, D.I. and Stallknecht, D.E. (1997) Cell-mediated immune response and IL-2 production in white-tailed deer experimentally infected with haemorrhagic disease viruses. *Vet. Immunol. Immunopathol.* **56**, 283–97.

Ramig, R.F. (2004) Pathogenesis of intestinal and systemic rotavirus infection. *J. Virol.* **78**, 10213–20.

Reisen, W.K. and Brault, A.C. (2007) West Nile virus in North America: perspectives on epidemiology and intervention. *Pest Manag. Sci.* 63(7): 641–646.

Reisen, W.K., Fang, Y., Lothrop, H.D., Martinez, V.M., Wilson, J., O'Connor, P., Carney, R., Cahoon-Young, B., Shafii, M. and Brault, A.C. (2006) Overwintering of west Nile virus in Southern California. *J. Med. Entomol.* **43**, 344–55.

Rohousova, I., Volf, P. and Lipoldova, M. (2005) Modulation of murine cellular immune response and cytokine production by salivary gland Lysate of three sand fly species. *Parasite Immunol.* **27**, 469–73.

Schneider, B.S., Soong, L., Girard, Y.A., Campbell, G., Mason, P. and Higgs, S. (2006) Potentiation of West Nile encephalitis by mosquito feeding. *Viral Immunol.* **19**, 74–82.

Schneider, B.S., Soong, L., Zeidner, N.S. and Higgs, S. (2004) *Aedes aegypti* salivary gland extracts modulate anti-viral and T(H)1/T(H)2 cytokine responses to Sindbis virus infection. *Viral Immunol.* **17**, 565–73.

Sellers, R.F. and Mellor, P.S. (1993) Temperature and the persistence of virus in *Culicoides* spp. During adverse conditions. *Off. Int. Epizoot. Sci. Tech. Rev.* **12**, 733–55.

Stott, J.L., Blanchard-Channell, M., Scibienski, R.J. and Stott, M.L. (1990) Interaction of bluetongue virus with bovine lymphocytes. *J. Gen. Virol.* **71**, 363–8.

Stott, J.L., Blanchard-Channell, M., Stott, M.L., Barrett-Boyes, S.M. and MacLachlan, N.J. (1992) Bluetongue virus tropism for bovine lymphocyte sub-populations. In: T.E. Walton and B.I. Osburn (eds), *Bluetongue, African Horse Sickness and Related Orbiviruses*. Boca Raton: CRC Press, pp. 781–7.

Stott, J.L. and Osburn, B.I. (1992) Immune response to bluetongue virus infection. In: P. Roy and B.M. Gorman (eds), *Current Topics in Microbiology and Immunology*, Vol. 162. Berlin-Heidelberg: Springer-Verlag, pp. 163–79.

Takamatsu, H.-H., Burroughs, J.N., Wade-Evans, A.M. and Mertens, P.P.C. (1992) Analysis of bluetongue virus serotype-specific and cross-reactive ovine T-cell determinants in virus structural proteins. In: T.E. Walton and B.I. Osburn (eds), *Bluetongue, African Horse Sickness and Related Orbiviruses*. Boca Raton: CRC Press, pp. 491–7.

Takamatsu, H-H., Mellor, P.S., Mertens, P.P.C., Kirkham, P.A., Burroughs, J.N. and Parkhouse, R.M.E. (2003) A possible overwintering mechanism for bluetongue virus in the absence of the insect vector. *J. Gen. Virol.* **84**, 227–35.

Takamatsu, H., Burroughs, J.H., Wadeevans, A.M. and Mertens, P.P.C. (1990) Identification of a bluetongue virus serotype-1 specific ovine helper T-cell determinant in outer capsid protein VP2. *Virology* **177**, 396–400.

Takamatsu, H. and Jeggo, M.H. (1989) Cultivation of bluetongue virus-specific ovine T-cells and their cross-reactivity with different serotype viruses. *Immunology* **66**, 258–63.

Titus, R.G. and Ribeiro, J.M.C. (1988) Salivary-gland lysates from the sand fly *Lutzomyia longipalpis* enhance *Leishmania* infectivity. *Science* **239**, 1306–8.

Villiers, E.-M. (1974) Comparison of the capsid polypeptides of various BTV serotypes. *Intervirology* **3**, 47–53.

Wanasen, N.R., Nussenzveig, H., Champagne, D.E., Soong, L. and Higgs, S. (2004) Differential modulation of murine host immune response by salivary gland extracts from the mosquitoes *Aedes aegypti* and *Culex quinquefasciatus*. *Med. Vet. Entomol.* **18**, 191–9.

Wasserman, H.A., Singh, S. and Champagne, D.E. (2004) Saliva of the yellow fever mosquito, *Aedes aegypti*, modulates murine lymphocyte function. *Parasite Immunol.* **26**, 295–306.

Wechsler, S.J. and Luedke, A.J. (1991) Detection of bluetongue virus by using bovine endothelial cells and embryonated chicken eggs. *J. Clin. Microbiol.* **29**, 212–14.

Wechsler, S.J. and McHolland, L.E. (1988) Susceptibilities of 14 cell lines to bluetongue virus infection. *J. Clin. Microbiol.* **26**, 2324–7.

Whetter, L.E., MacLachlan, N.J., Gebhard, D.H., Heidner, H.W. and Moore, P.F. (1989) Bluetongue virus infection of bovine monocytes. *J. Gen. Virol.* **70**, 1663–76.

White, D.M. and Mecham, J.O. (2004) Lack of detectable bluetongue virus in skin of seropositive cattle: implications for vertebrate overwintering of bluetongue virus. *Vet. Ital.* **40**, 513–19.

White, D.M., Wilson, W.C., Blair, C.D. and Beaty, B.J. (2005) Studies on overwintering of bluetongue viruses in insects. *J. Gen. Virol.* **86**, 453–62.

Wohlsein, P., Pohlenz, J.F., Davidson, F.L., Salt, J.S. and Hamblin, C. (1997) Immunohistochemical demonstration of African horse sickness viral antigen in formalin-fixed equine tissues. *Vet. Pathol.* **34**, 568–74.

Wohlsein, P., Pohlenz, J.F., Salt, J.S. and Hamblin, C. (1998) Immunohistochemical demonstration of African horse sickness viral antigen in tissues of experimentally infected equines. *Arch. Virol.* **14**, 57–65.

Yeruham, I., Braverman, Y. and Orgad, U. (1993) Field observations in Israel on hypersensitivity in cattle, sheep and donkeys caused by *Culicoides*. *Aust. Vet. J.* **70**, 348–52.

Yeruham, I., Braverman, Y. and Perl, S. (2000) Study of apparent hypersensitivity to *Culicoides* species in sheep in Israel. *Vet. Rec.* **147**, 360–3.

Clinical signs and pathology

13

N. JAMES MACLACHLAN* AND GEOFF GARD†

*Department of Pathology, Microbiology and Immunology, School of Veterinary Medicine, University of California, Davis, CA, USA
†Casey Crescent, Mystery Bay, NSW, Australia

Introduction

Bluetongue (BT) is a non-contagious, insect-transmitted disease of sheep and some species of wild ruminants (Spreull, 1905; Moulton, 1961; Erasmus, 1975; Gard, 1987; Parsonson, 1990; MacLachlan, 1994; Verwoerd and Erasmus, 2004) that was first recognized and comprehensively described in southern Africa, where it was initially called 'Malarial Catarrhal Fever' or 'Epizootic Catarrh of Sheep'. The name of 'bluetongue' was later adopted to describe the distinctive cyanotic tongue of some affected sheep. The original written descriptions of BT were published in the late nineteenth and early twentieth centuries, although farmers in southern Africa recognized the disease soon after the introduction of fine-wool European breeds of sheep to that region (Spreull, 1905; Verwoerd and Erasmus, 2004). Before the 1940s, BT was thought to be confined to southern Africa. The first well-documented epizootic of BT outside Africa occurred amongst sheep in Cyprus in 1943, although the disease had probably occurred there previously. BT was recognized in Texas soon thereafter, and an extensive epizootic occurred on the Iberian Peninsula in 1956–1960. The disease was subsequently recognized in the Middle East, Asia and southern Europe. Although BT virus (BTV), the causative agent of BT, has now been isolated from ruminants and/or vector insects from all continents except Antarctica (Gibbs and Greiner, 1994; Tabachnick, 2004), BT is either rare or non-existent amongst ruminants in many regions with enzootic BTV infection (Walton, 2004).

ISBN-13: 978-0-12-369368-6

Clinical signs

BT occurs principally in sheep and some species of wild ruminants, and BTV infection of cattle, goats and most wild ruminant species is typically asymptomatic or subclinical (Parsonson, 1990; Barratt-Boyes and MacLachlan, 1995; Verwoerd and Erasmus, 2004). The clinical signs of BTV infection are also highly variable even in susceptible species such as sheep, reflecting inherent differences in the susceptibility of different sheep breeds, as well as of individual animals, and external stressors such as solar irradiation that can exacerbate the clinical signs of BT. The signs of BT in sheep are the result of virus-mediated vascular injury that produces oedema, hyperaemia and vascular congestion, haemorrhage and tissue infarction (Gard, 1987; Moulton, 1961; MacLachlan, 1994; Verwoerd and Erasmus, 2004). Thus, sheep with acute BT have any combination of fever, anorexia and malaise, respiratory distress, excessive salivation, serous to bloody nasal and ocular discharge that becomes increasingly mucopurulent so that crusty exudates accumulate around the nostrils, petechial and ecchymotic haemorrhages in the mucous membranes of the oral and nasal cavities, oral erosions and ulcers (Figure 13.1), lameness and/or a stiff gait, hyperaemia and haemorrhage of the coronary band, oedema of the head and neck (including the ears) and congestion and focal haemorrhages in the conjunctiva and skin. The swollen and cyanotic tongue that gives the disease its name is uncommon. Mortality rates vary from 0% in mild outbreaks to 30% or even higher in outbreaks caused by virulent strains of BTV in highly susceptible breeds of sheep. Most animals that succumb to acute BT die within 14 days of infection.

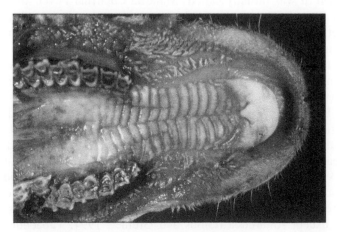

Figure 13.1   Ulcers in the dental pad and haemorrhages in the palate of a sheep with BT (See colour plate 24).

Sheep that survive the acute disease can have a prolonged convalescence and some will show substantial deterioration in body condition and become emaciated. BTV-induced muscle injury and necrosis prevent normal locomotion in some animals and can lead to torticollis (wryneck). The wool of convalescent sheep is frequently shed (wool break), and some sheep shed their hooves. These chronically affected sheep may succumb to other diseases such as bacterial pneumonia.

The post-mortem lesions of severe acute BT in sheep include vascular congestion, haemorrhage, erosion and ulceration of the mucosa of the upper gastrointestinal tract (oral cavity, oesophagus, forestomachs) and nasal cavity; sub-intimal haemorrhages in the pulmonary artery (Figure 13.2); pulmonary oedema with abundant froth in the trachea; pleural and/or pericardial effusion; oedema within the fascial planes of the muscles of the abdominal wall and necrosis of skeletal and cardiac muscle (Figure 13.3) with the

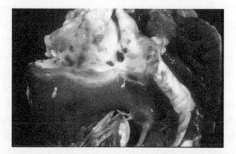

Figure 13.2 Sub-intimal haemorrhage of the pulmonary artery of a sheep with BT (See colour plate 25).

Figure 13.3 Acute necrosis and haemorrhage in the myocardium (papillary muscle of the left ventricle) of a sheep with BT (See colour plate 26).

papillary muscle of the left ventricle being an especially characteristic site (Spreull, 1905; Moulton, 1961; Erasmus, 1975; Gard, 1987; Verwoerd and Erasmus, 2004).

The pathogenesis of bluetongue virus infection

Despite marked differences in expression of the disease, the pathogenesis of BTV infection is similar in sheep and cattle, and most probably in all species of ruminants (Pini, 1976; Mahrt and Osburn, 1986; MacLachlan *et al.*, 1990; Barratt-Boyes and MacLachlan, 1995). After cutaneous instillation of the virus through the bite of a BTV-infected *Culicoides* vector, the virus travels to the regional lymph node where initial replication occurs. The virus then is disseminated to a variety of tissues throughout the body where replication occurs principally in mononuclear phagocytes and endothelial cells. Viraemia in BTV-infected ruminants is highly cell-associated and is prolonged but not persistent (Barratt-Boyes and MacLachlan, 1995; Singer *et al.*, 2001; Bonneau *et al.*, 2002b). The virus promiscuously associates with all blood cells, and titres of virus in each cell fraction are proportionate to the numbers of each cell type; specifically, BTV is quantitatively associated most with platelets and erythrocytes and, because of the short lifespan of platelets, the virus is most associated with erythrocytes late in the course of a BTV infection of ruminants (Barratt-Boyes and MacLachlan, 1995). BTV persists within invaginations of the erythrocyte cell membrane (Figure 13.4) where it apparently is protected from immune clearance (Brewer and MacLachlan, 1992; Brewer and

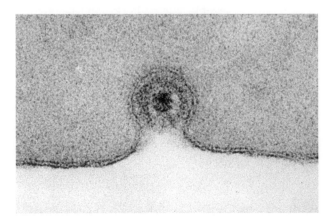

Figure 13.4   BTV virion in an invagination of erythrocyte cell membrane (Courtesy of veterinary pathology).

MacLachlan, 1994). Association of BTV with erythrocytes likely facilitates both the prolonged viraemia that is characteristic of BTV infection of ruminants and the infection of haematophagous insect vectors that feed on viraemic ruminants. Interestingly, BTV nucleic acid may be detected by polymerase chain reaction (PCR) in the blood of infected cattle and sheep for many months after it no longer can be detected by virus isolation in cell culture or inoculation of susceptible sheep. Furthermore, ruminant blood that contains BTV nucleic acid as determined by PCR assay, but not infectious BTV as determined by virus isolation, is not infectious to vector insects even by intrathoracic inoculation (MacLachlan *et al.*, 1994; Tabachnick *et al.*, 1996; Bonneau *et al.*, 2002b).

Clinical signs and lesions in BTV-infected sheep probably reflect virus-mediated endothelial injury, as BTV replicates in endothelial cells causing cell injury and necrosis (Pini, 1976; Mahrt and Osburn, 1986). Similarly, white-tailed deer, which are highly susceptible to BT, develop consumptive coagulopathy as a consequence of BTV-induced damage to endothelial cells (Howerth *et al.*, 1988). Consumptive coagulopathy (disseminated intravascular coagulation) in BTV-infected sheep and deer results in the bleeding tendency that characterizes fulminant BT. Endothelial injury is also likely to be responsible for increased vascular permeability leading to oedema in tissues such as the lung (pulmonary oedema), and vascular thrombosis leads to tissue infarction.

A fundamental question that has vexed scientists for many years is: Why virulent strains of BTV often produce disease in sheep but not in cattle (Barratt-Boyes and MacLachlan, 1995; Russell *et al.*, 1996)? The similar or identical pathogenesis of BTV infection of cattle and sheep further emphasizes this obvious paradox. DeMaula *et al.* (2001, 2002a, b) recently described inherent differences in the susceptibility of endothelial cells from cattle and sheep to BTV infection. To facilitate these studies, pure cultures of endothelial cells from the microvasculature of sheep and cattle were isolated and subsequently propagated in vitro, and then their responses to infection with BTV evaluated. Lung microvascular endothelial cells were selected because pulmonary oedema and microvascular injury are both highly characteristic of BT. Interestingly, whereas BTV infection of bovine endothelial cells resulted in endothelial activation, with the increased transcription of genes encoding a variety of vaso-active and inflammatory mediators, and increased expression of cell surface adhesion molecules, similar infection of sheep endothelial cells resulted in minimal activation of endothelial cells. Furthermore, the ratio of thromboxane to prostacyclin, which is indicative of enhanced coagulation and possible consumptive coagulopathy, was significantly greater in sheep than in cattle that were experimentally infected with BTV.

It is to be emphasized that BTV-infected sheep often develop mild or no obvious disease, especially in BTV-enzootic areas like those in the tropics. Outbreaks of BT typically occur either when susceptible sheep are introduced into BTV-enzootic regions or when the virus spreads into immunologically naïve sheep populations at the interface of BTV-enzootic and non-enzootic regions (such as the recent epizootic in the Mediterranean Basin). Expression of BT clearly reflects a variety of host, virus, vector and environmental factors. Specifically, field strains of BTV in enzootic areas exhibit remarkable genetic heterogeneity, even amongst strains that co-circulate and amongst viruses of the same serotype (de Mattos *et al.*, 1994; Bonneau *et al.*, 2002a; Pritchard *et al.*, 2004; Nikolakaki *et al.*, 2005). The genetic heterogeneity of field strains of BTV likely results from a combination of genetic drift and reassortment, the former as a consequence of quasispecies evolution of the virus in its insect and ruminant hosts (Bonneau *et al.*, 2001). It is logical that this genetic variability of BTV is reflected by differences in phenotypic properties of each virus strain, including their virulence to susceptible ruminants. Although sheep and certain species of wild ruminants such as the white-tailed deer are clearly most susceptible to expression of BT, the susceptibility of individual sheep breeds varies markedly and the fine-wool European breeds are most susceptible. Nutritional status, immune status and age also influence the severity of BT in individual sheep, as can environmental stresses such as high temperature and ultraviolet radiation.

Culicoides vectors are critical to the survival and transmission of BTV as the infection is not contagious, and there is no credible evidence of long-term maintenance of BTV in ruminants. Thus, BTV infection occurs only where competent vectors are present, although the species of *Culicoides* that transmit BTV in different regions of the world clearly are very different and often poorly defined (Meiswinkel *et al.*, 2004; Tabachnick, 2004; Savini *et al.*, 2005). Uncertain as yet is what role, if any, individual vector species might exert on the expression of BT in susceptible ruminants.

Persistent bluetongue virus infection of cattle

Luedke and colleagues (1977a, b) proposed a BTV carrier state following in utero infection of foetal cattle, a suggestion that has had a marked and highly adverse impact on animal trade from BTV-enzootic regions for several decades (Walton, 2004; MacLachlan and Osburn 2006). However, extensive studies by other investigators failed to confirm the findings of Luedke *et al.*; indeed, BTV infection of the ruminant foetus in early gestation leads to teratogenic defects and not persistent postnatal infection (MacLachlan, 2004). Thus, the concept of persistent BTV infection of cattle was abandoned because of a lack of credible evidence, and the OIE (i.e. World Animal Health Organization) Sanitary Code now lists the maximum infective period for BTV infection of ruminants as 60 days, reflection of extensive research that has confirmed prolonged but not persistent infection of ruminant erythrocytes (MacLachlan, 1994; Barratt-Boyes and MacLachlan, 1995; Tabachnick *et al.*, 1996; Singer *et al.*, 2001; Bonneau *et al.*, 2002b;; MacLachlan, 2004; Walton, 2004; MacLachlan and Osburn, 2006).

Bluetongue in other animal species

A variety of animal groups other than ruminants have been implicated in the life cycle of a BTV infection. Serological evidence indicates that large African carnivores are infected with BTV, whereas smaller predators that co-habit with them are not, suggesting that large carnivores are infected through feeding on BTV-infected ruminants (Alexander *et al.*, 1994). Inadvertent contamination of a canine vaccine with BTV confirmed that dogs are susceptible to BTV infection; indeed pregnant bitches that received this contaminated vaccine typically got aborted and died (Akita *et al.*, 1994). There is no evidence, however, that dogs or other carnivores are important to the natural cycle of a BTV infection.

References

Akita, G.Y., Ianconescu, M., MacLachlan, N.J., Osburn, B.I. and Greene, R.T. (1994) Bluetongue disease in dogs associated with contaminated vaccine. *Vet. Rec.* **135**, 283–4.

Alexander, K.A., MacLachlan, N.J., Kat, P.W., House, C., O'Brien, S.J., Lerche, N.W., Sawyer, M.M., Frank, L.G., Holekamp, K., Smale, J., McNutt, J.W., Laurenson, M.K., Mills, M.G. and Osburn, B.I. (1994) Evidence of natural bluetongue virus infection among African carnivores. *Am. J. Trop. Med. Hyg.* **51**, 568–76.

Barratt-Boyes, S.M. and MacLachlan, N.J. (1995) Pathogenesis of bluetongue virus infection of cattle. *J. Am. Vet. Med. Assoc.* **206**, 1322–9.

Bonneau, K.R., DeMaula, C.D., Mullens, B.A. and MacLachlan, N.J. (2002b) Duration of viraemia infectious to *Culicoides sonorensis* in bluetongue virus infected cattle and sheep. *Vet. Microbiol.* **88**, 115–25.

Bonneau, K.R., Mullens, B.A. and MacLachlan, N.J. (2001) Occurrence of genetic drift and founder effect during quasispecies evolution of the VP2 and NS3/3a genes of bluetongue virus upon passage between sheep, cattle and *Culicoides sonorensis*. *J. Virol.* **75**, 8298–305.

Bonneau, K.R., Topol, J.B., Gerry, A.C., Mullens, B.A., Velten, R.K. and MacLachlan, N.J. (2002a) Variation in the NS3/3a genes of bluetongue viruses contained in culicoides sonorensis collected from a single site. *Virus Res.* **84**, 59–65.

Brewer, A.W. and MacLachlan, N.J. (1992) Ultrastructural characterization of the interaction of bluetongue virus with bovine erythrocytes *in vitro*. *Vet. Pathol.* **29**, 356–9.

Brewer, A.W. and MacLachlan, N.J. (1994) The pathogenesis of bluetongue virus infection of bovine blood cells *in vitro*: ultrastructural characterization. *Arch. Virol.* **136**, 287–98.

DeMaula, C.D., Jutila, M.A., Wilson, D.W. and MacLachlan, N.J. (2001) Infection kinetics, prostacyclin release, and cytokine-mediated modulation of the mechanism of cell death during bluetongue virus infection of cultured ovine and bovine pulmonary artery and lung microvascular endothelial cells. *J. Gen. Virol.* **82**, 787–94.

DeMaula, C.D., Leutennegger, C.M., Bonneau, K.R. and MacLachlan, N.J. (2002b) The role of endothelial cell-derived inflammatory and vasoactive mediators in the pathogenesis of bluetongue. *Virology* **296**, 330–7.

DeMaula, C.D., Leutennegger, C.M., Jutila, M.A. and MacLachlan, N.J. (2002a) Bluetongue virus induced activation of primary bovine lung microvascular endothelial cells. *Vet. Immunol. Immunopathol.* **86**, 147–57.

de Mattos, C.A., de Mattos, C.C., Osburn, B.I. and MacLachlan, N.J. (1994) Heterogeneity of the L2 gene of field isolates of bluetongue virus serotype 17 from the San Joaquin valley of California. *Virus Res.* **31**, 67–87.

Erasmus, B.J. (1975) Bluetongue in sheep and goats. *Aust. Vet. J.* **51**, 165–70.

Gard, G.P. (1987) *Studies of Bluetongue Virulence and Pathogenesis in Sheep. Technical Bulletin No. 103.* Darwin: Department of Industries and Development.

Gibbs, E.P. and Greiner, E.C. (1994) The epidemiology of bluetongue. *Comp. Immunol. Microbiol. Infect. Dis.* **17**, 207–20.

Howerth, E.W., Greene, C.E. and Prestwood, A.K. (1988) Experimentally induced bluetongue virus infection in white-tailed deer: coagulation, clinical pathologic, and gross pathologic changes. *Am. J. Vet. Res.* **49**, 1906–13.

Luedke, A.J., Jochim, M.M. and Jones, R.H. (1977a) Bluetongue in cattle: effects of Culicoides variipennis-transmitted bluetongue virus on pregnant heifers and their calves. *Am. J. Vet. Res.* **38**, 1687–96.

Luedke, A.J., Jochim, M.M. and Jones, R.H. (1977b) Bluetongue in cattle: effect of vector-transmitted bluetongue virus on calves previously infected in utero. *Am. J. Vet. Res.* **38**, 1697–700.

MacLachlan, N.J. (1994) The pathogenesis and immunology of bluetongue virus infection of ruminants. *Comp. Immunol. Microbiol. Infect. Dis.* **17**, 197–206.

MacLachlan, N.J. (2004) Bluetongue: pathogenesis and duration of viraemia. *Vet. Ital.* **40**, 462–7.

MacLachlan, N.J., Jagels, G., Rositto, P.V., Moore, P.F. and Heidner, H.W. (1990) The pathogenesis of experimental bluetongue virus infection of calves. *Vet. Pathol.* **27**, 223–9.

MacLachlan, N.J. and Osburn, B.I. (2006) Impact of bluetongue virus infection on the international movement and trade of ruminants. *J. Am. Vet. Med. Assoc.* **228**, 1346–9.

MacLachlan, N.J., Nunamaker, R.A., Katz, J.B., Sawyer, M.M., Akita, G.Y., Osburn, B.I. and Tabachnick, W.J. (1994) Detection of bluetongue virus in the blood of inoculated calves: comparison of virus isolation, PCR assay, and *in vitro* feeding of *Culicoides variipennis*. *Arch. Virol.* **136**, 1–8.

Mahrt, C.R. and Osburn, B.I. (1986) Experimental bluetongue virus infection of sheep; effect of vaccination: pathologic, immunofluorescent, and ultrastructural studies. *Am. J. Vet. Res.* **47**, 1198–203.

Meiswinkel, R., Gomulski, L.M., Delcolle, J.C., Goffredo, M. and Gasperi, G. (2004) The taxonomy of *Culicoides* vector complexes – unfinished business. *Vet. Ital.* **40**, 151–9.

Moulton, J.E. (1961) Pathology of bluetongue in sheep. *J. Am. Vet. Med. Assoc.* **138**, 493–8.

Nikolakaki, S.V., Nomikou, K., Koumbati, M., Mangana, O., Papanastassopoulou, M., Mertens, P.P. and Papadopoulos, O. (2005) Molecular analysis of the NS3/3a gene of bluetongue virus isolates from the 1979 and 1998–2001 epizootics in Greece and their segregation into two distinct groups. *Virus Res.* **114**, 1–14.

Parsonson, I.P. (1990) Pathology and pathogenesis of bluetongue infections. *Curr. Top. Microbiol. Immunol.* **162**, 119–41.

Pini, A. (1976) A study on the pathogenesis of bluetongue: replication of the virus in the organs of infected sheep. *Onderstepoort J. Vet. Res.* **43**, 159–64.

Pritchard, L.I., Sendow, I., Lunt, R., Hassan, S.H., Kattenbelt, J., Gould, A.R., Daniels, P.W. and Eaton, B.T. (2004) Genetic diversity of bluetongue viruses in South East Asia. *Virus Res.* **101**, 193–201.

Russell, H., O'Toole, D.T., Bardsley, K., Davis, W.C. and Ellis, J.A. (1996) Comparative effects of bluetongue virus infection of ovine and bovine endothelial cells. *Vet. Pathol.* **33**, 319–31.

Savini, G., Goffredo, M., Monaco, F., Di Gennaro, A., Cafiero, M.A., Baldi, L., De Santis, P., Meiswinkel, R. and Caporale, V. (2005) Bluetongue virus isolations from midges belonging to the *Obsoletus* complex(*Culicoides*, diptera: ceratopogonidae) in Italy. *Vet. Rec.* **157**, 133–9.

Singer, R.S., MacLachlan, N.J. and Carpenter, T.E. (2001) Maximal predicted duration of viraemia in bluetongue virus-infected cattle. *J. Vet. Diag. Invest.* **13**, 43–9.

Spreull, J. (1905) Malarial catarrhal fever (bluetongue) of sheep in South Africa. *J. Comp. Pathol. Ther.* **18**, 321–37.

Tabachnick, W.J. (2004) *Culicoides* and the global epidemiology of bluetongue virus infection. *Vet. Ital.* **40**, 145–50.

Tabachnick, W.J., MacLachlan, N.J., Thompson, L.H., Hunt, G.J. and Patton, J.F. (1996) Susceptibility of *Culicoides variipennis sonorensis* to infection by polymerase chain reaction-detectable bluetongue virus cattle blood. *Am. J. Trop. Med. Hyg.* **54**, 481–5.

Verwoerd, D.W. and Erasmus, B.J. (2004) Bluetongue. In: J.A. Coetzer and R.C. Tustin (eds), *Infectious Diseases of Livestock*. Cape Town: Oxford University Press, pp. 1201–30.

Walton, T.E. (2004) The history of bluetongue and a current global overview. *Vet. Ital.* **40**, 31–8.

Bluetongue virus in the insect host

14

PHILIP S. MELLOR,* SIMON CARPENTER* AND
DAVID M. WHITE†

*Institute for Animal Health, Pirbright Laboratory, Pirbright, Woking, Surrey, UK
†Centres for Disease Control and Prevention, NCID/DVRD/Special Pathogens Branch,
Atlanta, GA 30329-4018, USA

Introduction

Bluetongue virus (BTV) is an arthropod-borne virus (arbovirus) that is transmitted biologically between its vertebrate hosts (ruminants) by certain species of *Culicoides* biting midge (Diptera: Ceratopogonidae). Du Toit (1944) in South Africa was the first to implicate a *Culicoides* species in the transmission of BTV when he showed that *Culicoides imicola* (=*pallidipennis*) was able to transmit the virus from infected to susceptible sheep. Since that time numerous authors have confirmed this original observation with additional species of *Culicoides* and different serotypes of BTV (see Mellor, 1990; Mellor et al., 2000).

Vector species of *Culicoides* are the only important means by which BTV transmission occurs, consequently the distribution of the virus is limited to those areas of the globe where these species occur and even within such regions the virus displays a seasonal incidence that reflects the abundance of the adult vectors at certain times of the year. In temperate areas, vectors are usually most abundant in the late summer and autumn; this is therefore the time when BT (i.e. the disease) is most frequently seen. Nevertheless, this should not be interpreted as meaning that the distribution of the virus is static, and indeed, recent observations suggest that in certain areas of the globe its distribution is rapidly expanding. The underlying causes and mechanisms of this phenomenon are considered elsewhere in this monograph.

ISBN-13: 978-0-12-369368-6

Culicoides biting midges

Culicoides biting midges are among the world's smallest blood sucking insects ranging in size from 1 to 3 mm in length. They generally have wings patterned with light and dark markings, and the adults of most species tend to be crepuscular, i.e. active at dawn and dusk. The four larval instars and pupal stage require a certain amount of moisture to develop; hence, the eggs are laid in a wide range of habitats that meet this criterion. Those *Culicoides* species involved in the transmission of pathogens of domestic livestock tend to breed in sites that bring them into close contact with the host animals such as irrigation pipe leaks, irrigated pasture, cattle trough overflows and even dung, the precise type of breeding site being dependent upon the species of vector. More than 1400 species of *Culicoides* have been named, and these insects occur on almost all large inhabited land masses, with the notable exceptions of New Zealand, Iceland and the Hawaiian Islands. They also range from sea level to over 4000 m in elevation (Mellor et al., 2000).

   Culicoides transmit a wide range of bacterial, protozoal and helminth parasites to humans and animals (Kettle, 1965; Linley, 1985), but it is as vectors of viruses that they attain their greatest importance. Worldwide, over 50 viruses have been isolated from *Culicoides* species (Meiswinkel et al., 1994). One of these, Oropouche virus, causes significant disease in humans in central and southern America (LeDuc and Pinheiro, 1988), but the remainder, if they cause disease at all, do so in animals. African horse sickness (AHS) and bluetongue (BT) are the two most important animal diseases caused by *Culicoides*-transmitted viruses and are of major international significance. AHS is usually confined to Africa south of the Sahara but periodically makes excursions into southern Europe and across Asia as far as Pakistan and India (Mellor and Hamblin, 2004). BTV occurs far more widely (Mellor and Boorman, 1995).

Bluetongue virus

BTV is the prototype of the genus *Orbivirus* in the family Reoviridae and exists as a number of serotypes, of which 24 have been identified to date. The virus infects all species of ruminant but severe disease usually occurs only in certain breeds of sheep (mainly the fine wool and mutton breeds) and some species of deer (Taylor, 1986; MacLachlan, 1994). The global distribution of BTV has traditionally been understood as lying between latitudes 35° S and 40° N, although in parts of North America, Europe and Asia it may extend up to around 50° N (Dulac et al. 1989; Guo et al., 1996; Qin et al., 1996; Mellor et al., 2000; Mellor and Wittmann, 2002). Within these areas, the virus has a virtually worldwide distribution being found in North, Central and South America, Africa, Europe, the Middle East, the Indian subcontinent, China, Southeast Asia and Australasia. The species of

Culicoides transmitting BTV in many of these regions have already been identified, and these will be discussed in the next section; however, in some regions the major vector species have yet to be identified.

Vector species of *Culicoides*

Africa

Culicoides imicola is the major vector of BTV throughout Africa, and many isolations of the virus have been made from this species, in South Africa (Du Toit, 1944; Meiswinkel et al., 1994), Zimbabwe (Blackburn et al., 1985), Kenya (Walker and Davies, 1971) and the Sudan (Mellor et al., 1984). BTV has also been isolated from pools of mixed species of *Culicoides* in Nigeria, in locations where *C. imicola* was the most common species (Lee et al., 1974; Herniman et al., 1983). In certain cooler, less arid locations in South Africa, *C. bolitinos*, a sibling species of *C. imicola* predominates and has also been implicated in BTV transmission being up to 20 times more susceptible to infection under laboratory conditions than *C. imicola* (Venter et al., 1998; 2004). This species also supports virus replication to higher titres than those recorded in *C. imicola*, leading to the suggestion that in regions where it is common it may be the primary BTV vector. Paweska et al. (2002) also reported the laboratory infection of small numbers of several other species of *Culicoides* in South Africa (*C. magnus*, *C. bedfordi*, *C. leucostictus*, *C. pycnostictus*, *C. gulbenkiani* and *C. milnei*). Individuals of these species sustained BTV replication for at least 10 days post infection (dpi) and in a small proportion of the infected *C. magnus*, *C. bedfordi*, *C. leucostictus* and *C. milnei*, virus titres reached levels suggesting that transmission would be possible. A further 14 species of *Culicoides* were not able to be infected with BTV.

In Kenya, in addition to *C. imicola*, BTV has been isolated from *C. tororoensis* and *C. milnei* (Walker and Davies, 1971). Further evidence linking these two species with BTV transmission, however, has not been forthcoming.

Culicoides imicola is one of the most widespread *Culicoides* species in the world occurring across the whole of Africa and extending through the Middle East to the Indian subcontinent and stretching as far east as Laos and Vietnam (Dyce and Wirth, 1983; Howarth, 1985; Wirth and Hubert, 1989; Mellor, 1990). It is also found in many countries of southern Europe where its northern boundary seems to be extending, driven at least in part by climate change (Tatem et al., 2003; Purse et al., 2005). Across the whole of this vast region virtually every country that has been shown to support populations of *C. imicola* has also reported serological or clinical evidence of BTV at one time or another – indicating that the mere presence of *C. imicola* carries with it a high risk of BTV exposure.

Asia

Many countries in Asia have recorded BTV activity, and it is likely that the virus is present at times in most countries in the tropical, subtropical and temperate parts of the region. The northernmost areas of BTV occurrence in this region are in northern China, where the virus has been recorded up to 48° N (Guo et al., 1996; Qin et al., 1996). However, information on the identity of the vectors across Asia is patchy. BTV has been isolated from *C. imicola* in many of the countries of the Middle East (Braverman and Galun 1973; Braverman et al., 1976; Jennings et al., 1983; Boorman, 1989; Al-Busaidy and Mellor, 1991), and this species is also known to occur in India (=*C. minutus*), Sri Lanka, China, Thailand, Laos and Vietnam (Howarth, 1985; Muller and Li, 1996), but to date no virus isolations have been made from the species in this area. BTV has been isolated from *Culicoides* in India and China (Yunnan Province), but the insects were not identified to the species level (Jain et al., 1988; Zhang et al., 1996). The most important vectors in Yunnan Province, circumstantially, are considered to be *C. schultzei*, *C. gemellus*, *C. peregrinus*, *C. arakawae* and *C. circumscriptus*; however, no data are published to justify this selection (Bi et al., 1996). *Culicoides actoni* is mentioned as 'being able to transmit BTV in China', but again no data are presented to support this assertion (Zhou, 1996). In Anhui Province, slightly more information is available and *C. homotomus* is considered to be the most likely BTV vector on the basis of its high abundance and peak activity periods which coincide with the occurrence of the virus, but so far there are no published data on BTV isolations from this species to confirm this association (Zhou, 1996). It is apparent, therefore, that the identity of the major BTV vectors in China is still uncertain and further work is required to resolve this situation. As *C. imicola*, *C. fulvus* and *C. actoni* have all been confirmed as vectors elsewhere, it seems prudent to allocate a high priority to a study of these species.

In Indonesia, more research has been carried out on the vectors than in any other country in the east Asian region. Almost 50 species of *Culicoides* have been identified, including four proven BTV vectors in other regions (*C. actoni*, *C. brevitarsis*, *C. fulvus*, *C. wadai*) (Sukarsih et al., 1993, 1996), and these probable vector species are also present in neighbouring Malaysia (Daniels et al., 2004). It is notable that *C. imicola* was absent from surveys carried out in both of these countries. Several other potential vectors in this region (*C. brevipalpis*, *C. peregrinus*, *C. oxystoma*, *C. nudipalpis*, *C. orientalis*) have also been identified (Sukarsih et al., 1996), and BTV-1 has been isolated from pools of *C. fulvus* (Sendow et al., 1996) while BTV-21 has been isolated both from pools of *C. peregrinus* (Sendow et al., 1996) and from mixed pools of *C. fulvus* and *C. orientalis* (Sendow et al., 1996).

Australasia

At least eight serotypes of BTV occur in Australia (Gard, 1987; Kirkland, 2004), three in Papua New Guinea (Puana, 1996) and one in the Solomon Islands (St George, 1996). Several *Culicoides* species, including *C. brevipalpis*, *C. peregrinus*, *C fulvus*, *C. wadai*, *C. actoni* and *C. brevitarsis*, are thought to be capable of transmitting the virus in the region with the latter four considered to be the principal vectors (Standfast et al., 1985; St George, 1996; Kirkland, 2004).

Of these four species, *C. fulvus* is thought to be the most efficient vector (Standfast et al., 1985) but is restricted to areas with high summer rainfall and does not occur in the drier sheep-rearing areas of Australia. *Culicoides wadai*, also an efficient BTV vector, was first recorded in Australia in 1971, probably having been introduced by being blown on the prevailing winds from Indonesia, where it is common (Sukarsih et al., 1996). This species was originally restricted to the Darwin area but from 1978 to 1988 extended its range southwards into the Kimberleys and eastwards to the Queensland coast and from there southwards again into New South Wales. This expansion of some 2000 miles in around 10 years put this efficient BTV vector into close proximity to some of the major sheep-rearing areas, causing concern for the Australian authorities at the time (Doyle, 1992). Fortunately, a long succession of dry summers appears to have halted its southwards expansion (Muller, 1995). Changing climate patterns could support this halt in the expansion of *C. wadai*, or could reverse it depending on the local conditions created.

Culicoides brevitarsis, although a relatively inefficient transmitter of BTV compared with *C. fulvus* and *C. wadai*, is considered to be the major vector in the Australasian region because of its high abundance in many areas and its very wide distribution which, in Australia, approximates the distribution of BTV seropositive animals (St George, 1996; Kirkland, 2004). Much work has been carried out into the ecology of *C. brevitarsis* leading to a detailed understanding of the environmental conditions controlling its survival and dispersal (see Kirkland, 2004). Such data are being used to develop models predicting its seasonal incidence and dispersal (Bishop et al., 1995, 1996).

The Americas

Until relatively recently, the major vector throughout almost the whole of the BTV zone in North America was considered to be *C. variipennis*. However, during the 1990s convincing evidence was put forward to show that *C. variipennis* is complex of at least three species (*C. variipennis*, *C. occidentalis* and *C. sonorensis*) (Tabachnick, 1992; Tabachnick and Holbrook, 1992). Of these, it is *C. sonorensis* that has been shown to be the primary vector on the basis of field isolations of virus, a close relationship between the distribution

of BTV-specific antibodies and the range of this species, and vector competence studies (Tabachnick, 1992; Tabachnick and Holbrook, 1992). It is also
the case that populations of *C. variipennis* have been shown to be significantly more resistant than *C. sonorensis* to oral infection with BTV (Lopez
et al., 1992; Tabachnick, 1996). On the basis of these findings, it has been
suggested that this is the major reason why BTV does not seem to be
transmitted in northeastern USA, a region where *C. variipennis* is common
but *C. sonorensis* is absent.

As almost all of the earlier published work on BTV and the *C. variipennis*
complex has been carried out using *C. sonorensis* under its earlier names of
C. variipennis or *C. variipennis sonorensis*, for the remainder of this chapter
the name *C. sonorensis* will be used in their stead.

In southeastern USA, BTV is transmitted by *C. sonorensis* and, probably,
other species of *Culicoides*. In parts of Alabama, where *C. sonorensis* is scarce
and *C. stellifer* very common, BTV sero-conversion rates of almost 90% in
cattle and white-tailed deer have been reported and BTV RNA has been detected
in both species (Mullen and Anderson, 1998), suggesting that *C. stellifer* could
act as a vector in these areas. Further south in the USA, *C. sonorensis* becomes
less common and in this and adjacent regions *C. insignis*, a mainly Central and
South American species, appears to be the most important vector with BTV-2
isolations being made from this species (Kramer et al., 1985; Greiner et al.,
1985). *Culicoides* of the *variipennis* complex do not seem to extend further
south than central Mexico (Wirth and Jones, 1957).

In Central and South America, *C. insignis* is considered to be the major
vector of BTV. This species is widespread throughout most of the region
occurring as far south as Argentina (Blanton and Wirth, 1979; Lager, 2004;
Lager et al., 2004), and it is by far the most abundant biting midge near
ruminant livestock (Mo et al., 1994; Lager et al., 2004). BTV-3 and 6 have
both been isolated from *C. insignis* in Central America (Mo et al., 1994).
Additional isolations of BTV, serotypes 3 and 4, have been made from
C. pusillus, but this species is considered to be less important as a vector as
its distribution is more restricted than that of *C. insignis* and it is significantly
less abundant (Mo et al., 1994).

Europe

Until recently, Europe had experienced only sporadic incursions of BTV, into
Spain and Portugal (1956–1960) and the Greek Islands of Lesbos and Rhodes
in 1979 and 1980, respectively (Vassalos, 1980; Dragonas, 1981; Mellor et al.,
1983). However, there was no information on the identity of the vectors in
these regions. Consequently, during the 1980s attempts were made to identify
BTV vectors in many areas across southern Europe, including the Iberian
Peninsula, Italy and Greece. As a result, *C. imicola* was identified in Portugal,
SW Spain and certain Greek islands adjacent to the Anatolian Turkish coast

01 (Mellor et al., 1983, 1985; Boorman, 1986). This distribution was virtually
02 identical to the areas affected by BTV, which led to the assumption that
03 *C. imicola* was the only important vector in Europe and that areas at risk of
04 BT were limited to the regions where it had been identified.
05   However, starting in 1998 and continuing until the time of writing (2006), the
06 continent has experienced the most severe and widespread incursion of BTV
07 ever seen, with a dozen countries affected, 5 serotypes of the virus involved and
08 over 1.5 million sheep having died, and with the virus extending some 800 km
09 further north than ever before (Mellor and Wittmann, 2002; Purse et al., 2005).
10 The vectors responsible are now known to be *C. imicola* and also members of
11 the *C. obsoletus* and *C. pulicaris* species complexes (Caracappa et al., 2003;
12 Savini et al., 2003; Purse et al., 2005; Savini et al., 2005). This unprecedented
13 spread of BTV seems to have been driven by recent changes in the European
14 climate (Purse et al., 2005). This has facilitated the northwards expansion of
15 *C. imicola* across most of southern Europe, which itself has significantly increased
16 the areas at risk to BT. In addition, these movements have also dramatically
17 increased overlap between *C. imicola* and the more northerly, *C. obsoletus* and
18 *C. pulicaris* groups of midges, some of which have also been shown to be
19 competent BTV vectors – thus extending the areas at risk well beyond the range
20 of *C. imicola* itself (Mellor and Wittmann, 2002; Purse et al., 2005).

Infection and replication of bluetongue virus in vector *Culicoides*

27 After ingestion of a viraemic blood meal, the titre of virus in a competent
28 insect vector usually falls for a variable period of time (eclipse or partial
29 eclipse phase) and then increases significantly to reach a plateau that is
30 maintained for the duration of the insect's life. Transmission becomes possible
31 after the completion of the extrinsic incubation period (EIP), which is the
32 interval between ingestion of virus and the earliest time at which virus is
33 released in the saliva. All of these times and titres vary, according to the virus,
34 the vector and the ambient temperature.
35   For example, females of the North American vector *C. sonorensis* can
36 ingest approximately 10^{-4} ml of blood; therefore, when feeding upon viraemic
37 blood containing 10^6 TCID$_{50}$ (tissue culture infectious doses 50%) of BTV
38 per ml, each midge will on average ingest about 100 TCID$_{50}$ of virus (day 0
39 value). Subsequently, over the first 1–2 days after ingestion, the viral titre per
40 midge decreases in the eclipse phase (if virus seems to disappear totally) or the
41 partial eclipse phase (if some virus is still detectable). This decrease is due to
42 the inactivation of some virus particles in the hostile environment of the gut
43 (mesenteron) lumen and/or to excretion of virus before viral proliferation in
44 infected gut cells becomes evident. After the trough of the eclipse phase is

reached, replication begins to exceed inactivation and the viral concentration rises to a plateau (5–6 $\log_{10}$ TCID$_{50}$ of virus per midge) between 7 and 9 dpi at 25° C. This represents a 1000- to 10 000-fold increase in virus per midge over the day 0 value, a concentration that is maintained for the remainder of the insect's life. Transmission of BTV from midges maintained at about 25° C becomes possible at 7–14 dpi, by which time the virus will have reached and replicated in the salivary glands thus completing the EIP (Foster et al., 1963; Bowne and Jones, 1966; Luedke et al., 1967; Foster and Jones, 1973; Chandler et al., 1985; Fu et al., 1999). Fu et al. (1999), working with *C. sonorensis* maintained at 24 ± 1° C, showed that BTV disseminates sequentially through the body of competent individuals with midgut cells infected first, followed by fat body cells at 1 dpi, neural tissue by 3 dpi and salivary gland cells by 5 dpi, transmission becoming possible from 7 dpi. Fu (1995) also reported that BTV particles infected the salivary gland acinar cells and, after replication, progeny virus particles were released into the terminal acini from where they were transported through the intermediate ducts to accumulate in para-crystalline arrays in the lumen of the major secretary ducts. The titre of BTV present in the injected saliva during a bite by a single *C. sonorensis* was estimated to lie between 0.32 and 7.79 TCID$_{50}$ (Fu et al., 1999). Despite this apparently small amount of virus, Foster et al. (1968) reported that the bite of a single *C. sonorensis* was sufficient to infect a susceptible sheep. More recent work by O'Connell 2002 and Baylis et al. (see Chapter 15 this volume) who fed variable numbers of infected *C. sonorensis* on groups of naïve sheep has confirmed and extended this observation showing that the bite of a single infected midge almost invariably gives rise to infection in susceptible sheep.

When Foster and Jones (1979) infected *C. sonorensis* with BTV, they observed a complete eclipse phase at 3 dpi. Thereafter, a two-phase viral replication cycle ensued before the titre reached a plateau [10^7 egg lethal doses 50% (ELD$_{50}$)] per infected midge. They interpreted the initial decrease in viral titre as being due to digestion of the infecting blood meal with 'attachment, penetration and uncoating' in the midgut cells of infected midges. They further suggested that the first increase in titre corresponded to viral growth 'through' the gut wall, and that the second increase was due to further cycles of replication in the salivary glands and other secondary target organs. The plateau region, beyond 14 dpi, was thought to represent either a cessation of virus replication with retention of infectivity or a steady state of virus replication and inactivation. No explanation was offered to explain why BTV infectivity was apparently limited to 10^7 ELD$_{50}$ per midge. It is possible that the overall controlling factor may merely be the number of susceptible cells available within each infected midge. This proposition is supported by the fact that smaller vector species of *Culicoides*, such as *C. imicola* and *C. obsoletus*, usually support lower peak viral titres than *C. sonorensis* (Carpenter et al., 2006; Venter et al., 2005). In this context Fu et al. (1999) showed that BTV replicates in midgut cells, neural tissues, fat body cells and salivary glands of infected midges but not in hindgut

01 cells, muscle cells, Malpighian tubule cells or oocyte/nurse cells. However,
02 based upon mosquito work, Murphy (1975), Murphy et al. (1975) and Hardy
03 et al. (1983) suggested that the levelling off of arbovirus titres in infected vectors
04 might be related to an innate ability of the vector to control or modulate viral
05 titres and thus avoid the pathological effects of infection.
06   The reason why arboviruses rarely produce the catastrophic damage in suscep-
07 tible insect cells that they do in mammalian cells is not entirely clear, but it may
08 have to do with the mechanisms the virus uses to exit cells. Fu (1995) showed
09 that BTV is released from mammalian cells both by a process of membrane
10 budding and by extrusion (exocytosis). Particle release by extrusion was accom-
11 panied by the co-release of 'debris' from the cell surface, indicating damage to
12 the cell membrane; such damage is likely to be a major factor in the lysis of
13 infected mammalian cells. Release of virus from vector insect cells, however,
14 was observed to occur only via membrane budding and without obvious cellular
15 damage. Furthermore, Mertens et al. (1984) working with BTV showed that shut-
16 off of host-cell protein translation in mammalian cells occurs within 10 h of
17 infection but no such shut-off occurs in infected insect cells (Mertens, personal
18 communication). These findings may help to explain why arboviruses such as
19 BTV usually cause little or no damage in the vector host, even though they may
20 be responsible for severe damage in the vertebrate host.

Bluetongue virus infection and susceptibility rates in vector species of *Culicoides*

26 The infection rate (IR) of *C. sonorensis* has been shown to be dependent upon
27 the concentration of virus in the blood meal (Jones and Foster, 1971a, 1974).
28 By feeding *C. sonorensis* a series of blood meals containing low titres of BTV
29 (3×10^5 ELD$_{50}$/ml), these authors progressively increased the IR (% of indi-
30 viduals actually infected) to a level where it equalled the susceptibility rate
31 (SR – i.e. the % of individuals in a population able to be infected). They also
32 showed that the ingestion of a single blood meal containing a high titre of BTV
33 produced an IR that equalled the SR.
34   The oral SR of any vector species of arthropod for an arbovirus rarely
35 attains 100%, and the susceptibility of *C. sonorensis* to BTV is no exception.
36 Jones and Foster (1978a) showed not only that different field populations of
37 this midge vary in their susceptibility to several serotypes of BTV but also that
38 the susceptibility of a single population to different serotypes may vary. The
39 first part of this work has recently been confirmed by Carpenter et al. (2006)
40 when working with different geographical populations of *C. obsoletus* in the UK.
41 Jennings and Mellor (1987) found that even within an established laboratory
42 colony of *C. sonorensis*, the response to oral infection with a single serotype
43 of BTV could vary widely, and they recorded IRs ranging from 0 to 51.6%.
44 Jones and Foster (1974) considered that oral susceptibility was under the control

of a single gene with a dominant allele for resistance. Completely susceptible and highly refractory populations of *C. sonorensis* were apparently derived by selective breeding over the course of a single generation from a parent population with an intermediate SR. This suggests a simple form of Mendelian inheritance. However, breeding experiments by other workers, while significantly enhancing the rate of occurrence of one or other of the desired traits, failed to produce completely susceptible or refractory populations of *C. sonorensis*, even after many generations of selective breeding (Mellor et al., unpublished observations). It would seem, therefore, that the inheritance mechanisms may be somewhat more complex than originally suggested. Tabachnick (1991) achieved results which suggest the following: (1) Resistance and susceptibility are controlled by a single locus that has resistant and susceptible alleles. (2) The maternal genotype determines the phenotype of the offspring. (3) The paternal gene is the dominant allele in the genotype of the daughter. His work also suggests that there may be a second locus that is able to modify the 'degree' of infection, i.e. the level of dissemination of the virus within individual infected insects. The physical manifestations of these heritable traits presumably find expression as one or another of the 'barriers' that will be discussed in the next section.

The information discussed in this section relates mainly to the vector competence of *Culicoides* species which is under genetic, and as will be seen later, environmental control (see sections on the effect of temperature on infection and transmission). However, it should be borne in mind that vector competence is only one element of vectorial capacity which includes also vector abundance, biting rates, host preferences, survival rates and duration of the EIP. All of these factors, together, provide a much more comprehensive measure of the disease risk posed by each vector species or population. For most vector species of *Culicoides*, many or most of the variables incorporated in vectorial capacity remain unmeasured and, indeed, standard techniques for measuring them may not exist or may not be widely available. Accordingly, Mellor et al. (2004) recently recommended that

a. Standard techniques for measuring the variables of vectorial capacity should be developed and adopted to facilitate comparison of data and data sharing.
b. Trapping methods should be evaluated against a 'gold standard' (e.g. drop trap, and the Onderstepoort-type light trap).
c. Biases in trapping methods should be measured.
d. Improved methods for reliably aging *Culicoides* should be developed.
e. Improved methods for recording host preferences should be developed.
f. The effects of the environment, host demography and climate on vectorial capacity should be investigated.
g. Measures of vectorial capacity should be correlated with other indicators of disease risk, such as host disease status.

Barriers to the infection, dissemination and transmission of arboviruses in insect vectors

Over 70 years ago, Storey (1933) demonstrated that if the gut wall of the leaf hopper *Circadulina mbila* was disrupted by puncture, strains of the insect that previously would not transmit maize-streak virus became transmitters. This suggested that the gut wall itself could provide a barrier to viral transmission. Demonstration of this same phenomenon with haematophagous Diptera and their associated arboviruses has now also been extensively documented. However, it has become apparent that the insect gut is not the only site at which interference with arboviral infection, dissemination through the body of the vector and transmission may occur. Neither is the gut itself necessarily a simple single-stage barrier. In fact, it has become evident that vector competence of insects for arboviruses is a complex subject associated with multiple barrier systems. Presence or absence of these barriers seem to be hereditary traits; consequently, within a vector species of insect, the proportion of individuals expressing one or other of the barriers or expressing no barriers at all can be manipulated by selective breeding.

Barriers to the infection of vector *Culicoides* with bluetongue virus

Some female *C. sonorensis* exhibit a midgut infection barrier (MIB) to BTV, and Jones and Foster (1974, 1978a, 1979) and Jennings and Mellor (1987) have described how different populations of this species contained variable proportions of individuals that are refractory to oral infection with this virus even though 100% of each population could be infected parenterally (Jones and Foster, 1966; Foster and Jones, 1973; Jochim and Jones, 1966).

Jennings and Mellor (1987) also demonstrated the presence of a midgut escape barrier (MEB) to BTV in *C. sonorensis*. They showed that persistently infected *C. sonorensis* containing less than $10^{2.5}$ TCID$_{50}$ of virus consistently failed to transmit whereas transmission was regularly demonstrated by midges containing $10^{2.7}$–$10^{5.1}$ TCID$_{50}$ of virus. Furthermore, Mellor and Jennings (unpublished data 1987) dissected the midguts from female *C. sonorensis* orally infected with BTV, at intervals ranging from 10 to 14 dpi. In midges containing less than $10^{2.5}$ TCID$_{50}$ of BTV, the virus was completely restricted to the midgut cells and had failed to disseminate to other target organs. These midges, although persistently infected with BTV, were therefore incapable of operating as vectors and exhibited an MEB. Up to 43.6% of persistently infected *C. sonorensis* were found to express such a barrier.

Interestingly, Riegler (2002) on the basis of work with AHSV in *C. sonorensis* and *C. imicola* suggested that in addition to being insect-genetically controlled, as suggested by Tabachnick (1991), the MIB and MEB in these vectors could also be strongly influenced by viral factors. Using two reassortants of AHSV-3 that varied only in the origin of Segment 10, he showed that oral IRs and dissemination rates from the midgut cells varied significantly between insects infected with one or other of these two viruses. This leads to the intriguing possibility that the observed vector competence may be an expression of the genetic make-up of both the insect vector and the infecting virus.

Fu et al. (1999) showed that in *C. sonorensis* orally-infected with BTV, dissemination of virus through the haemocoel to the salivary glands was usually a rapid process in insects expressing neither an MIB nor an MEB. However in a proportion of such individuals, BTV dissemination was restricted to abdominal fat body cells while all other tissues except the midgut remained uninfected. Because fat body is known to play a part in insect immune responses (Rees et al., 1997), Fu et al. suggested that their observations reflected such a response operating to limit viral dissemination through the haemocoel, which they termed a dissemination barrier. The same phenomenon was not observed in *Culicoides* inoculated intrathoracically with BTV. Fu et al. suggested that this was so because in their system, intrathoracic inoculation always resulted in large amounts of virus being introduced into the haemocoel, thereby exceeding the level at which fat bodies could clear it. On the contrary, some orally infected individuals were presumed to release virus into the haemocoel at concentrations below this level.

Although salivary gland infection and escape barriers have been demonstrated for a number of mosquito/virus combinations, these have yet to be identified in any species of *Culicoides* biting midge, which suggests that once virus reaches and infects the salivary glands of these insects, transmission invariably becomes possible (Fu, 1995).

Transovarian transmission (TOT) barriers have been the subject of much study. There is some evidence to indicate TOT of BTV does not occur in laboratory-adapted colonies of *C. sonorensis* when it was not detected in a systematic examination of approximately 1000 progeny of 72 infected females (Jones and Foster, 1971b; Nunamaker et al., 1990). However, as will be seen below, this result could have been influenced by the (vertebrate) assay system used which, though it was the best available at the time, could have failed to identify virus present in insects. Furthermore, in the same colony insects examined with electron microscopy, BTV antigen was detected in the vitelline membrane of infected, adult females and in the proteoid yolk bodies of their oocytes but not in the ovarian tissue itself. However, this study does not preclude the possibility of vertical transmission (VT) of the virus through some other structure. In fact, demonstration of 'dense' BTV antigen in the reproductive structures in which it was found (Nunamaker et al., 1990) would suggest such a possibility. For example, VT of flaviviruses in mosquitoes occurs via the micropyle as the egg passes through the oviducts during oviposition (Rosenstock et al., 2003). A similar mechanism could function for VT of BTV.

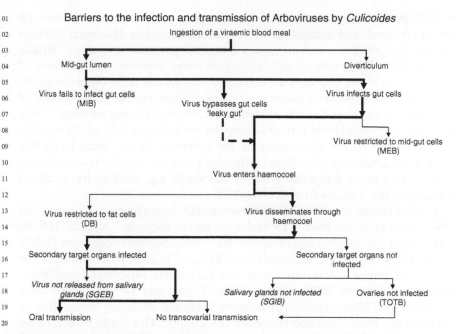

Barriers to the infection and transmission of Arboviruses by *Culicoides*

Ingestion of a viraemic blood meal

Mid-gut lumen — Diverticulum

Virus fails to infect gut cells (MIB) — Virus bypasses gut cells 'leaky gut' — Virus infects gut cells

Virus restricted to mid-gut cells (MEB)

Virus enters haemocoel

Virus restricted to fat cells (DB) — Virus disseminates through haemocoel

Secondary target organs infected — Secondary target organs not infected

Virus not released from salivary glands (SGEB) — Salivary glands not infected (SGIB) — Ovaries not infected (TOTB)

Oral transmission — No transovarial transmission

Figure 14.1   Barriers to the infection and transmission of arboviruses in *Culicoides* biting midges and mosquitoes. MIB, mid-gut infection barrier; MEB, mid-gut escape barrier; DB, dissemination barrier; SGIB, salivary gland infection barrier; SGEB, salivary gland escape barrier; TOTB, transovarial transmission barrier (Adapted from Hardy et al., 1983).

Figure 14.1 summarizes the barriers to viral infection, dissemination and transmission in *Culicoides* biting midges and mosquitoes from the stage of ingestion of a viraemic blood meal to transmission (oral and/or transovarial).

A possible overwintering mechanism of bluetongue virus in vector *Culicoides*

Overwintering of BTV has been a contentious issue for several years, with scientific and regulatory implications that confound the emergence of a clear answer to what might otherwise be an easily answered question. Theoretically, there are several ways in which BTV might survive in locations where the harsh winters mean that there are no adult vectors for several months of the year. The most likely possibility, and one that matches the observed epidemiology of BT, is overwintering via TOT through the developmental stages of the vectors which do persist throughout the winter period, even in the absence of the adults. An alternative overwintering hypothesis postulates the survival of BTV in the vertebrate animal, either through persistent infection or through lateral (venereal) and vertical (transplacental) transmission. However, although data on the generation and maintenance of viral persistence in sheep (Brodie et al., 1998; Takamatsu et al., 2003) have been reported on the basis of laboratory experiments, such a mechanism has not yet been observed in the field. Another hypothesis postulates not overwintering but re-incursions of the virus via the long-distance movement of infected insects on wind. While the movement of infected *Culicoides* spp. by prevailing winds and/or weather patterns can certainly be responsible for outbreaks at the local level, this mechanism is unlikely to account for the successive and regular re-emergences of BTV into many temperate areas of the world, e.g. such as has occurred annually in the Balkans from 2000 to 2004.

To examine the role of the insect vector in BTV overwintering, *C. sonorensis* larvae were collected from BTV-endemic sites in northern Colorado, U.S.A., and assayed for the presence of several BTV RNA segment sequences (White et al., 2005). Segment 7 sequences were detected in roughly one-third of the larvae and pupae – less often with less favourable processing conditions. However, infectious BTV was not isolated from any of the insects using vertebrate cell-based, traditional isolation methods, including inoculation of embryonated hens' eggs and vertebrate cell culture. This could be due to the fact that segment 2 RNA was detected at roughly half the rate of segment 7 RNA, in the field-collected larvae. Segment 2 encodes the vertebrate cell receptor ligand VP2, and a reduction in the expression of that gene product would decrease the amount of 'vertebrate infective' BTV in an insect. Interestingly, upon examination of a persistently-infected *Culicoides* cell line, this pattern was repeated: segments necessary for invertebrate cell infection (3, 7 and 10) were found consistently, whereas segment 2 sequences were found significantly less often, and only when using one of two primer sets of similar sensitivity, perhaps indicating a fragmentation of 'unnecessary' genome segments (White et al., 2005). These data may best be interpreted in the light of the properties of the BTV virion. It would appear that there is a structural and functional differentiation of the outer and inner capsids of the virion, where the outer capsid mediates vertebrate cell infection and the inner, invertebrate cell infection. It can be hypothesized that a downregulation or elimination of 'unnecessary' genome segments occurs during the persistent infection of the insect vector – possibly to decrease the metabolic burden of BTV infection on the overwintering insect. The different infectivities of intact virions, infectious subviral particles (ISVP) and core particles for insect and mammalian cells provides further support for this hypothesis (see below, and Mertens et al., 1996). Such a situation could explain the relatively low rate of viral isolation from insects (using vertebrate cell-based systems) and also emphasizes the close relationship between insect and viral components in the epidemiology of BT. The epidemiological significance of the presence of viral antigen and nucleic acid in field-collected larvae remains to be confirmed; however, this finding may have provided the first hard evidence that TOT may be a major strategy in the transeasonal maintenance of BTV.

Effects of temperature on arbovirus infection and transmission by insect vectors

Temperature is probably the most important extrinsic variable affecting arbovirus transmission via the vector and is likely, therefore, to have a significant effect on the epidemiology of many arboviral diseases. However, its precise effects upon the dynamics of arbovirus transmission may be difficult to predict and are likely to vary with each virus–vector combination. Generally speaking, at higher temperatures a vector may blood-feed more frequently and the rate of virogenesis within a vector is usually faster, leading to an enhanced probability of transmission. On the contrary, increase in temperature may shorten the lifespan of the vector, which would lessen the transmission potential. As temperature decreases, virogenesis usually slows and at some point (specific for each arbovirus) may cease altogether; however, at lower temperatures the lifespan of the vector may be extended. The likelihood of arbovirus transmission by arthropod vectors is therefore a function of the interaction of these two opposing trends (McMichael et al., 1996). Additionally, there may be other, so far ill-defined but more subtle, effects of temperature upon the ability of arthropods to transmit arboviruses, for example, upon (1) the ability of the vector to modulate viral replication within its cells (Hardy et al., 1983; Kramer et al., 1983) or (2) the threshold or efficacy levels of the various barriers to viral replication and dissemination that exist within a variable proportion of most vector species.

Effect of temperature on bluetongue virus infection by vector *Culicoides*

Mullens et al. (1995) when working with C. *sonorensis* and BTV-11 showed that virogenesis in the vector proceeded significantly faster at higher temperatures and peak levels of viral antigen were detected after 5–7, 7–13 and 18–22 days in midges held at 32, 27 and 21°C, respectively. These authors failed to detect viral replication in any midges held at 15°C for 22 days; however, when such midges were transferred to 27°C, latent virus replicated and became detectable within 4–10 days. On the basis of their work, Mullens et al. (1995) suggested that high summer temperatures should favour BTV transmission through more efficient virus replication and a decreased duration of the EIP. Although high temperatures might also reduce the daily survival rate of vectors, Mullens et al. (1995) considered that this would be more than offset by the increased feeding frequency which would enhance the likelihood of infection and transmission.

Working with the related AHS virus (AHSV), which can also be transmitted by *C. sonorensis*, Wellby et al. (1996) and Mellor and Wellby (1998) achieved results similar to those of Mullens et al. (1995) and showed that at elevated temperatures, IRs tended to be higher, rates of virogenesis faster and transmission earlier. At constant temperatures lower than 15°C, AHSV failed to replicate in or be transmitted, though insect survival was extended. When insects maintained at 10–15°C were moved to higher, virus permissive, temperatures, previously undetected latent virus became evident in many individuals, just as had been shown by Mullens et al. (1995) with BTV. Under variable temperature regimes chosen to simulate nocturnal–diurnal variations, Mellor and Wellby (1999) showed that AHSV infection and viral replication rates were proportional to the time spent at temperatures permitting virogenesis (i.e. above 15°C) and the cumulative total of time spent at these temperatures was the major factor controlling transmission potential. More recently, Paweska et al. (2002), when working with BTV-1 and wild-caught *C. imicola* and *C. bolitinos* at temperatures varying from 10 to 30°C, also observed that IRs and rates of virus development increased significantly at the higher temperatures. Wittmann et al. (2002), who worked with both BTV and AHSV in *C. sonorensis*, obtained similar though rather more detailed results to those outlined above and stated that 'the transmission potential of *C. sonorensis* for AHSV-4, BTV-10 and BTV-16 was greater at higher temperatures, because although vector survival was reduced, this was more than compensated for by the accompanying decrease in duration of the EIP', and they considered that temperatures between 27 and 30°C were probably most favourable for maximum transmission potential. Interestingly, 28–29°C is also the optimal temperature range for BTV in vitro viral synthesis, based upon the activity of the RNA-dependent viral polymerase (van Dijk and Huismans, 1982).

Effects of temperature on bluetongue virus infection and transmission by 'non-vector' *Culicoides*

Culicoides nubeculosus, a species that occurs through much of Europe, is considered to be incapable of transmitting BTV (and AHSV) because of a MIB. However, when the immature stages of this species were reared at a range of temperatures, the BTV oral IR increased from 0% (at 25 and 30°C) to 13.4% (at 33°C), and virus replicated to high titre in infected individuals. Indeed, exposure of the pupal stage alone to high rearing temperatures also caused similar increases in IR (Wittmann, 2000). The results of this work are very similar to those achieved in earlier studies with *C. nubeculosus* and AHSV (Mellor et al., 1998). Wittmann (2000) suggested that at the higher rearing temperatures the integrity of the gut wall might be compromised, thus enabling virus to bypass the gut barriers (the so-called 'leaky gut' phenomenon). The

01 critical period for this effect appeared to be during the time of development of
02 the adult midgut epithelium cells, i.e. during the pupal stage. Wittmann (2000)
03 considered that at a time of climate change this phenomenon could result in the
04 appearance of BTV (and presumably AHSV) competent *C. nubeculosus* in
05 the field. Taking into account the reduced survival rates observed when rearing
06 the immature stages of this species at high temperature, she calculated that for
07 every 100 individuals reared at 33°C, two would be competent to transmit
08 BTV whereas none would be competent at 25–32°C. Mellor (1998) suggested
09 that the isolations of AHSV made from *C. obsoletus* and *C. pulicaris* in Spain
10 in 1988 (see Mellor et al., 1990), species that had not previously been
11 connected with this virus, could be examples of the leaky gut phenomenon
12 in operation potentiated by the unusually warm conditions prevailing in the
13 area at the time. Also, he suggested that with increasing temperature due to
14 climate change further such 'unusual' isolations were likely. The subsequent
15 identification of *C. obsoletus* and *C. pulicaris* as likely vectors of BTV in the
16 Balkans, Bulgaria, NW Greece, European Turkey and some areas of Italy
17 during the period 1999–2004 and isolation of BTV from these species may be
18 further examples of the sort that Mellor was envisaging (Caracappa et al.,
19 2003; Savini et al., 2003; De Liberato et al., 2003; Mellor, 2004; Torina et al.,
20 2004; Purse et al., 2005).

Other factors influencing the infection of *Culicoides* with bluetongue virus

25 The existence of the various barriers to infection of *Culicoides* with BTV
26 provides a means whereby virus infection can be modulated. However, few
27 studies have been carried out in this area and the basis of these barriers is poorly
28 understood. In this context, Mertens et al. (1996) showed that modification of
29 the viral particles themselves by digestive enzymes within the insect gut can
30 affect the efficacy of the MIB. Mertens et al. (1987) originally showed that
31 treatment of BTV particles with proteases (chymotrypsin and trypsin) cleaves
32 virus protein 2 in the outer capsid to give rise to an ISVP. A third particle type
33 (core) may also be produced by the uncoating of either ISVPs or intact viral
34 particles. In mammalian cells, cores have a very low level of infectivity while
35 ISVPs and intact non-aggregated viral particles have a similar but much
36 higher ($\times 10^4$) level of infectivity. All three particle types are highly infective
37 for BTV-susceptible *Culicoides* (Mertens et al., 1996), but ISVPs are over
38 100 times more infectious than intact aggregated viral particles or core particles.
39 Consequently, it would be advantageous for the virus if the intact particles
40 ingested from a vertebrate host were to be converted into ISVPs in the insect's
41 gut lumen. The presence of proteases such as trypsin and chymotrypsin, which
42 occur naturally in the gut of many haematophagous insects, might well accom-
43 plish this conversion. Indeed, the conversion of merely 1% of ingested particles
44 into ISVPs would account for all of the apparent infectivity of the intact

particles. Further, the oral infection of vector *Culicoides* with core particles of BTV showed that such particles are as infective as intact particles for vector insects (Mertens et al., 1996). Because cores, unlike intact particles and ISVPs, have no outer capsid, Mertens et al. (1996) considered it likely that the initial stages of infection of these particles (interaction between cores and gut cells, and entry) must require receptors or mechanisms different from those utilized by the other two particle types. Also, because core particles are virtually non-infectious for mammalian cells, it also suggests that this mechanism is peculiar to inter- actions with insect cells. More recently it has been shown that VP7 is the invertebrate cell receptor ligand (Xu et al., 1997; Tan et al., 2001). Clearly, this suggests that VP7 must be accessible from the surface of the intact virion and ISVPs, in order to be involved in insect cell infection, and antibody studies with intact virions support this suggestion (Eaton et al., 1991; Wang et al., 1994). Further studies, at the molecular level, are required to explore these complex arbovirus–vector interactions in greater depth. One such study could examine the role of viral genome segment reassortment observed in nature (White et al., 2005), as different S7 genotypes could be responsible for some of the variability in vectorial capacity mentioned above even though initial *in silico* attempts at delineating this relationship have failed.

Summary

BTV is maintained in nature by an endless series of alternating cycles of replication in vector species of *Culicoides* biting midge and ruminant mam- mals. Experimentation and field studies have shown that the ability of the virus to infect *Culicoides* and to be transmitted by them is restricted to a relatively small number of species. Experimentation has also shown that the virus can only be transmitted by competent *Culicoides* between certain, rather ill-defined, temperature limits. In essence, therefore, the world distribution of BTV is little more than a reflection of the distribution of vector species of biting midge within virus-permissive temperature limits.

Once ingested by a competent vector, BTV attaches to the luminal surface of the midgut cells, infects these cells and replicates in them. Progeny virus particles then escape through the abluminal surface and the basement lamina into the haemocoel from where the secondary target organs including the salivary glands are infected. Subsequent to virus replication in the salivary glands and release into the salivary ducts, transmission can take place. The whole cycle from infection to transmission takes between 10 and 15 days at 25°C but in a lesser time at higher temperatures and a longer time at lower temperatures. The minimum temperature for BTV replication in vectors seems to lie between 10 and 15°C.

Not all female midges within a vector species are susceptible to infection with BTV or if infected are competent to transmit the virus. A series of barriers exists within certain individuals which either prevent virus infection or else restrict it in such a way as to preclude transmission. Every population of a vector species of *Culicoides* has a variable proportion of these refractory individuals. The susceptible and refractory traits for BTV are under genetic control so that by selective breeding vector populations that are of high, low or medium susceptibility can be produced. However, the molecular basis of these barriers and their mode of operation are poorly understood. It is known, however, that temperature influences their efficacy so that at higher temperatures a greater proportion of a vector population becomes susceptible to infection and transmission. The ideal temperature for transmission seems to be in the region of 27–30°C because most competent individuals survive long enough to transmit at these temperatures and viral polymerase activity is also optimal. At these higher temperatures, ability to transmit BTV may also extend to additional species of *Culicoides*. The genetic make-up of the infecting virus may also influence infection and transmission so the observed level of vector competence may be an expression of the genetic make-up of both the insect vector and the infecting virus.

Transovarial transmission of BTV through its vectors has not previously been reported, but recent work has now shown that such a mechanism may exist. If confirmed, this could explain the apparent ability of BTV to survive, in certain regions, over vector-free periods (i.e. overwinter).

References

Al-Busaidy, S.M. and Mellor, P.S. (1991) Isolation and identification of arboviruses from the Sultanate of Oman. *Epidemiol. Infect.* **106**, 403–13.

Bi, Y., Li, C., Li, S., Qing, B., Zhong, N., Hu, J. and Yang, R. (1996) An epidemiological survey of bluetongue disease in Yunnan Province, China In: T.D. St George, B.H. Kay and J. Blok (eds), *Bluetongue Disease in Southeast Asia and the Pacific*. Canberra: ACIAR, pp. 51–6.

Bishop, A.L., Kirkland, P.D., McKenzie, H.J. and Barchia, I.M. (1996) The dispersal of *Culicoides brevitarsis* in Eastern New South Wales and associations with the occurrences of arbovirus infections in cattle. *Aust. Vet. J.* **73**, 174–8.

Bishop, A.L., McKenzie, H.J., Barchia, I.M. and Harris, A.M. (1995) The effects of habitat on the distribution of *Culicoides brevitarsis* Kieffer (Diptera, Ceratopogonidae) during its resting phase. *Aust. J. Zool.* **43**, 531–9.

Blackburn, N.K., Searle, L. and Phelps, R.J. (1985) Viruses isolated from *Culicoides* (Diptera: Ceratopogonidae) caught at the veterinary research farm, Mazowe, Zimbabwe. *J. Entomol. Soc. S. Afr.* **48**, 331–6.

Blanton, F.S. and With, W.W. (1979) Arthropods of Florida and neighbouring land areas. Gainesville, Fl: *The Sand Flies (Culicoides) of Florida (Diptera, Ceratopogonidae)*, Vol. 10. Department of Agriculture and Consumer services, p. 204.

Boorman, J. (1986) Presence of bluetongue virus vectors on Rhodes. *Vet. Rec.* **118**, 21.

Boorman, J. (1989) *Culicoides* (Diptera: Ceratopogonidae) of the Arabian Peninsula with notes on their medical and veterinary importance. *Fauna Saudi Arabia* **10**, 160–224.

Bowne, J.G. and Jones, R.H. (1966) Observations on bluetongue virus in the salivary glands of an insect vector, *Culicoides variipennis. Virology* **30**, 127–33.

Braverman, Y. and Galun, R. (1973) The occurrence of *Culicoides* in Israel with reference to the incidence of bluetongue. *Refu. Vet.* **30**, 121–7.

Braverman, Y., Kremer, M. and Boorman, J. (1976) Faunistic list of *Culicoides* (Diptera, Ceratopogonidae) from Israel. *Cah. O.R.S.T.O.M. Ser. Ent. Med. et Parasitol.* **14**, 179–85.

Brodie, S.J., Wilson, W.C., O'Hearn, P.M., Muthui, D., Diem, K. and Pearson, L.D. (1998) The effects of pharmacological and lentivirus-induced immune suppression on orbivirus pathogenesis: assessment of virus burden in blood monocytes and tissues by reverse transcription-*in situ* PCR. *J. Virol.* **72**, 5599–609.

Caracappa, S., Torina, A., Guercio, A., Vitale, F., Calabro, A., Purpari, G., Ferrantelli, V., Vitale, M. and Mellor, P.S. (2003) Identification of a novel bluetongue virus vector species of *Culicoides* in Sicily. *Vet. Rec* **153**, 71–4.

Carpenter, S., Lunt, H.L., Venter, G.J. and Mellor, P.S. (2006) Oral susceptibility to bluetongue virus of *Culicoides* (Diptera: Ceratopogonidae) from the United Kingdom. *J. Med. Entomol.* **43**, 73–8.

Chandler, L.J., Ballinger, M.E., Jones, R.H. and Beaty, B.J. (1985) The virogenesis of bluetongue virus in *Culicoides variipennis*. In: T.L. Barber and M.M. Jochim (eds), *Progress in Clinical and Biological Research. Bluetongue and Related Orbiviruses.* New York: Liss, pp. 245–53.

Daniels, P.W., Sendow, I., Prichard, L.I., Sukarsih and Eaton, B.T. (2004) Regional overview of bluetongue viruses in South-East Asia: viruses, vectors and surveillance. *Vet. Ital.* **40**, 94–100.

De Liberato, C., Purse, B.V., Goffredo, M., Scholl, F. and Scaramozzino, P. (2003) Geographical and seasonal distribution of the bluetongue virus vector, *Culicoides imicola*, in central Italy. *Med. Vet. Entomol.* **17**, 388–94.

Doyle, K.A. (1992) An overview and perspective on orbivirus disease prevalence and occurrence of vectors in Australia and Oceania. In: T.E. Walton and B.I. Osburn (eds), *Bluetongue, African Horse Sickness, and Related Orbiviruses.* Boca Raton, Fl: CRC Press, pp. 44–57.

Dragonas, P.N. (1981) Evolution of bluetongue in Greece. *OIE Monthly Epizoot. Circ.* **9**, 10–11.

Dulac, G.C., Dubuc, C., Myers, D.J., Afshar, A. and Taylor, E.A. (1989) Incursion of bluetongue virus type 11 and epizootic haemorrhagic disease of deer virus type 2 for two consecutive years in the Okanagan valley. *Can. Vet. J.* **30**, 351.

Du Toit, R.M. (1944) The transmission of bluetongue and horsesickness by *Culicoides. Onderstepoort. J. Vet. Sci. Anim. Ind.* **19**, 7–16.

Dyce, A.L. and Wirth, W.W. (1983) Reappraisal of some Indian *Culicoides* species in the subgenus Avaritia (Diptera: Ceratopogonidae). *Int. J. Entomol.* **25**, 221–5.

Eaton, B.T., Gould, A.R., Hyatt, A.D., Coupar, B.E., Martyn, J.C. and White, J.R. (1991) A bluetongue serogroup-reactive epitope in the amino terminal half of the major core protein VP7 is accessible on the surface of bluetongue virus particles. *Virology* **180**, 687–96.

Foster, N.M. and Jones, R.H. (1973) Bluetongue virus transmission with *Culicoides variipennis* via embryonating chicken eggs. *J. Med. Entomol.* **10**, 529–32.

Foster, N.M. and Jones, R.H. (1979) Multiplication rate of bluetongue virus in the insect vector *Culicoides variipennis* infected orally. *J. Med. Entomol.* **15**, 302–303.

Foster, N.M., Jones, R.H. and Luedke, A.J. (1968) Transmission of attenuated and virulent bluetongue virus with *Culicoides variipennis* infected orally via sheep. *Am. J. Vet. Res.* **29**, 275–9.

01 Foster, N.M., Jones, R.H. and McCrory, B.R. (1963) Preliminary investigations on insect
02 transmission of bluetongue virus in sheep. *Am. J. Vet. Res.* **24**, 1195–200.
03 Fu, H. (1995) Mechanisms controlling the infection of *Culicoides* biting midges with
04 bluetongue virus. PhD Thesis, University of Hertfordshire, pp. 154.
05 Fu, H., Leake, C.J., Mertens, P.P.C. and Mellor, P.S. (1999) The barriers to bluetongue
06 virus infection, dissemination and transmission in the vector, *Culicoides variipennis*
07 (Diptera: Ceratopogonidae). *Arch. Virol.* **144**, 747–61.
07 Gard, G.P., Shorthose, J.E., Weir, R.P. and Erasmus, B.J. (1987) The isolation of a
08 bluetongue serotype new to Australia. *Aust. Vet. J.* **64**, 87–8.
09 Greiner, E.C., Barber, T.C, Pearson, J.E., Kramer, W.L. and Gibbs, E.P.J. (1985)
10 Orbiviruses from *Culicoides* in Florida. In: T.L. Barber and M.M. Jochim (eds), *Progress
11 in Clinical and Biological Research. Bluetongue and Related Orbiviruses* New York:
12 Liss, pp. 195–200.
13 Guo, Z., Hao, J., Chen, J., Li, Z., Zhang, K., Hu, Y., Li, G. and Pu, L. (1996) Investigation
14 of bluetongue disease in the Bayannur Meng of inner Mongolia. In: T.D. St George, B.H.
15 Kay, and J. Blok (eds), *Bluetongue Disease in Southeast Asia and the Pacific.* Canberra:
16 ACIAR, pp. 80–3.
17 Hardy, J.L., Houk, E.J., Kramer, L.D. and Reeves, W.C. (1983) Intrinsic factors affecting
17 vector competence of mosquitoes for arboviruses. *Annu. Rev. Entomol.* **28**, 229–62.
18 Herniman, K., Boorman, J. and Taylor, W. (1983) Bluetongue virus in a Nigerian dairy
19 cattle herd. 1. Serological studies and correlation of virus activity to vector population.
20 *J. Hyg.* **90**, 177–93.
21 Howarth, F.G. (1985) Biosystematics of the *Culicoides* of Laos (Diptera: Ceratopogonidae).
22 *Int. J. Entomol.* **27**, 1–96.
23 Jain, N.C., Prasad, G., Gupta, Y. and Mahajan, N.K. (1988) Isolation of bluetongue virus
24 from *Culicoides* sp. in India. *Rev. Sci. Tech. OIE.* **7**, 275–8.
25 Jennings, M., Boorman, J.P.T. and Ergun, H. (1983) *Culicoides* from western Turkey in
26 relation to bluetongue disease of sheep and cattle. *Rev. Elev. Med. Vet. Pays Trop.* **36**, 67–70.
27 Jennings, D.M. and Mellor, P.S. (1987) Variation in the responses of *Culicoides variipennis*
28 (Diptera, Ceratopogonidae) to oral infection with bluetongue virus. *Arch. Virol.* **95**,
177–82.
29 Jochim, M.M. and Jones, R.H. (1966) Multiplication of bluetongue virus in *Culicoides
30 variipennis* following artificial infection. *Am. J. Epidemiol.* **84**, 241–6.
31 Jones, R.H. and Foster, N.M. (1966) The transmission of bluetongue virus to embryonating
32 chicken eggs by *Culicoides variipennis* (Diptera: Ceratopogonidae) infected by intrathor-
33 acic inoculation. *J. Am. Mosq. Control Assoc.* **26**, 185–9.
34 Jones, R.H. and Foster, N.M. (1971a) The effect of repeated blood meals infective for
35 bluetongue on the infection rate of *Culicoides variipennis*. *J. Med. Entomol.* **8**, 499–501.
36 Jones, R.H. and Foster, N.M. (1971b) Transovarian transmission of bluetongue virus
37 unlikely for *Culicoides variipennis*. *J. Am. Mosq. Control Assoc.* **31**, 434–7.
38 Jones, R.H. and Foster, N.M. (1974) Oral infection of *Culicoides variipennis* with bluetongue virus: development of susceptible and resistant lines from a colony population.
39 *J. Med. Entomol.* **11**, 316–23.
40 Jones, R.H. and Foster, N.M. (1978a) Heterogeneity of *Culicoides variipennis* field popula-
41 tions to oral infection with bluetongue virus. *Am. J. Trop. Med. Hyg.* **27**, 178–83.
42 Jones, R.H. and Foster, N.M. (1978b) Relevance of laboratory colonies of the vector in
43 arbovirus research *Culicoides variipennis* and bluetongue. *Am. J. Trop. Med. Hyg.* **27**,
44 168–77.

Jones, R.H. and Foster, N.M. (1979) *Culicoides variipennis*: Threshold to infection for bluetongue virus. *Ann. Parasitol.* **54**, 250.

Kettle, D.S. (1965) Biting ceratopogonidae as vectors of human and animal diseases. *Acta Trop.* **22**, 356–62.

Kirkland, P.D. (2004) Bluetongue viruses, vectors and surveillance in Australia – the current situation and unique features. *Vet. Ital.* **40**, 47–50.

Kramer, L.D., Hardy, J.L. and Presser, S.B. (1983) Effect of temperature of extrinsic incubation on the vector competence of *Culex tarsalis* for western equine encephalomyelitis virus. *Am. J. Trop. Med. Hyg.* **32**, 1130–9.

Kramer, W.L., Greiner, E.C. and Gibbs, E.P.J. (1985) A survey of *Culicoides* midges (Diptera, Ceratopogonidae) associated with cattle operations in Florida, USA. *J. Med. Entomol.* **22**, 153–62.

Lager, I.A. (2004) Bluetongue virus in South America: overview of viruses, vectors, surveillance and unique features. *Vet. Ital.* **40**, 89–93.

Lager, I.A., Duffy, S., Miquet, J., Vagnozzi, A., Gorchs, C., Draghi, M., Cetra, B., Soni, C., Hamblin, C., Maan, S., Samnel, A.R., Mertens, P.P.C., Ronderos, M. and Ramirez, V. (2004) Incidence and isolation of bluetongue virus infection in cattle of the Santo Tome Department, Corrientes Province, Argentina. *Vet. Ital.* **40**, 141–44.

LeDuc, J.W. and Pinheiro, F.P. (1988) Oropouche fever. In: T.P. Monath (ed), *The Arboviruses: Epidemiology and Ecology, Vol. IV.* Boca Raton, Fl: CRC Press, pp. 1–14.

Lee, V.H., Causey, O.R. and Moore, D.L. (1974) Bluetongue and related viruses in Ibadan, Nigeria: isolation and preliminary identification of viruses. *Am. J. Vet. Res.* **35**, 1105–8.

Linley, J.R. (1985) Biting midges (Diptera: Ceratopogonidae) as vectors of non-viral animal pathogens. *J. Med. Entomol.* **22**, 589–99.

Lopez, J.W., Dubovi, E.J., Cupp, E.W. and Lein, D.H. (1992) An examination of the bluetongue virus status of New York state. In: T.E. Walton (ed), *Bluetongue, African Horse Sickness, and Related Orbiviruses.* Boca Raton, FL: CRC Press, pp. 140–6.

Luedke, A.J., Jones, R.H. and Jochim, M.M. (1967) Transmission of bluetongue between sheep and cattle by *Culicoides variipennis*. *Am. J. Vet. Res.* **28**, 457–60.

MacLachlan, N.J. (1994) The pathogenesis and immunology of bluetongue virus infections of ruminants. *Comp. Immunol. Microbiol. Infect. Dis.* **17**, 197–206.

McMichael, A.J., Haines, A., Slooff, R. and Kovats, S. (1996) *Climate Change and Human Health.* Geneva: World Health Organisation, p. 297.

Meiswinkel, R., Nevill, E.M. and Venter, G.J. (1994) Vectors: *Culicoides* spp. In: J.A.W. Coetzer, G.R. Thomson,   R.C. Tustin (eds), *Infectious Diseases of Livestock with Special Reference to Southern Africa, Vol. 1.* Cape Town: Oxford University Press, pp. 68–89.

Mellor, P.S. (1990) The replication of bluetongue virus in *Culicoides* vectors. *Curr. Top. Microbiol. Immunol.* **162**, 143–61.

Mellor, P.S. (1998) Climate change and the distribution of vector-borne diseases with special reference to African horsesickness virus. In: Towards livestock disease diagnosis and control in the 21st century. Proceedings of the IAEA/FAO symposium on diagnosis and control of livestock diseases using nuclear and related techniques. Vienna: IAEA. pp. 439–54.

Mellor, P.S. (2004) Infection of the vectors and bluetongue epidemiology in Europe. *Vet. Ital.* **40**, 167–74.

Mellor, P.S., Boned, J., Hamblin, C. and Graham, S.D. (1990) Isolations of African horse sickness virus from vector insects made during the 1988 epizootic in Spain. *Epidemiol. Infect.* **105**, 447–54.

Mellor, P.S. and Boorman, J. (1995) The transmission and geographical spread of African horse sickness and bluetongue viruses. *Ann. Trop. Med. Parasitol.* **89**, 1–15.

Mellor, P.S., Boorman, J. and Baylis, M. (2000) *Culicoides* biting midges: their role as arbovirus vectors. *Ann. Rev. Entomol.* **45**, 307–40.

Mellor, P.S., Boorman, J.P.T., Wilkinson, P.J. and Martinez-Gomez, F. (1983) Potential vectors of bluetongue and African horse sickness viruses in Spain. *Vet. Rec.* **112**, 229–30.

Mellor, P.S. and Hamblin, C. (2004) African horse sickness. *Vet. Res.* **35**, 445–66.

Mellor, P.S., Jennings, D.M., Wilkinson, P.J. and Boorman, J.P.T. (1985) *Culicoides imicola* a bluetongue virus vector in Spain and Portugal. *Vet. Rec.* **116**, 589–90.

Mellor, P.S., Osborne, R. and Jennings, D.M. (1984) Isolation of bluetongue and related viruses from *Culicoides* species in the Sudan. *J. Hyg.* **93**, 621–8.

Mellor, P.S., Rawlings, P., Baylis, M. and Wellby, M.P. (1998) Effect of temperature on African horse sickness virus infection in *Culicoides*. *Arch. Virol.* 155–63.

Mellor, P.S., Tabachnick, W., Baldet, T., Baylis, M., Bellis, G., Calistri, P., Delecolle, J.-C., Goffredo, M., Meiswinkel, R., Mullens, B., Scaramozzino, P., Torina, A., Venter, G. and White, D. (2004) Conclusions, working group 2, vectors. *Vet. Ital.* **40**, 715–7.

Mellor, P.S. and Wellby, M.P. (1999) Effect of temperature on African horse sickness virus serotype 9 infection of vector species of *Culicoides* (Diptera: Ceratopogonidae). In: U. Wernery, J.F. Wade, J.A. Mumford and O-R. Kaaden (eds) *Equine Infectious Diseases VIII*. Newmarket, UK: R & W Publications, pp. 246–51.

Mellor, P.S. and Wittmann, E.J. (2002) Bluetongue virus in the Mediterranean basin (1998–2001). *Vet. J.* **164**, 20–37.

Mertens, P.P.C., Brown, F. and Sangar, D.V. (1984) Assignment of the genome segments of bluetongue virus type 1 to the proteins which they encode. *Virology* **135**, 207–17.

Mertens, P.P.C., Burroughs, J.N. and Anderson, J. (1987) Purification and properties of virus-particles, infectious subviral particles, and cores of bluetongue virus serotype-1 and serotype-4. *Virology* **157**, 375–86.

Mertens, P.P., Burroughs, J.N., Walton, A., Wellby, M.P., Fu, H., O'Hara, R.S., Brookes, S.M. and Mellor, P.S. (1996) Enhanced infectivity of modified bluetongue virus particles for two insect cell lines and for two *Culicoides* vector species. *Virology* **217**, 582–93.

Mo, C.L., Thompson, L.H., Homan, E.J., Oviedo, M.T., Greiner, E.C., Gonzales, J. and Saenz, M.R. (1994) Bluetongue virus isolation from vectors and ruminants in Central America and the Caribbean. *Am. J. Vet. Res.* **55**, 211–5.

Mullen, G.R. and Anderson, R.R. (1998) Transmission of Orbiviruses *by Culicoides* Latreille species (Ceratopogonidae) among cattle and white-tailed deer in the south-eastern United States. In: 4th International Congress of Dipterology, Oxford, UK, pp. 155–6.

Mullens, B.A., Tabachnick, W.J., Holbrook, F.R. and Thompson, L.H. (1995) Effects of temperature on virogenesis of bluetongue virus serotype-11 in *Culicoides variipennis-sonorensis*. *Med. Vet. Entomol.* **9**, 71–6.

Muller, M.J. (1995) Veterinary arbovirus vectors in Australia – a retrospective study. *Vet. Microbiol.* **46**, 101–16.

Muller, M.J. and Li, H. (1996) Preliminary results of trapping for *Culicoides* in South China: future bluetongue vector studies. In: T.D. St George, B.H. Kay and J. Blok (eds), *Bluetongue Disease in Southeast Asia and the Pacific*. Canberra: ACIAR, pp. 295–8.

Murphy, F.A. (1975) Cellular resistance to arbovirus infection. *Ann. N. Y. Acad. Sci.* **266**, 197–203.

Murphy, F.A., Whitfield, S.G., Sudia, W.D. and Chamberlain, R.W. (1975) Interaction of vector with vertebrate pathogenic viruses. In: K. Maramorosch and R.E. Shope (eds), *Invertebrate Immunity*. New York: Academic Press, Inc, pp. 25–53.

Nunamaker, R.A., Sieburth, P.J., Dean, V.C., Wigington, J.G., Nunamaker, C.E. and Mecham, J.O. (1990) Absence of transovarial transmission of bluetongue virus in *Culicoides variipennis*: immunogold labelling of bluetongue virus antigen in developing oocytes from *Culicoides variipennis* (coquillett). *Comp. Biochem. Physiol. A* **96**, 19–31.

O'Connell, L. (2002) Entomological aspects of the transmission of arboviral diseases by Culicoides biting midges. PhD Thesis, University of Bristol, pp. 176.

Paweska, J.T., Venter, G.J. and Mellor, P.S. (2002) Vector competence of South African *Culicoides* species for bluetongue virus serotype 1 (BTV-1) with special reference to the effect of temperature on the rate of virus replication in *C. Imicola* and *C. Bolitinos*. *Med. Vet. Entomol.* **16**, 10–21.

Puana, I. (1996) Bluetongue virus status in Papua New Guinea. In: T.D. St George, B.H. Kay and J. Blok (eds) *Bluetongue Disease in Southeast Asia and the Pacific*. Canberra: ACIAR, p. 41.

Purse, B.V., Mellor, P.S., Rogers, D.J., Samuel, A.R., Mertens, P.P.C. and Baylis, M. (2005) Climate change and the recent emergence of bluetongue in Europe. *Nat. Rev. Microbiol.* **3**, 171–81.

Qin, Q., Tai, Z., Wang, L., Luo, Z., Hu, J. and Lin, H. (1996) Bluetongue epidemiological survey and virus isolation in Xinjiang. In: T.D. St George, B.H. Kay and J. Blok (eds), *Bluetongue Disease in Southeast Asia and the Pacific*. Canberra: ACIAR, pp. 67–71.

Rees, J.A., Moniatte, M. and Bulet, P. (1997) Novel antibacterial peptides isolated from a European bumblebee, *Bombus pascuorum* (Hymenoptera, Apoidea). *Insect Biochem. Mol. Biol.* **27**, 413–22.

Riegler, L. (2002) Variation in African horse sickness virus and its effect on the vector competence of Culicoides biting midges. PhD thesis, University of Surrey, pp. 175.

Rosenstock, S.S., Ramberg, F., Collins, J.K. and Rabe, M.J. (2003) *Culicoides mohave* (Diptera: Ceratopogonidae): New occurrence records and potential role in transmission of hemorrhagic disease. *J. Med. Entomol.* **40**, 577–9.

Savini, G., Goffredo, M., Monaco, F., de Santis, P. and Meiswinkel, R. (2003) Transmission of bluetongue virus in Italy. *Vet. Rec.* **152**, 119.

Savini, G., Goffredo, M., Monaco, F., Di Gennaro, A., Cafiero, M.A., Baldi, L., de Santis, P., Meiswinkel, R. and Caporale, V. (2005) Bluetongue virus isolations from midges belonging to the *Obsoletus* complex (*Culicoides*, Diptera: Ceratopogonidae) in Italy. *Vet. Rec.* **157**, 133–9.

Sendow, I., Sukarsih, Soleha, E., Pearce, M., Bahri, S. and Daniels, P.W. (1996) Bluetongue virus research in Indonesia. In: T.D. St. George and K. Peng (eds), *Bluetongue Disease in Southeast Asia and the Pacific*. Canberra: ACIAR, pp. 28–32.

Standfast, H.A., Dyce, A.L. and Muller, M.J. (1985) Vectors of bluetongue virus in Australia. In: T.L. Barber and M.M. Jochim (eds), *Bluetongue and Related Orbiviruses*. New York, Liss, pp. 177–86.

St. George, T.D. (1996) The history of bluetongue in Australia and the Pacific Islands. In: T.D. St. George and K. Peng (eds), *Bluetongue Disease in Southeast Asia and the Pacific*. Canberra: ACIAR, pp. 33–40.

Storey, H.H. (1933) Investigations of the mechanism of transmission of plant viruses by insect vectors. *Proc. R. Soc. Lond. Ser. B* **113**, 463–85.

Sukarsih, Daniels, P.W., Sendow, I. and Soleha, E. (1933) Longitudinal studies of *Culicoides* spp. associated with livestock in Indonesia. In: M.F. Uren and B.H. Kay (eds), *Arbovirus Research in Australia, 6th edn.* Brisbane: CSIRO/QIMR, pp. 203–9.

Sukarsih, Sendow, I., Bahari, S., Pearce, M. and Daniels, P.W. (1996) *Culicoides* survey in Indonesia. In: T.D. St. George and K. Peng (eds), *Bluetongue Disease in Southeast Asia and the Pacific.* Canberra: ACIAR, pp. 123–8.

Tabachnick, W.J. (1991) Genetic control of oral susceptibility to infection of *Culicoides variipennis* for bluetongue virus. *Am. J. Trop. Med. Hyg.* **45**, 666–71.

Tabachnick, W.J. (1992) Genetic differentiation among populations of *Culicoides variipennis* (Diptera, Ceratopogonidae), the North American vector of bluetongue virus. *Ann. Entomol. Soc. Am.* **85**, 140–7.

Tabachnick, W.J. and Holbrook, F.R. (1992) The *Culicoides variipennis* complex and the distribution of the bluetongue viruses in the United States. *Proc. US Anim. Health. Assoc.* **96**, 207–12.

Tabachnick, W.J. (1996) *Culicoides variipennis* and bluetongue virus epidemiology in the United States. *Annu. Rev. Entomol.* **41**, pp. 23–43.

Takamatsu, H., Mellor, P.S., Mertens, P.P.C., Kirkham, P.A., Burroughs, J.N. and Parkhouse, R.M.E. (2003) A possible overwintering mechanism for bluetongue virus in the absence of the insect vector. *J. Gen. Virol.* **84**, 227–35.

Tan, B.H., Nason, E., Staeuber, N., Jiang, W.R., Monastryrskaya, K. and Roy, P. (2001) RGD tripeptide of bluetongue virus VP7 protein is responsible for core attachment to *Culicoides* cells. *J. Virol.* **75**, 3937–47.

Tatem, A.J., Baylis, M., Mellor, P.S., Purse, B.V., Capela, R., Pena, I. and Rogers D.J. (2003) Prediction of bluetongue vector distribution in Europe and North Africa using satellite imagery. *Vet. Microbiol.* **97**, 13–29.

Taylor, W.P. (1986) The epidemiology of bluetongue. *Rev. Sci. Tech. OIE.* **5**, 351–6.

Torina, A., Caracappa, S., Mellor, P.S., Baylis, M. and Purse, B.V. (2004) Spatial distribution of bluetongue virus and its *Culicoides* vectors in Sicily. *Med. Vet. Entomol.* **18**, 81–9.

van Dijk, A.A. and Huismans, H. (1982) The effect of temperature on the *in vitro* transcriptase reaction of bluetongue virus, epizootic hemorrhagic disease virus and African horsesickness virus. *Onderstepoort J. Vet. Res.* **49**, 227–32.

Vassalos, M. (1986) Cas de fievre catarrhale du mouton dans l'ile de Lesbos (Grece). *Bull. Off. Int. Epizoot.* **92**, 547–55.

Venter, G.J., Paweska, J.T., van Dijk, A.A., Mellor, P.S. and Tabachnick, W.J. (1998) Vector competence of *Culicoides bolitinos and C. imicola* for South African bluetongue virus serotypes 1, 3 and 4. *Med. Vet. Entomol.* **12**, 378–85.

Venter, G.J., Gerdes, G.H., Mellor, P.S. and Paweska, J.T. (2004) Transmission potential of South African *Culicoides* species for live-attenuated bluetongue virus. *Vet. Ital.* **40**, 198–202.

Venter, G.J., Paweska, J.T., Lunt, H.L., Mellor, P.S. and Carpenter, S. (2005) An alternative method of blood-feeding *Culicoides imicola* and other haematophagous *Cuclciodes* spp. for vector competence studies. *Vet. Parasitol.* **131**, 331–35.

Walker, A.R. and Davies, F.G. (1971) A preliminary survey of the epidemiology of bluetongue in Kenya. *J. Hyg. Camb.* **69**, 47–60.

Wang, L.F., Scanlon, D.B., Kattenbelt, J.A., Mecham, J.O. and Eaton, B.T. (1994) Fine mapping of a surface-accessible, immunodominant site on the bluetongue virus major core protein VP7. *Virology* **204**, 811–4.

Wellby, M.P., Baylis, M., Rawlings, P. and Mellor, P.S. (1996) Effect of temperature survival and rate of virogenesis of African horse sickness virus in *Culicoides variipennis sonorensis* (Diptera: Ceratopogonidae) and its significance in relation to the epidemiology of the disease. *Bull. Entomol. Res.* **86**, 715–20.

White, D.M., Wilson, W.C., Blair, C.D. and Beaty, B.J. (2005) Studies on overwintering of bluetongue viruses in insects. *J. Gen. Virol.* **86**, 453–62.

Wirth, W.W. and Hubert, A.A. (1989) The *Culicoides* of Southeast Asia (Diptera: Ceratopogonidae). **44**, *Mem. Am. Entomol. Inst.*, Gainesville, FL. Am. Entomol. Inst. p. 508.

Wirth, W.W. and Jones, R.H. (1957) The North American subspecies of *Culicoides variipennis* (Diptera: Heleidae). *U.S. Dept. Agr. Tech. Bull.* **1170**, 1–35.

Wittmann, E.J. (2000) Temperature and the transmission of arboviruses by Culicoides biting midges. PhD Thesis, University of Bristol, p. 159.

Wittmann, E.J., Mellor, P.S. and Baylis, M. (2002) Effect of temperature on the transmission of orbiviruses by the biting midge, *Culicoides sonorensis. Med. Vet. Entomol.* **16**, 147–56.

Xu, G., Wilson, W., Mecham, J., Murphy, K., Zhou, E.M. and Tabachnick, W. (1997) VP7: an attachment protein of bluetongue virus for cellular receptors in *Culicoides variipennis. J. Gen. Virol.* **78**, 1617–23.

Zhang, N., Li, Z., Zhang, K., Hu, Y. and Li, G. (1996) Bluetongue history, serology and virus isolations in China. In: T.D. St George, B.H. Kay and J. Blok (eds), *Bluetongue Disease in Southeast Asia and the Pacific.* Canberra: ACIAR, pp. 43–50.

Zhou, W. (1996) An epidemiological study of bluetongue in Anhui Province, China. In: T.D. St George, B.H. Kay and J. Blok (eds), *Bluetongue Disease in Southeast Asia and the Pacific.* Canberra: ACIAR, pp. 61–4.

Rates of bluetongue virus transmission between *Culicoides sonorensis* and sheep

15

MATTHEW BAYLIS, LELIA O'CONNELL AND PHILIP S. MELLOR

Institute for Animal Health, Pirbright Laboratory, Pirbright, Surrey, UK

Introduction

Bluetongue (BT) is an insect-borne disease of ruminants caused by BT virus (BTV). Infection is not contagious between its mammalian hosts and is not transmitted transovarially (from mother to offspring) by its *Culicoides* vectors. Accordingly, viral transmission is restricted to events of *Culicoides* feeding on hosts, with virus passing either from an infected vector to a susceptible host (hereafter, V → H transmission), or from an infected host to a susceptible vector (H → V). Feeding events between hosts/vectors that are infected/susceptible (whichever way around) do not, however, invariably lead to transmission. The proportion of such events in which transmission occurs successfully is described as the *transmission rate*. The transmission rates (V → H, H → V) of any vector-borne pathogen are fundamentally important to the epidemiological dynamics of the disease.

One of the most significant concepts in epidemiology is that of R_0, the basic case reproduction ratio (Anderson and May, 1991). This is a measure of the

ISBN-13: 978-0-12-369368-6

number of new infections that arise, on average, from the introduction of one infected case into a susceptible population. As such, it provides an indication of the possible scale of an outbreak and the effort required to control it. The R_0 of a 'simple' vector-borne disease, such as malaria, with just single host and vector species, is of the form:

$$R_0 = m\, a^2\, b\, \beta\, p / \mu\, (r + c) \tag{1}$$

where m is the ratio of vector numbers to host numbers; a is the vector-biting rate (the daily probability of a vector feeding on a susceptible host); b and β are the $V \rightarrow H$ and $H \rightarrow V$ transmission rates; p is the proportion of vectors that survive the latent period and become infectious; μ is the vector daily mortality rate ($1/\mu$ is the longevity in days) and $r + c$ is the rate of loss of infected hosts through recovery and mortality (Lord *et al.*, 1996a). Equation (1) shows that R_0 and, therefore, estimates of outbreak risk, scale and necessary control effort, are related linearly to both the $V \rightarrow H$ and $H \rightarrow V$ transmission rates.

Models have been developed for the R_0 of the *Culicoides*-borne disease, African horse sickness (AHS) (Lord *et al.*, 1996a), and are currently in development for BT (David Munyinyi, Personal Communication). These models must allow for a minimum of two host types (horses and donkeys for AHS, cattle and sheep for BT) and, quite possibly, should in some circumstances allow for multiple vectors [e.g. BTV transmission in Italy; Purse *et al.* (2004); Torina *et al.* (2004)]. Accordingly, the models are significantly more complex than equation (1), despite retaining its general form. In the models for AHS, R_0 is once again linearly related to the $V \rightarrow H$ and $H \rightarrow V$ transmission rates. It is important to note, however, that current models allow the two host types to differ in certain ways, such as the duration of infectiousness and recovery rates, but assume that the $V \rightarrow H$ and $H \rightarrow V$ transmission rates are equal (e.g. donkeys and horses are infected with equal probability by the bite of an infected midge; midges are infected with equal probability after feeding on an infected horse or donkey). The truth of these assumptions is not clear. If the transmission rates differ for different host types, or if they differ for different vector species (when multiple vectors occur), then current models are likely to be over-simplistic. Finally, it is worth noting that in many situations there may be more than two host types (e.g. mules as additional hosts of AHS; goats or other ruminants, or sheep breeds of differential susceptibility to BT).

The central importance of transmission rates to the epidemiology of *Culicoides*-borne disease is underlined by simulation modelling, of AHS epidemic likelihood, following viral introduction into a susceptible host population. In the work of Lord *et al.* (1996a, b) the $V \rightarrow H$ and $H \rightarrow V$ transmission rates were both among the six or seven most important independent variables (out of 15) in models of the size of an epidemic (peak prevalence in horses) and the timing of the peak.

Accordingly, there are sound epidemiological reasons for empirical measurement of the $V \rightarrow H$ and $H \rightarrow V$ transmission rates for a *Culicoides*-borne disease such as BT. First, they are central to the estimation of R_0 and are important determinants of the likely scale and timing of an epidemic. Second, data are required on the transmission rates on different hosts, and with different vectors, to help evaluate the validity of current models. Therefore, in this paper we present an attempt to quantify the $V \rightarrow H$ and $H \rightarrow V$ transmission rates between a laboratory colony of *Culicoides sonorensis* and sheep of the Poll Dorset breed.

Materials and methods

Two experiments were undertaken: (I) a study of the transmission of BTV from infected *C. sonorensis* to susceptible sheep, and (II) the transmission of BTV from infected sheep to susceptible *C. sonorensis*. Each of the two experiments was carried out separately using a total of twenty (18 for Experiment I and 2 for Experiment II) 6-month-old Poll Dorset sheep. The sheep were housed in two separate boxes in an insect-free isolation unit, in compliance with the terms and conditions of the project license issued by the UK Home Office [Home Office Animals (Scientific Procedures) Act 1986]. The *C. sonorensis* were from the PIRB-s-3 strain (Wellby *et al.*, 1996) reared in the Pirbright colony (Boorman, 1974). This PIRB-s-3 strain was first selected for oral susceptibility to AHSV-9 (Wellby *et al.*, 1996); however, they have been used successfully for vector competency studies for both BTV and AHSV (Fu *et al.*, 1999; Wittmann, 2000). The virus used was an attenuated strain of BTV serotype 1, designated S_4BHK_2 (= four passages in sheep and two in baby hamster kidney cells) at a titre of 7.3 $\log_{10}/TCID_{50}/ml$.

Experiment I: vector to host ($V \rightarrow H$) transmission

Midge inoculation
Two hundred, 2- or 3-day-old, female *C. sonorensis* were inoculated with the BTV-1-attenuated virus using the parenteral (intrathoracic) technique, described by Mellor *et al.* (1974) and equipment described by Boorman (1975). This procedure ensures that all midges develop fully disseminated infections and are able to transmit the virus within a few days (Mellor, 1990). Inoculated females were maintained in waxed cardboard pillboxes, with fine mesh tops at one end and clear plastic disks at the other, at $25 \pm 1°C$ for 4 days, and without feeding, until the start of the experiment. A pad of cotton wool soaked in 10% sucrose solution medicated with 100 μg/ml of penicillin/streptomycin was placed on the mesh and replaced daily. This provided a source of energy, while the presence of antibiotics in the sucrose solution reduces bacterial contamination and has been shown to prolong survival of midges.

Experimental procedure

Female *C. sonorensis* were divided into three groups consisting of six pillboxes each: Group A with ~20 females per box, Group B with ~5 females per box and Group C with just one female per box. The pillboxes were then randomly assigned to one of 18 sheep (uh5–uh22; Table 15.1) on which the *C. sonorensis* were allowed to feed for a 20-min period. Feeding was conducted in the groin area of the sheep as, there being no wool, the midges had easy access to the epidermis. Blood-fed midges were then titrated individually within 1 h of feeding to confirm engorgement and to measure the virus titres of each one. Feeding rates on each sheep are summarized in Table 15.1.

Sheep

The sheep were monitored for clinical signs of BT (e.g. fever, anorexia, oedema, coronitis, lameness) for 21 days. Blood samples were taken daily in the form of 5 ml heparinized blood (for virus detection) and 5 ml non-heparinized blood (for serum collection and antibody detection). ELISAs for BTV antibody were

Table 15.1   V → H transmission: number of BTV-inoculated *C. sonorensis* exposed to 18 uninfected sheep

Group	Sheep	Number per pillbox	Blood-fed
A	uh5	23	2
	uh6	22	16
	uh7	19	15
	uh8	21	21
	uh9	20	6
	Uh10	19	15
	Mean	20.6	12.5
B	Uh11	6	6
	Uh12	6	5
	Uh13	5	2
	Uh15	5	5
	Uh17	6	6
	Uh18	6	6
	Mean	5.6	5.0
C	Uh14	1	1
	Uh16	1	1
	Uh19	1	1
	Uh20	1	1
	Uh21	1	1
	Uh22	1	1
	Mean	1	1

done on serum samples for Day 0, to confirm that each sheep was BTV serologically negative at the start of the experiment, and Day 12 post infection (p.i.) to confirm BTV infection.

Sheep uh5, uh6, uh9 (from Group A), uh11, uh15, uh17 (from Group B), uh14 and uh19 (from Group C) were euthanased on Day 16 p.i. because of the severity of their condition.

Data analysis

The relationship between BTV-infected *C. sonorensis* and the responses of the sheep on which they fed were investigated using linear regression analyses of four midge-related independent variables (no. 1–4) on nine sheep-related dependent variables (no. 5–13):

1. M = Number of midges exposed to each sheep (e.g. 23 midges for sheep uh5).
2. M_B = Number of midges that fed per sheep (e.g. only 2 midges blood-fed out of 23 exposed to uh5).
3. t_{sum} = Sum of BTV midge titre per sheep (e.g. the 2 midges that fed on sheep uh5 had a total BTV titre of 5.32 $\log_{10}$/TCID$_{50}$/midge).
4. t_{max} = Maximum midge titre per sheep (e.g. the highest titre of the 2 midges that fed on sheep uh5 was 5.25 $\log_{10}$/TCID$_{50}$/midge).
5. DT_{peak} = Days for sheep to reach peak temperature (>40°C)
6. T_{peak} = Days of sheep at peak temperature (>40°C)
7. T_{max} = Maximum body temperature of sheep
8. DT_{max} = Days for sheep to reach maximum body temperature
9. DV_{peak} = Days for sheep to reach peak BTV viraemia (>4 $\log_{10}$/TCID$_{50}$/ml)
10. V_{peak} = Days of sheep at peak viraemia (>4 $\log_{10}$/TCID$_{50}$/ml)
11. V_{max} = Maximum titre of BTV viraemia of sheep
12. DV_{max} = Days for sheep to reach maximum viraemia
13. A = Percentage inhibition matrix recorded from antibody detection ELISAs conducted on sera collected from sheep on day 12 p.i.

Experiment II: host to vector (H → V) transmission

Experimental procedure

Two sheep (uh23 and uh24) were inoculated with a total of 1.5 ml (1.1 ml subcutaneous; 0.2 ml intradermal at two sites) of a 1 in 10 dilution (6.3 $\log_{10}$/ TCID$_{50}$/ml) of the attenuated strain of BTV-1 described previously. The sheep were monitored daily for 19 days for clinical signs of BT. Blood samples were taken at 3, 5, 7, 8, 9, 12, 14, 16 and 19 days p.i. in the form of 5 ml heparinized blood (for virus detection) and 5 ml non-heparinized blood (for sera collection and antibody detection). BTV antibody detection ELISAs were undertaken on sera from day 0, to confirm that each sheep was

BTV serologically negative at the start of the experiment, to day 9 p.i. to confirm BTV infection. Sheep uh23 was euthanased on day 14 p.i. because of the severity of its condition.

Cohorts of *C. sonorensis* were produced every 2 days to provide ~2200 midges for every experimental day. Two pillboxes, containing ~550 *C. sonorensis* (males and females) each, were placed on the groin area of each of the two sheep (~1100 midges per sheep) and left in place for 20 min. This was undertaken daily during the period of peak viraemia in sheep (days 5–9 p.i.), and every 2 days thereafter. The experiment was terminated after 19 days.

After each feeding event, 20 engorged females from each sheep were placed into 1.5 ml Eppendorf tubes and stored at −70°C until required for virus assay. The remaining blood-engorged females were transferred into new, clean pillboxes and maintained as described earlier, and without further feeding, for 7 days. This is the extrinsic incubation period (EIP) of BTV in midges kept at the holding temperature. After 7 days, surviving females were also placed into 1.5 ml Eppendorf tubes and kept at −70°C until being assayed for virus.

Midge virus titration assay

Female midges from both experiments were assayed for virus as follows. Females were individually homogenized in 1 ml of Glasgow Minimum Essential Medium (MEM; Invitrogen Life Technologies Ltd.) containing 200 µg/ml of penicillin/streptomycin and 2.5 µg/ml of fungizone, using motor-driven 1.5 ml polypropylene pestles (Anachem Ltd.). Two hundred microlitres of the midge suspensions were then diluted 1 in 10 in the MEM. Seven additional ten-fold dilutions for each midge solution were prepared (i.e. down to dilutions of 10^{-7}).

Virus titrations were done in 96-well microtitre plates. One hundred microlitres of each midge solution and associated dilutions was inoculated onto each of four microtitre plate wells, so as to obtain four replicates per dilution. Each well contained a monolayer of BHK-21 cells and 100 µl of MEM supplemented with 2% foetal calf serum, 100 µg/ml of penicillin/streptomycin and 2.5 µg/ml of fungizone. The inoculated microtitre plates were then incubated at 37°C in a 5% carbon dioxide incubator and observed microscopically for cytopathic effects (CPE) for 5 days. CPE observations were used as a positive indicator of the presence of virus. Viral concentrations, calculated by the method adapted from Spearman (1908) and Kärber (1931), were expressed as tissue culture infectious doses ($TCID_{50}$)/midge. The assay could detect viral titres between 0.75 and 8.5 $\log_{10}/TCID_{50}$/midge. Samples with viral titres <0.75 $\log_{10}/TCID_{50}$/midge were classed as negative.

Blood virus titration assays

Each 5 ml heparinized blood sample from sheep was washed three times in PBS (Dulbecco's Phosphate Buffer Solution; CSU-IAH Pirbright) to remove any BTV antibody and was then sonicated with a Soniprep-150 (MSE) at 18 amplitude microns for 30 s. This was done to lyse the red blood cells and free the virus as most BTV is cell associated. To assay the blood samples, 200 µl of each one was diluted in 1 ml MEM. Two hundred microlitre of the blood dilutions was then diluted 1 in 10 in the MEM. Six additional ten-fold dilutions for each blood sample were prepared (i.e. down to dilutions of 10^{-6}).

Blood virus titrations were undertaken as per midge virus titrations. However, as CPE would not be visible in undiluted blood, the assay was only able to detect viral titres between 1.5 and 8.5 $\log_{10}/TCID_{50}/ml$. Samples with viral titres <1.5 $\log_{10}/TCID_{50}/ml$ were classed as negative.

Antibody detection ELISA

Serum decanted from the 5 ml clotted non-heparinized blood was stored at $-20°C$ until being tested by ELISA using standard methods (Anderson, 1984).

Results

Experiment I: vector to host (V → H) transmission

BTV titres of the intrathoracically infected midges, shortly after feeding on the 18 sheep, are shown in Figure 15.1. All engorged midges were positive for BTV, with titres ranging from 1.7 to 6.5 $\log_{10}/TCID_{50}/midge$.

All sheep developed a fever (>40°C) (Figures 15.2a–c), usually around day 6 p.i., but in some cases as early as day 4 p.i., apart from sheep uh22 (from Group C) whose body temperature remained at below 40°C throughout. Peak temperatures generally occurred during days 6–10 p.i. Group C's mean temperature peaked approximately 1 day before those of Groups A and B (day 8 vs day 9 p.i.) (Figure 15.2d).

Aside from temperature, the first clinical sign of BT to appear was hyperaemia of the oral mucus membranes, and conjunctivitis. By day 12 p.i., most sheep were showing signs of anorexia, coronitis and lameness. On day 16 p.i., the sheep with the most severe clinical signs were euthanased (uh5, uh6, uh9 from Group A; uh11, uh15, uh17 from Group B; uh14 and uh19 from Group C). Sheep uh22 showed no clinical signs of BT.

The BTV viraemia titrations for each sheep are shown in Figures 15.3a–c. As expected all eighteen sheep had no identifiable virus titres on day 0. Nearly half of the sheep (8/18) had detectable viraemia by day 3 p.i., and more than 80% (15/18) had detectable viraemia by day 5 p.i., but in some sheep virus was

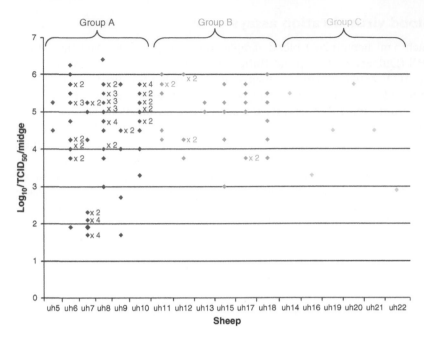

Figure 15.1 V → H transmission: BTV titres of each blood-fed *C. sonorensis*. Overlapping titres are highlighted. Groups correspond to those summarized in Table 15.1.

not detected until later (notably uh7 in Group A; day 7). Peak viraemia (defined here as >4.0 $\log_{10}/TCID_{50}/ml$) tended to occur between days 6 and 10 p.i.. The highest viraemia of any sheep was 6.5 $\log_{10}/TCID_{50}/ml$, achieved by sheep uh8 (Group A, reached on day 7 p.i.) and sheep uh15 (Group B, reached on days 7 and 12 p.i.). Following the period of peak viraemia, virus levels tended to decline. By the end of the experiment (day 21 p.i.), some sheep had become aviraemic while, in others, viraemia was still ongoing. No virus was detected at all during the course of this experiment in sheep uh22.

Considering averages per group (Figure 15.3d), the mean peak viraemia for Group A lasted ~6 days (days 6–11 p.i. inclusive), for Group B ~11 days (day 6 to day 16 p.i. inclusive) and for Group C ~7 days (days 7–13 p.i. inclusive). The maximum viraemia (averaged by group) was higher for Groups A and B (~5.2 $\log_{10}/TCID_{50}/ml$) than for Group C (~4.2 $\log_{10}/TCID_{50}/ml$).

Antibody detection ELISA undertaken on sera collected on day 12 p.i. confirmed seroconversion (i.e. BTV infection) of all 18 sheep. Seventeen sheep had positive inhibition matrix scores of $>90\%$; sheep uh22, however, had a score of only 54%. Thus, sheep uh22 appeared at first not to have become infected with BTV as no significant rise in temperature ($>40°C$) was recorded, no clinical signs were detected and no virus was isolated from the

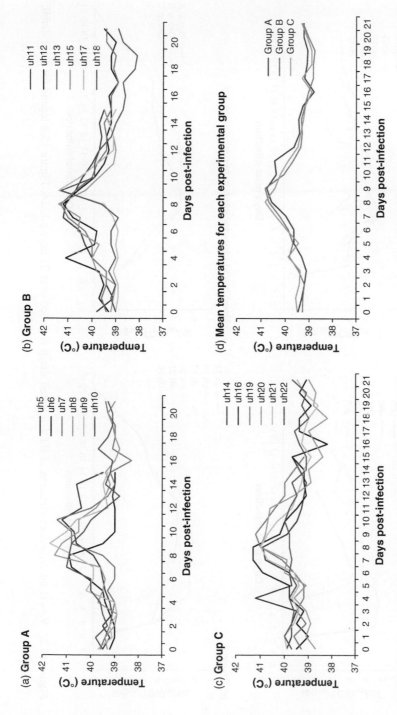

Figure 15.2 V → H transmission rectal temperatures of BTV-infected sheep recorded over 21 days p.i. (a) sheep from group A, (b) sheep from group B, (c) sheep from group C and (d) mean temperatures for each experimental group. Sheep uh5, uh6, uh9, uh11, uh15, uh17, uh14 and uh19 were euthanased on day 16 p.i.

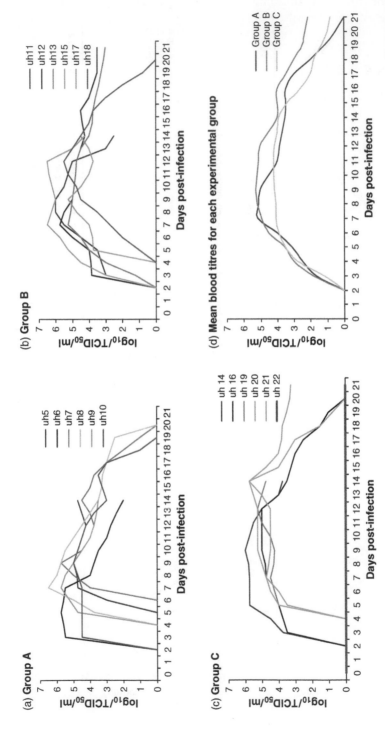

Figure 15.3 V → H transmission: viraemia of BTV-infected sheep recorded over 21 days p.i. See Figure 15.2 legend for more detail.

blood. However, the weak-positive response in the antibody detection ELISA showed, nevertheless, that a low level of infection had occurred.

Linear regression analysis

The results for one animal (uh22) tended to be different from all others; it lacked clinical signs, fever and viraemia, and its antibody response was low. We have excluded this animal from all analyses of temperature, as it was impossible to assign values to DT_{peak}, T_{peak}, DT_{max} and T_{max}. This animal was also excluded from one set of analyses of viraemia (DV_{peak}, V_{peak}, DV_{max} and V_{max}). However, the absence of detectable viraemia in this animal is, nevertheless, useful information that should not be entirely lost from analyses (e.g. the information that its viraemia was low enough to be undetected). Therefore, the viraemia analyses were repeated, with data for uh22 winsorized (Sokal and Rohlf, 1981). In winsorization, the outliers in an ordered array are replaced by the next closest values so that they still contribute to the sample size but are less extreme and have less leverage. The antibody response of sheep uh22 is valid, but low. Data for A are therefore analysed both including and excluding data for this sheep.

The results from linear regression analysis undertaken on combinations of midge (x) and sheep (y) variables are summarized in Table 15.2. There were no associations between any midge variable and sheep temperature, or the number of days to, or at, peak viraemia (DV_{peak}, V_{peak}). However, there were associations with the days to maximum viraemia, the level of this viraemia and the antibody response. Sheep took significantly less time to reach maximum viraemia (DV_{max}) when exposed to (M) or fed on (M_B) by a larger number of midges (Figure 15.4a). Sheep also took less time to reach maximum viraemia (DV_{max}) when exposed to a greater titre of virus in feeding midges, as measured by the total titre of all fed midges (t_{sum}) and the highest titre of any fed midge (t_{max}) (Figure 15.4b). For DV_{max}, all associations were significant with data for uh22 winsorized (Table 15.2); with data for uh22 excluded, a negative association between t_{sum} and DV_{max} approached but did not reach significance ($F_{1,15} = 3.1$, $P < 0.1$).

There was no association between midge numbers (M or M_B) and the maximum level of the viraemia (V_{max}) (Table 15.2). However, sheep achieved a higher maximum viraemia when exposed to a greater titre of virus in feeding midges, as measured by t_{sum} or t_{max} (Figure 15.5). Both associations were significant with data for sheep uh22 winsorized, and for t_{max} when data for uh22 were excluded (Table 15.2); for t_{sum} however, with uh22 excluded, the association with V_{max} approached but did not reach significance ($F_{1,15} = 3.2$, $P < 0.1$).

A sheep's antibody response at 12 days appeared to be stronger when fed on (M_B) by a larger number of midges (Figure 15.6a) or when exposed to a greater titre of virus in feeding midges, as measured by the total titre of all fed midges (t_{sum}) and the highest titre of any fed midge (t_{max}) (Figure 15.6b). The

Table 15.2 V → H transmission: F-values[a] of linear regression analyses undertaken on combinations of midge (x) and sheep (y) variables

Midge-Variables[b]	Sheep variables[c] (sheep uh22 excluded[d])										Sheep variables[c] (uh22 winsorized or included[d])			
	DT_{peak}	T_{peak}	DT_{max}	T_{max}	DV_{peak}	V_{peak}	DV_{max}	V_{max}	A	DV_{peak}	V_{peak}	DV_{max}	V_{max}	A*
M	1.2	0.0	2.9	1.0	0.0	2.9	**6.1**	0.0	3.3	0.2	3.6	**7.8**	0.1	1.9
M_B	0.1	0.2	0.0	0.4	0.2	3.3	**4.5**	1.6	**5.2**	0.4	4.0	**5.7**	2.3	1.8
t_{sum}	0.1	0.4	0.0	0.0	0.0	0.9	3.1	3.2	10.0	0.3	1.7	**7.1**	**5.8**	15.4
t_{max}	0.0	0.8	0.0	0.1	0.2	0.8	2.4	**5.6**	**6.3**	0.1	1.7	**6.1**	**8.3**	16.1

[a] **Bold**, $F \geq 8.40$, $P < 0.01$; Italic, $F \geq 8.40$, $P < 0.01$; Underline, $F \geq 15.8$, $P < 0.001$

[b] M, number of midges exposed on each sheep; M_B, number of blood-fed midges per sheep; t_{sum}, sum of BTV midge titre per sheep; t_{max}, maximum midge titre per sheep

[c] DT_{peak}, days to reach peak temperature (>40°C); T_{peak}, days of peak temperature (>40°C); T_{max}, maximum body temperature; DT_{max}, days to reach maximum body temperature; DV_{peak}, days to reach peak BTV viraemia (>4 $\log_{10}$TCID$_{50}$/ml); V_{peak}, days of peak viraemia (>4 $\log_{10}$TCID$_{50}$/ml); DV_{max}, days to reach maximum viraemia; V_{max}, maximum BTV viraemia; A, percentage inhibition matrix recorded from antibody detection ELISA on sera collected on day 12 p.i.

[d] As sheep uh22 gave unusual results, analyses were undertaken both excluding and winsorizing its data or, for A, included the data.

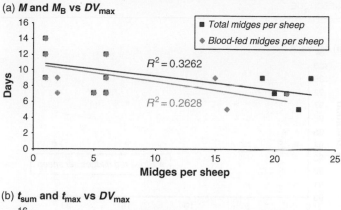

(a) *M* and M_B vs DV_{max}

■ Total midges per sheep
♦ Blood-fed midges per sheep

$R^2 = 0.3262$

$R^2 = 0.2628$

Days

Midges per sheep

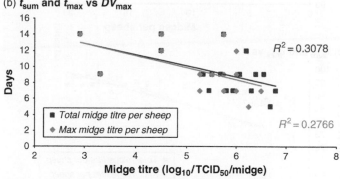

(b) t_{sum} and t_{max} vs DV_{max}

$R^2 = 0.3078$

■ Total midge titre per sheep
♦ Max midge titre per sheep

$R^2 = 0.2766$

Days

Midge titre ($\log_{10}/TCID_{50}$/midge)

Figure 15.4 V → H transmission: effects of feeding BTV-infected *C. sonorensis* on the time period until sheep reached maximum viraemia (DV_{max}). (a) The number of BTV-infected midges exposed to a sheep (*M*) and the number that blood-fed (M_B). (b) The sum of BTV titres of midges that fed (t_{sum}) and the maximum titre of any one midge that fed (t_{max}). The DV_{max} data point for sheep uh22 is winsorized (set to 14 days), and R^2 values correspond accordingly.

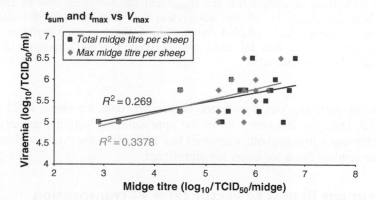

t_{sum} and t_{max} vs V_{max}

■ Total midge titre per sheep
♦ Max midge titre per sheep

$R^2 = 0.269$

$R^2 = 0.3378$

Viraemia ($\log_{10}/TCID_{50}$/ml)

Midge titre ($\log_{10}/TCID_{50}$/midge)

Figure 15.5 V → H transmission: effects of feeding BTV-infected *C. sonorensis* on the maximum viraemia developed by sheep (V_{max}). Association is shown for the sum of BTV titres of midges that fed (t_{sum}), and the maximum titre of any one fed midge (t_{max}). The V_{max} data point for sheep uh22 is winsorized (set to 5 $\log_{10}/TCID_{50}$/ml), and R^2 values correspond accordingly.

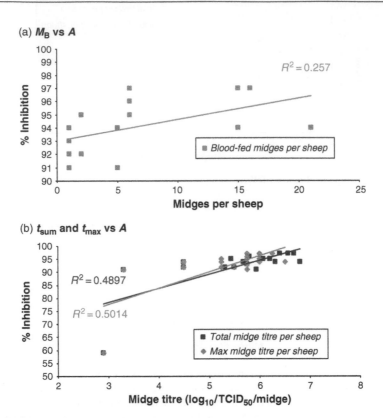

(a) M_B vs A

(b) t_{sum} and t_{max} vs A

Figure 15.6   V→H transmission: effects of feeding BTV-infected *C. sonorensis* on the level of antibody response of sheep after 12 days (A). (a) Association with the number of BTV-infected midges that blood-fed (M_B). (b) Association with the sum of BTV titres of midges that fed (t_{sum}) and the maximum titre of any one midge that fed (t_{max}). In (a), the association is significant with the outlying datapoint for sheep uh22 excluded (inhibition = 54%), but not significant when it is included. In (b), data for uh22 are included, but the association is also significant if uh22 is excluded.

association with M_B was significant only when data for sheep uh22 were excluded. Data for this sheep lie in the appropriate direction (i.e. only one fed midge and a low antibody response) but are so extreme that the additional variance renders the association not significant.

Experiment II: host to vector (H→V) transmission

The two sheep developed fever (>40°C) on day 6 p.i. (Figure 15.7), which peaked at 41°C on day 6 p.i. (uh24) and day 7 p.i. (uh23). After the peak, temperature declined at a faster rate for uh24 than for uh23. The first signs of

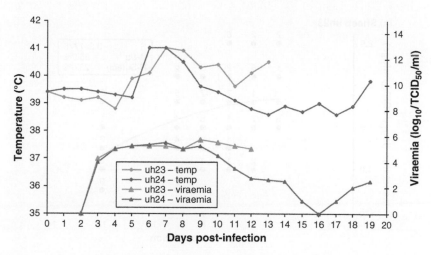

Figure 15.7 H → V transmission: daily temperature and viraemia profile over 19 days for the two BTV-infected sheep. Sheep uh23 was euthanased on day 14 p.i.

BT were hyperaemia of the oral mucus membranes and conjunctivitis. Both sheep showed signs of anorexia by day 8 p.i. and coronitis and lameness by day 12 p.i. The clinical signs of uh23 were sufficiently severe by day 14 p.i. that it was euthanased.

BT virus was first detected on day 3 p.i. (Figure 15.7). Titres for both sheep proceeded to rise to more than 5 $\log_{10}$/TCID$_{50}$/ml. Viraemia of uh23 remained high until its death; that of uh24 peaked between days 4–7 and began to decline after day 9 p.i. On day 16, no BTV was detected in the blood of sheep uh24; however, the BTV titre started to rise again from day 17 p.i. until the end of the experiment (day 19 p.i.) when blood titre was measured at 2.5 $\log_{10}$/TCID$_{50}$/ml. Antibody detection ELISA undertaken on sera collected on day 9 p.i. confirmed BTV infection of both sheep, with inhibition scores of 97% (uh23) and 88% (uh24).

BTV was successfully isolated from batches of *C. sonorensis* that were killed immediately after feeding on the infected sheep between days 5–9 p.i. (uh24) and days 5–12 p.i. (uh23) (Figure 15.8). During the period of peak viraemia (days 5–9 p.i.), BTV was successfully isolated from 94% (uh23) and 75% (uh24) of midges; titres were, however, quite low and variable (0.75–2.5 $\log_{10}$/TCID$_{50}$/ midge). The titres of virus ingested by midges appeared to decline with time. Linear regressions fitted to both data sets are significant (uh23, $r^2 = 23.5\%$, $F_{1,110} = 33.9$, $P < 0.001$; uh24, $r^2 = 15.9\%$, $F_{1,72} = 13.6$, $P < 0.001$). However, a quadratic relationship fitted the data better, significantly so for uh24 (uh23, $r^2 = 25.2\%$, $F_{2,109} = 18.3$, $P < 0.001$; uh24, $r^2 = 35.6\%$, $F_{2,71} = 19.7$, $P < 0.001$), suggesting that the titres ingested by midges increased to a peak at day ~7 p.i. and declined thereafter. It is noteworthy that the decline over time in the BTV

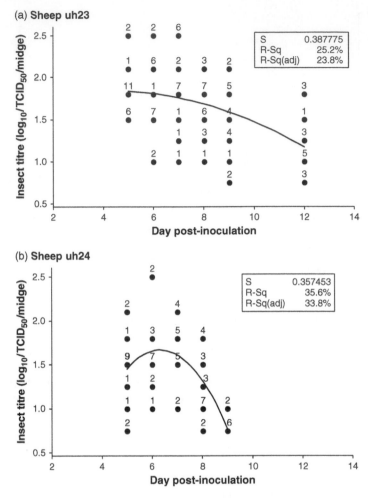

Figure 15.8 H → V transmission: individual BTV titres for *C. sonorensis* on the day of feeding on infected sheep (a) uh23 and (b) uh24. Labels above points show the frequency of identical data points. A quadratic regression line is plotted and the R^2 shown. Insects were fed on days 5, 6, 7, 8, 9, 12, 14 (uh23) or 5, 6, 7, 8, 9, 12, 14 and 19 (uh24) p.i. Sheep uh23 was euthanased on day 14 p.i.

titre ingested by *C. sonorensis* fed on sheep uh23 was not accompanied by a detectable decline in the BTV titre in uh23 itself (Figure 15.7); equally, the BTV titre in the *C. sonorensis* that fed on uh24 declined substantially between days 7 and 9 p.i., whereas the BTV titre in uh24 itself declined after day 9 p.i. The failure to detect BTV in midges that fed on uh24 after day 9 p.i. cannot, therefore, be attributed solely to the declining viraemia of the host.

Table 15.3 H → V transmission: oral transmission of BT infection from two infected sheep to uninfected *C. sonorensis*

Days post-inoculation[a]	Sheep uh23			Sheep uh24		
	Number analysed	BTV positive[b]	Titre range	Number analysed	BTV positive[b]	Titre range
5	216	7	0.75–4.75	133	2	0.75–4.5
6	444	5	0.75–4.5	245	3	1.5–4.75
7	236	1	1.75	218	0	–
8	304	0	–	381	0	–
9	310	0	–	255	2	0.75–4.75
12	404	3	1–2.25	257	0	–
14[c]	–	–	–	185	0	–
16	–	–	–	176	0	–
19[d]	–	–	–	165	0	–
Total	1914	16	0.75–4.75	2015	7	0.75–4.75

[a] Day post-inoculation of sheep on which midges were fed.
[b] Midges titrated 7 days after feeding.
[c] uh23 was euthanased on day 14 p.i.
[d] End of experiment.

A total of 3929 *C. sonorensis* females, which successfully blood-fed on sheep uh23 and uh24 (excluding the ones blood-fed and titrated on the day of feeding) and survived for 7 days, were titrated individually for BTV. Twenty-three (0.6%) were positive for BTV: 16 (0.8%) for sheep uh23 and 7 (0.3%) for sheep uh24 (Table 15.3). For both sheep, individual midge titres at 7 days p.i. varied between 0.75 and 4.75 $\log_{10}$/TCID$_{50}$/midge. There were no trends in titre over time (uh23, $r^2 = 3.1\%$, n.s.; uh24, $r^2 = 0\%$, n.s.).

Discussion

A motivation for this work was to investigate whether there is a threshold number of bites by BTV-infected *Culicoides* below which a host sheep does not develop infection. No threshold was found; instead, our results show that the bite of a single infected *C. sonorensis* is sufficient to infect a sheep. Five of the six sheep fed on by just a single infected *C. sonorensis* developed full infections; the sixth had no clinical signs and viraemia was not detected but did produce a weak antibody response. Accordingly, the V → H transmission rate for the *C. sonorensis*–BTV–sheep system appears to be extremely high, at 80–100% per midge infected with a BTV titre ≥ 2.9 $\log_{10}$/TCID$_{50}$/midge.

Foster *et al.* (1968) also report that the bite of a single infected female *C. sonorensis* is sufficient to transmit BTV to sheep. In that work, pools of 12 or more midges were allowed to feed on sheep and, in three cases, the sheep developed infection when just a single midge fed. The authors infer that a single bite is sufficient to transmit infection but cannot exclude the possibility that some midges transmitted virus while probing the sheep but did not proceed to engorge. Our results extend the work of Foster *et al.* (1968) by proving that a single bite is sufficient. This observation lends support to the concept of wind-dispersed epizootics of *Culicoides*-borne disease (Sellers, 1992; Mellor *et al.*, 2000); only a single infected midge needs to survive carriage on winds for the possible spread of an outbreak to a new country or region.

Some differences were discernable between the sheep fed on by one, ~5 or 20 infected midges. Sheep fed on by fewer infected *C. sonorensis* tended to take longer to reach maximum viraemia, of lower titre, than counterparts fed on by larger numbers of infected *C. sonorensis*. One might expect the titre of virus in engorged midges to be a better explanatory variable than, simply, the number that fed or that were in the pot placed on the sheep (many of which did not engorge). Our data provide some support: midge virus titres were better than midge numbers as predictors of the maximal level of viraemia and the antibody response. In contrast, however, midge numbers were at least as good, if not better, than midge virus titres as predictors of the time to maximum viraemia in sheep.

The results for sheep uh22 complicated statistical analysis, but are of interest in themselves. This animal showed no clinical signs of BT and no viraemia was detected, but there was a BTV antibody response, albeit a low one, confirming that infection occurred. Sheep uh22 was in Group C, where one midge fed on each sheep, and the midge that fed on uh22 had the lowest BTV titre (2.9 $\log_{10}$/TCID$_{50}$/midge) of its group. In previous, laboratory-based studies, flies infected with <2.5 $\log_{10}$/TCID$_{50}$/midge have failed to transmit BTV into clean blood through a chick-skin membrane [unpublished results described by Jennings and Mellor (1987)]. Accordingly, there may be a threshold of epidemiological significance: a bite by a single *C. sonorensis* of BTV titre ≤ 2.9 $\log_{10}$/TCID$_{50}$/midge may be insufficient to lead to demonstrable viraemia in a sheep.

A key finding is that there are significant associations between the highest titre of BTV in any one engorged midge and the subsequent pattern of viraemia in the host. A logical deduction is that midges of higher BTV titre must inoculate more virus upon feeding than midges of lower BTV titre. This deduction, if valid, would be novel, as most previous workers have assumed that any midge with a fully disseminated infection is equally likely to infect a host upon feeding (Philip Mellor, Personal Communication). However, this deduction could be premature as there is confounding with the number of midges that feed. Clearly, more virus is likely to be inoculated if more midges feed, but also, the maximum virus titre of any one midge is likely to be greater as more midges are tested. Therefore, it remains possible that the significant association of t_{max} with host viraemia is a function of other variables, such as the total virus inoculated (t_{sum}) or the number of midges that feed (M_B). It is worth noting that in at least one analysis (V_{max}), however, t_{max} was a significant predictor while these other variables were not.

Our second experiment shows that the H → V transmission rate, from BTV-infected sheep to *C. sonorensis*, is <1%, even though at least 70% of midges acquired detectable levels of virus on the day of feeding. This is consistent with the concept of barriers to infection in midges, such as the midgut infection barrier, midgut escape barrier and dissemination barrier, which prevent many *Culicoides* exposed to virus from becoming infected (Mellor *et al.*, 2000). The majority of *C. sonorensis* that became persistently infected did so during peak viraemia, suggesting a direct link between the presence of virus in sheep blood and the infection of feeding midges. This proposition is complicated, however, by the observation that the titres of BTV in midges just after feeding declined significantly with time after about day 6 p.i., without any measurable decline in the viraemia of the sheep on which they fed until day 9 p.i. This observation is currently unexplained. Bearing in mind that the sheep blood was collected from the jugular vein, but the midges fed through skin on capillary blood, it could relate to a decreased concentration of virus in capillary blood after day 6 p.i. but not in jugular blood, or to a decreased availability of virus to midges feeding through skin attributable to immune or non-immune responses being mounted by the host.

There are some notable differences between our results and those of Bonneau *et al.* (2002) who fed *C. sonorensis* on three sheep infected with BTV sero-type 17. *Culicoides sonorensis* that fed on two of three sheep on day 7 p.i. (during the period of sheep viraemia) became infected; BTV titre was measured in one of these sheep at 2.4 $\log_{10}$/TCID$_{50}$/ml well below the threshold we suggest above. Furthermore, transmission was not 100% successful as midges fed on the third viraemic sheep did not become infected. Interpretation of this negative result is complicated, however, by the small number of midges (only 13) that survived 10 days until being assayed for virus. A particularly important finding, however, was the isolation of BTV from one midge that fed on one of the sheep on day 21 p.i. despite no isolation of virus from the sheep since day 11 p.i. While our data are consistent with midges only developing infection if they feed on a viraemic host, the work of Bonneau *et al.* (2002) shows that a small proportion of midges can be infected at least a week after viraemia in the host has become undetectable. This finding also extends the time span during which midges can be infected to at least 21 days post host-infection, compared with a maximum of 12 days in our work.

What, then, can we conclude about BTV viraemia in sheep and infection of midges that fed on them? We, and other workers, have observed that midges are most likely to become infected if the host is at peak viraemia at the time of feeding. This could (and probably does) relate to the presence of detectable virus in the blood but it could also be a simple function of time: midges may be most infectable early on during host infection, when the host is most likely to have detectable viraemia. Later during infection virus may still be detectable in host blood, but the likelihood of midge infection may decline (c.f. our failure to detect virus in midges that fed on viraemic hosts after day 12 p.i., either immediately after feeding or after the EIP). Equally, a small proportion of midges may still be infectable after viraemia is no longer detectable [c.f. the results of Bonneau *et al.* (2002)], either because virus is in the blood but at a titre below the minimum sensitivity of the diagnostic test being used, or through some alternative mechanism of infection such as that proposed by Takamatsu *et al.* (2003). Clearly, more work is required to clarify the relationship between the infection of hosts and the midges that feed on them.

Acknowledgements

We are grateful to Eric Denison, Sue Graham, Chris Hamblin, Lutz Riegler and Richard Wall for help. This work was funded by an IAH studentship awarded to MB and PSM.

References

Anderson, J. (1984) Use of monoclonal-antibody in a blocking Elisa to detect group-specific antibodies to bluetongue virus. *J. Immunol. Methods* **74**, 139–49.

Anderson, R.M. and May, R.M. (1991) *Infectious Diseases of Humans: Dynamics and Control.* Oxford: Oxford University Press.

Bonneau, K.R., DeMaula, C.D., Mullens, B.A. and MacLachlan, N.J. (2002) Duration of viraemia infectious to *Culicoides sonorensis* in bluetongue virus-infected cattle and sheep. *Vet. Microbiol.* **88**, 115–25.

Boorman, J. (1974) The maintenance of laboratory colonies of *Culicoides variipennis* (Coq.), *C. nubeculosus* (Mg.) and *C. riethi* kieff. (diptera, ceratopogonidae). *Bull. Entomol. Res.* **64**, 371–7.

Boorman, J. (1975) Semi-automatic device for inoculation of small insects with viruses. *Lab. Pract.* **24**, 90.

Foster, N.M., Jones, R.H. and Luedke, A.J. (1968) Transmission of attenuated and virulent bluetongue virus with *Culicoides variipennis* infected orally via sheep. *Am. J. Vet. Res.* **29**, 275–9.

Fu, H., Leake, C.J., Mertens, P.P.C. and Mellor, P.S. (1999) The barriers to bluetongue virus infection, dissemination and transmission in the vector, *Culicoides variipennis* (diptera: ceratopogonidae). *Arch. Virol.* **144**, 747–61.

Jennings, D.M. and Mellor, P.S. (1987) Variation in the responses of *Culicoides variipennis* (diptera, ceratopogonidae) to oral infection with bluetongue virus. *Arch. Virol.* **95**, 177–82.

Kärber, G. (1931) Beitrag zur kolllektiven behandlung pharmakologischer reihenversuche. *Archiv Für Experimentelle Pathologie Und Pharmakologie* **162**, 480–3.

Lord, C.C., Woolhouse, M.E.J., Heesterbeek, J.A.P. and Mellor, P.S. (1996a) Vector-borne diseases and the basic reproduction number: A case study of African horse sickness. *Med. Vet. Entomol.* **10**, 19–28.

Lord, C.C., Woolhouse, M.E.J., Rawlings, P. and Mellor, P.S. (1996b) Simulation studies of African horse sickness and *Culicoides imicola* (diptera: ceratopogonidae). *J. Med. Entomol.* **33**, 328–38.

Mellor, P.S. (1990) The replication of bluetongue virus in *Culicoides* vectors. *Curr. Top. Microbiol. Immunol.* **162**, 143–61.

Mellor, P.S., Boorman, J. and Baylis, M. (2000) *Culicoides* biting midges: their role as arbovirus vectors. *Ann. Rev. Entomol* **45**, 307–40.

Mellor, P.S., Boorman, J. and Loke, R. (1974) The multiplication of main drain virus in two species of *Culicoides* (diptera, ceratopogonidae). *Arch. Gesamte Virusforsch* **46**, 105–10.

Purse, B.V., Tatem, A.J., Caracappa, S., Rogers, D.J., Mellor, P.S., Baylis, M. and Torina, A. (2004) Modelling the distributions of *Culicoides* bluetongue virus vectors in Sicily in relation to satellite-derived climate variables. *Med. Vet. Entomol.* **18**, 90–101.

Sellers, R.F. (1992) Weather, culicoides, and the distribution and spread of bluetongue and African horse sickness viruses. Bluetongue, African horse sickness and related orbiviruses. In: T.E. Walton and B.I. Osburn (eds), *Bluetongue, African Horse Sickness and Related Orbiviruses*. Boca Raton: CRC Press, pp. 284–90.

Sokal, R.R. and Rohlf, F.J. (1981) *Biometry*. New York: W. H. Freeman & Company.

Spearman, C. (1908) The method of 'right and wrong cases' ('constant stimuli') without Gauss's formulae. *Br. J. Psychol.* **2**, 227–42.

Takamatsu, H., Mellor, P.S., Mertens, P.P.C., Kirkham, P.A., Burroughs, J.N. and Parkhouse, R.M.E. (2003) A possible overwintering mechanism for bluetongue virus in the absence of the insect vector. *J. Gen. Virol.* **84**, 227–35.

Torina, A., Caracappa, S., Mellor, P.S., Baylis, M. and Purse, B.V. (2004) Spatial distribution of bluetongue virus and its *Culicoides* vectors in Sicily. *Med. Vet. Entomol.* **18**, 81–9.

Wellby, M., Baylis, M., Rawlings, P. and Mellor, P.S. (1996) Effect of temperature on survival and rate of virogenesis of African horse sickness virus in *Culicoides variipennis sonorensis* (diptera: ceratopogonidae) and its significance in relation to the epidemiology of the disease. *Bull. Entomol. Res.* **86**, 715–20.

Wittmann, E. (2000) Temperature and the transmission of arboviruses by culicoides biting midges. PhD Thesis, University of Bristol, pp. 159.

Bluetongue virus and climate change

16

BETHAN V. PURSE AND DAVID J. ROGERS

TALA Research Group, Department of Zoology, South Parks Rd., Oxford OX13PS, United Kingdom

Introduction

Vector-borne pathogens are particularly sensitive to climate because they are transmitted between vertebrate hosts by small, poikilothermic, blood-sucking insects or ticks (Rogers and Randolph, 2003). This has led to widespread and continued speculation that anthropogenic climate change will increase the incidence and intensity of their transmission (Cook, 1992; Martens and Moser, 2001; Patz et al., 2005). There are many examples of recent changes in vector borne and other diseases in the literature but there is little direct evidence that any of these changes were caused by climate change (Patz, 2002; Kovats et al., 2001; Rogers and Randolph, 2003) and a growing realisation that other non-climatic abiotic and biotic factors may also affect disease distributions (Kuhn et al., 2003, 2004; Randolph, 2004; Reiter, 1998). Against this background, the recent unprecedented emergence of bluetongue (BT) in Europe constitutes a rare example of a clear impact of climate change on a vector-borne disease (Purse et al., 2005).

Bluetongue virus (BTV) replicates in all ruminant species but severe disease is mostly restricted to certain breeds of sheep and some species of deer (Taylor, 1986). In cattle, BTV causes long-lived sub-clinical infections making these ruminants the main reservoir host. Bluetongue virus is transmitted between its ruminant hosts, primarily by certain species of *Culicoides* biting midges (Diptera: Ceratopogonidae). *Culicoides* populations can build up to high abundances under suitable conditions, and adults can be transported by wind for several kilometres within one night, leading to rapid spread of the

ISBN-13: 978-0-12-369368-6

diseases they carry (Sellers *et al.*, 1977, 1978; Sellers, 1992). Bluetongue virus is restricted to areas where these competent vector species occur – broadly the tropical and subtropical parts of the world, between latitudes 35°S and 40°N. Thus, BTV reaches its northern range margin in Europe, and this region contained, until recently, large populations of naïve and susceptible ruminants (particularly fine wool and mutton breeds of sheep).

In this chapter, we describe the shifted pattern of BT epidemics in Europe (Mellor and Wittmann, 2002; Purse *et al.*, 2005) and summarise briefly the strands of evidence that link this emergence to climate change. The biological mechanisms underlying this response to climate change may include increased virus persistence over winter, the northward expansion of the primary Old World vector – *Culicoides imicola* – and, beyond this vector's range, transmission by indigenous European *Culicoides* species. This chapter considers these potential mechanisms in more detail, focusing on questions that remain to be answered and their consequences for the spread of BT in Europe. Finally we will speculate on the features of the European BTV–*Culicoides* 'episystem' that made it particularly sensitive to climate change. Understanding this sequence of events may help us predict the emergence of BT and other vector-borne pathogens across other continents.

The shifting pattern of European bluetongue epidemics

Bluetongue virus has circulated on Europe's fringes for decades – in sub-Saharan Africa, Turkey and the Middle East (Taylor *et al.*, 1985; Hassan, 1992; Reda *et al.*, 1992; Giangaspero *et al.*, 1995). Throughout this period, these fringe areas have been connected to Europe by synoptic wind systems and by traditional livestock trade routes. The potential for BTV to enter Europe has therefore long existed either by the movement of infected ruminants or by the wind dispersal of infected midges.

Yet, historically, this disease has made only brief sporadic incursions into Europe. Outbreaks were confined to southern Portugal, Spain and Greek islands and occurred wholly within the range of the major Old World vector *C. imicola* Kieffer (Figure 16.1, purple hatched area and blue line). During these incursions, only one or two countries were affected at a time and only a single BTV serotype was involved.

Since 1998, however, six strains (of five serotypes) of BTV have entered Europe more-or-less simultaneously from at least two directions (Figure 16.2, East and West) and spread across 12 countries up to 800 km further north in Europe, than ever before (Mellor and Wittmann, 2002; Purse *et al.*, 2005). Not only has the distribution of BTV expanded dramatically northwards, but so also has that of *C. imicola* (Figure 16.1, red line, into the Balearic Islands (Miranda

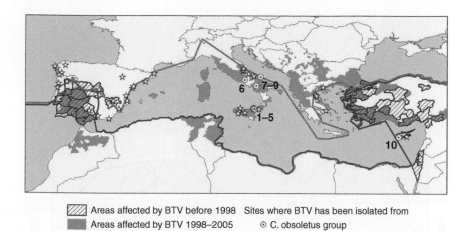

Areas affected by BTV before 1998   Sites where BTV has been isolated from
Areas affected by BTV 1998–2005
— Known limit of C. imicola to 2005   ⊙ C. obsoletus group
— Known limit of C. imicola to 1997   ⊚ C. pulicaris s.s.
   ☆ Sites without C. imicola before 1998

Figure 16.1 The changed distribution of bluetongue (BTV) and its vectors in Europe: Map showing the distribution of BTV prior to 1998 and that of BTV since 1998 (up to October 2004). The distribution of BTV prior to 1998 in North Africa and the Middle East is likely to have been extensive but much transmission in these endemic areas occurs silently in disease-resistant host animals. The reported outbreaks mapped here therefore vastly underestimate the extent of historical transmission in these fringe areas. Lines indicate the known northern range limit of the major Old World vector midge species *Culicoides imicola* up to 1998 (blue) and up to the present day (red). Dotted circles indicate sites where BTV has been isolated from wild–caught, non-engorged individuals from Palearctic vectors groups – from *C. pulicaris s.s.* in green and from *C. obsoletus* group in yellow. All these isolations were made during 2002 [Sites 1–5 (Caracappa *et al.*, 2003), Site 6 (De Liberato *et al.*, 2005), Sites 7–9 (Savini *et al.*, 2003, 2005)] except for site 10 in Cyprus (Mellor and Pitzolis, 1979). Stars indicate sites where *C. imicola* was found to be absent before 1998 (Mellor *et al.*, 1984; Gallo *et al.*, 1984; Gloukova *et al.*, 1991; Dilovski *et al.*, 1992; Scaramozzino *et al.*, 1996; Rawlings *et al.*, 1997; Ortega *et al.*, 1998); sites in Bulgaria were surveyed but not georeferenced, so are not shown (See colour plate 27).

et al., 2003), mainland France (J.-C. Delacolle, Personal Communication, 2004), Switzerland, eastern Spain (Sarto I Monteys and Saiz-Ardanaz, 2003; J. Delgado and P. Collantes, Personal Communication), mainland Greece (Patakakis, 2003), Sicily (Torina *et al.*, 2004) and mainland Italy (Calistri *et al.*, 2003)). Transmission has now persisted for 8 years in the region. Furthermore, and early in the epidemic, transmission occurred beyond even the expanded range of *C. imicola* (in NW Greece, Bulgaria and the Balkans – Figure 16.1), indicating a vector role for other *Culicoides* species in these areas.

The primary candidate species for such a vector role are from the *Culicoides obsoletus* and *Culicoides pulicaris* species' complexes, which are abundant and widespread in northern Europe and also extend southwards into North

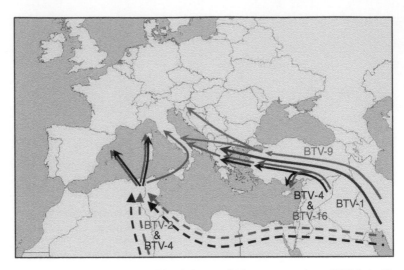

Figure 16.2   The molecular epidemiology of bluetongue virus (BTV) in Europe. Sequence analysis of the five European BTV serotypes has identified six lineages, which have arrived from at least two sources (Mertens *et al.*, this volume). The European strain of BTV-1 (Greece2001/01) belongs to an 'eastern' group of viruses and is similar to viruses isolated in India. The European strain of BTV-2, which first appeared in Tunisia in 1998, belongs to a 'western' group of viruses and is similar to strains from South Africa, Nigeria, Sudan and the United States and probably entered Europe from the south. Both European BTV-9 and -16 belong to 'eastern' groups while the European type 4, initially isolated in Greece (in 2000), is very similar to viruses periodically isolated in the region since 1969. This suggests that it may have been circulating in the fringes of Europe for many years. However, in late 2003, a new strain of BTV-4 arrived in Corsica and the Balearics, which is distinct from that seen in Greece and Turkey, and therefore may have arrived from North Africa. Importantly, distinct strains are still entering Europe on an annual/biannual basis (See colour plate 28).

Africa (Baylis *et al.*, 1997), Turkey (Jennings, 1983) and the Middle East Braverman *et al.*, 1974). Despite their extensive geographical overlap with both the major vector *C. imicola* and areas of historical BTV incursions, there is little evidence, either from the timing or the fine-scale spatial distribution of historical outbreaks (Mellor *et al.*, 1984, 1985), that these complexes played a major role in transmission before the 1990s (Mellor and Boorman, 1995). Their current involvement in transmission in Europe is well substantiated by fine-scale overlap of their distributions with outbreaks (Torina *et al.*, 2004; De Liberato *et al.*, 2005), by virus isolation from wild-caught *C. pulicaris* and *C. obsoletus* group adults in several sites (dotted circles in Figure 16.1; Savini *et al.*, 2003, 2005; De Liberato *et al.*, 2005) and by laboratory competence studies (Carpenter *et al.*, 2006).

Evidence linking patterns in bluetongue to climate change

Examining responses to recent, unprecedented climate change across a range of biological systems, Walther (2002) stated that the 'clearest evidence for climate trigger occurs where a suite of species, with different histories of introduction, spread *en masse* during periods of climate amelioration' – a situation shown, we suggest, by the more-or-less simultaneous entry of six BTV strains into Europe (Figure 16.2). What are the competing explanations for this dramatic change in BT epidemiology in Europe that has involved several BTV strains?

Non-climatic (biotic and abiotic) factors unlikely to influence bluetongue distribution in Europe

Some biotic factors (pathogen characteristics, host distribution or movements) can easily be ruled out. The opening up of trade routes between Europe and the Middle East (Riddle, 1998) may have slightly increased the number of host animal movements, but the total number of ruminants has actually declined in Europe since the 1980s, particularly in central areas (FAO, 2001). When considering whether BT spread is due to circulation of new, perhaps more virulent strains of BTV, it is important to remember that the large, naïve populations of European sheep are likely to have been highly susceptible to the entry of any BTV strains in the late 1990s – not only novel ones. In fact the strains involved are closely related to those that have been circulating in the region for decades (Mertens *et al.*, Chapter 6). Although small genetic changes in RNA arboviruses can have dramatic changes on viral phenotype (Weaver and Barrett, 2004), it is difficult to envisage a factor that would select for such 'small changes' simultaneously and independently in six lineages of BTV and to select for them now rather than at any other time in previous decades.

Non-climatic, abiotic factors (socio-economy, land use and animal health systems) also appear unlikely to be responsible. Any recent changes in agricultural practice or land use are unlikely to have had a substantial impact on *Culicoides* vector distributions. These species are habitat generalists, breeding in a range of moist microhabitats (irrigation channels, drainage pipes and dung heaps) that are ubiquitous across many farmyard types. Regardless of any improvements in disease surveillance, any previous incursions would not have gone unreported due to the high susceptibility of European ruminants for BTV mentioned above. Control strategies against BTV in fringe regions have also remained relatively unchanged for decades, with only a handful of countries using the polyvalent, live attenuated vaccines on an ad hoc (Turkey and Cyprus) or an annual (Israel) basis.

Biological sensitivity of bluetongue virus and *Culicoides* to climate

A direct causal link has been suggested between BT emergence and climate change (Purse *et al.*, 2005) because of close adherence to criteria set out by earlier workers (Kovats *et al.*, 2001; Rogers and Randolph, 2003; Randolph, 2004), namely the following:

1. Research along several axes (theoretical model systems, laboratory experiments, field manipulations and observations) demonstrates the biological sensitivity of both *Culicoides* vectors and BTV to climate (cf. Parmesan and Yohe, 2003).
2. There is meteorological evidence of climate change with sufficient measurements in the study region.
3. Significant changes in the climatic drivers of infection in Europe have occurred at the same time and in the same places as the changes in the incidence of BT.

In common with vector-borne disease systems, key events, both in the BTV transmission cycle and in the life cycle of its *Culicoides* vectors, are modulated by temperature and moisture availability. These phenomena are reviewed extensively elsewhere (Mellor and Leake, 2000; Mellor *et al.*, 2000; Wittmann and Baylis, 2000). Significantly, the competence of *Culicoides* vectors, both the degree of transmission by 'traditional vectors' and the extension of transmission to historically 'non-vector' species, is enhanced by warm temperatures (Pawesak *et al.*, 2002; Wittmann *et al.*, 2002). Within traditional vectors, warm temperatures increase viral replication rates (optimal temperatures 28–29 °C, Van Dijk and Huismans, 1982) and may reduce the efficiency of heritable barrier mechanisms that constrain virus dissemination through a vector individual at various stages following oral infection (Tabachnick, 1991). Bluetongue virus can persist at low temperatures (say <10 °C), for up to 35 days inside adult vectors, and later replicate and be transmitted when the temperature increases (Mellor *et al.*, 1998). In a 'non-vector' species, *C. nubeculosus*, competence can be induced when larvae are reared at high temperatures, with 10% of emerging adults being infectible when reared at 33–35 °C compared to 0% at 30 °C (Mellor *et al.*, 1998; Wittmann *et al.*, 2002). This phenomenon has been attributed to the leakage of virus directly into the haemocoel, bypassing the midgut barriers, allowing virus replication and dissemination. Considering both vector and non-vector species together then, an increase in the cumulative frequency of either warm or hot periods in summer/autumn may increase their transmission potential for BTV.

Broadly speaking, warm temperatures also enhance the recruitment, development, activity and survival rates of *Culicoides* vectors' (Mellor *et al.*, 1998; Wittmann *et al.*, 2002) (taking 6–8 days at 25 °C but only 4 days at 30 °C) and daily survival probability of adult *Culicoides* (optimal between 10 °C and 20 °C, vastly reduced at 30 °C).

Moisture availability is the second most important extrinsic variable affecting *Culicoides* vectors and, in turn, transmission. Precipitation governs the size and persistence of semi-aquatic breeding sites (Mellor *et al.*, 2000). *Culicoides Imicola*, for example, breeds in wet but not flooded, organically enriched soil and mud (Braverman *et al.*, 1974; Braverman, 1978). Precipitation also determines the availability and duration of humid microhabitats in summer/autumn where adults can carry out key activities and shelter from desiccation (Murray, 1991). These interacting effects mean that the spatial and temporal patterns of *Culicoides* vectors can often be matched successfully to patterns in climate (Nevill, 1971; Walker and Davies, 1971; Baylis and Rawlings, 1998; Baylis *et al.*, 1999, 2001; Wittmann *et al.*, 2001; Tatem *et al.*, 2003; Purse *et al.*, 2004a, b).

Consideration of the various (independent and sometimes opposing) responses of these biological processes to climate allows us to pin down the climatic drivers of BTV transmission. Increases in both temperature (particularly at night-time and in winter) and precipitation (particularly in summer/autumn) could lead to an increased geographical and seasonal incidence of BTV transmission by (1) increasing the range, abundance and seasonal activity of vectors; (2) increasing the proportion of a vector species that is competent and (3) increasing the development rates of the virus within vectors and extending transmission ability to additional *Culicoides* species.

Spatio-temporal correspondence between changes in European climate and changes in bluetongue virus

Coincident in time and space with the emergence of BTV in Europe (Purse *et al.*, 2005), there have been pronounced increases in night-time and winter temperatures (Fomby and Vogelsang, 2003; Klein Tank and Konnen, 2003), fewer frost days and changes in moisture conditions (Klein Tank *et al.*, 2002; Haylock and Goodess, 2004; Klein Tank and Konnen, 2003). Bluetongue incidence has increased most markedly in areas where temperature increases are greatest (yellow, orange and red areas in Figure 16.3), in both south-central (Italy, Corsica and the Balearic Islands) and south-eastern Europe (West Bulgaria, northern Greece, Albania, F.Y.R. Macedonia, Bosnia and Herzegovina, Serbia and Montenegro, Croatia). Areas such as central Iberia, northern Morocco and Algeria – where the distribution of BT has broadly remained stable – have cooled. What are the mechanisms that might underlie these responses of BTV to climate change in Europe and what are their consequences for future spread?

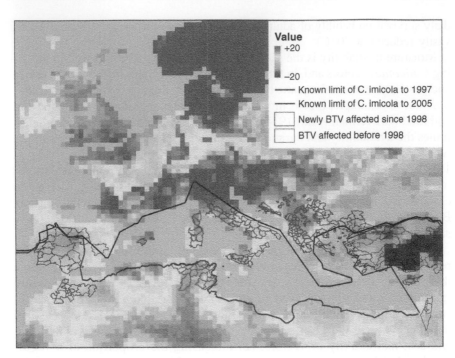

Figure 16.3 Spatial variation in recent climate change in Europe. These images show the changes in annual minimum temperatures between the 1990s and 1980s for each 0.5° square of longitude and latitude on a sliding colour scale ranging from a reduction of 2.0 °C (dark blue and –20 on inset legend) to an increase of 2.0 °C (dark red and +20 on inset legend). Temperature increases are most marked in both central (Italy, Corsica, the Balearic Islands) and eastern Europe (West Bulgaria, northern Greece, Albania, F.Y.R. Macedonia, Bosnia and Herzegovina, Serbia and Montenegro, Croatia) whilst central Iberia and the zone around the border between northern Morocco and Algeria have cooled. This is overlaid with the areas historically affected by BT in grey outline and the newly affected areas in red outline and the distribution limits of *C. imicola* from Figure 16.1. (This image was produced by temporal Fourier processing the raw time series of data and reconstituting it by summing the annual, biannual and triannual harmonics – essentially smoothing the data. The minimum values here are the minima of the reconstituted series for the period in question.) (See colour plate 29)

Extended distribution of the major Old World vector, *C. imicola*

Vector surveillance efforts have been dramatically stepped up in Europe during the recent BT epidemic such that people are looking for the major Old World vector, *C. imicola*, more often and in more places than before. This vastly increased sampling effort might result in an extension of the recorded northern range limit of *C. imicola* regardless of any actual extension on the ground. However, *C. imicola* has now been found in or near sites in which it had been searched for and found to be absent before 1998 (Mellor *et al.*, 1984; Gallo *et al.*, 1984; Scaramozzino *et al.*, 1996; Rawlings *et al.*, 1997; Ortega *et al.*, 1998), suggesting that the actual distribution limit really has extended, at least in places.

What is the reason for this extension in *C. imicola's* northern range limit? Has climate change increased the extent of suitable habitat in Europe? Has *C. imicola*, in the absence of climate change, simply been filling suitable habitat in Europe that was always available to it, but unoccupied by it until recently? Has the environmental envelope of *C. imicola* itself changed over this time period? These alternatives are extremely difficult to tease apart without extensive historical and current distribution data for *C. imicola*. Broadly speaking, however, the pattern of *C. imicola's* range extension during the 1990s shows considerable spatial correspondence with the pattern of warming (Figure 16.4). This species has expanded most into warmed areas in central and eastern Europe (e.g. north-eastern Spain, south France, northern Italy and north-eastern parts of Greece) but has remained stable in central and western Iberia (Capela *et al.*, 2003). This is consistent with temperature determining the distributional limits of BTV and its vectors in cooler places (Baylis and Rawlings, 1998; Rawlings *et al.*, 1998), which are generally wet enough to support larval development, whereas moisture limits vector abundance in warmer areas.

Even in the absence of a clear picture of the current (or previous) extent of suitable habitat for *C. imicola* in Europe, it is evident that this species is capable of rapid colonisation of new habitat. Colonisation distances of *C. imicola* of around 200–300 km were recorded in the 1990s (200 km between the Balearic Islands and North Africa, 160 km between Corsica and southern France and 300 km between Lesvos and mainland Greece) and were comparable to range shifts documented for other vectors such as the mosquito, *Aedes albopictus*, that moved southwards at a rate of 65 km a year in Florida following its introduction into the United States (Mellor and Leake, 2000). Highly dispersive vectors such as *C. imicola*, for which habitats are fairly ubiquitous (see above), are likely to track shifts in their climatic envelopes relatively rapidly, whilst high reproductive rates may permit more rapid adaptation to novel environments (Lounibos, 2002).

Involvement of Palearctic vectors in bluetongue virus transmission in Europe

Palearctic vector groups are now substantially involved in BTV transmission in Europe and have facilitated this virus's expansion of BTV into areas where

C. imicola is rare or absent as set out above (in Bulgaria, the Balkans, north west Greece, European Turkey, some areas of Sicily, and Lazio and Tuscany). It has been hypothesised that the recent climate warming in Europe may have increased their importance – by increasing their population sizes and survival rates to compensate for their low competence levels and by increasing their individual susceptibility through the developmental temperature effects mentioned above (Purse *et al.*, 2005). In support of this hypothesis, the areas where they are involved in transmission again coincide with those areas of Europe that have warmed the most.

It is important to note, however, that a similar temporal pattern in their involvement in transmission could have been produced, in the absence of climate change, given geographical variation in these aspects of their vector capacity, if, for example, the populations of the *C. obsoletus* and *C. pulicaris* groups in the Balkans, Bulgaria, northern Greece, European Turkey have (or had) higher vectorial capacity than do populations in areas historically affected by outbreaks (Spain, Portugal, Morocco and the Greek Islands). In either scenario, their involvement in transmission would has been facilitated when *C. imicola* recently expanded sufficiently to overlap with these 'high vectorial capacity' populations, allowing 'handover' of the virus between the traditional and novel vectors (cf. Mellor, 1996). Evidence for the importance of such hand-over events during the current epidemic is substantial. In both Sicily and Lazio and Tuscany provinces, BTV was transmitted initially in lowland areas by *C. imicola* but was then handed over and spread inland by *C. pulicaris* and *C. obsoletus* groups (De Liberato *et al.*, 2003; Torina *et al.*, 2004; De Liberato *et al.*, 2005).

Vectorial capacity is a complex phenomenon for these Palearctic groups. Geographical variation in vector competence – one aspect of vectorial capacity – has, for example, been noted within *C. obsoletus* group within the UK (Carpenter *et al.*, 2006). Further work is ongoing to pinpoint whether competence is environmentally determined or whether it is heritable, and if heritable, whether particular species of *C. obsoletus s.l.* or genotypes within species are highly competent. Further data on geographical and temporal variation in vectorial capacity of these Palearctic vector groups may later permit a finer understanding of the mechanism governing their recent enhanced role in transmission.

Regardless of the mechanisms involved, what are the consequences of BTV being transmitted by Palearctic vectors that are common and widespread across the whole of central and northern Europe? As expected when any pathogen moves to a new vector, the distributions of the new vectors extend the environmental space that can be occupied by the pathogen. To what extent and along what environmental axes has the environmental envelope of BTV extended by the end of the 1990s due to its enhanced transmission by Palearctic vector species?

For a preliminary investigation of this question, we took the current-day distribution of BT in Europe at a resolution of 0.5° latitude and longitude and

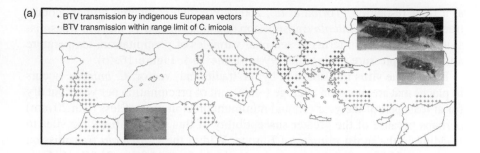

(a)

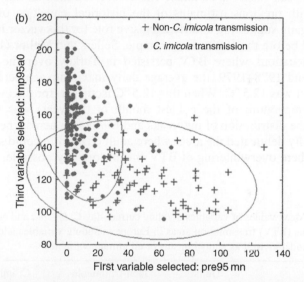

(b)

Figure 16.4 (a) 'Current-day' (1998–2005) distribution of bluetongue (BT) in Europe at a 0.5° latitude and longitude resolution, divided crudely into *C. imicola* and non-*C. imicola* transmission areas, defined by the northern range limit of *C. imicola*. (Modified from Purse *et al.* (2006) State of Science Review: Hindsight and foresight on the spread of bluetongue virus in Europe, UK Government's Foresight Project, Detection and Identification of Infectious Diseases). (b) Bivariate plot of mean annual temperature versus mean annual precipitation for areas of non-*C. imicola* (grey plus) and *C. imicola* transmission in Europe.

divided it crudely into areas where *C. imicola* was mainly responsible for transmission and areas where Palearctic vector groups must be responsible for transmission, using the known northern range limit of *C. imicola* (Figure 16.4a, Purse *et al.*, 2006). We are aware that Palearctic vectors are involved in transmission in some areas below the northern range limit of *C. imicola*, but insufficient reliable data are available to tackle this question at a finer resolution (Torina *et al.*, 2004). We then compared climate conditions in areas of *C. imicola* transmission and areas of non-*C. imicola* transmission – specifically in terms of seasonal variables of precipitation, temperature and vapour pressure during the 1990s (Table 16.1, Figure 16.4b).

Bluetongue virus transmission by its 'traditional' vector, *C. imicola*, occurs in places that are moist on average (~500 mm of precipitation per year) but dry for at least part of the year (annual minimum precipitation of less than 20 mm) perhaps because of the greater susceptibility of this species' breeding sites to flooding in the moist climate of Europe. Within the *C. imicola* transmission envelope of BTV, mean temperatures are high (from 11 °C to 21 °C, 16 °C on average) and peak earlier in summer. These observed thermal limits are fairly consistent with previous estimates of the historical envelope of BTV and *C. imicola* again, suggesting a fairly exclusive role for this vector in outbreaks that occurred before the 1990s. For example, Sellers and Mellor (1993) found that across locations where BTV persisted in Turkey over the winters of 1977–1978 and 1978–1979, the average daily maximum temperature of the coldest month was 12.5 °C. When this 12.5 °C isotherm (for the average daily maximum temperature of the coldest month of the year) was then superimposed on the distribution of outbreaks and *Culicoides* across the Mediterranean, it broadly delineated the areas where adult *C. imicola* could survive year round and where overwintering of BTV was historically possible.

Table 16.1 Mean values of variables dividing current-day *C. Imicola* and non-*C. Imicola* bluetongue virus (BTV) transmission areas in Europe including variables added to a discriminant model as well as annual means of all climate variables

Variables	Units	Non-C. imicola transmission areas		C. imicola transmission areas	
		Mean	s.e.	Mean	s.e.
Minimum annual precipitation	mm	48.6	3.5	6.3	0.8
Timing of annual peak of temperature	Decimal months	7.2 (early August)	0.0	7.4 (mid August)	0.0
Mean annual temperature	°C	12.0	0.2	15.9	0.1
Mean annual amplitude of precipitation	mm	280.2	23.2	379.2	8.7
Mean annual precipitation	mm	681.9	19.6	506.4	8.1
Mean annual vapour pressure	hp	99.2	2.0	122.7	1.3

Source: Modified from Purse *et al.* (2006). *State of Science Review: Hindsight and Foresight on the Spread of Bluetongue Virus in Europe*, UK Government's Foresight Project, Detection and Identification of Infectious Diseases.

By contrast, the current areas of non-*C. imicola* transmission occupy cooler (average annual temperatures of 12 °C), wetter habitats (average precipitation level of 681 mm) than either the current *C. imicola* transmission areas or the historical distribution of BTV or of *C. imicola* delineated in the literature. Figure 16.4b shows that although non-*C. imicola* transmission occurs across a similar range of temperatures as *C. imicola* transmission (albeit lower absolute values) from 8 °C to 16 °C (versus 11–21 °C for *C. imicola*), this envelope spans a much larger range of precipitation levels than current *C. imicola* transmission – from the dry habitats in which *C. imicola* transmits to very wet habitats (where levels of precipitation in a month never drop below 120 mm). Thus, due to the movement into vectors that occupy cooler and moister habitats than *C. imicola*, BTV transmission now occurs in places where annual average temperatures reach as low as 10 °C or where levels of precipitation in a month never drop below 120 mm during a year. Further investigation of the extent to which BTV has filled the environmental space opened up by the distributions of these Palearctic vectors is essential to delimit the areas of Europe that should now be considered at risk of BTV transmission.

Why is the bluetongue virus–*Culicoides* episystem responsive to climate change?

At least six strains of five serotypes have been involved in the recent unprecedented emergence of BTV across much of southern and central Europe. Changes in BT incidence in Europe have been matched by spatio-temporal changes in regional climates, including the specific climatic drivers of BTV infection, and match particularly closely the spatial pattern of warming between the 1980s and 1990s. Changes in other factors (such as agricultural land use changes, changes in animal health systems, increases in livestock trade and increases in host density) could not be shown to follow a similar geographical pattern (Purse *et al.*, 2005).

When considering whether other vector-borne diseases in Europe may undergo a similar emergence, it is pertinent to ask whether there are features of the European BTV–*Culicoides* 'episystem' that predisposed BTV to be able to skip vectors easily and exploit an extended climatic envelope. Bluetongue virus is promiscuous in its choice of ruminant hosts; susceptible host populations are likely to be spatially continuous across agricultural systems. Vector populations of *Culicoides* are similarly widespread and continuous since they are highly dispersive and occupy a wide range of moist soil/dung microhabitats in farmyards, ubiquitous across different agricultural systems. *Culicoides* can rapidly build up large populations under suitable climatic conditions and are catholic in their biting habits, feeding upon any available large mammal. Since most hosts are sub-clinically infected, most infections are neither identified nor removed rapidly from the population and so persist as sources of infection for these biting vectors. Considering Europe in particular, indigenous *Culicoides* vector groups (specifically *C. pulicaris* and *C. obsoletus) are* widespread and abundant and have a wide zone of overlap with the major Old World vector *C. imicola*, providing the opportunity for frequent and widespread 'hand-over' events of the virus between major and novel vector groups. In this zone, climate change, by extending the duration of their seasonal activity, may have additionally increased the temporal overlap of these vectors.

Finally, the vector competence of *Culicoides* for BTV is dynamic and climate-mediated. The temperature effects of both traditional vectors and extensions of competence to 'non-vectors' on the vectorial capacity are described above. Bluetongue virus also has additional mechanisms for coping with harsh winters that commonly occur in northern Europe. The virus can persist at low titres inside adult vectors for up to 35 days and is later able to replicate and be transmitted when the temperature increases (as would occur in spring). In addition, in areas where adult vectors are unable to persist in substantial numbers over winter (such as Bulgaria), it has been hypothesised that the virus can instead persist covertly in the ruminant host itself inside the $\gamma\delta$ T cells of the immune system (Takamatsu *et al.*, 2003).

Similarly, the recent dramatic emergence of West Nile Virus (a flavivirus) into northern temperate United States was undoubtedly assisted by its promiscuity between more than 30 vector species of mosquito and at least 150 bird species – each with different residence ranges or migration routes across North and Central America (Morens *et al.*, 2004). In the year 2000, strains of Rift Valley fever virus (RVFV) (probably originating in East Africa), another virus that is promiscuous between vertebrate hosts (ranging from rodents to hippopotamus) and vectors (including 23 mosquito species, a *Simulium* sp. and a *Rhipicephalus* tick), escaped from Africa for the first time and invaded the Arabian Peninsula, an area well-connected to Europe by a 'ruminant street' (Shoemaker *et al.*, 2002).

A global view of climate change and bluetongue epidemiology

But is there evidence that climate change is currently influencing the epidemiology of BT in other parts of the world? The current consensus seems to be that the BT–*Culicoides* 'episystem' in Europe is unusually dynamic, with regard to the changing role of its *Culicoides* vectors and rapid extension of the virus that occurred in this region. It has been postulated that BTV has evolved worldwide within stable continental episystems defined by the competence (or capacity) of vectors to transmit only indigenous viruses despite movements of hosts (due to trade) and vectors between episystems

(Gibbs and Greiner, 1994; Tabachnick, 2003; Walton, 2003). For example, in the New World, there are thought to be two different episystems – *C. sonorensis* is responsible for the transmission of serotypes in North America and *C. insignis* for the transmission of serotypes in South America (Tabachnick, 1996). Since there is little clinical disease in tropical and subtropical areas, changes in distributional limits of BTV and its vectors are likely to be much harder to detect in most continents than in Europe, which had large populations of naïve susceptible sheep. In fact, there is evidence for a degree of dynamism in these other continental episystems though no particular associations with climate changes have been noted. Long periods of BTV faunal stability have been punctuated in some continents by the intermittent introduction and establishment of foreign or novel strains. For example, several South East Asian strains of BT have become established in northern Australia during the 1990s (Kirkland, 2003; Daniels *et al.*, 2004) and, very recently in 2004, BTV-1 was isolated for the first time on the Gulf Coast of the United States (Johnson*et al.*, unpublished data (2005), described in Report of the committee on BT and bovine retroviruses, 2005).

In the Europe–Africa–Middle East episystem, a similarly long period of stability, despite frequent host and vector movements around the episystem, was followed by an emergence involving the entry and establishment of several new virus strains and transmission by Palearctic vectors in Europe. We demonstrate in this chapter that this period of instability was largely precipitated by climate change – probably acting on critical determinants of vectorial capacity. The major lesson to be drawn from this emergence is that vectorial capacity should be considered as temporally and geographically variable, within and between *Culicoides* vectors, and the genetic and environmental (particularly climatic) factors underpinning this variation should be thoroughly investigated (Tabachnick, 1991; Carpenter *et al.*, 2006). Just because a *Culicoides* species or genotype has been historically refractory to BTV in one area, it does not mean that it will remain so in other places or at other times. Given the global nature of climate change and the fact that short-term changes in climate can produce short-term changes in vector or virus distributions across most continental BT episystems (Walker, 1977; Wright *et al.*, 1993; Bishop *et al.*, 1996a, 2000, Purse *et al.*, 2004a), it should be considered as one of the suite of potential abiotic and biotic factors when investigating the mechanisms for dynamism in BT epidemiology worldwide.

Considering future change in BT epidemiology in Europe, it is conceivable that European livestock may quickly develop immunity to extant strains of BTV in Europe. However, molecular epidemiological evidence indicates that novel strains (to which the population will be naïve) are still entering Europe on an annual basis (Mertens *et al.*, Chapter 6) and that BTV represents a continuing threat, at least in the short term. Within Europe, the environmental envelope of BTV has now expanded as expected when any pathogen moves into new vectors with different geographical distributions. Due to the movement into Palearctic vectors (*C. obsoletus* and *C. pulicaris* groups) that occupy cooler and moister habitats than *C. imicola*, BTV transmission now occurs in places where annual average temperatures reach as low as 10 °C or where levels of precipitation in a month never drop below 120 mm during a year. Further investigation of the extent to which BTV has filled the environmental space opened up by the distributions of these Palearctic vectors is essential to delimit the areas of Europe that should now be considered at risk of BTV transmission.

Acknowledgements

BVP was supported by Epidemiology & Control of Orbiviral Diseases in the UK, with particular reference to bluetongue & African horse sickness (BBSRC Research Grant BBS/B/00603; Defra Research Grant SE4104). The authors would like to thank Philip Mellor, Simon Carpenter, Matthew Baylis, Peter Mertens, Alan Samuel, James MacLachlan, Peter Kirkland and Irene Lager for useful discussions about climate change and bluetongue. Benjamin McCormick gave invaluable assistance with spatial analyses and comments on the manuscript.

Glossary

Environmental envelope   The range of environmental variation in which a species can currently persist in the face of competitors, predators and disease.
Episystem   The species (vectors, hosts and pathogens) and environmental aspects of an epidemiological system within a particular ecosystem.
Extrinsic incubation period   The time interval between oral infection of the vector and parasite transmission, involving entry of virus into the midgut, replication, dissemination through the haemocoel and infection of the salivary glands.
Fourier processing   A mathematical technique that expresses a time series as the sum of a series of sine waves, used by climatologists to summarise the seasonal cycles (annual, biannual and tri-annual) of multi-temporal climate data.
Haemocoel   The main body cavity of many invertebrates, including insects, formed from an expanded 'blood' system.
Overwintering   The persistence of a virus in a location between one vector transmission season and the next: by persistence within surviving adult vectors themselves, within the juvenile stages of the vectors following transovarial transmission or by prolonged/persistent infection in viraemic or aviraemic vertebrate hosts.
Palaearctic   One of the eight ecozones (or biogeographical realms) into which the world is divided, which extends across Europe, North Africa and North Asia, north of the tropics.

Ruminant street The 'corridor' between South Asia and Europe formed from connected ruminant populations of Pakistan, Afghanistan, Iran and Turkey.

Vector competence The (innate) ability of a vector to acquire a pathogen and to successfully transmit it to another susceptible host.

γδ T cells A subset of T cells predominant in skin and mucosal tissues. They probably act as a first line of defence against infection/cancer and have immunoregulatory functions.

References

Baylis, M., El Hasnaoui, H., Bouayoune, H., Touti, J. and Mellor, P.S. (1997) The spatial and seasonal distribution of African horse sickness and its potential *Culicoides* vectors in Morocco. *Med. Vet. Entomol.* **11**(3), 203–12.

Baylis, M., Meiswinkel, R. and Venter, G.J. (1999) A preliminary attempt to use climate data and satellite imagery to model the abundance and distribution of *Culicoides imicola* (Diptera: Ceratopogonidae) in southern Africa. *J. S. Afr. Vet. Assoc.* **70**, 80–9.

Baylis, M., Mellor, P.S., Wittmann, E.J. and Rogers, D.J. (2001) Prediction of areas around the Mediterranean at risk of bluetongue by modelling the distribution of its vector using satellite imaging. *Vet. Rec.* **149**, 639–43.

Baylis, M. and Rawlings, P. (1998) Modelling the distribution and abundance of *Culicoides imicola* in Morocco and Iberia using climatic data and satellite imagery. *Arch. Virol.* (Suppl) **14**, 137–53.

Bishop, A.L., Barchia, I.M. and Spohr, L.J. (2000) Models for the dispersal in Australia of the arbovirus vector, *Culicoides brevitarsis* Kieffer (Diptera: Ceratopogonidae). *Prev. Vet. Med.* **47**, 243–54.

Bishop, A.L., Kirkland, P.D., McKenzie, H.J. and Barchia, I.M. (1996a) The dispersal of *Culicoides brevitarsis* in eastern New South Wales and associations with the occurrences of arbovirus infections in cattle. *Aust. Vet. J.* **73**, 174–8.

Bishop, A.L., McKenzie, H.J., Barchia, I.M. and Harris, A.M. (1996b) Effect of temperature regimes on the development, survival and emergence of *Culicoides brevitarsis* Kieffer (Diptera: Ceratopogonidae) in bovine dung. *Aust. J. Entomol.* **35**, 361–8.

Braverman, Y. (1978) Characteristics of *Culicoides* (Diptera, Ceratopogonidae) breeding places near Salisbury, Rhodesia. *Ecol. Entomol.* **3**, 163–70.

Braverman, Y., Galun, R. and Ziv, M. (1974) Breeding sites of some *Culicoides* species (Diptera: Ceratopogonidae) in Israel. *Mosq. News* **34**, 303–8.

Calistri, P., Goffredo, M., Caporale, V. and Meiswinkel, R. (2003) The distribution of *Culicoides imicola* in Italy: Application and Evaluation of Current Mediterranean models based on Climate. *J. Vet. Med.* **50**, 132–8.

Capela, R., Purse, B.V., Pena, I., Wittmann, E.J., Margarita, Y., Capela, M., Romo, L., Mellor, P.S. and Baylis, M. (2003) Spatial distribution of *Culicoides* species in Portugal in relation to the transmission of African horse sickness and bluetongue viruses. *Med. Vet. Entomol.* **17**, 165–77.

Caracappa, S., Torina, A., Guercio, A., Vitale, F., Calabro, A., Purpari, G., Ferrantelli, V., Vitale, M. and Mellor, P.S. (2003) Identification of a novel bluetongue virus vector species of *Culicoides* in Sicily. *Vet. Rec.* **153**, 71–4.

Carpenter, S.T., Lunt, H.L., Arav, D., Venter, G.J. and Mellor, P.S. (2006) The oral susceptibility of UK *Culicoides* species complexes to bluetongue virus infection. *J. Med. Entomol.* **43**, 73–8.

Cook, G. (1992) Effect of global warming on the distribution of parasitic and other infectious diseases: a review. *J. R. Soc. Med.* **85**, 688–91.

Daniels, P.W., Sendow, I., Prichard, L.I., and Sukarsih Eaton, B.T. (2004) Regional overview of bluetongue viruses in South East Asia: viruses, vectors and surveillance. *Vet. Ital.* **40**, 94–100.

De Liberato, C., Purse, B.V., Goffredo, M., Scholl, F. and Scaramozzino, P. (2003) Geographical and seasonal distribution of the bluetongue virus vector, *Culicoides imicola*, in central Italy. *Med. Vet Entomol.* **17**, 388–94.

De Liberato, C., Scavia, G., Lorenzetti, R., Scaramozzino, P., Amaddeo, D., Cardeti, G., Scicluna, M., Ferrari, G. and Autorino, G.L. (2005) Identification of *Culicoides obsoletus* (diptera: ceratopogonidae) as a vector of bluetongue virus in central Italy. *Vet. Rec.* **156**, 301–4.

Dilovski, M., Nedelchev, N. and Petkova, K. (1992) Studies on the species composition of *Culicoides* – potential vectors of the virus of the bluetongue in Bulgaria. *Vet. Sci.* **25**, 63–6.

FAO (2001) *Livestock Geography: An Introductory Atlas of Animal Resources,* Rome, Italy: Food and Agricultural Organisation of the United Nations, Animal Health and Production Division.

Fomby, T.B. and Vogelsang, T.J. (2003) Test of common deterministic trend slopes applied to quarterly global temperature data. Maximum likelihood estimation of misspecified models: Twenty years later. *Adv. Econometr. Res. Annu.* **17**, 29–43.

Gallo, C., Guercio, V., Caracappa, S., Boorman, J. and Wilkinson, P.J. (1984) Indagine siero-entomologica sulla possibile presenza del virus bluetongue nei bovini in Sicilia. *Atta della Societa Italiana di Buiatria* **16**, 393–8.

Giangaspero, M., Vanopdenbosche, E., Clercq, K.D., Tabbaa, D. and Nishikawa, H. (1995) Seroepidemiological survey for bluetongue virus, orf virus, caprine herpes virus type 1 and foot and mouth disease virus type A, O and C in Awassi sheep in Syria. *Vet. Ital.* **31**, 24–9.

Gibbs, P.J. and Greiner, E.C. (1994) The epidemiology of Bluetongue. *Comp. Immunol. Microbiol. Infect. Dis.* **17**, 207–20.

Gloukova, V.M., Nedelchev, N., Rousev, I. and Tanchev, T. (1991) On the fauna of bloodsucking midges of the genus *Culicoides* (diptera: ceratopogonidae) in Bulgaria. *Vet. Sci.* **25**, 63–6.

Hassan, A. (1992). Status of bluetongue in the Middle East and Africa. In: T.E. Walton and B.I. Osburn (eds), *Bluetongue, African horse sickness and related orbiviruses, Proceedings of the Second International Symposium,* Paris, 17–21 June 1991. Boca Raton: CRC Press, pp. 38–41.

Haylock, M.R. and Goodess, C.M. (2004) Inter-annual variability of European extreme winter rainfall and links with mean large-scale circulation. *Int. J. Climatol.* **24**, 759–76.

Jennings, D.M. (1983) *Culicoides* from Western Turkey in relation to bluetongue disease of sheep and cattle. *Rev. Elev. Med. Vet. Pays Trop.* **36**, 67–70.

Johnson, D.J., Mecham, J.O., Ostlund, E.N. and Stallknecht, D.E. unpublished data (2005) In *Report of the Committee on Bluetongue and Bovine Retroviruses.* Committee on bluetongue and bovine retroviruses. US Animal Health Association. http://www.usaha.

org/committees/btbr/btbr.shtml   http://www.usaha.org/committees/reports/2005/report-btbr-2005.pdf

Kirkland, P.D. (2003) Bluetongue viruses, vectors and surveillance in Australia – the current situation and unique features. *Vet. Ital.* **40**, 47–50.

Klein Tank, A.M.G. and Konnen, G.P. (2003) Trends in indices of daily temperature and precipitation extremes in Europe, 1946–1999. *J. Clim.* **16**, 3665–80.

Klein Tank, A.M.G., Wijngaard, J.B. and Van Engelen, A.F.V. (2002) *Climate of Europe; Assessment of observed daily temperature and precipitation extremes.* De Bilt, The Netherlands: KNMI, pp. 36.

Kovats, R.S., Campbell-Lendrum, D.H., McMichael, A.J., Woodward, A. and Cox, J.S. (2001) Early effects of climate change: do they include changes in vector-borne disease. *Philos. Trans. R. Soc. Lond. Ser. B* **356**, 1057–68.

Kuhn, K.G., Campbell-Lendrum, D.H., Armstrong, B. and Davies, C.R. (2003) Malaria in Britain: Past, present and future. *Proc. Natl. Acad. Sci. U.S.A* **100**, 9997–10001.

Kuhn, K.G., Campbell-Lendrum, D.H. and Davies, C.R. (2004) Tropical diseases in Europe? How we can learn from the past to predict the future. *EpiNorth J.* **5**, 6–13.

Lounibos, L.P. (2002) Invasions by insect vectors of human disease. *Annu. Rev. Entomol.* **47**, 233–66.

Martens, P. and Moser, S.C. (2001) Health impacts of climate change. *Science* **292**, 1065–6.

Mellor, P.S. (1996) *Culicoides*: vectors, climate change and disease risk. *Vet. Bull.* **66**, 301–6.

Mellor, P.S. and Boorman, J. (1995) The transmission and geographical spread of African horse sickness and bluetongue viruses. *Ann. trop. Med. Parasitol.* **89**(1), 1–15.

Mellor, P.S., Boorman, J. and Baylis, M. (2000) *Culicoides* biting midges: Their role as Arbovirus Vectors. *Annu. Rev. Entomol.* **45**, 307–40.

Mellor, P.S., Jennings, M. and Boorman, J.P.T. (1984) *Culicoides* from Greece in Relation to the spread of bluetongue virus. *Rev. Elev. Med. Vet. Pays Trop.* **37**, 286–9.

Mellor, P.S., Jennings, D.M., Wilkinson, P.J. and Boorman, J.P.T (1985) *Culicoides imicola*: a bluetongue virus vector in Spain and Portugal. *Vet. Rec.* **116**, 589–590.

Mellor, P.S. and Leake, C.J. (2000) Climatic and geographic influences on arboviral infections and vectors. *OIE Sci. Tech. Rev.* **19**, 41–54.

Mellor, P.S. and Pitzolis, G. (1979) Observations on breeding sites and light-trap collections of *Culicoides* during an outbreak of bluetongue in Cyprus. *Bull. Entomol. Res.* **69**, 229–34.

Mellor, P.S., Rawlings, P., Baylis, M. and Wellby, M.P. (1998) Effect of temperature on African horse sickness virus infection in *Culicoides*. *Arch. Virol.* (Suppl) **14**, 55–164.

Mellor, P.S. and Wittmann, E.J. (2002) Bluetongue virus in the Mediterranean basin, 1998–2001. *Vet. J.* **164**, 20–37.

Miranda, M.A., Borras, D., Rincon, C. and Alemany, A. (2003) Presence in the Balearic Islands (Spain) of the midges *Culicoides imicola* and *Culicoides obsoletus* group. *Med. Vet. Entomol.* **17**, 52–4.

Morens, D.M., Folkers, G.K. and Fauci, A.S. (2004) The challenge of emerging and re-emerging infectious diseases. *Nature* **430**, 242–9.

Murray, M.D. (1991) The seasonal abundance of female biting-midges, *Culicoides brevitarsis* Kieffer (diptera: ceratopogonidae), in coastal south-eastern Australia. *Aust. J. Zool.* **39**, 333–42.

Nevill, E.M. (1971) Cattle and *Culicoides* biting midges as possible overwintering hosts of bluetongue virus. *Onderstepoort J. Vet. Res.* **38**, 65–72.

Ortega, M.D., Mellor, P.S., Rawlings, P. and Pro, M.J. (1998) The seasonal and geographical distribution of *Culicoides imicola*, *C.pulicaris* group and *C. obsoletus* biting midges in central and southern Spain. *Arch. Virol.* (Suppl) **14**, 85–91.

Parmesan, C. and Yohe, G. (2003) A globally coherent fingerprint of climate change impacts across natural systems. *Nature* **421**, 37–42.

Patakakis, J.M. (2003) *Culicoides imicola* in Greece. *Vet. Ital.* **40**, 232–4.

Patz, J.A. (2002) A human disease indicator for the effects of recent global climate change. *Proc. Natl. Acad. Sci.U.S.A.* **99**, 12506–08.

Patz, J.A., Campbell-Lendrum, D., Holloway, T. and Foley, J.A. (2005) Impact of regional climate change on human health. *Nature* **438**, 310–17.

Pawesak, J.T., Venter, G.J. and Mellor, P.S. (2002) Vector competence of South African *Culicoides* species for bluetongue virus serotype 1 (BTV-1) with special reference to the effect of temperature on the rate of virus replication in *C. imicola* and *C. bolitinos*. *Med. Vet. Entomol.* **16**, 10–21.

Purse, B.V., Baylis, M., McCormack, B.J.J. and Rogers, D.J. (2006) State of Science Review: Hindsight and foresight on the spread of bluetongue virus in Europe. In *Foresight, Infectious Diseases: Preparing for the Future*. London: Office of Science and Innovation, UK Government, pp. 1–46.

Purse, B.V., Baylis, M., Tatem, A.J., Rogers, D.J., Mellor, P.S., Van Ham, M., Chizov-Ginsburg, A. and Braverman, Y. (2004a) Predicting the risk of bluetongue through time: climate models of temporal patterns of outbreaks in Israel. *OIE Rev. Sci. Tech.* **23**, 761–75.

Purse, B.V., Mellor, P.S., Rogers, D.J., Samuel, A.R., Mertens, P.P.C. and Baylis, M. (2005) Climate change and the recent emergence of bluetongue in Europe. *Nature Rev. Microbiol.* **3**, 171–81.

Purse, B.V., Tatem, A.J., Caracappa, S., Rogers, D.J., Mellor, P.S., Baylis, M. and Torina, A. (2004b) Modelling the distributions of *Culicoides* bluetongue virus vectors in Sicily in relation to satellite-derived climate variables. *Med. Vet. Entomol.*, **18**, 90–101.

Randolph, S.E. (2004) Evidence that climate change has caused 'emergence' of tick-borne diseases in Europe. *Int. J. Med. Microbiol.*, **37**, 5–15.

Rawlings, P., Capela, R., Pro, M.J., Ortega, M.D., Pena, I., Rubio, C., Gasca, A. and Mellor, P.S. (1998) The relationship between climate and the distribution of *Culicoides imicola* in Iberia. *Arch. Virol.* (Suppl) **14**, 93–102.

Rawlings, P., Pro, M.J., Pena, I., Ortega, M.D. and Capela, R. (1997) Spatial and seasonal distribution of *Culicoides imicola* in Iberia in relation to the transmission of African Horse sickness virus. *Med. Vet. Entomol.* **11**, 49–57.

Reda, I.M., Bastawisi, I.M., Ismail, I.M. and Agag, A.E. (1992) Is bluetongue virus still circulating in the blood of ruminants in Egypt? *Egypt. J. Agric. Res.* **70**, 1333–9.

Reiter, P. (1998) Global warming and vector-borne disease in temperate regions and at high altitudes. *Lancet* **351**, 839–40.

Riddle, J. (1998) *FAO: Europe still Threatened by Animal Epidemics – Better Disease Control Required*. Brussels: FAO Press Release 08/05.

Rogers, D.J. and Randolph, S.E. (2003) Studying the global distribution of infectious diseases using GIS and RS. *Nature Rev. Microbiol.* **1**, 231–6.

Sarto i Monteys, V. and Saiz-Ardanaz, M. (2003) *Culicoides* midges in Catalonia (Spain), with special reference to likely bluetongue virus vectors. *Med. Vet. Entomol.* **17**, 288–93.

Savini, G., Goffredo, M., Monaco, F., De Santis, P. and Meiswinkel, R. (2003) Transmission of bluetongue virus in Italy. *Vet. Rec.* **152**, 119.

Savini, G., Goffredo, M., Monaco, F., Di Gennaro, A., De Santis, P., Meiswinkel, R., Caporale, V., Cafiero, M.A. and Baldi, L. (2005) Bluetongue virus isolations from midges belonging to the Obsoletus complex (Culicoides, diptera: ceratopogonidae). Italy. Vet. Rec. **157**, 133–9.

Scaramozzino, P., Boorman, J., Semproni, G., Vitale, F., Mellor, P.S. and Caracappa, S. (1996) An entomological survey of Ceratopogonidae in central-southern Italy. Parassitologia **38**, 1.

Sellers, R.F. (1992) Weather, Culicoides and the distribution and spread of bluetongue and African horse sickness viruses. In: T.E. Walton and B.I. Osburn (eds), Bluetongue, African horse sickness and related orbiviruses, Proceedings of the Second International Symposium, Paris, 17–21 June 1991. Boca Raton: CRC Press, pp. 284–90.

Sellers, R.F. and Mellor, P.S. (1993) Temperature and the persistence of viruses in Culicoides spp during adverse conditions. Rev. Sci. Tech. **12**(3), 733–55.

Sellers, R.F., Pedgley, D.E. and Tucker, M.R. (1977) Possible spread of African horse sickness on the wind. J. Hyg. **79**, 279–98.

Sellers, R.F., Pedgley, D.E. and Tucker, M.R. (1978) Possible windborne spread of bluetongue to Portugal, June–July 1956. J. Hyg. **81**, 189–98.

Shoemaker, T., Boulianne, C., Vincent, M.J., Pezzanite, L., Al-Qahtani, M.M., Al-Mazrou, Y., Khan, A.S., Rollin, P.E., Swanepoel, R., Ksiazek, T.G. and Nichol, S.T. (2002) Genetic analysis of viruses associated with emergence of Rift Valley fever in Saudi Arabia and Yemen, 2000–2001. Emerg. Infect. Dis. **8**, 1415–20.

Tabachnick, W.J. (1991) Genetic control of oral susceptibility to infection of Culicoides variipennis with bluetongue virus. Am. J. Trop. Med. Hyg. **45**, 666–71.

Tabachnick, W.J. (1996) Culicoides variipennis and bluetongue virus epidemiology in the United States. Annu. Rev. Entomol. **41**, 23–43.

Tabachnick, W.J. (2003) Culicoides and the global epidemiology of bluetongue virus infection. Vet. Ital. **40**, 145–50.

Takamatsu, H-H., Mellor, P.S., Mertens, P.P.C., Kirkham, P.A., Burroughs, J.N. and Parkhouse, R.M.E. (2003) A possible overwintering mechanism for bluetongue virus in the absence of the insect vector. J. Gen. Virol. **84**, 227–35.

Tatem, A.J., Capela, R., Pena, I., Mellor, P.S., Baylis, M., Purse, B.V. and Rogers, D.J. (2003) Prediction of bluetongue vector distribution in Europe and North Africa using satellite imagery. Vet. Microbiol. **97**, 13–29.

Taylor, W.P. (1986) The epidemiology of bluetongue. OIE Rev. Sci. Tech. **5**, 351–6.

Taylor, W.P., Sellers, R.F., Gumm, I.D., Herniman, K.A.J. and Owen, L (1985). Bluetongue epidemiology in the Middle East. In: T.L. Barber, M.M. Jochim, and B.I. Osburn (eds), Bluetongue and Related Orbiviruses. New York: Alan R. Liss, pp. 527–30.

Torina, A., Caracappa, S., Mellor, P.S., Baylis, M. and Purse, B.V. (2004) Spatial distribution of bluetongue and its vectors in Sicily. Med. Vet. Entomol. **18**, 81–9.

Van Dijk, A.A. and Huismans, H. (1982) The effect of temperature on the in vitro transcriptase reaction of bluetongue virus, epizootic haemorrhagic disease virus and african horse sickness virus. Onderstepoort J. Vet. Res. **49**, 227–32.

Walker, A.R. (1977) Seasonal fluctuations of Culicoides species (diptera: ceratopogonidae) in Kenya. Bull. Entomol. Res. **67**, 217–33.

Walker, A.R. and Davies, F.G. (1971) A preliminary survey of the epidemiology of bluetongue in Kenya. J. Hyg. **69**, 47–60.

Walther, G.R., Post, E., Convey, P., Menzel, A., Parmesian, C., Beebee, T.J.C., Fromentin, J.M., Hoegh-Guldberg, O. and Bairlein, F. (2002) Ecological responses to recent climate change. Nature **416**, 389–95.

Walton, T.E. (2003) The history of bluetongue and a current global overview. *Vet. Ital.* **40**, 31–8.

Weaver, S.C. and Barrett, A.D.T. (2004) Transmission cycles, host range, evolution and emergence of arboviral disease. *Nat. Rev. Microbiol.* **2**, 789–801.

Wittmann, E.J., Mellor, P.S. and Baylis, M. (2001) Using climate data to map the potential distribution of *Culicoides imicola* (diptera: ceratopogonidae) in Europe. *OIE Rev. Sci Tech.* **20**, 731–40.

Wittmann, E.J. and Baylis, M. (2000) Climate change: Effects on *Culicoides*-transmitted viruses and implications for the UK. *Vet. J.* **160**, 107–17.

Wittmann, E.J., Mellor, P.S. and Baylis, M. (2002) Effect of temperature on the transmission of orbiviruses by the biting midge, *Culicoides sonorensis*. *Med. Vet. Entomol.* **16**, 147–56.

Wright, J.C., Getz, R.R., Powe, T.A., Nusbaum, K.E., Stringfellow, D.A., Mullen, G.R. and Lauerman, L.H. (1993) Model based on weather variables to predict seroconversion to bluetongue virus in Alabama cattle. *Prev. Vet. Med.* **16**, 271–8.

Bluetongue virus diagnosis

17

PETER P.C. MERTENS, SUSHILA MAAN, CARRIE BATTEN, KARIN E. DARPEL, ANDREW SHAW, NARENDER S. MAAN, KYRIAKI NOMIKOU, SIMON J. ANTHONY, EVA VERONESI, CHRIS A.L. OURA, SIMON CARPENTER AND PHILIP S. MELLOR

Department of Arbovirology and the Community Reference Laboratory for Bluetongue, Pirbright Laboratory, Institute for Animal Health, Pirbright, Woking GU24 0NF, UK

Introduction

'Diagnosis of bluetongue' (BT) is the identification of animals that are or have previously been infected with the bluetongue virus (BTV). A positive diagnosis usually involves detection and identification of BTV-specific antigens, antibodies or RNA in diagnostic samples taken from animals that are potentially infected using virus isolation and serological or molecular assays to identify the virus serogroup and serotype.

Each *Orbivirus* species/serogroup exists as a complex spectrum of serotypes, topotypes and strains, the identification and differentiation of which is essential for a full understanding of virus movement and epidemiology during the distribution and spread of disease outbreaks (see Chapter 7). The BT 'serogroup' reflects the presence of antigens or RNAs that are both conserved and cross-reactive between the different members of the virus species (this includes most of the BTV non-structural and structural proteins of the virus core – particularly VP7). Consequently, serogroup-specific assays can detect any BTV strain [e.g. by enzyme-linked immunoadsorbent assay (ELISA) or reverse transcriptase-polymerase chain reactions (RT-PCR)] and can be used to distinguish it from the other orbiviruses.

ISBN-13: 978-0-12-369368-6

In contrast, BTV serotype is determined by the BTV outer capsid proteins VP2 and VP5, particularly VP2, which primarily controls the specificity of interactions with neutralising antibodies in serum neutralisation assays. Twenty-four distinct serotypes of BTV have been identified so far, originally using serum neutralisation/virus neutralisation tests (SNT or VNT – see 'Serum neutralisation tests' and 'Virus neutralisation tests', respectively). In recent years, multiple strains of different BTV serotypes have been introduced into Europe (and other areas), either by natural incursion or by the use of live attenuated vaccines. It is particularly important to know which serotype and strain of BTV is circulating in a region to ensure the use of an appropriate vaccine. These different BTVs can now also be detected and differentiated by RT-PCR and phylogenetic analyses targeting segment 2 (Seg-2) of the virus genome, which codes for outer capsid protein VP2 (Maan *et al.*, 2007a; Mertens *et al.*, 2007a, b).

However, BTV also includes a number of distinct geographical variants (topotypes) and subtypes, which show distinctive nucleotide sequence variations in each of their genome segments (including segment 2) (Maan *et al.*, 2007a). Any of these serotypes, topotypes and subtypes has the potential to cause disease. The identification of any individual 'type' of BTV can be used to demonstrate conclusively that the virus belongs to the BTV serogroup/ species and can therefore be used to confirm an initial diagnosis. Indeed, the identification of a specific BTV serotype is cited as one of the most reliable methods of BTV diagnosis (Hamblin, 2004).

Two major geographic groups of BTVs have been identified and have been designated as 'eastern' or 'western' topotypes. They include viruses from Australia and the Middle/Far East, or Africa and the Americas, respectively (Maan *et al.*, 2008). However, initial identification of a novel isolate as BTV can potentially be more difficult if the virus belongs to a group that is only distantly related to the strains already characterised. One example of this is 'Toggenburg orbivirus', which was detected during routine surveillance of goats in Switzerland during 2008 (Martin Hofman and Barbara Thür, personal communication).

The International Committee for the Taxonomy of Viruses (ICTV) has recognised certain parameters for the identification of different species of orbivirus and other reoviruses (see Chapter 3; Mertens *et al.*, 2005a, b). These include the serological and phylogenetic relationships that exist between different viruses, which can be detected using serological or molecular assays (as described in 'Molecular assays'). However, the more distantly related members of the same virus species may cross-react only poorly in some of these assays. In such cases, a demonstration that two viruses can exchange/reassort genome segments during co-infection of the same host cell would provide a definitive confirmation that they belong to the same virus species (Mertens *et al.*, 2005a).

Many of the different assays developed to identify BTV, or animals that have been infected with the virus, are commonly referred to as 'group-specific'.

Table 17.1 Differential diagnoses of bluetongue (BT): agents or conditions causing clinical signs similar to BT that can be distinguished using a series of laboratory-based assay systems to positively identify BTV-specific antigens, antibodies or RNA

Infectious diseases	Most common hosts
Foot-and-mouth	Cattle, sheep and goats, camelids, pigs
Vesicular stomatitis	Horses, pigs, cattle, sheep and goats
Malignant catarrhal fever	cattle
Mucosal disease and/or bovine viral diarrhoea	cattle
Orf	sheep
Sheep and goat pox	Sheep and goats
Peste des petite ruminants	Sheep and goats
Rinderpest	cattle
Lumpy skin disease	cattle
Infectious bovine rhinotrachaeitis	cattle
Bovine popular stomatitis/pseudo-cowpox	cattle
Bovine herpes mammillitis	cattle
Contagious ovine digital dermatitis	sheep
Foot rot	sheep
Clostridial enterotoxaemia	sheep
Non-infectious diseases	
Photosensitisation	Cattle and sheep
Cobalt deficiency	sheep
Poising (e.g. Bracken)	Cattle and sheep

To avoid false-negative results, these group-specific assays must be able detect all known serotypes and strains of the virus and must be sufficiently sensitive to detect even low level, early, or late infections. To avoid false-positive results, these assays must also be able to distinguish any BTV from other viruses that cause similar clinical signs, such as epizootic haemorrhagic disease virus (EHDV), foot-and-mouth disease virus (FMDV), or malignant catarrhal fever virus (MCFV) (Table 17.1).

BTV group-specific assays include serological methods to detect BTV-specific antibodies generated during infection of the mammalian host; serological assays to detect and identify BTV-specific protein antigens and molecular techniques such as RT-PCR and cDNA sequencing/phylogenetic analyses that can be used to detect and identify BTV RNA extracted from diagnostic samples (e.g. blood, spleen and insects). These assays may also involve or depend on the 'isolation' of the virus and its growth in cell culture.

An understanding of the strengths and limitations of each diagnostic method is required to ensure that it is used correctly and with the appropriate samples to give meaningful results. RT-PCR assays (particularly nested or real-time assays) can be very sensitive and can give positive results at an earlier stage of post-infection (p.i.) (e.g. after 1–3 days) than methods that detect the production of

BTV-specific antibodies in the infected host (e.g. 7–10 days) (Figure 17.1). However, the extreme sensitivity of RT-PCR assays can also render them sensitive to low levels of cross-contamination, potentially generating false-positive results. BTV-specific antibodies are also detectable for long periods, possibly for the life of the ruminant host, whereas virus particles and viral RNA can be cleared much more quickly from the blood of the infected animals.

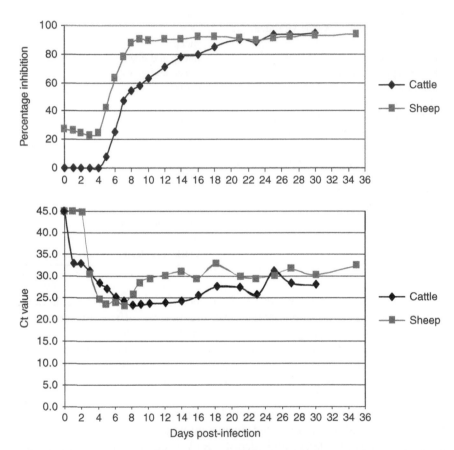

Figure 17.1 Duration of BTV-specific antibodies and RNA in the blood of animals experimentally infected with BTV-8. Cattle (♦) and sheep (■) were experimentally infected with the northern European strain of BTV-8 (NET2006/01) on day 0 (Darpel *et al.*, 2007). Samples of serum and EDTA-treated blood were taken over a 36-day period and analysed for the presence of BTV antibodies by ELISA (top panel) and BTV RNA by real-time RT-PCR (bottom panel). Antibodies were detected in ELISA from 8 dpi and reached a plateau at day 9 in sheep but in cattle continued to rise until day 21. BTV RNA was detected in blood 1 dpi in cattle and 3 dpi in sheep. High levels of viraemia (indicated by low Ct value <25) were observed in cattle from 6–16 dpi and in sheep 4–8 dpi.

The duration of BTV viraemia (detected by the isolation of infectious virus) is usually quite short (<60 days p.i.) in sheep, cattle and goats although it has occasionally been detected for up to 100 days in cattle (Gibbs and Greiner, 1988, Bonneau *et al.*, 2002). In some cases, BTV RNA has been detected by conventional (gel based) RT-PCR assays in sheep and cattle blood, for only 14–20 days p.i. (Darpel *et al.*, 2007), although it often persists for much longer periods, with an average of ~160 days p.i. (as detected by real-time RT-PCR) in both cattle and sheep. BTV RNA has sometimes been detected for as long as to 220 days p.i. (by conventional nested RT-PCR) (Bonneau *et al.*, 2002).

BTV can be detected in vector insects by virus isolation and subsequent characterisation although field strains of the virus are not usually well adapted to infection of cell cultures and may fail to grow. The extreme sensitivity of real-time RT-PCR methods makes them well suited for the detection of BTV RNA in individuals or pools of vector insects although these assays cannot by themselves confirm the presence of infectious virus.

Clinical diagnosis

Clinical signs as an early indicator of BTV infection

Recognition of the clinical signs of BT can provide an early indication of infection and forms a basis for 'passive surveillance', particularly in areas where BTV is exotic and not the subject of routine 'active surveillance'. Indeed, the observation of clinical signs of disease in the mammalian host (sheep), in the Maastricht region of the Netherlands during August 2006, provided the first indication of the first ever northern European outbreak, which was caused by a western strain of BTV-8 (Backx *et al.*, 2007; Elbers *et al.*, 2008; also see Chapters 7 and 11). Clinical diagnosis can therefore be of critical importance, as BTV like many other exotic diseases can spread rapidly and may quickly become established within a naive and susceptible host population. Early diagnosis leading to a rapid implementation of control measures is therefore vital. The likelihood of an early clinical diagnosis can be enhanced if farmers and veterinarians are familiar with the clinical signs of the disease, possibly through awareness campaigns. However, as seen in northern Europe during 2006, BTV transmission tends to occur primarily in bovines that frequently exhibit few, if any, clinical signs of infection. Consequently, the infection may be widespread before the first clinical signs are detected. The incursion of BTV-8 in northern Europe was first detected in the Netherlands although retrospective investigations indicate that the disease had been present in the region for several weeks, with many animals already infected in the Netherlands, Belgium and Germany (Dercksen and Lewis, 2007).

The typical clinical signs of BT in sheep and cattle are described elsewhere (e.g. see Chapters 8, 12, and 13). However, none of these signs are pathognomonic and any disease that causes haemorrhage or oedema could be confused with BT. The severity and range of clinical signs caused by BTV in its hosts can also vary significantly and may be influenced by a number of factors. These include the following:

- The species, breed and even individual animal variations within a host species. BTV infection in cattle, in particular, is frequently asymptomatic.
- The age of an animal. There is some indication that older animals may be more severely affected.
- Differences between infecting virus strains can influence the severity of the clinical signs elicited, ranging from inapparent to fatal.
- The immune status of the host. Antibodies generated by previous infections with the same, or even different strain/serotypes of the virus, or maternal antibodies (colostrum) may have particular significance in determining the outcome of the infection.
- Environmental stress (e.g. high levels of solar radiation) can significantly influence the severity and outcome of disease caused by BTV infection.

Consequently, the clinical diagnosis of BT always requires confirmation by laboratory testing. Diagnostic samples should be sent to a local or regional reference laboratory for analysis where the appropriate diagnostic assay systems can be used to detect and identify the virus, its RNA or proteins and/or specific antibodies. With the first samples from a new outbreak in a region, these tests are likely to include group-specific serological and molecular assays and virus isolation and 'typing' assays. Table 17.1 lists some of the infectious agents that need to be considered in differential diagnoses of BT in cattle and sheep (Bexiga *et al.*, 2007; Watson, 2001, 2004; Williamson *et al.*, 2008).

Virus isolation

BTV can be isolated from blood, semen and various other tissue samples including liver, spleen, brain, lymph nodes and mucosal epithelium. The virus is very stable at neutral pH, in solutions with a high protein content (Mertens *et al.*, 2005b), and it has been isolated successfully from blood samples stored in the presence of EDTA at 4°C for more than a year (Veronesi, unpublished results). The cells in fresh blood samples (with EDTA to prevent clotting) can be washed to remove humoral neutralising antibodies, thereby improving the chances of successful virus isolation. Traditionally, BTV isolation has been performed by inoculation of embryonated chicken eggs (ECE), either by yolk sac inoculation (Mason *et al.*, 1940) or more effectively via an intravenous route (Goldsmit and Barzilai, 1985). Sheep inoculation and cerebral inoculation of suckling mouse brains have also been used for BTV isolation (Afshar, 1994). After initial isolation (e.g. by passage in ECE), the virus is usually adapted to mammalian cell (usually BHK-21, or Vero cells) or insect cell culture [e.g. *Culicoides sonorensis* (KC) cells] (Wechsler and McHolland, 1988; Wechsler and Luedke, 1991). Wechsler and McHolland (1988) evaluated the susceptibility of various cell lines for isolation of BTV and concluded that calf pulmonary artery endothelium (CPAE) cells were also highly sensitive.

Cells of insect origin (e.g. *Aedes albopictus* cells – C6-36 or *C. sonorensis* cells – KC cells) can also be used for the initial isolation of BTV, directly from blood/tissue samples, thereby reducing the use of experimental animals (ECE or mice). These cells seem to be more sensitive to infection with field strains of BTV than mammalian cell lines (Mertens *et al.*, 1996).

Isolation and adaptation of the virus to grow in cell culture can provide larger amounts of material for further study and characterisation of the virus, e.g. by sequence analysis of the whole or part of the viral genome. Such isolates can be stored in a virus reference collection (see www.reoviridae.org/dsRNA_virus_proteins/ReoID/BTV-isolates.htm), providing a basis for further comparisons and studies.

Recent developments have made it possible to amplify and sequence cDNA copies of individual BTV genome segments directly from RNA extracted from viraemic blood or tissue samples. However, virus isolates still provide an important source of viral RNA for full genome analyses or for work on novel or poorly characterised strains that may not amplify well with existing primers. Virus isolates are also essential to provide material for the determination of virus serotype in VNT (see 'Virus neutralisation tests'), for use in antigen detection ELISA and for studies of the biological properties of the virus (see 'Group-specific antigen detection assays'). It is important to note that adaptation to growth in cell culture can itself affect the biological properties of the virus often leading to a reduction in virulence for the mammalian host. It appears likely that this effect is due to selection of a particular subpopulation of the virus, and it is therefore likely to lead to some variation in the consensus sequence of the virus genome.

Antibody/antigen-based assays

Group-specific antibody detection assays

Agar gel immuno-diffusion tests

Agar gel immuno-diffusion (AGID) tests, historically, have been widely used for the detection of group-specific antibodies against BTV (Pearson *et al.*, 1985) although the development of more rapid, sensitive, specific and higher throughput antibody-detection ELISA has led to a decline in their use. The AGID test relies on the availability of purified soluble antigens, derived from BTV-infected cell cultures and positive control serum from hyper-immunised animals (Pearson and Jochim, 1979). The major BTV group-specific antigen, which is detected in these tests as a precipitin line, has been identified as the BTV outer core protein VP7 (Gumm and Newman, 1982). Although the AGID test is simple and rapid to perform, it lacks sensitivity and is not quantitative. Results are often difficult to interpret and are further complicated by cross-reactions with other orbiviruses (Della-Porta *et al.*, 1983; 1985) or by reaction of antibodies against host cell proteins with components of the AGID antigen preparation, making it essential to include uninfected cell-antigens as negative controls.

Complement fixation tests

Complement fixation (CF) tests have been used to identify BTV or to detect a rise in BTV-specific antibody titre following infection (Van Den Ende *et al.*, 1954; Boulanger *et al.*, 1967). These assays that primarily detect early antibodies, IgM, depend on inhibition of the complement-mediated lysis of activated erythrocytes by BTV antigen/antibody complexes that can also fix the available complement. However, they may only be effective for a relatively short period of time p.i. and have largely been superseded by the use of the ELISA.

Competition and blocking ELISA

These ELISAs can be used to detect the presence of BTV-specific humoral antibodies indicating that the animal has either been infected with BTV or has been vaccinated against the virus. During the 1980s, several indirect ELISAs were described for the detection of BTV-specific antibodies (Hubschle *et al.*, 1981) although these assays lacked specificity due to the use of semi-purified antigens. However, Anderson (1984) described a blocking ELISA, which utilised a group-specific monoclonal antibody (identified as 'A3') that is directed against a conformational epitope on the VP7 trimers that form the surface of the BTV core particle. Briefly, immobilised BTV antigen is reacted with the test serum, followed by reaction with the A3 monoclonal antibody. After washing to remove excess, bound A3 can be detected using conjugated anti-mouse antibodies. Antibodies to BTV in the test serum can block the binding of the monoclonal antibody reducing the signal detected. This 'blocking ELISA' is specific for the detection of ovine, bovine and guinea pig antibodies to BTV, and no cross-reactions are observed with antiserum to EHDV, or against cellular proteins present in the antigen preparation (unlike the AGID test). The method has since been developed into a 'dot-blocking' ELISA, in which the antigen is bound to nitrocellulose, to increase its ease and rapidity of use (Afshar *et al.*, 1987a). The blocking ELISA was also adapted in to give a 'competition-ELISA' (Afshar *et al.*, 1987b). In this format, the test antiserum and the BTV-specific monoclonal antibody are reacted with the antigen simultaneously, competing for antigen-binding sites. The competition ELISA is reported to have a comparable sensitivity to the AGID test, CF test and plaque neutralisation tests but is more efficient than the AGID test or the indirect ELISA for antibody detection during early infection (Afshar *et al.*, 1989). The competition ELISA has also been shown to be specific for

Table 17.2 Commercially available BTV ELISA for detection of BTV antibodies

Assay supplier	Available from
BDSL	www.BDSL2000.com
VMRD	www.vmrd.com
ID –Vet	www.id-vet.com
Pourquier	www.institut-pourquier.fr
Ingenasa	www.ingenasa.es
IZS	www.izs.it

Each of these assays has been evaluated in a ring trial and was shown to be capable of detecting antibodies to the BTV strains circulating in Europe since 1998 although their sensitivities vary (Batten *et al.*, 2008).

antibodies to BTV, and interpretation of results is unaffected by the presence of antibodies to EHDV (Afshar *et al.*, 1987b).

The current edition of the OIE *Manual of Standards for Diagnostic Tests and Vaccines* (2008) cites the competition ELISA as a prescribed test for the detection of BTV-specific antibodies (OIE, 2008). Competitive ELISAs have also been described for the detection of antibodies against other orbiviruses, including EHDV (Thevasagayam *et al.*, 1995, 1996a, b) and african horse sickness virus (AHSV) (Hamblin *et al.*, 1990). These ELISAs are specific for their respective virus species/serogroup and show no cross-reactivity with BTV making them useful for differential diagnosis.

A new indirect ELISA for the detection of BTV-specific antibodies in bulk milk samples is reported to be robust, specific and sensitive (Kramps *et al.*, 2008). This assay may provide a valuable non-invasive method for wide-scale surveillance and may be particularly useful for detection of new incursions into non-endemic areas.

ELISAs are robust, rapid assays that can be carried out in a 96-well format. They can also be carried out with a high degree of standardisation and automation (robotics), making them well suited for high throughput, and wide-scale surveillance and diagnostic testing (Table 17.2).

Assays to differentiate infected from vaccinated animals (DIVA)

The implementation of a non-DIVA vaccination campaign using either live attenuated vaccines (as in the Mediterranean region: see Chapters 7 and 11) or partially purified inactivated vaccines (as in both southern and northern Europe) (see Chapter 11; Savini *et al.*, 2008) inevitably compromises the use of ELISA to detect BTV-specific antibodies as a marker for natural infection. Anderson *et al.* (1993) developed a competitive ELISA for the detection of antibodies to the BTV non-structural protein 1 (NS1) as a means of differentiating between BTV-infected and BTV-vaccinated animals. This ELISA used a monoclonal antibody against NS1 to detect antibodies to the BTV NS1 (tubule) protein in infected animals. Serum from animals vaccinated with an inactivated vaccine showed significantly lower reactions in the assay. However, serum from animals vaccinated with live attenuated vaccines gave a positive reaction. This assay is no longer available. Recent studies have focussed on the development and evaluation of similar assays to distinguish between animals infected with the northern European strain of BTV-8 from those vaccinated with commercially available inactivated vaccines (Pourquier and Lesceu, 2008).

Group-specific antigen detection assays

Antigen ELISA
Sandwich ELISAs have been described for the detection of BTV antigens in infected cell cultures or adult *Culicoides* midges (Elhussein *et al.*, 1989; Mecham 1993). Thevasagayam *et al.* (1995; 1996a, b) also described two sandwich ELISAs for the detection of BTV- and EHDV-specific antigens in tissue culture fluids or adult *Culicoides*. These ELISAs use polyclonal rabbit and guinea pig antiserum raised against purified core particles of BTV-1 or EHDV-1, respectively. Both assays are specific for their target virus species/serogroup and gave no detectable cross-reactions with related orbiviruses. Similar sandwich ELISAs have been described for AHSV (Hamblin *et al.*, 1991) and equine encephalosis virus (EEV) (Crafford *et al.*, 2003), both of which show no cross-reactivity with other related orbiviruses. However, although antigen ELISA are specific, they are insensitive requiring relatively large amounts of antigen (equivalent to ≥ 2.5–$3.0 \log_{10}$ infectious units of virus) to give a positive result and consequently are rarely used as a front line test for the detection of BTV.

Fluorescent antibody staining and electron microscopy
These methods have traditionally been used to identify BTV infection (Gould *et al.*, 1989). Electron microscopy of the infected cell cultures tissues or insects can be used to identify viral inclusion bodies, NS1 tubules or virus particles, all of which are characteristic of BTV infection and replication in cell culture (Fu *et al.*, 1999). Immuno-gold labelling of individual BTV proteins using monospecific antibodies can be used to identify specific BTV antigens/proteins, increasing the specificity of the method (Brookes *et al.*, 1993, 1994). Fluorescent antibody staining of either infected tissues or cell cultures can also be used to differentiate between EHDV and BTV in cell culture (Gould *et al.*, 1989).

Serotype-specific antibody/antigen assays

The serotype of each BTV strain is determined by the specificity of reactions between components of the outer capsid (proteins VP2 and VP5) of the virus particle and neutralising antibodies that are generated during infection of the mammalian host (Mertens *et al.*, 1989; Erasmus, 1990; Gould & Eaton, 1990; Mertens, 1999; DeMaula *et al.*, 2000; Maan *et al.*, 2007a). These reactions can be analysed and measured using a variety of micro-titre, plaque reduction or other neutralisation assays (see 'Serum neutralisation tests' and 'Virus neutralisation tests'). Most of the antibodies that neutralise intact BTV particles are specific for VP2 although VP5 can also influence the specificity of the reaction, probably through its interactions with VP2 (Mertens *et al.*, 1989; DeMaula *et al.*, 2000). Antibodies to VP7 can also neutralise the infectivity associated with BTV core particles although in this case the specificity of neutralisation is independent of virus serotype. The rapid and reliable identification of individual BTV strains is particularly important to ensure that a vaccine of the appropriate serotype is used. Before the advent of molecular epidemiology studies and associated databases (see Chapter 7), serotype determination was the only rapid and effective method used to distinguish between different strains of BTV. It has therefore played an important role in elucidating the movements and spread of the virus. Twenty-four serotypes of BTV are currently recognised although it is always possible that a new type or types will be discovered, or will emerge in the future.

Serotype-specific antibody detection assays

Serum neutralisation tests

SNTs can be used to detect neutralising antibodies that are specific for each BTV serotype in diagnostic serum samples. Although the test methodologies can vary, the tests are conventionally based on interaction of known amounts/concentrations of test serum with fixed amounts of virus of each of the different BTV serotypes. Neutralisation is then detected biologically using susceptible cell cultures as an indicator of infectivity, in either micro-titre or plaquing assays (Hamblin, 2004). Conventionally, a dilution series of the serum is reacted with a specific concentration of a reference strain of each BTV serotype (e.g. 100 $TCID_{50}$ or 100 plaque forming units of each virus) to determine the concentration at which the serum reduces or neutralises virus titre in comparison to known negative control sera (Pearson *et al.*, 1985). Neutralisation is considered to have occurred if $2\log_{10}$ infectious units of virus can be neutralised as a result of reaction with the test serum. The antibody titre of the serum is 1, the highest dilution at which it prevents replication of $2\log_{10}$ infectious units (in 50% of the wells micro-titre assay or 50% plaque reduction). Each test serum needs to be evaluated with reference strains of all 24 BTV serotypes to confirm its specificity and level of cross-reaction with different serotypes.

SNTs are highly sensitive and are usually specific for each BTV serotype although circulation of more than one serotype in a region, leading to sequential infections with different serotypes, is likely to cause cross-reactions with multiple additional serotypes (Jeggo *et al.*, 1986). However, even where multiple serotypes have been co-circulating in a region, it may still be possible to identify the response to individual serotypes, as 'common features' of the immune response in different animals, the so-called clustering technique. This approach was used successfully in the early identification of BTV-2 in the United States and in the identification of BTV serotypes in Turkey (Taylor *et al.*, 1985; Taylor and Mellor, 1994). The requirement to use all 24 serotypes in SNTs makes them time consuming and labour intensive. They also require laboratory conditions that are suitable for the routine use of cell cultures, with appropriate biosecurity and access to both reference virus stocks and high-quality biological reagents (Afshar, 1994; Hamblin, 2004). Consequently, SNTs are not widely used for routine identification of BTV infections or surveillance.

Serotype-specific antigen detection assays

Virus neutralisation tests

VNTs can be used to identify the serotype of BTV isolates. These tests are based on the inactivation of virus infectivity by standardised preparations of polyclonal neutralising antibodies for each of the 24 BTV serotypes. Each BTV serotype can be neutralised by antibodies generated against other isolates of the homologous serotype. The results generated are therefore primarily dependent on the antigenic specificity of the BTV outer capsid proteins VP2 and to a lesser extent on VP5.

VNTs require an initial adaptation of the virus strain being tested, to infect and replicate in mammalian tissue culture cells. They usually involve incubation of a dilution series of the 'test virus', with a fixed amount of reference antisera for each of the 24 BTV serotypes, and 'negative' control sera (Afshar, 1994). After incubation, the residual infectivity of the virus is determined using tissue culture cells, and a quantitative assessment is made of the reduction in infectivity caused by each antiserum. These assays can be carried out using a micro-titre format with the infectivity of each treated sample calculated as described by Karber (1931). A reduction of $>2 \log_{10}$ beyond that caused by the negative control serum can usually be regarded as evidence of serotype specificity. The sensitivity of the assay is therefore dependent on the initial titre of the virus being tested (Hamblin, 2004).

It has been suggested that the identification of a specific BTV serotype is itself one of the most reliable methods for 'group-specific' identification of infection by BTV. However, these assays are very time consuming, and it may take several weeks to isolate/adapt the virus to cell culture and then carry out the test (at least one further week). These assays may also have difficulty 'typing' virus isolates containing more than one serotype and they are entirely dependent on the availability of high-quality and standardised antisera for all 24 BTV serotypes (Hamblin, 2004). Such antisera can be difficult to produce/obtain and may not therefore be available in all diagnostic laboratories. The serotyping of BTV isolates is also possible by plaque reduction assays

(Thomas *et al.*, 1976) although it can be difficult to induce new virus isolates to reliably produce suitable plaques in cell cultures, and these assays have become less popular (Afshar, 1994).

Alternative serotyping methods have used small filter-paper discs soaked in serotype-specific neutralising antisera which are placed on an agar overlay to create a zone of protection in lawns of tissue culture cells that are challenged with the test virus isolate (Stott *et al.*, 1978; OIE, 2008). Although not strictly quantitative, the size of the protected area provides some estimate of the degree of interaction between virus and antisera. Fluorescence inhibition tests for the typing of BTV serotypes have also been described (Blacksell and Lunt, 1996) although their use has not been widely reported, and these assays have been largely superseded by molecular RT-PCR-based typing methods.

Molecular assays

The term 'Molecular assay' is used here to describe methods that can detect and identify the viral RNA of BTV or related viruses. These may involve RT-PCR, sequence analyses and phylogenetic comparisons, molecular probes and electrophoresis.

The BTV genome

Electropherotype
The BTV genome is composed of 10 segments of double-stranded RNA (dsRNA), which code for at least 10 distinct viral proteins (Mertens *et al.*, 1984; Firth, 2008). The genome segments can be analysed by polyacrylamide or agarose gel electrophoresis (PAGE or AGE), separating them into 10 distinct bands, representing the different size classes of dsRNA molecules. The genome segment migration pattern generated by polyacrylamide gel electrophoresis (PAGE) depends in part on the primary sequences of the RNAs and therefore varies significantly between different strains of BTV (Figure 17.2). In some cases, the position and even the migration order of the different segments can vary between isolates, with segment 5 and segment 6 usually migrating in the reverse order (Pedley *et al.*, 1988). In contrast, the migration pattern of the different genome segments during AGE correlates with their molecular weight (Figure 17.3) generating a pattern (electropherotype) which is in itself characteristic of BTV and can therefore be used to identify members of the virus species, although not usually for routine diagnostic purposes.

BTV genomic RNA extracted from diagnostic blood or tissue samples (e.g. using guanidinium isothiocyanate-based kits) can be detected using oligonucleotide primers that are complimentary to certain regions/segments

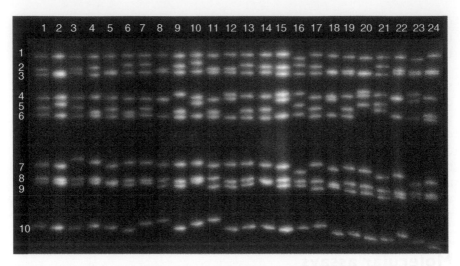

Figure 17.2 Polyacrylamide gel electrophoresis (PAGE) of genomic dsRNAs from reference strains of the twenty-four BTV serotypes. Purified BTV dsRNA were radio-actively labelled (^{32}P-pCp) (Mertens and Sangar, 1985) and analysed by 11% SDS–PAGE (Laemmli, 1970). The bands were detected by autoradiography. The majority of individual genome segments separated by this technique migrate in order of decreasing molecular weight (top to bottom) although genome segments 5 and 6 of most strains migrate in a reverse order (Pedley *et al.*, 1988). Some of the more intense bands contain two genome segments which have co-migrated. Reproduced from Maan *et al.* (2007b).

of the virus genome. This allows the synthesis of cDNA copies of the bound template RNA by a reverse transcriptase enzyme. By using two primers (up stream and down stream) that generate overlapping copies from opposite strands (in opposite directions), it is possible to transcribe individual genome segments into complimentary dsDNA (cDNA). These cDNAs can then serve as templates for multiple further rounds of denaturation and amplification by a DNA-dependent DNA polymerase in a PCR, using either the same or different primers. The amplified cDNA products of 'conventional' RT-PCR can be visualised after agarose gel electrophoresis, whereas those from 'real-time assays' can be detected by the emission of light generated during the amplification reaction. Several types of real-time RT-PCR exist, most being based on either SYBR (where an intercalating molecule fluoresces upon binding to double stranded DNA) or on fluorescence resonance energy transfer (FRET) (e.g. TaqMan and molecular beacon assays). In the FRET-based assays, a fluorescent 'probe-oligonucleotide' binds specifically to the region between the two primers. A positive signal is generated when the probe is degraded by the polymerase as it synthesises new complimentary DNA strands.

The extreme sensitivity of RT-PCR, particularly real-time or nested techniques, coupled with the use of primers that specifically target regions of the

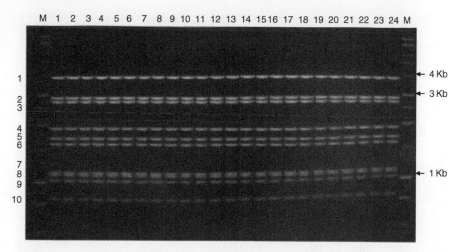

Figure 17.3   Agarose gel electrophoresis of genomic dsRNAs from reference strains of the 24 BTV serotypes. Unlabelled BTV dsRNA preparations were analysed by electrophoresis in a 1% agarose gel containing 0.5 μg/mL ethidium bromide and visualised by exposure to UV light. Genome segments are numbered, in order of decreasing molecular weight. DNA markers were run (lanes 'M') to allow an estimation of molecular weights. Lanes 1–24 (in Figures 17.2 and 17.3) contain RNAs of the reference strains of BTV serotypes 1–24, respectively (see www.iah.bbsrc.ac.uk/dsRNA_virus_proteins/ReoID/ BTV-isolates.htm). Reproduced from Maan *et al.* (2007b).

virus genome that are conserved within the BTV serogroup/species makes them ideally suited for use in diagnostic assays. Conventional RT-PCR assays generate a cDNA product, the 'expected size' of which can be confirmed after AGE. These cDNA products can be sequenced and compared to existing data for known BTV isolates, confirming an initial positive result. Phylogenetic analysis of the sequences generated can also be used to identify BTV topotype and even identify individual lineages of the genome segment (and potentially of the whole virus) being analysed.

The detection of amplification in real-time RT-PCR assays (via the fluorescence produced during the reaction) eliminates the need to open the reaction tube post-amplification. This significantly reduces the risk of contaminating the diagnostic laboratory with the cDNA that has just been synthesised and which may be present in relatively large amounts. Because of the extreme sensitivity of the assays, contamination even at a very low level could give rise to false-positive results. However, the products of real-time RT-PCR assays are not usually suitable for confirmation by definitive sequence analyses/ comparisons. The results of real-time RT-PCR assays are expressed as a cycle-threshold (Ct) value. This represents the number of amplification cycles that are required under standard test conditions to cross a certain threshold level of fluorescence (Figure 17.4), and higher Ct values therefore indicate that

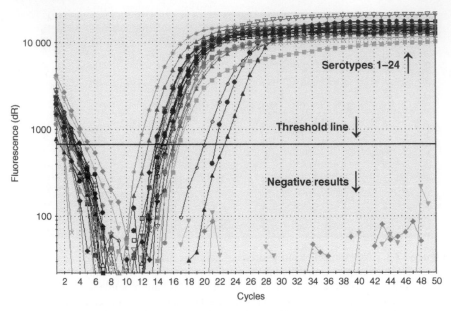

Figure 17.4 Typical real-time RT-PCR results for reference strains of all 24 BTV serotypes. The figure shows amplification plots for 24 reference strains of BTV representing serotypes 1–24, by real-time RT-PCR as described by Shaw *et al.* (2007). Clear 'curves' are generated during the logarithmic phase of PCR amplification. The cycle at which the level of fluorescence crosses the threshold line defines the cycle-threshold (Ct) value. The Ct value is proportional to the input copy number of RNA allowing a subjective (without standard curve) or quantitative (with standard curve) evaluation of the level of RNA present in the sample and therefore an estimation of the level of infection.

smaller amounts of the target gene are present in the test sample than do lower Ct values. Blood samples taken from an infected animal at the peak of viraemia may give Ct values of <20, whereas a low level viraemia could still be detectable with a Ct value >35. Negative control samples should not achieve a positive Ct value although assays are rarely conducted for more than 45 cycles. Standardisation of the assay, described by Shaw *et al.* (2007) and using a dilution series of viral RNA, indicated that a change of 3 Ct units was approximately equivalent to a 10-fold dilution of the sample.

Group-specific molecular assays

Many of the earliest BTV group-specific 'molecular assays' are based on conventional gel-based RT-PCR methods (Dangler *et al.*, 1990; Wade-Evans, *et al.*, 1990; McColl and Gould, 1991; Akita *et al.*, 1992; Shad *et al.*, 1997; Aradaib *et al.*, 1998, 2003, 2005; Billinis *et al.*, 2001; Anthony *et al.*, 2007) and were developed before the advent of real-time RT-PCR. Real-time diagnostic assays have rapidly become popular because they are less labour intensive, quicker, can be more sensitive and have a lower risk of cross-contamination in the laboratory. However, the equipment and reagents required are more expensive and may not be universally available, and their high sensitivity increases the potential to generate false positives due to low-level cross-contamination. Segments 1, 5 and 10 have all been targeted for the detection of the BTV serogroup often with high levels of sensitivity (Batten *et al.*, 2008). In addition, segments 5 and 10 have been used to differentiate between co-circulating strains of BTV.

Conventional RT-PCR assays

These assays have been developed for detection of BTV by targeting genome segments that code for serologically cross-reactive proteins including segments 1, 3 and 7 (Wade-Evans *et al.*, 1990; Pearson *et al.*, 1992; Afshar, 1994; Tabachnick *et al.*, 1996; Bonneau *et al.*, 2000; Wilson *et al.*, 2000; Billinis *et al.*, 2001; Breard *et al.*, 2003; Mecham *et al.*, 2003; Anthony *et al.*, 2007). Such assays have been extensively validated to ensure their optimal performance (Anthony *et al.*, 2007) and to provide a 'front-line' diagnostic technique in laboratories where real-time PCR technology is not available, as well as providing back-up methods in diagnostic laboratories that do have real-time capabilities. Perhaps surprisingly, genome segment 7, which encodes the most immunodominant of the BTV serogroup-specific antigens VP7 (Gumm and Newman, 1982) is also the third most variable of the genome segments (Wilson *et al.*, 2000; Mecham *et al.*, 2003; Anthony *et al.*, 2007; Maan *et al.*, 2008). The variations in segment 7 reflect the geographic origins (topotype) of the virus, but additionally, isolates from within each geographic region can also form a number of distinct clades, which at the time of writing are of uncertain significance. These clades could potentially relate to the role of VP7 in infection of insect cells and the different insect vector populations that are found in different geographic regions (Mertens *et al.*, 1996).

The most conserved regions of the BTV genome are segments 1, 3, 4, 5, 8 and 9 (Maan *et al.*, 2008) (coding for the proteins: VP1[Pol], VP3[T2], VP4[CaP], NS1[Tup], NS2[VIP], VP6[Hel], respectively – see Chapter 7). These genome segments therefore represent potential targets for the development of additional BTV species/serogroup-specific RT-PCR assays.

Real-time (quantitative) pan-BTV RT-PCR assays

Such assays have been reported to target BTV genome segments 1 (Shaw *et al.*, 2007; Toussaint *et al.*, 2007) and 5 (Jimenez-Clavero *et al.*, 2006; Toussaint *et al.*, 2007). They appear to be very sensitive to mis-matches in the target sequences, particularly in the fluorescent probe target region. Segment 1 encodes the viral RNA polymerase and is the most highly conserved segment across the orbiviruses. Fears that such a high level of conservation could result in cross-reaction with closely related non-BTV viruses have proved to be unfounded. However, even the most conserved of the BTV genome segments do show sequence variations that correlate with the geographic origins of the virus isolate (Maan *et al.*, 2008) which can also affect the specificity and sensitivity of RT-PCR assays. To circumvent these problems, Shaw *et al.* (2007) describe a duplex assay, using two separate sets of primers and probes, that target a highly conserved region of genome segment 1 from the eastern and western BTV topotypes, respectively. This assay is currently used by the BTV reference laboratories at IAH Pirbright and elsewhere although assays targeting genome segments 5 and 10 are also in regular use by other BTV reference laboratories (Jimenez-Clavero *et al.*, 2006; Orrù *et al.*, 2006; Toussaint *et al.*, 2007).

BTV genome segment 5 (encoding NS1) is widely regarded as being sufficiently conserved across the serogroup for diagnosis of BTV. Indeed, the OIE manual itself recommends Seg 5-specific assays (OIE, 2000). Despite this high level of conservation, one published assay targeting this segment failed to detect some eastern strains of BTV (Jimenez-Clavero *et al.*, 2006). In contrast, another assay targeting the same segment, which is capable of detecting any BTV, has been widely adopted by several national laboratories (Toussaint *et al.*, 2007).

Genome segment 9, coding for minor core protein VP6 (the core associated helicase of BTV), is also conserved across the BTV serogroup (Maan *et al.*, 2008; see Chapter 7). This genome segment has recently been targeted for development of a pan-BTV real-time RT-PCR assay. This is intended to provide a double target for the detection and identification of BTV in combination with the assays for segment 1 (Shaw *et al.*, 2007). Such an approach would help to circumvent any potential problems that could be caused by mutation in the virus genome within a primer or probe 'foot-print', resulting in a mis-match and reduction in test efficiency for any single segment (Maan *et al.*, – unpublished).

Genome segment 10 of BTV (encoding the small membrane glycoprotein NS3/NS3A) is the fourth most variable region of the BTV genome segments compared by Maan *et al.* (2008), with variations that only show a partial correlation with the geographic origins of the virus isolate (see Chapter 7) (Billinis *et al.*, 2001; Balasuriya *et al.*, 2008). However, the small size of genome segment 10 has made it a target for sequencing and phylogenetic studies leading to the development of a number of both real time and conventional RT-PCR targeting segment 10. Recently, a molecular beacon assay has been developed based on a conserved region of segment 10 (Orrù *et al.*, 2006).

Real-time RT-PCR assays have also been developed with the aim of differentiating between strains of BTV. The first real-time RT-PCR assay for BTV, targeted genome segment 5, and was used to differentiate between a field strain and the South African vaccine strain of BTV serotype 2 (Orrù *et al.*, 2004). Segment 10 has subsequently been used for differentiation between strains of BTV serotypes 2 and 9 in duplex real-time RT-PCR reactions

(Elia *et al.*, 2008). However, it is important to note that because different BTV strains can exchange genome segments by reassortment, only genome segments 2 and 6, coding for the outer capsid proteins VP2 and VP6, can be used to reliably identify virus serotype (see Serotype-specific molecular assays).

Serotype-specific molecular assays

The outer capsid proteins VP2 and VP5 of BTV are encoded by segments 2 and 6 (Mertens *et al.*, 1984), which represent the first and second most variable regions of the virus genome, respectively (Maan *et al.*, 2008). Sequence comparisons of genome segments 2 and 6 from isolates of the 24 BTV serotypes have shown variations in segment 2 that correlate primarily with virus serotype although they also show differences within each serotype that correlate with the geographic origins of the virus isolate (Maan *et al.*, 2007a). Variations in segment 6 also show some correlation with virus serotype (Singh *et al.*, 2004) although unlike segment 2 this is not an absolute or exclusive relationship, with isolates of type 9 appearing in two distinct groups and isolates of type 7 and 19 containing segment 6 with very similar sequences (see Chapter 7). These studies and sequence comparisons of full length segment 2 from over 300 virus strains belonging to different BTV serotypes have demonstrated that BTV isolates can be reliably 'typed' by sequence analyses and phylogenetic comparisons of segment 2 (Maan *et al.*, 2007a).

Conventional RT-PCR assays (targeting segment 2)

These assays have been used to detect and identify a limited range of BTV serotypes within certain geographical regions, including Australia and the United States (Gould and Pritchard, 1990; McColl and Gould, 1991; Wilson and Chase, 1993; Pritchard and Gould, 1995; Johnson *et al.*, 2000). However, these studies were hampered by the absence of segment 2 sequences for multiple isolates of each serotype from geographically distinct origins, and the specificity of the assays was not fully evaluated. The development of a sequence data set for the reference strains of all 24 BTV serotypes (Maan *et al.*, 2007a) confirmed the earlier studies showing that VP2 is the primary determinant of BTV serotype (Mertens *et al.*, 1989; DeMaula *et al.*, 2000). These sequencing studies also provided a basis for the development of RT-PCR assays to identify the serotype of RNA extracted from virus isolates or directly from diagnostic samples (e.g. blood) (Zientara *et al.*, 2006; Mertens *et al.*, 2007a, b). Such assays do not require prior isolation of the virus in cell culture, and they have the potential to significantly reduce the use of experimental animals (ECE) in BTV diagnosis.

RT-PCR assays are also much faster than conventional antigen-based 'typing' methods giving a positive serotype identification within 12 h of sample receipt, compared with conventional VNT typing methods which can take over a month. These molecular assays are also independent of the standardised preparation of neutralising antisera or reference virus strains that are needed for VNTs and SNTs. The ready availability of synthetic oligonucleotide primers from commercial suppliers has removed one of the major bottlenecks in BTV diagnosis putting typing assays within the reach of any competent laboratory. RT-PCR assays also give a positive identification of segment 2 for each serotype and are consequently able to detect and confirm the identity and presence of multiple serotypes within a single sample, which can be difficult or impossible by SNT or VNT.

The conventional RT-PCRs that have been developed to identify the different serotypes of BTV can also be used to distinguish between field and vaccines strains as well as different topotypes of genome segment 2, in a manner that is impossible using antibody/antigen-based methods alone. Primers for the detection of the European BTV serotypes, differentiation between eastern and western topotypes and identification of vaccine/field strains have been reported by Mertens *et al.*, (2007b), and their sequences are available at www.reoviridae.org/dsRNA_virus_proteins/ReoID/BTV-S2-Primers-Eurotypes.htm.

The conventional (gel based) format of these typing assays also generates cDNA products from segment 2 of the BTV strain being analysed. These products can be sequenced and compared to databases of previously characterised BTV genome segment 2s. This not only confirms virus serotype but can also be used to identify individual virus lineages for molecular epidemiology studies (see Chapter 7). These methods have formed a basis for the identification of the different BTV serotypes in Europe and the United States since 1998 (Mertens *et al.*, 2007a, b) including the initial identification of BTV-8 in the Netherlands during 2006, demonstrating that it was not derived from a South African live attenuated vaccine strain that had previously been used in Bulgaria (Maan *et al.*, 2008).

The RNA of different orbiviruses can also be detected and identified using nucleic acid probe-based methods. By targeting these to the serotype-specific genome segments, it is possible to identify individual virus serotypes. Such assays have been described for both AHSV and BTV (Gould, 1988; Koekemoer *et al.*, 2000), but their use has not been widely reported. The wider acceptance and specificity of RT-PCR-based methods have largely superseded other molecular techniques.

Detection of BTV in *Culicoides* midges

The detection of BTV in field collected populations of adult *Culicoides* biting midges is most commonly attempted in areas where outbreaks are occurring or from endemic regions during periods of intense transmission. The techniques used are, with a few exceptions, the same as those used for detection of the virus in ruminants (see above) although the results may be more difficult to interpret and can be assay specific. The BTV infection rates in wild populations of adult *Culicoides* is usually very low, making attempts to identify individual infected *Culicoides* impractical as a method of surveillance of virus incursion or resurgence. The probability of finding any BTV-positive *Culicoides* is low, even during periods of high transmission. Sampling of ruminants, particularly cattle, is far more likely to generate positive results.

The primary reason for trying to detect BTV in biting midges is therefore to implicate a particular species of *Culicoides* in transmission. These assessments can provide valuable information concerning the epidemiology of BT in the field, along with estimates of the both seasonality and distribution of the virus. Knowledge of such BTV transmission mechanisms is vital to enable accurate assessments of the likely effect of *Culicoides* control interventions on BTV transmission.

The detection of BTV in *Culicoides* is most often carried out using ECE, followed by passage (often blind) in cell culture, to isolate the virus from pools of parous midges. As with samples from ruminants, the isolation of BTV is not always successful and virus strains which fail to grow under these conditions will remain undetectable. In these cases, other techniques (particularly RT-PCR) can be used to identify the presence of viral RNA in the pooled insects providing evidence that the virus itself is present. With this supporting evidence, multiple cell lines from different origins can be used to increase chances of virus isolation from any RNA-positive insect populations.

Although useful in this role, RT-PCR has significant limitations when used to attempt to implicate vector species as they will detect both inactivated and replicating viral RNA, with apparently equal efficiency. It is therefore possible that any *Culicoides* (or indeed any other biting arthropod) that has fed on a viraemic animal will be detected as 'positive', even if they contain no infectious virus (Veronesi *et al.*, 2008; Carpenter *et al.*, 2008). Although real-time RT-PCR methods could be used to determine the amount of viral RNA present in a sample of *Culicoides* biting midges, they do not assess the amount of live virus that is present, as the relationship between Ct value and infectivity is still uncertain. The level of virus infectivity in a midge correlates with its degree of dissemination in the insect and allows persistent and transmissible BTV infections to be differentiated from persistent but non-transmissible infections (see Chapter 14). Studies of the relationship between the amounts of viral RNA, as measured by real-time RT-PCR, and virus titre as detected using cell culture could improve our understanding of virus replication within and transmission by the insect vector. However, current investigations into the vector competence of *Culicoides* species should always include virus isolation as part of their methodology.

These difficulties have been addressed in part through the use of rapid, tissue lyser-based systems that can be used to homogenise midges individually (Kato and Mayer, 2007; Veronesi *et al.*, 2008). These techniques have been used successfully with laboratory-infected *Culicoides* but have not yet been used to process field-caught individuals. While the techniques remain slower and hence more expensive than pool-based isolations, they have the advantage of identifying infected individuals and could be used to estimate the level of viral dissemination within the insect itself (e.g. by testing the heads of potentially infected individuals). The vector competence of *Culicoides* species implicated using these screening techniques can be assessed in laboratory-based transmission experiments to confirm their status.

Ring-trials and proficiency testing

The EU Community Reference Laboratory is responsible for sending a proficiency panel of samples to all Member States (MS) on a yearly basis to determine the ability of National Reference Laboratories to detect BTV.

In 2006, two proficiency tests were performed to evaluate the sensitivity and specificity of ELISA and RT-PCR assays for the detection of BTV antibodies and RNA. The first ring trial evaluated the ability of each laboratory to detect antibodies to all 24 serotypes of BTV. The second ring trial, which included both antisera and EDTA-treated blood samples from animals experimentally infected with the northern European strain of BTV-8, evaluated the diagnostic sensitivity of the assays used by each laboratory and their ability to detect antibodies to BTV-8 and viral RNA. A total of six C-ELISAs, six real-time RT-PCR and three conventional RT-PCR assays were used. All the C-ELISAs were capable of detecting antibodies to the BTV serotypes currently circulating in Europe (BTV-1, 2, 4, 8, 9 and 16). However, some of the assays displayed inconsistencies in the detection of other serotypes, particularly BTV-19. All C-ELISAs detected BTV-8 antibodies in cattle and sheep by 21 dpi, whereas the majority of assays detected antibodies by 9 dpi in cattle and 8 dpi in sheep. All the RT-PCR assays were able to detect BTV-8 although real-time assays were more sensitive than conventional assays. The majority of the real-time RT-PCR assays detected BTV RNA as early as 2 dpi in cattle and 3 dpi in sheep (Batten *et al.*, 2008).

In 2007, the proficiency test evaluated the sensitivity of the assays used by each participating laboratory and their ability to detect antibodies to a series of BTV serotypes. EDTA blood samples from animals infected with the northern European strain of BTV-8 were diluted in BTV negative blood to determine the sensitivity of RT-PCR assays and to mimic pooling of blood samples. A total of six C-ELISAs, seven real-time RT-PCR and two conventional RT-PCR assays were used.

All the C-ELISAs were shown to be 'fit for purpose' detecting antibodies to all the BTV serotypes circulating in Europe. The real-time RT-PCR assays were also all capable of detecting BTV-8 RNA although with varying sensitivities. However, inconsistency in the ability to detect BTV RNA was observed between the two gel-based PCR assays (Batten *et al.*, 2008).

The results of the proficiency tests were collated by the CRL for BT and disseminated to the European Commission and participating laboratories. The results will help the Commission to formulate recommendations concerning testing regimes and help member states to choose which diagnostic assays to use in their laboratories. Individual countries usually perform further national proficiency tests to evaluate the ability of their own laboratories to carry out diagnostic assays for BTV.

Diagnostic samples

One important aspect of BTV diagnosis is the preparation, transportation and storage of diagnostic samples. If these are being sent to a reference laboratory for testing, this must be done in accordance with local regulations concerning notifiable disease agents, and samples must be transported in appropriate containers with the corrected documentation. In case of any uncertainty, it is important to contact local veterinary authorities and the test laboratory, before samples are dispatched.

Although BTV can be detected in many tissue types, one of the easiest and least invasive samples for diagnosis is blood. As a result, the majority of routine diagnostic methods have been developed primarily for use with either whole blood (collected in the presence of heparin or EDTA) or serum (collected from blood samples that have been allowed to clot in the absence of heparin or EDTA). EDTA-treated blood can be used both for extraction of viral RNA for molecular assays and for virus isolation. It should therefore be held unfrozen at 4°C since freeze-thawing of BTV will reduce infectivity by approximately 10-fold and will cause lysis of red blood cells thereby releasing the cell-associated BTV which may then be neutralised by any humeral antibody that may be present. Sterile blood samples can be transported effectively at room temperature for short periods (a few days).

The transportation of virus isolates is highly regulated and requires particular care to avoid escape or contamination. In these cases, appropriate licenses and permissions need to be obtained from relevant government departments and samples need to be packed in accordance with relevant guidelines. Samples of live viruses should only be sent to reference laboratories after they have been contacted for advice on the protocols required.

Collections of vector insect catches can also be tested for the presence of BTV or other orbiviruses. These materials need to be stored in a suitable manner that will avoid degradation of virus infectivity or RNA – depending on the assay systems that will be used for their analysis. Samples should only be sent to the reference laboratory after the recipient has been contacted to ensure that the correct protocols are used and to confirm that the samples can be dealt with on arrival.

It is particularly important that multiple samples are identified indelibly, that all container tops are tightly sealed to avoid any chance of cross-contamination or escape of potentially infectious agents. In addition, information describing the sample and including the date and location of collection, host species, details of clinical signs and any other relevant observation should accompany any consignment of samples.

The future

It is clear that although there have been many significant advances and developments in the diagnosis of BTV in recent years, driven by the expansion of BTV into new areas and the need for better/faster assays, further developments continue to be required

The advent of real-time RT-PCR assays has already increased sensitivity, and the speed with which BTV can be detected and identified, and these technologies are currently being applied to the typing of BTV in diagnostic samples. However, it is important to remember that the segmented nature of the BTV genome and the ability of different virus strains to reassort their genome segments makes genome segment 2 (and possibly segment 6) the only valid targets for serotyping assays. Variations in the other genome segments do not reliably correlate with virus serotype (Maan *et al.*, 2008). Real-time RT-PCR assays are commercially available to detect and identify BTV serotypes 1 and 8. Although such assays, to detect a specific strain of BTV, can be rapidly designed and produced, the design and validation of assays that are reliably serotype-specific is a much longer and more difficult undertaking but the development of such monoplex and multiplex real-time assays that can be used to detect and distinguish all 24 BTV serotypes is currently underway.

Eradication of a BTV incursion is most likely to be achieved the earlier the control strategies are implemented. Consequently, the development of robust and reliable 'pen-side' assays for BTV based either on antibody/antigen detection methods or on RT-PCR could offer significant and possible vital time savings in diagnosis. The further development of probe-based diagnostic systems, using libraries of oligonucleotides printed on chips, could enable the BTV detection and other orbiviruses even to serotype level. Printing of different peptide or protein antigens into ELISA plates also has the potential to allow testing of antibodies against a wide range of pathogens in a single plate, or even a single well.

However, distinguishing between animals that have received live-attenuated or semi-purified inactivated BTV vaccines and animals that have been naturally infected is likely to remain problematic. The development of the next generation of subunit vaccines will ease the problems of designing and producing an effective DIVA assay.

Sources of information

BTV reference collection: www.reoviridae.org/dsRNA_virus_proteins/ReoID/BTV-isolates.htm
Recent BTV outbreaks: www.reoviridae.org/dsRNA_virus_proteins/outbreaks.htm
European outbreaks: www.reoviridae.org/dsRNA_virus_proteins/ReoID/BTV-mol-epidem.htm

BTV RNA and proteins: www.reoviridae.org/dsRNA_virus_proteins/BTV.htm
BTV sreotype distribution: www.reoviridae.org/dsRNA_virus_proteins/btv-serotype-distribution.htm
BTV serotype specific primers: www.reoviridae.org/dsRNA_virus_proteins/ReoID/BTV-S2-Primers-Eurotypes.htm

References

Afshar, A. (1994) Bluetongue – laboratory diagnosis. *Comp. Immunol. Microbiol. Infect. Dis.* **17**, 221–42.

Afshar, A., Thomas, F.C., Wright, P.F., Shapiro, J.L. and Anderson, J. (1989) Comparison of competitive ELISA, indirect ELISA and standard AGID tests for detecting bluetongue virus antibodies in cattle and sheep. *Vet. Rec.* **124**, 136–41.

Afshar, A., Thomas, F.C., Wright, P.F., Shapiro, J.L., Anderson, J. and Fulton, R.W. (1987a) Blocking dot-elisa, using a monoclonal-antibody for detection of antibodies to bluetongue virus in bovine and ovine sera. *J. Virol. Methods* **18**, 271–9.

Afshar, A., Thomas, F.C., Wright, P.F., Shapiro, J.L., Shettigara, P.T. and Anderson, J. (1987b) Comparison of competitive and indirect enzyme-linked immunosorbent assays for detection of bluetongue virus- antibodies in serum and whole-blood. *J. Clin. Microbiol.* **25**, 1705–10.

Akita, G.Y., Chinsangaram, J., Osburn, B.I., Ianconescu, M. and Kaufman, R. (1992) Detection of bluetongue virus serogroup by polymerase chain reaction. *J. Vet. Diagn. Invest.* **4**, 400–5.

Anderson, J. (1984) Use of monoclonal-antibody in a blocking ELISA to detect group-specific antibodies to bluetongue virus. *J. Immunol. Methods* **74**, 139–49.

Anderson, J., Mertens, P.P.C. and Herniman, K.A.J. (1993) A competitive ELISA for the detection of anti-tubule antibodies using a monoclonal-antibody against bluetongue virus non-structural protein NS1. *J. Virol. Methods* **43**, 167–76.

Anthony, S., Jones, H., Darpel, K.E., Elliott, H., Maan, S., Samuel, A., Mellor, P.S. and Mertens, P.P.C. (2007) A duplex RT-PCR assay for detection of genome segment 7 (VP7 gene) from 24 BTV serotypes. *J. Virol. Methods* **141**, 188–97.

Aradaib, I.E., Mohamed, M.E., Abdalla, T.M., Sarr, J., Abdalla, M.A., Yousof, M.A., Hassan, Y.A. and Karrar, A.R. (2005) Serogrouping of united states and some African serotypes of bluetongue virus using RT-PCR. *Vet. Microbiol.* **111**, 145–50.

Aradaib, I.E., Schore, C.E., Cullor, J.S. and Osburn, B.I. (1998) A nested PCR for detection of north American isolates of bluetongue virus based on NS1 genome sequence analysis of BTV-17. *Vet. Microbiol.* **59**, 99–108.

Aradaib, I.E., Smith, W.L., Osburn, B.I. and Cullor, J.S. (2003) A multiplex PCR for simultaneous detection and differentiation of North American serotypes of bluetongue and epizootic hemorrhagic disease viruses. *Comp. Immunol. Microbiol. Infect. Dis.* **26**, 77–87.

Backx, A., Heutink, C.G., Van Rooij, E.M.A. and van Rijn, P., A. (2007) Clinical signs of bluetongue virus serotype 8 infection in sheep and goats. *Vet. Rec.* **161**, 591–3.

Balasuriya, U.B., Nadler, S.A., Wilson, W.C., Pritchard, L.I., Smythe, A.B., Savini, G., Monaco, F., De Santis, P., Zhang, N., Tabachnick, W.J. and MacLachlan, N.J. (2008) The NS3 proteins of global strains of bluetongue virus evolve into regional topotypes through negative (purifying) selection. *Vet. Microbiol.* **126**, 91–100.

Batten, C.A., Bachanek-Bankowska, K., Bin-Tarif, A., Kgosana, L., Swain, A.J., Corteyn, M., Darpel, K., Mellor, P.S., Elliott, H.G. and Oura, C.A. (2008) Bluetongue virus: European community inter-laboratory comparison tests to evaluate ELISA and RT-PCR detection methods. *Vet. Microbiol.* **129**, 80–8.

Bexiga, R., Guyot, H., Saegerman, C., Mauroy, A., Rollin, F., Thiry, E., Philbey, A.W., Logue, D.N., Mellor, D.J., Barrett, D.C. and Ellis, K. (2007) Clinical differentiation of malignant catarrhal fever, mucosal disease and bluetongue. *Vet. Rec.* **161**, 858–9.

Billinis, C., Koumbati, M., Spyrou, V., Nomikou, K., Mangana, O., Panagiotidis, C.A. and Papadopoulos, O. (2001) Bluetongue virus diagnosis of clinical cases by a duplex reverse transcription-PCR: a comparison with conventional methods. *J. Virol. Methods* **98**, 77–89.

Blacksell, S.D. and Lunt, R.A. (1996) A simplified fluorescence inhibition test for the serotype determination of Australian bluetongue viruses. *Aust. Vet. J.* **73**, 33–4.

Bonneau, K.R., DeMaula, C.D., Mullens, B.A. and MacLachlan, N.J. (2002) Duration of viraemia infectious to *Culicoides sonorensis* in bluetongue virus-infected cattle and sheep. *Vet. Microbiol.* **88**, 115–25.

Bonneau, K.R., Zhang, N.Z., Wilson, W.C., Zhu, J.B., Zhang, F.Q., Li, Z.H., Zhang, K.L., Xiao, L., Xiang, W.B. and MacLachlan, N.J. (2000) Phylogenetic analysis of the S7 gene does not segregate Chinese strains of bluetongue virus into a single topotype. *Arch. Virol.* **145**, 1163–171.

Boulanger, P., Ruckerbauer, G.M., Bannister, G.L., Gray, D.P. and Girard, A. (1967) Studies on bluetongue. 3. Comparison of two complement-fixation methods. *Can. J. Comp. Med. Vet. Sci.* **31**, 166–70.

Breard, E., Sailleau, C., Coupier, H., Ravaud, K.M., Hammoumi, S., Gicquel, B., Hamblin, C., Dubourget, P. and Zientara, S. (2003) Molecular epidemiological analysis of genome segments 2, 7 and 10 of bluetongue virus in Corsica, and differentiation between field isolates and the vaccine strain by RT-PCR. *Virus Res.* **34**, 1–13.

Brookes, S.M., Hyatt, A.D. and Eaton, B.T. (1993) Characterization of virus inclusion bodies in bluetongue virus-infected cells. *J. Gen. Virol.* **74**, 525–30.

Brookes, S.M., Hyatt, A.D. and Eaton, B.T. (1994) The use of immuno-gold silver staining in bluetongue virus adsorption and neutralisation studies. *J. Virol. Methods* **46**, 117–32.

Carpenter, S., McArthur, C., Selby, R., Ward, R., Nolan, D.V., Mordue (Luntz), A.J., Dallas, J.F., Tripet, F. and Mellor, P.S. (2008) Artificial infection studies of UK *Culicoides species* with bluetongue virus serotypes 8 and 9. *Vet. Rec.* In press.

Crafford, J.E., Guthrie, A.J., van Vuuren, M., Mertens, P.P.C., Burroughs, J.N., Howell, P.G. and Hamblin, C. (2003) A group-specific, indirect sandwich ELISA for the detection of equine encephalosis virus antigen. *J. Virol. Methods* **112**, 129–35.

Dangler, C.A., de Mattos, C.A., de Mattos, C.C. and Osburn, B.I. (1990) Identifying bluetongue virus ribonucleic acid sequences by the polymerase chain reaction. *J. Virol. Methods* **28**, 281–92.

Darpel, K.E., Batten, C.A., Veronesi, E., Shaw, A.E., Anthony, S., Bachanek-Bankowska, K., Kgosana, L., Bin-Tarif, A., Carpenter, S., Mueller-Doblies, U.U., Takamatsu, H.H., Mellor, P.S., Mertens, P.P.C. and Oura, C.A.L. (2007) Clinical signs and pathology shown by British sheep and cattle infected with bluetongue virus serotype 8 derived from the 2006 outbreak in northern Europe. *Vet. Rec.* **161**, 253–61.

Della-Porta, A.J., Parsons, I.M. and McPhee, D.A. (1985). Problems in the interpretation of diagnostic tests Due To cross-reactions between orbiviruses and broad serological responses in animals. In: T.L., Barber, M.M. Jochim, and B.I. Osburn, (eds), *Bluetongue and Related Orbiviruses*. New York: Alan R. Liss, Inc, 445–53.

Della-Porta, A.J., Sellers, R.F., Herniman, K.A., Littlejohns, I.R., Cybinski, D.H., St George, T.D., McPhee, D.A., Snowdon, W.A., Campbell, J., Cargill, C., Corbould, A., Chung, Y.S. and Smith, V.W. (1983) Serological studies of Australian and Papua New Guinean cattle and Australian sheep for the presence of antibodies against bluetongue group viruses. *Vet Microbiol.* **8**, 147–62.

DeMaula, C.D., Bonneau, K.R. and MacLachlan, N.J. (2000) Changes in the outer capsid proteins of bluetongue virus serotype ten that abrogate neutralization by monoclonal antibodies. *Virus Res.* **67**, 59–66.

Dercksen, D. and Lewis, L. (2007) Bluetongue virus serotype 8 in sheep and cattle: a clinical update. *In Practice* **29**(6).

Elbers, A.R.W., Backx, A., Ekker, H.M., van der Spek, A.N. and van Rijn, P.A. (2008) Performance of clinical signs to detect bluetongue virus serotype 8 outbreaks in cattle and sheep during the 2006-epidemic in The Netherlands. *Vet. Microbiol.* **129**, 156–2.

Elhussein, A., Calisher, C.H., Holbrook, F.R., Schoepp, R.J. and Beaty, B.J. (1989) Detection of bluetongue virus-antigens in Culicoides variipennis by Enzyme-immunoassay. *J. Clin. Microbiol.* **27**, 1320–3.

Elia, G., Savini, G., Decaro, N., Martella, V., Teodori, L., Casaccia, C., Di Gialleonardo, L., Lorusso, E., Caporale, V. and Buonavoglia, C. (2008) Use of real-time RT-PCR as a rapid molecular approach for differentiation of field and vaccine strains of bluetongue virus serotypes 2 and 9. *Mol. Cell. Probes* **22**, 38–46.

Erasmus, B.J. (1990) Bluetongue virus. In: Z. Dinter and B. Morein (eds) *Virus Infections of Ruminants.* New York: Elsevier, pp. 227–37.

Firth, A.E. (2008) Bioinformatic analysis suggests that the orbivirus VP6 cistron encodes an overlapping gene. *Virol. J.* **5**, 48.

Fu, H., Leake, C.J., Mertens, P.P.C. and Mellor, P.S. (1999) The barriers to bluetongue virus infection, dissemination and transmission in the vector, *Culicoides variipennis* (Diptera: Ceratopogonidae). *Arch Virol.* **144**, 747–61.

Gibbs, E.P.J. and Greiner, E.C. (1988) Bluetongue and epizootic haemorrhagic disease. In: T.P. Monath, (ed), *The Arboviruses; Epidemiology and Ecology, Vol. II.* Boca Raton: CRC Press, pp. 39–70.

Goldsmit, L. and Barzilai, E. (1985) Isolation and propagation of bluetongue virus in embryonating chicken eggs. *Prog. Clin. Biol. Res.* **178**, 307–18.

Gould, A.R. (1988) The use of recombinant DNA probes to group and type orbiviruses. A comparison of Australian and south African isolates. *Arch. Virol.* **99**, 205–20.

Gould, A.R. and Eaton, B.T. (1990) The amino acid sequence of the outer coat protein VP2 of neutralizing monoclonal antibody-resistant, virulent and attenuated bluetongue viruses. *Virus Res.* **17**, 161–72.

Gould, A.R. and Pritchard, L.I. (1990) Relationships amongst bluetongue viruses revealed by comparisons of capsid and outer coat protein nucleotide sequences. *Virus Res.* **17**, 31–52.

Gould, A.R., Hyatt, A.D., Eaton, B.T., White, J.R., Hooper, P.T., Blacksell, S.D. and Le Blanc Smith, P.M. (1989) Current techniques in rapid bluetongue virus diagnosis. *Aust. Vet. J.* **66**, 450–4.

Gumm, I.D. and Newman, J.F. (1982) The preparation of purified bluetongue virus group antigen for use as a diagnostic reagent. *Arch. Virol.* **72**, 83–93.

Hamblin, C. (2004) Bluetongue virus antigen and antibody detection, and the application of laboratory diagnostic techniques. *Vet. Ital.* **40**, 538–45.

Hamblin, C., Graham, S.D., Anderson, E.C. and Crowther, J.R. (1990) A competitive Elisa for the detection of group-specific antibodies to African horse sickness virus. *Epidemiol. Infect.* **104**, 303–12.

Hamblin, C., Mertens, P.P.C., Mellor, P.S., Burroughs, J.N. and Crowther, J.R. (1991) A serogroup specific enzyme-linked-immunosorbent-assay for the detection and identification of African horse sickness viruses. *J. Virol. Methods* **31**, 285–92.

Hubschle, O.J.B., Lorenz, R.J. and Matheka, H.D. (1981) Enzyme-linked immunosorbent-assay for detection of bluetongue virus antibodies. *Am. J. Vet. Res.* **42**, 61–5.

Jeggo, M.H., Wardley, R.C., Brownlie, J. and Corteyn, A.H. (1986) Serial inoculation of sheep with two bluetongue virus types. *Res. Vet. Sci.* **40**, 386–92.

Jimenez-Clavero, M.A., Aguero, M., San Miguel, E., Mayoral, T., Lopez, M.C., Ruano, M.J., Romero, E., Monaco, F., Polci, A., Savini, G. and Gomez-Tejedor, C. (2006) High throughput detection of bluetongue virus by a new real-time fluorogenic reverse transcription-polymerase chain reaction: application on clinical samples from current Mediterranean outbreaks. *J. Vet. Diagn. Invest.* **18**, 7–17.

Johnson, D.J., Wilson, W.C. and Paul, P.S. (2000) Validation of a reverse transcriptase multiplex PCR test for the serotype determination of U.S. isolates of bluetongue virus. *Vet. Microbiol.* **76**, 105–15.

Karber, G. (1931) Beitrag zur kollektiven behandlung pharmakologischer reihenversche. *Arch. Exp. Pathol. Pharmakol.* **162**, 480–3.

Kato, C.Y. and Mayer, R.T. (2007) An improved, high-throughput method for detection of bluetongue virus RNA in culicoides midges utilizing infrared-dye-labelled primers for reverse transcriptase PCR. *J. Virol Methods* **140**, 140–7.

Koekemoer, J.J., Potgieter, A.C., Paweska, J.T. and Van Dijk, A.A. (2000) Development of probes for typing African horsesickness virus isolates using a complete set of cloned VP2-genes. *J. Virol. Methods* **88**, 135–44.

Kramps, J.A., van Maanen, K., Mars, M.H., Popma, J.K. and van Rijn, P.A. (2008) Validation of a commercial ELISA for the detection of bluetongue virus (BTV)-specific antibodies in individual milk samples of Dutch dairy cows. *Vet. Microbiol.* **130**, 80–7.

Laemmli, U.K. (1970) Cleavage of structural proteins during the assembly of the head of bacteriophage T4. *Nature* **227**, 680–5.

Maan, S., Maan, N.S., Ross-Smith, N., Batten, C.A., Shaw, A.E., Anthony, S.J., Samuel, A.R., Darpel, K.E., Veronesi, E., Oura, C.A., Singh, K.P., Nomikou, K., Potgieter, A.C., Attoui, H., van Rooij, E., van Rijn, P., De Clercq, K., Vandenbussche, F., Zientara, S., Bréard, E., Sailleau, C., Beer, M., Hoffman, B., Mellor, P.S. and Mertens, P.P.C. (2008) Sequence analysis of bluetongue virus serotype 8 from the Netherlands 2006 and comparison to other European strains. *Virology* **377**, 308–18.

Maan, S., Maan, N.S., Samuel, A.R., Rao, S., Attoui, H. and Mertens, P.P.C. (2007a) Analysis and phylogenetic comparisons of full-length VP2 genes of the 24 bluetongue virus serotypes. *J. Gen. Virol.* **88**, 621–30.

Maan, S., Rao, S., Maan, N.S., Anthony, S.J., Attoui, H., Samuel, A.R. and Mertens, P.P.C. (2007b) Rapid cDNA synthesis and sequencing techniques for the genetic study of bluetongue and other dsRNA viruses. *J. Virol. Methods* **143**, 132–9.

Mason, J.H., Coles, J.D.W.A. and Alexander, R.A. (1940) Cultivation of bluetongue virus in fertile eggs produced on vitamin deficient diet. *Nature* **145**, 1022–3.

McColl, K.A. and Gould, A.R. (1991) Detection and characterisation of bluetongue viruses using the polymerase chain reaction. *Virus Res.* **21**, 19–34.

Mecham, J.O. (1993) Detection of bluetongue virus from blood of infected sheep by use of an antigen-capture enzyme-linked-immunosorbent-assay after amplification of the virus in cell-culture. *Am. J. Vet. Res.* **54**, 370–2.

Mecham, J.O., Stallknecht, D.E. and Wilson, W.C. (2003) The S7 gene and VP7 protein are highly conserved among temporally and geographically distinct American isolates of epizootic hemorrhagic disease virus. *Virus Res.* **94**, 129–33.

Mertens, P.P.C. (1999). Orbiviruses and coltiviruses – general features. In: R.G. Webster and A. Granoff (eds) *Encyclopaedia of Virology,* 2nd edn., Academic Press, London, pp. 1043–61.

Mertens, P.P.C. and Sangar, D.V. (1985) Analysis of the terminal sequences of the genome segments of four orbiviruses. *Virology* **140**, 55–67.

Mertens, P.P.C., Brown, F. and Sangar, D.V. (1984) Assignment of the genome segments of bluetongue virus type 1 to the proteins which they encode. *Virology* **135**, 207–17.

Mertens, P.P.C., Burroughs, J.N., Walton, A., Wellby, M.P., Fu, H., O'Hara, R.S., Brookes, S.M. and Mellor, P.S. (1996) Enhanced infectivity of modified bluetongue virus particles for two insect cell lines and for two culicoides vector species. *Virology* **217**, 582–93.

Mertens, P.P.C., Duncan, R., Attoui, H. and Dermody, T.S. (2005a) Reoviridae. In: C.M. Fauquet, M.A. Mayo, J. Maniloff, U. Desselberger, and L.A. Ball, (eds), *Virus Taxonomy, VIIIth Report of the ICTV.* London: Elsevier/Academic Press, pp. 447–54.

Mertens, P.P.C., Maan, N.S., Johnson, D.J., Ostlund, E.N. and Maan, S (2007a). Sequencing and RT-PCR assays for genome segment 2 of the 24 bluetongue virus serotypes: identification of exotic serotypes in the southeastern USA (1999-2006). Abstract: World Association of Veterinary Laboratory Diagnosticians (WAVLD) – 13th International Symposium Proceeding. Melbourne, Australia, 12-14 November 2007, p. 55.

Mertens, P.P.C., Maan, N.S., Prasad, G., Samuel, A.R., Shaw, A.E., Potgieter, A.C., Anthony, S.J. and Maan, S. (2007b) Design of primers and use of RT-PCR assays for typing European bluetongue virus isolates: differentiation of field and vaccine strains. *J. Gen. Virol.* **88**, 2811–23.

Mertens, P.P.C., Maan, S., Samuel, A. and Attoui, H. (2005b). Orbivirus, Reoviridae. In: C.M. Fauquet, M.A. Mayo, J. Maniloff, U. Desselberger and L.A. Ball (eds), *Virus Taxonomy, VIIIth Report of the ICTV.* London: Elsevier/Academic Press, pp. 466–83.

Mertens, P.P.C., Pedley, S., Cowley, J., Burroughs, J.N., Corteyn, A.H., Jeggo, M.H., Jennings, D.M. and Gorman, B.M. (1989) Analysis of the roles of bluetongue virus outer capsid proteins VP2 and VP5 in determination of virus serotype. *Virology* **170**, 561–5.

OIE (2000) Bluetongue. In: *Manual of Standards for Diagnostic Tests and Vaccines for Terrestrial Animals,* 4th edn., Paris: OIE, pp. 153–67.

OIE (2008) Bluetongue. In: *Manual of Standards for Diagnostic Tests and Vaccines for Terrestrial Animals,* 6th edn., Chapter 2.1.3, Paris: OIE, pp. 158–74.

Orrù, G., Ferrando, M.L., Meloni, M., Liciardi, M., Savini, G. and De Santis, P. (2006) Rapid detection and quantitation of bluetongue virus (BTV) using a molecular beacon fluorescent probe assay. *J. Virol. Methods* **137**, 34–42.

Orrù, G., Santis, P.D., Solinas, F., Savini, G., Piras, V. and Caporale, V. (2004) Differentiation of Italian field and south African vaccine strains of bluetongue virus serotype 2 using real-time PCR. *J. Virol. Methods* **122**, 37–43.

Pearson, J.E., Carbrey, E.A. and Gustafson, G.A. (1985) Bluetongue and related orbivirus diagnosis in the United States. In: T.L. Barber, M.M., Jochim and B.I. Osburn (eds), *Bluetongue and Related Orbiviruses.* New York: Alan R. Liss, Inc, pp. 469–75.

Pearson, J.E., Gustafson, G.A., Shafer, A.L. and Alstad, A.D. (1992) Diagnosis of blue-tongue and epizootic hemorrhagic disease. In: T.E. Walton and B.I. Osborne (eds), *Proceedings of the 2nd International Symposium on Bluetongue, African Horsesickness and Related Orbiviruses*. Boca Raton, Florida: CRC Press, Inc, pp. 533–46.

Pearson, J.E. and Jochim, M.M. (1979) Protocol for the immunodiffusion test for blue-tongue. *Am. Assoc. Vet. Lab. Diagn. Proced.* **22**, 463–71.

Pedley, S., Mohamed, M.E. and Mertens, P.P.C. (1988) Analysis of genome segments from six different isolates of bluetongue virus using RNA-RNA hybridisation: a generalised coding assignment for bluetongue viruses. *Virus Res.* **10**, 381–90.

Pourquier, P. and Lesceu, S. (2008) Preliminary validation of the non-structural protein-based ID screen NS BTV ELISA to be used for the differentiation of infected and vaccinated animals (DIVA) Abstracts, Epizone Bluetongue Satellite Symposium, 7th June 2008 Brescia Italy. pp35.

Pritchard, L.I. and Gould, A.R. (1995) Phylogenetic comparison of the serotype-specific VP2 protein of bluetongue and related orbiviruses. *Virus Res.* **39**, 207–20.

Savini, G., MacLachlan, N.J., Sanchez-Vizcaino, J.M. and Zientara, S. (2008) Vaccines against bluetongue in Europe. *Comp. Immunol. Microbiol. Infect. Dis.* **31**, 101–20.

Shad, G., Wilson, W.C., Mecham, J.O. and Evermann, J.F. (1997) Bluetongue virus detection: a safer reverse-transcriptase polymerase chain reaction for prediction of viraemia in sheep. *J. Vet. Diagn. Invest.* **9**, 118–24.

Shaw, A.E., Monaghan, P., Alpar, H.O., Anthony, S., Darpel, K.E., Batten, C.A., Guercio, A., Alimena, G., Vitale, M., Bankowska, K., Carpenter, S., Jones, H., Oura, C.A., King, D.P., Elliott, H., Mellor, P.S. and Mertens, P.P.C. (2007) Development and initial evaluation of a real-time RT-PCR assay to detect bluetongue virus genome segment 1. *J. Virol. Methods* **145**, 115–26.

Singh, K.P., Maan, S., Samuel, A.R., Rao, S., Meyer, A. and Mertens, P.P.C. (2004) Phylogenetic analysis of bluetongue virus genome segment 6 (encoding VP5) from different serotypes. *Vet. Ital.* **40**, 479–83.

Stott, J.L., Barber, T.L. and Osburn, B.I. (1978) Serotyping bluetongue virus: a comparison of plaque inhibition (disc) and plaque neutralization methods. *Am. Assoc. Vet. Lab. Diagn., Proc. Ann. Meet.* **21**, 399–410.

Tabachnick, W.J., MacLachlan, N.J., Thompson, L.E., Hunt, G.J. and Patton, J.F. (1996) Susceptibility of *Culicoides variipennis sonorensis* to infection by polymerase chain reaction-detectable bluetongue virus in cattle blood. *Am. J. Trop. Med. Hyg.* **54**, 481–5.

Taylor, W.P., Gumm, I.D., Gibbs, E.P.J. and Homan, J. (1985). The use of serology in bluetongue epidemiology. In: T.L. Barber, M.M. Jochim and B.I. Osburn, (eds), *Bluetongue and Related Orbiviruses*. New York: Alan R. Liss, pp. 461–8.

Taylor, W.P. and Mellor, P.S. (1994) Distribution of bluetongue virus in turkey, 1978–1981. *Epidemiology and Infection.* **112**, 623–3.

Thevasagayam, J.A., Mertens, P.P.C., Burroughs, J.N. and Anderson, J. (1995) Competitive ELISA for the detection of antibodies against epizootic haemorrhagic disease of deer virus. *J. Virol. Methods* **55**, 417–25.

Thevasagayam, J.A., Wellby, M., Mertens, P.P.C., Burroughs, J.N. and Anderson, J. (1996a) Detection and differentiation of epizootic haemorrhagic disease of deer and bluetongue viruses by serogroup-specific sandwich ELISA. *J. Virol. Methods* **56**, 49–57.

Thevasagayam, J.A., Woolhouse, T.R., Mertens, P.P.C., Burroughs, J.N. and Anderson, J. (1996b) Monoclonal antibody based competitive ELISA for the detection of antibodies against epizootic haemorrhagic disease of deer virus. *J. Virol. Methods* **57**, 117–26.

Thomas, F.C., Girard, A., Boulanger, P. and Ruckerbauer, G. (1976) A comparison of some serological tests for bluetongue virus infection. *Can. J. Comp. Med.* **40**, 291–7.

Toussaint, J.F., Sailleau, C., Breard, E., Zientara, S. and De Clercq, K. (2007) Bluetongue virus detection by two real-time RT-qPCRs targeting two different genomic segments. *J. Virol. Methods* **140**, 115–23.

Van Den Ende, M., Linder, A. and Kaschula, V.R. (1954) Experiments with the cyprus strain of blue-tongue virus: multiplication in the central nervous system of mice and complement fixation. *J. Hyg. (Lond)* **52**, 155–64.

Veronesi, E., Mertens, P.P.C., Shaw, A.E., Brownlie, J., Mellor, P.S. and Carpenter, S.T. (2008) Quantifying bluetongue virus in adult culicoides biting midges (diptera: cerato-pogonidae). *J. Med. Entomol.* **45**, 129–32.

Wade-Evans, A.M., Mertens, P.P.C. and Bostock, C.J. (1990) Development of the polymerase chain reaction for the detection of bluetongue virus in tissue samples. *J. Virol. Methods* **30**, 15–24.

Watson, P. (2001) Clinical diagnosis of FMD in sheep. *Vet. Rec.* **149**(16), 499.

Watson, P. (2004) Differential diagnosis of oral lesions and FMD in sheep. *In Practice* **26**, 182.

Wechsler, S.J. and Luedke, A.J. (1991) Detection of bluetongue virus by using bovine endothelial cells and embryonated chicken eggs. *J. Clin. Microbiol.* **29**, 212–14.

Wechsler, S.J. and McHolland, L.E. (1988) Susceptibilities of 14 cell lines to bluetongue virus infection. *J. Clin. Microbiol.* **26**, 2324–7.

Williamson, S., Woodger, N. and Darpel, K. (2008) Differential diagnosis of bluetongue in cattle and sheep. *In Practice* **30**, 242–51.

Wilson, W.C. and Chase, C.C. (1993) Nested and multiplex polymerase chain reactions for the identification of bluetongue virus infection in the biting midge, *Culicoides variipennis*. *J. Virol. Methods* **45**, 39–47.

Wilson, W.C., Ma, H.C., Venter, G.J., Van Dijk, A.A., Seal, B.S. and Mecham, J.O. (2000) Phylogenetic relationships of bluetongue viruses based on gene S7. *Virus Res.* **67**, 141–51.

Zientara, S., Bréard, E. and Sailleau, C. (2006) Bluetongue: characterization of virus types by reverse transcription polymerase chain reaction. *Dev. Biol. (Basel)* **126**, 187–96.

Bluetongue virus vaccines past and present

18

H. OYA ALPAR,* VINCENT W. BRAMWELL,* EVA VERONESI,† KARIN E. DARPEL,† PAUL-PIERRE PASTORET‡ AND PETER P.C. MERTENS†

*Centre for Drug Delivery Research, London School of Pharmacy, 29–39 Brunswick Square, London, UK

†The Arbovirology Department, Pirbright Laboratory, Institute for Animal Health, Ash Road, Pirbright, Woking, Surrey, UK

‡Publications Department, World Organisation for Animal Health (OIE), Paris, France and 7, rue de la Clissure, 4130 FONTIN (Esneux), Belgium

Introduction

Bluetongue virus (BTV) is endemic in many regions of the world, between the latitudes of 40°N and 30°S, where adults of competent *Culicoides* vector species are present throughout the year, although not all of the 24 serotypes are found in each location (see www.reoviridae.org/dsrna_virus_proteins/btv-serotype-distribution.htm). Areas at the margins of the endemic regions tend to experience seasonal variations in the population densities of adult insect vectors (usually because of colder winters), leading to periods when BTV transmission cannot occur. This results in major seasonal variation in the incidence of bluetongue (BT), the disease caused by this virus. In regions still further from the equator, BT is usually absent, although occasional outbreaks can occur after introduction of the virus, that may also be linked to variations or changes in local climate (particularly temperature) and/or the distribution of vector species of *Culicoides* (Purse *et al.*, 2005).

BTV infections can cause heavy losses in domesticated livestock, particularly in the naïve populations of susceptible breeds of sheep that are present in non-endemic areas. Although other ruminants can also be infected, disease

ISBN-13: 978-0-12-369368-6

is usually less severe and may be sub-clinical in cattle, goats and most other ruminant species. For example, during the outbreak of BT caused by BTV-8 in northern Europe during 2006–2007, the case fatality rate in sheep was estimated as >30%. Fewer cattle showed clinical signs of infection, and these had a lower case fatality rate (estimated at < 5%) (Szmaragd *et al.*, 2007). However, infected animals that do not show clinical signs of the disease can also serve as 'reservoir hosts' for the virus, infecting vector insects that feed on them, leading to transmission of the virus, and disease.

During an outbreak, BTV is transmitted in the field primarily by a continuous cycle of infection between the insect vectors and the susceptible ruminant hosts. However, the virus can also survive for periods (up to 9–12 months) in the absence of adult vectors or detectable transmission/disease, by an undetermined 'over-wintering' mechanism. As a result of overwintering, BT outbreaks can last for several years, even in non-endemic areas. For example, the first recorded BT outbreak in Portugal and Spain lasted from 1956 till 1960 (Roberts, 1990); more recent outbreaks in Greece, caused by BTV types 1, 4, 9 and 16, collectively lasted from 1998 to 2001; and BTV-8 survived the northern European winter (2006–2007), reappearing almost simultaneously in many areas of Germany, France, Belgium and the Netherlands in 2007. Takamatsu *et al.* (2003) suggested a possible mechanism for overwintering of BTV in the ruminant host, involving persistent infection of cellular components of the host's immune system (see Chapter 12). The inflammation caused by insect saliva also results in recruitment of these cells to the biting site, potentially facilitating transmission to the vector insect during feeding (Darpel, 2007). It has been reported that several other arboviruses can overwinter in the developmental stages of their invertebrate vectors (transovarial or vertical transmission). Recent data suggest that BTV may also exploit this strategy (White *et al.*, 2005) (see Chapter 14), although attempts to recover infectious virus from the next generation of adult insects have so far been unsuccessful.

Control measures such as the application of insecticides and larvicides, or protecting susceptible hosts against exposure to biting insects are discussed elsewhere (see Chapters 14 and 19) but are considered unlikely to be entirely successful by themselves in preventing the occurrence, persistence and spread of BTV. Although the BTV transmission cycle can also be interrupted by immunization of ruminant hosts, including sheep, goats and cattle, the full impact of a vaccine control strategy is difficult to predict as the immunization of ruminant species and breeds that rarely suffer from the disease may be seen as undesirable or unnecessary by farmers and therefore difficult to achieve. Although not fully understood, the process of overwintering clearly represents an important factor in the epidemiology of BTV that also needs to be considered in the design, implementation and timing of control strategies.

Vaccination can be used to protect susceptible animals against the clinical signs of BT and prevent onward spread of the virus, by restricting virus replication and neutralizing the virus within the infected animals. Until recently, the only commercial BTV vaccines that were widely available were live virus strains that had been attenuated by multiple passage in embryonated hens' eggs and passage or plaque cloning in mammalian cell cultures. Most of these vaccines were developed for use in southern Africa (Table 18.1 – where multiple serotypes of BTV are endemic), primarily to protect exotic breeds of sheep against clinical signs of the disease. Indeed, the vaccination of sheep with live BTV vaccines (Onderstepoort Biological Product OPB – South Africa) has proved to be an effective control measure for disease in South Africa, for almost a century (Chapter 2). These live vaccines are inexpensive and easy to produce, and represent a 'tried and tested' technology for the control of clinical BT in sheep in South Africa. Most South African sheep breeds have some innate resistance to BT disease and are therefore unlikely to be adversely affected by infection and replication of the attenuated vaccine strains of the virus. However, it is important to note that the primary aim of vaccination with these strains is to protect against clinical disease in areas where BTV is already endemic, rather than to prevent onward transmission of the virus. A live attenuated BTV-10 vaccine is also currently licensed nationally, for use in sheep in the USA (Kemeny and Drehle, 1961; Luedke and Jochim, 1968). Attenuated BTV 10, 11 and 17 vaccines are also available for use in sheep in the USA (see Chapter 9), and attenuated vaccine - containing

Table 18.1 Details of Onderstepoort bluetongue virus, attenuated virus vaccine strains

Virus type	Strain	Origin	Passage History
BTV-1	Biggarsberg/8012	RSA, 1958	50E 3P 4BHK
BTV-2	Vryheid/5036	RSA, 1958	50E 3P 4BHK
BTV-3	Cyprus/8231	Cyprus, 1944	45E 2BHK 3P 5BHK
BTV-4	Theiler/79043	RSA, ~1900	60E 3Pa 9BHK
BTV-5	Mossop/4868	RSA, 1953	50E 2BIIK 3Pa 6BHK
BTV-6	Strathene/5011	RSA, 1958	60E 3Pa 7BHK
BTV-7	Utrecht/1504	RSA, 1955	60E 3Pa 7BHK
BTV-8	Camp/8438	RSA, 1937	50E 3BHK 10Pa 7BHK
BTV-9	University Farm/2766	RSA, 1942	70E 2BHK pp 3BHK 7P 6BHK
BTV-10	Portugal/2627	Portugal, 1956	E81
BTV-11	Nelspoort/4575	RSA, 1944	35E 3P 5BHK
BTV-12	Estantia/75005	RSA, 1941	55E 3P 4BK
BTV-13	Westlands/7238	RSA, 1959	45E 2BHK 3Pa 4BHK
BTV-14	Kolwani/89/59	RSA, 1959	60E 3Pa 4BHK
BTV-16	Pakistan/7766	Pakistan	37E 3P 2BHK 1 Vero
BTV-19	143/76	RSA, 1976	29E 3Pa 3BHK

E = Number of passages in eggs; BHK = Number of passages in baby hamster kidney cells; Vero = Number of passages in green monkey kidney cells; P = Number of large plaque selections; p = Number of small plaque selections; a = Small plaque variant.

Data derived from: OIE Manual of Standards for Diagnostic Tests and Vaccines (1996). Onderstepoort Biological Products, Products Manual (1996): National Department of Agriculture, ISBN 062116142 X.

serotypes 10, 11, 13 and 17 - have been used to vaccinate wildlife species, in California (McConnell *et al.*, 1985) (see Chapter 9).

Several live attenuated BTV vaccine strains have been shown to replicate efficiently in the vaccinated host, which results in the development of a strong immune response. The fact that vaccinated animals may develop viraemia and can potentially transmit the attenuated virus may therefore be considered to be relatively unimportant in endemic areas. However, the major purposes of BTV vaccination within Europe (an area where the BT is epidemic, rather than endemic, and where local sheep and cattle have little, if any, innate resistance) are not only to prevent infection and protect against the clinical signs of the disease in individual animals, but also to prevent the replication, spread and persistence of the virus. Consequently, there are entirely justifiable concerns about the risks associated with the use of live BTV vaccine strains in naïve animal populations such as those in Europe. Indeed, it has been demonstrated that vaccination can result in severe clinical signs in certain European breeds of sheep that would prove fatal under field conditions (Veronesi *et al.*, 2005, 2008). Live vaccine strains can cause high levels of viraemia that are sufficient for transmission in the field, leading to outbreaks of 'vaccine derived' disease, providing opportunities for exchange (reassortment) of genome segments between different strains (e.g. BTV-2 and 16 in mainland Italy and Sardinia in 2002 and 2004; – Ferrari *et al.*, 2005; Batten *et al.*, 2008a, b). Live attenuated vaccines can also be teratogenic, and their use in pregnant animals should therefore be avoided (see Chapter 13). With live virus vaccines there is also a risk that the vaccine virus could become established or even endemic in the region, representing a continuing threat of disease, as well as longer term economic losses because of restrictions in trade. Nevertheless, a decision not to vaccinate sheep in a region that is at risk, or a failure to vaccinate all or most of the susceptible animals in the region, could result in considerable losses as a result of severe disease caused by field strains of BTV.

During the recent outbreaks of BT in the Mediterranean region which started in 1998, Greece made a decision not to use live attenuated BTV vaccines but attempted control by relying only on the traditional zoosanitary measures (see Chapter 19). Although this led, initially, to some outbreaks of disease, after four years the virus had died out, allowing the country to become disease-free. In contrast, Italy adopted a programme of active vaccination, using the attenuated vaccines that were currently available and has experienced continuing though reduced outbreaks of disease caused by transmission of both field strains and the vaccine viruses themselves (Ferrari, *et al.*, 2005; Batten *et al.*, 2008a, b). The problems associated with live vaccines have led to a demand for safer vaccines that can be used to effectively control the severity, spread and persistence of the disease. Alternative safer vaccines, generated by inactivation of tissue culture-derived virus, are now becoming available for at least some of the European BTV types (including type 8), and inactivated BTV-2, 4 and 16 vaccines have recently been used successfully in southern Europe (Savini *et al.*, 2007). There is also ongoing research to develop the next generation of subunit, recombinant virus and nucleic acid-based vaccines, as well as more advanced delivery and adjuvant systems. It has been suggested that 80% of the total susceptible population (cattle, sheep and goats) need to be immunized in order to provide adequate protection against BTV transmission (Giovannini *et al.*, 2004). Consequently, although prophylactic immunization of sheep may still be the most easily implemented and effective measure to combat disease, this is unlikely to control the spread or recrudescence (overwintering) of the virus. Certain problems also exist that relate specifically to vaccination against BTV. These include the existence of 24 different serotypes of the virus, which do not usually cross-neutralize or cross-protect and each of which has the potential to cause outbreaks of disease. Each serotype involved in an outbreak therefore needs to be targeted specifically by use of a vaccine of the same serotype. Multiple serotypes can also be simultaneously involved in an outbreak, so that more than one vaccine serotype may be required at the same time.

Live attenuated BTV vaccines are currently available for most of the 24 BTV serotypes and have been used in southern Europe against BTV types 2, 4, 9 and 16, contributing to the genetic diversity of the circulating viruses (Veronesi *et al.*, 2005; Savini *et al.*, 2007). However, it is now widely recognized that the virulence of these strains for some European breeds of sheep, their ability to be transmitted in the field and their involvement in the generation of novel progeny virus strains by exchange of genome segments (reassortment), compromises their suitability for use in non-endemic regions (Ferrari *et al.*, 2005; Veronesi *et al.*, 2005, 2008; Savini *et al.*, 2007; Batten *et al.*, 2008a, b). Alternative vaccination strategies, including recombinant expressed virus-like or core-like particles (VLP or CLP), subunit vaccines composed of individual expressed proteins, recombinant heterologous virus vectors (e.g. vaccinia virus or canary pox) or vaccines based on inactivated whole virus (some of which are now available), have all shown experimental promise with BTV and/or related orbiviruses. In the longer term, synthetic delivery systems, subunit vaccines and novel adjuvant combinations have the potential to generate much safer and more effective BTV vaccines. A downside is that many of the replacements for the attenuated virus vaccines will be more expensive, but this is likely to be outweighed by increased safety and an inability to be transmitted.

Multiple BTV serotypes

Outbreaks of BT can occur involving a single BTV serotype, as in the northern European outbreak during 2006 and 2007, caused by BTV-8 (Darpel *et al.*, 2007; Mertens *et al.*, 2007; Maan *et al.*, 2008). However, outbreaks of BT can often also occur that involve more than one virus serotype. For example, three serotypes (4, 9 and 16) were detected during the BT outbreaks in Greece during 1999 (see Chapters 11 and 19). It is also evident that multiple serotypes are frequently present in endemic areas (see www.reoviridae.org/dsrna_virus_proteins//btv-serotype-distribution.htm), making reassortment between distinct serotypes and strains, an additional concern (Stott *et al.*, 1982; Oberst, 1985; Samal *et al.*, 1987a, b; Pritchard *et al.*, 2004; Batten *et al.*, 2008a). The absence of significant cross protection subsequent to infection with a single serotype of BTV indicates that effective vaccines either need to be the same serotype as the outbreak virus or will require multivalent serotype components. Some cross-protection can be generated by immune responses to the more conserved viral antigens, and sequential inoculation with multiple serotypes can raise a broad spectrum neutralizing antibody response against several of the 24 BTV serotypes including some not present in the inoculum, although this is usually less enduring (Jeggo *et al.*, 1983a, b, 1984, 1986; Jones *et al.*, 1997; Wade-Evans *et al.*, 1996; Ghosh *et al.*, 2002a, b). Problems involving interference between virus strains, differences in the replication efficiency and immunogenicity of different virus strains, and variations in the responses of animals to individual vaccine components, may also prevent the development of an effective response to at least some of the component strains used in attenuated multivalent vaccines (Jeggo *et al.*, 1983a, b, 1986). On this basis the South African multivalent vaccines are designed to be used as three sequential inoculations, each of which involves multiple serotypes, with the intention of generating a broad protective response (Onderstepoort Biological Products (OBP) Products Manual).

Reliable diagnostic systems that can rapidly identify the BTV type(s) circulating in a region and/or those involved in an outbreak are necessary for the design and implementation of appropriate control strategies that target the virus strains involved (see Chapter 17). Vaccines that can be used to safely elicit a cross-reactive and protective immune response against multiple BTV strains/types would be of particular value in regions that are under the threat of incursions by multiple BTV serotypes.

Attenuated 'live' vaccines

History and efficacy

A monotypic blood vaccine (BTV-4) was introduced and used with some success for approximately 40 years, in the Republic of South Africa, but was withdrawn later because of safety and efficacy concerns (see Chapter 2). The discovery that BTV can be grown in embryonated hens' eggs led to the development of safer and immunogenic vaccine strains that were attenuated by multiple passage of field strains (Table 18.1). When cell culture technology became widespread, the egg-attenuated vaccine strains were subsequently plaque purified. For the last 50 years the polyvalent live attenuated BTV vaccines that were produced in this way have been used to prevent clinical disease in the endemic regions of South Africa, with considerable success. These vaccines generate effective and long-lasting immunity in the vaccinated animal (Verwoerd and Erasmus, 1994).

Live attenuated BTV vaccines have also been used in a more limited way in a few other countries [including the USA and more recently in Italy, Bulgaria, Israel, France (Corsica) and Spain]. The present, standard, OBP BTV vaccine is produced in cell culture and supplied as a freeze-dried polyvalent preparation (Reg. No. G 358 Act No. 36/1947) that is administered subcutaneously. The vaccine comprises three bottles and includes the following serotypes of BTV – Bottle A: BTV serotypes 1, 4, 6, 12 and 14; Bottle B: BTV serotypes 3, 8, 9, 10 and 11; Bottle C: BTV serotypes 2, 5, 7, 13 and 19.

The three components of the multivalent vaccine are administered separately, at 3-week intervals. A monovalent BTV vaccine or combination of vaccines can be produced on special request by OBP but requires two months for production and quality control. Details of the vaccine strains are summarized in Table 18.1. These Onderstepoort attenuated vaccine strains were initially validated for use in sheep only and their efficacy in other ruminants was not determined. Following the 1999 outbreaks of BT in Greece, further research was recommended in order to establish the efficacy and safety of live attenuated vaccines in cattle and goats. However, four different monovalent BTV vaccines (types 2, 4, 9 and 16) were also used extensively in both sheep and cattle in Italy from 2002 onwards (Caporale *et al.*, 2004), although this strategy did not lead to the eradication of BTV from the affected areas.

Concerns associated with attenuated BTV vaccines

Concerns about the use of live attenuated BTV vaccines, particularly in non-endemic regions, can be summarized into seven main areas: (1) The virulence of vaccine strains in susceptible breeds of sheep; (2) The possibility that (single or multiple) passage through the insect vector and /or mammalian host could lead to an increase in or reversion to virulence; (3) Teratogenicity of the attenuated virus for the developing foetus; (4) Vertical transmission of the vaccine strain in the mammalian host (e.g. by infection of the Foetus *in utero*, or by excretion of virus in the semen of bulls and rams); (5) The possibility that vaccine virus will infect vectors, becoming established and transmitted in the field; (6) The possibility of disease outbreaks caused by transmission of the BTV vaccine strains; (7) The possibility that reassortant viruses will be generated, containing genome segments derived from vaccine and field strains, with altered biological properties that might include novel serological characteristics, or different (enhanced) virulence (Murray and Eaton, 1996; Veronesi *et al.*, 2005; Batten *et al.*, 2008a).

The application of attenuated vaccines therefore needs to be well managed. Although it has been suggested that breeding ewes and rams should be vaccinated before mating, maternally derived (passive) immunity in young sheep can interfere with replication of the vaccine-virus strains and the development of active immunity. There is also a consensus that the live virus vaccines should not be used during the vector seasons because adult *Culicoides* vectors may become infected and transmit the virus to other animals.

The South African BTV vaccine strains have been attenuated by multiple passage in embryonated hens' eggs and plaquing/passage in cell culture (Table 18.1). However, selection of virus strains that can replicate effectively in heterologous culture systems does not preclude their replication in sheep or other ruminants. Indeed, replication of the vaccine strains may be advantageous or even essential for the development of an effective immune response. BTV is transmitted from one host to another through the bites of the insect vector (adult female *Culicoides*), which become infected after ingesting the virus as part of a blood meal. Any significant level of replication by BTV vaccine strains in the ruminant host may also result in a significant level of viraemia (see Chapter 12) and the potential for infection and transmission of the vaccine strain by feeding *Culicoides*. It is considered possible that transmission of the vaccine strain in the field, potentially involving repeated serial passage through the mammalian host and vector insect, would lead to selection of variants that are better adapted to transmission and survival. This selection could also be accompanied by a reversion to virulence, generating a distinct and potentially dangerous strain of the virus.

It has been suggested that a viraemia of less than 1000 plaque-forming units (PFU) per millilitre of blood will ensure that the virus cannot be transmitted by bloodsucking insects (OBP Products Manual: National Department of Agriculture, 1996). However, recent studies of BTV replication in the sheep skin suggest that the replication and level of BTV in the skin itself may be more significant and the risk of transmission does not correlate accurately with the level of virus detected in the bloodstream (viraemia) (Darpel, 2007; Chapter 12). Titres of less than 1000 PFU per millilitre of blood at the height of viraemia, are still considered likely to elicit an effective neutralizing antibody response, and may reduce the likelihood of effective transmission. This has therefore been suggested as a parameter for vaccine strain selection (OBP Products Manual: National Department of Agriculture, 1996). However, the level of viraemia is also likely to vary between individual animals, and between different breeds and species of ruminant host.

Some of the concerns relating to live BTV vaccines have recently been investigated under laboratory conditions (Veronesi *et al.*, 2005) using a highly susceptible European breed of sheep (Dorset Poll). Four different attenuated monovalent BTV vaccines (BTV2, 4, 9 and 16) all caused viraemia in each of the sheep that were vaccinated, with titres ranging from 4.25 to 6.83 $\log_{10}$ $TCID_{50}$/mL. The longest duration of viraemia was 27 days post infection with the BTV-16 vaccine strain (Table 18.2). These high levels of viraemia suggest that feeding *Culicoides* would become infected and could potentially transmit the virus, helping to explain the seroconversion and isolation of vaccine strains from unvaccinated sentinel cattle in Italy (Ferrari *et al.*, 2005; Batten *et al.*, 2008b).

Table 18.2 Pyrexia and viraemia duration recorded in Dorset poll sheep after inoculation with bluetongue virus vaccines and passaged vaccines

Vaccine	Serotype	No. of sheep showing pyrexia	Pyrexia duration range (days)	Maximum viraemia titre recorded[a]	Maximum viraemia duration (days)	Minimum viraemia duration (days)
Vaccine	BTV2	2/4	1	4.75	17	17
	BTV4	2/4	2	4.25	16	11
	BTV9	4/4	1–5	5.25	15	7
	BTV16	4/4	6	6.83	19	17
	BTV16	3/3	1	6.0	23	10
C. sonorensis	BTV2	0/4	0	4.5	15	5
1st passage	BTV4	2/4	2	4.25	13	11
	BTV9	4/4	1–5	5.5	9	5
C. sonorensis	BTV2	1/4	1	3.75	19	13
2nd passage	BTV4	4/4	3	5.0	10	6
	BTV9	4/4	1–3	6.0	13	7
C. nubeculosus	BTV2	0/4	0	3.5	15	9
1st passage	BTV4	1/4	1	4.25	14	9
	BTV9	2/4	1–4	6.25	17	9

Four different serotypes of BTV live attenuated vaccine viruses were individually tested in sheep: **BTV2**, **4** and **16** produced by Onderstepoort Biological Products (OPB) S. Africa and **BTV9** labelled as 'IZS Istituto Zooprofilattico Sperimentale dell'Abruzzo e del Molise G. Caporale – Via Campo Boario 64100 Teramo (Italy)'. Separate groups of sheep were also inoculated with homogenised adults of a vector species (*C. sonorensis*) and a non vector species (*C. nubeculosus*) previously infected (once or twice), with same vaccine viruses. All of the attenuated viruses caused viraemia in all the vaccinated sheep, with peak titres ranging between 4.25 and 6.83 log10 TCID50/mL. The longest duration period of viraemia detected in vaccinated sheep was 27 days post infection with BTV16 serotype. A variable period of pyrexia was detected (that lasted for up to 6 days) in at least two out of four sheep vaccinated with each of the BTV serotypes – data from Veronesi et al. (2005, 2008).

[a] $\log_{10}$TCID50/mL.

Separate groups of sheep were also inoculated with homogenized adult *Culicoides sonorensis* that had previously been infected with the same vaccine strains, although there was no conclusive evidence for reversion to virulence after passaged in vector *Culicoides* (Table 18.2).

One of the major concerns is that different vaccine strains, or vaccine and field strains may reassort, giving rise to progeny virus strains, that could have different or enhanced biological characteristics (Cowley and Gorman, 1989; Mertens *et al.*, 1989; O'Hara *et al.*, 1998; Monaco *et al.*, 2006; Batten *et al.*, 2008a). The strain of BTV-16 that was isolated in Italy during 2002 represents the first BTV reassortant that was identified in the field in Europe (Monaco *et al.*, 2006; Batten *et al.*, 2008a). This virus is a reassortant between the BTV-16 vaccine (which contributed genome segment 2) and the BTV-2 vaccine (which contributed genome segment 5 [*NS1* gene]) (Batten *et al.*, 2008a). However, the BTV-16 vaccine had not been used in Europe during or prior to 2002, indicating a different origin for the BTV-16 parental strain that gave rise to this reassortant. Although BTV-16 isolates were also made in the eastern Mediterranean region (Turkey, Bulgaria and Cyprus from 1998 onwards, see www.iah.bbsrc.ac.uk/dsRNA_virus_proteins/ReoID/btv-16.htm), these strains were more distantly related to the vaccine, indicating that the Italian field strain had not arrived from that direction. However, the BTV-16 vaccine had been used for many years, as part of an annual vaccination campaign in Israel, giving rise to vaccine-derived BTV-16 field strains in the region (Batten *et al.*, 2008a). Since 2000, strains of BTV-1, 2 and 4 have entered Italy and the western Mediterranean region, from North Africa (see Chapter 11). It is possible that a field strain of BTV-16 recently derived from the vaccine in Israel, also arrived in Italy via this route, and had already reassorted (e.g. in Israel) or subsequently acting as one of the parental strains for reassortment in Italy 2002. There is also evidence for the generation of other reassortant viruses within Europe (Maan *et al.*, 2008). Previous studies in other areas of the world where the disease is endemic have demonstrated high-frequency reassortment of BTV genome segments during dual infection of sheep and cattle (Stott *et al.*, 1982; Samal *et al.*, 1987b) and in *Culicoides* species experimentally infected with different BTV serotypes (Samal *et al.*, 1987a).

Reassortant strains of BTV are generated by replication of two or more distinct parental strains within the same cell. The emergence of a single reassortant strain of BTV from one infected cell, to dominate the population of virus causing an outbreak, suggests that it is particularly well suited to the local ecosystem and may have a selective advantage over both the parental strains and the other reassortant viruses that would also be generated (e.g. more easily or efficiently transmitted) (Batten *et al.*, 2008a).

Concerns over the safety of the live attenuated BTV vaccines have implications for their use (particularly in non-endemic regions), with recommendations that provide farmers with logistical considerations as well as limitations. In South Africa, it is recommended that OBP's attenuated vaccines are not used in sheep between August and October, with immunization of ewes commencing 9–12 weeks before mating. Vaccination of pregnant ewes is not advised during the first half of pregnancy, and rams should be inoculated after the mating season. Lambs from immunized ewes should be vaccinated at the age of six months and older. If animals are vaccinated at an earlier age in heavily infected areas, it is recommended that they are revaccinated at the age of six months. If sheep need to be vaccinated annually, it is recommended that exposure to wild-type strains and virus vector are limited. There is clear evidence in Europe that the attenuated vaccine viruses can be transmitted by insects from vaccinated to unvaccinated animals (Ferrari *et al.*, 2005; Monaco *et al.*, 2006; Savini *et al.*, 2007; Batten *et al.*, 2008a, b) and consequently the possibility of reversion to virulence becomes more than a simple theoretical possibility. Transmission of vaccine strains in the field also dramatically increases the possibility of reassortment with other field or vaccine strains.

There are internationally accepted rules restricting the movement of animals for sixty days after they have been vaccinated with the live attenuated viruses. Primarily, this is to allow any viraemia that may be generated to subside, removing or very significantly reducing the risks of subsequent virus transmission (OIE, 2007).

Live vaccine 'platforms'

Jeggo *et al.*, (1984) observed that the degree of protection conferred by a BTV infection did not correlate with the level of neutralizing antibody. They concluded that although neutralizing antibodies can play a significant role in protection against BTV, the outcome of challenge with a virulent strain of the virus will depend on several interacting factors that may include a cell-mediated immune response.

Expression of heterologous antigens in replication competent, heterologous viral carriers, has the potential to generate potent immune responses, and would remove concerns about reassortment of genes from wild and vaccine strains of BTV. A recombinant capripox virus (CPV) expressing VP7 of BTV-1 has previously been used to generate at least partial cross-protection against BTV-3 (Wade-Evans *et al.*, 1996). Eight sheep receiving the recombinant CPV recovered fully in comparison with a control group where all of the animals died following challenge with a virulent strain of BTV-3. Interestingly, this study did not involve VP2 (the major target for BTV-specific neutralizing antibodies) and no neutralizing antibody response was generated, or consequently involved in the protection observed. Viral proteins, other than the type-specific structural components of the outer capsid layer, can therefore play an important role in protection against BTV, possibly through a cell-mediated component of the immune response (Andrew *et al.*, 1995). This protection was also not serotype specific.

CPV has a restricted host range (cattle, sheep and goats), and it has been suggested that it may therefore be more useful for vaccination than a vaccinia virus vector that would potentially infect a much broader host range (Jeggo *et al.*, 1984). However, more recent studies have involved the development of a vaccinia virus, Modified Vaccinia Ankara (MVA) strain that is replication defective in any normal mammalian cell but is still an effective vector for expression of heterologous antigens in infected cells (Julyr *et al.*, 1975; Chiam *et al.*, 2008). A recombinant canary pox vector, expressing BTV outer capsid proteins, that generates a high level of protection, has also recently been reported (Boone *et al.*, 2007). CPV, MVA and canary pox virions contain one molecule of linear double-stranded DNA (unlike the segmented dsRNA genome of BTV). The mRNAs that are transcribed from the genomic DNA of these viral carriers, for expression of BTV proteins, have modified terminal regions (compared with native BTV mRNAs) removing any possibility of 'rescue' and development of novel BTV strains by interaction with wild-type BTV. Recent studies with African horse sickness virus (Chiam *et al.*, 2008) have also shown development of an antibody response to AHSV-4 NS1, and a neutralizing antibody response to VP2. Similar studies are in progress for other serotypes of BTV.

Inactivated BTV vaccines

Preparations of BTV are commonly inactivated for use as vaccines, at 37°C by treatment with 0.02 M binary ethylenimine (BEI) prepared from 2-bromoethylamine hydrobromide (e.g. Odeon *et al.*, 1999). Lefevre and Desoutter (1988) cite early trials of inactivated vaccines against BTV prepared from whole virus that were conducted as early as 1975 (Parker *et al.*, 1975). The advantages were obvious; such vaccines could be given to pregnant female ruminants, could be prepared quickly in case of the emergence of a new serotype, would avoid the risk of reassortment and reversion to virulence and could be administered to cattle without risk.

Several attempts were made (Parker *et al.*, 1975; Campbell, 1985; Stott *et al.*, 1985) to develop an effective inactivated vaccine but results were disappointing. The first commercially available vaccine was developed by the company Merial under the name 'BTVPUR AISap'. It contains inactivated BTV, and is an adjuvanted liquid vaccine. It is recommended that vaccination is applied before, or as soon as, an outbreak is detected. This suggests that inactivated vaccines could be used, either as an emergency or ring vaccination of infected flocks to avoid spread of infection, or as a preventive vaccination campaign (possibly in winter) to avoid new outbreaks during the following year.

The Merial vaccine contains inactivated, chromatography-purified and concentrated particles of BTV-2, with aluminium hydroxide and saponin adjuvants.

Only moderate local reactions are observed at the injection sites. Vaccinated animals are protected against challenge with the homologous BTV serotype and virus 'shedding' was prevented. Although this vaccine is monovalent (BTV-2), similar monovalent or polyvalent inactivated vaccines are also now available for BTV-4, and are under development for other BTV serotypes, particularly those currently circulating in and near to Europe (i.e. BTV-1, 8, 9, 15 and 16). Recently, the development of an inactivated BTV-16 vaccine to be used in Italy has been reported (Savini *et al.*, 2007).

In 2008, several of the northern European governments also developed vaccination strategies, using inactivated vaccines (as produced by Intervet, Merial and Fort Dodge), to control the disease outbreak caused by BTV-8, which started in the Netherlands and Belgium during 2006 (see Chapters 7 and 11).

Subunit vaccines

Virus-like particles

The BTV particle is composed of three concentric layers, composed of seven structural proteins (VP1-VP7) (colour plates 1, 2 and 15 (virus cross-section cartoon and images of particles from cryomicroscopy and X-ray crystallography) see Chapters 4 and 6). The innermost complete shell of the virus capsid (the 'subcore') is composed of VP3 and surrounds the three minor enzyme proteins [VP1(Pol), VP4(Cap) and VP6(Hel)] and the 10 dsRNA segments of the BTV genome. When VP3 is expressed by itself, it can self assemble to form sub-CLP and the sub-cores of some orbiviruses are sufficiently stable so that they can be purified. These observations indicate that the initial formation of the subcore provides a scaffold, for the assembly of the BTV particle. The sub-core is subsequently decorated with trimers of VP7, which forms the 'core-surface layer' of the virus capsid, which gives strength and rigidity to the capsid structure. The BTV 'outer-capsid' is composed of trimers of VP2, and trimers VP5, which are added to the VP7 surface of the completed core particle. (Grimes *et al.*, 1998: Gouet *et al.*, 1999; Mertens and Diprose, 2004).The innermost structural proteins of BTV, other orbiviruses and even other more distantly related dsRNA viruses, tend to be more highly conserved and therefore antigenically cross-reactive within each virus species than the components of the outer capsid layers, which tend to be serotype specific (Mertens, 2004).

CLP and VLP can be made by expression of BTV structural proteins VP3 and VP7, or VP2, VP3, VP5 and VP7 (respectively) using recombinant baculoviruses (French and Roy, 1990). This indicates that the genomic dsRNA and non-structural proteins (NS1, NS2 or NS3) are not required for assembly of the structural-proteins into BTV particles (French and Roy, 1990;

Balyaev and Roy, 1993). Their structural similarity to native BTV virus or core particles, led to early recognition that VLP or CLP have potential as safe vaccine components. These synthetic particles do not contain viral nucleic acids or cellular components, removing concerns about BTV vaccine replication or reassortment, and reducing the chance of adverse reactions in the host. Vaccination of sheep with as little as 10 μg of VLP (containing the four major structural proteins of BTV: VP2, VP3, VP5 and VP7) has been reported to give protection against challenge with the homologous BTV serotype (Roy, 2003). Some cross-protection was also achieved, depending on the challenge virus and the amount of antigen used for vaccination. Development of a system for simultaneous expression for the four major structural proteins of BTV from the same baculovirus genome was seen as a step forward. This removes problems potentially caused by incomplete expression of all four BTV proteins from different recombinant viruses, if one or more of them fail to infect every cell.

VLP are morphologically similar to native BTV particles (Belyaev and Roy, 1993). However, the absence of the packaged genome segments reduces the outward pressure exerted on the CLP and VLP structure (Gouet *et al.*, 1999; Mertens and Diprose, 2004), which may result in a slightly reduced diameter. This appears to make it more difficult for the full complement of outer capsid proteins to fit onto the VP7 surface of the core and may reduce the stability of the VLP structure. Although the major protein components of CLP (VP3 and VP7) are not involved in the neutralizing antibody response against BTV, limited vaccination trials indicate that these synthetic particles can give partial protection against homologous and heterologous virus challenges (Roy, 2003). The BTV core proteins are antigenically conserved across the 24 BTV serotypes, and VP7 raises a very strong antibody response in infected or vaccinated animals. CLP could therefore form a component of a 'cross-reactive' BTV vaccine.

The lack of neutralizing antibodies generated to inactivated BTV (Stott *et al.*, 1985) provides evidence for the importance of maintaining viral protein conformation. VLP present the major BTV structural proteins in the correct conformation (including components of the outer capsid), mimicking native virus particles and raising a neutralizing antibody response. They are also readily recognized by the immune system, so that smaller amounts of VLP are needed to elicit a similar level of protection to the same BTV structural proteins expressed individually (Noad and Roy, 2003). However, in order to remove contaminating cellular components, CLP and VLP must still be purified. Unfortunately their assembly from expressed proteins in the heterologous insect cell systems used may also be suboptimal, and there are indications that they are less stable than native virus particles, potentially reducing their efficiency of production/assembly and increasing losses during purification. The expression vectors/recombinant viruses used for synthesis of VLP may also suffer losses of coding capacity during preparation, again reducing yield and increasing costs. Indeed, the costs of VLP production are considered to be relatively high compared with the value of the individual host animal, and as yet, there are no commercially available vaccines for BTV based on CLP or VLP, despite the technology having been available for almost twenty years.

Individual BTV proteins as vaccine candidates

The majority of neutralizing epitopes of BTV are situated on VP2 (Huismans and Bremer, 1981; Appleton and Letchworth, 1983). VP2 has also been identified as the major determinant of BTV serotype by cross hybridization, studies of reassortant viruses (Mertens *et al.*, 1987, 1989; Cowley and Gorman, 1989) and in sequencing studies (Maan *et al.*, 2007; Mertens *et al.*, 2007). However, these studies also showed that the other outer-capsid protein 'VP5' can also have some influence on virus serotype. VP5 is thought to exert an influence over the structure of the highly conformationally dependent VP2 molecule thereby affecting neutralizing antigenic sites, although VP5 may also contain a small number of neutralization sites in its own right (DeMaula *et al.*, 2000). This may help to explain why the use of a combination of VP2 and VP5 as vaccine subunits is more effective than VP2 alone. Inoculation of BTV outer-capsid protein VP2 is necessary for production of neutralizing antibodies and makes a significant contribution to the generation a protective response against viral challenge in sheep. In studies by Roy *et al.* (1990), doses of greater than 50 µg of baculovirus expressed VP2 per sheep were required for full protection. A dose of 50 µg was partially protective, but inoculation in combination with VP5 resulted in protection and higher BTV-neutralizing antibody titre. This enhancement of the neutralizing antibody response was not observed during co-inoculation with any of the other viral proteins. Monomeric VP2 has a molecular weight of around 111 kDa, but VP2 appears as a trimer in the virus particle (Hewat *et al.*, 1992a, b, 1994) and multimers of VP2 are thought to be stabilized by disulfide linkages (Hassan and Roy, 1999). VP2 is responsible for cell attachment and is involved in cell-entry in both mammalian and insect cell systems, emphasizing the potential for antibodies to VP2 to prevent initiation of infection and block transmission (Hassan and Roy, 1999).

A neutralizing serotype-specific determinant of VP2 from BTV-13 has been cited as representing the major part of a conformational epitope (Hwang and Li, 1993) (the linear epitope EMDDDETEYE located at residues 642–651). However, the majority of neutralizing epitopes on VP2 appear to be highly conformationally dependent (Gould and Eaton, 1990; White and Eaton, 1990; DeMaula *et al.*, 2000) and some may therefore be discontinuous in nature. A loss in reactivity with neutralizing monoclonal antibodies has been reported when viral proteins are solubilized (Grieder and Schultz, 1989). Although detergent treatment can also result in stronger reactivity of BTV-specific antibodies, this may be due to unmasking epitopes usually located in less-accessible regions of the protein (Nagesha *et al.*, 2001).

Binding of monoclonal antibodies to VP2 can influence neutralization determinants elsewhere on the protein (Gould and Eaton, 1990; White and Eaton, 1990), and it was concluded that the conformation of the protein itself determines whether epitopes that are conserved across the BTV serogroup are involved in neutralization of the different virus serotypes. Sequencing studies of BTV genome segment 2 from different serotypes have only identified a maximum region of six conserved nucleotides across all BTV strains, and these are the conserved hexanucleotide termini within the non-coding regions of the segment. The longest stretch of conserved amino acids within VP2, across all isolates, was found to be only 5 residues in length (Maan et al., 2007). If there are relatively more conserved sequences or epitopes within VP2, these will exist more frequently between the members of the same genome segment 2 'nucleotype', as identified by Maan et al. (2007). The existence of conserved sites on VP2 that are not always active neutralization epitopes within the different BTV serotypes, could potentially help to explain why sequential infection with different BTV types generates a broad neutralizing antibody response, which even recognizes types not previously encountered by the host animal. (Jeggo et al., 1986).

The BTV outer-capsid protein VP2 contains antigenic sites recognized by B cells that are involved in the neutralizing antibody response. However, there is also a cell-mediated immune response following BTV infection and at least one site is recognized by ovine helper (CD4$^+$) T cells, and epitopes for cytotoxic T cells have been identified in VP2 (Takamatsu et al., 1990, Andrew et al., 1995). The mechanism involved in the cell-mediated response against BTV and its role in protection are only partly understood (see Chapter 12). However, it has been demonstrated that it can contribute to the overall level of protective immunity. Anti-BTV cytotoxic T lymphocytes (CTL) have been identified in mice (Jeggo and Wardley, 1982a, b, c) and in sheep (Jeggo et al., 1984, 1985). Maximum CTL activity was observed from 7 to 21 d.p.i. and CTLs were capable of heterotypic lysis (Jeggo and Wardley, 1982a).

Recipient sheep, which received thoracic duct lymphocytes (TLDs) from donor sheep previously infected with BTV were partly protected (reduced viraemia) against challenge with homologous virus challenge (Jeggo et al., 1984). Partial protection was also achieved when T-cell-enriched TLDs were used in similar transfer experiments, demonstrating that the observed effect was not due to transferred B cells. A boostable 'memory-component' of the cellular immune response (specifically CTLs) was also reported (Jeggo et al., 1986; Andrew et al., 1995). Sheep that received TLDs (taken from donor sheep at 14 d.p.i. with BTV-3) were also partially protected against challenge with a heterologous virus serotype (BTV-4) (Jeggo et al., 1984) indicating that at least part of the CTL response involved is serotype cross-reactive, suggesting that it may involve proteins other than outer-capsid components VP2 and VP5. Later studies also demonstrated that some BTV-specific T-cell lines were able to reduce homologous and heterologous BTV replication in autologous skin fibroblast cultures.

The cross-reactivity observed in CTLs against heterologous BTV strains did not correlate with cross-reactions shown by neutralizing antibodies in virus neutralization assays (VNT) of close related serotypes (Takamatsu and Jeggo, 1989). It was suggested that this also indicates that different viral proteins are responsible for the induction of cellular immune response, as compared with neutralizing antibody responses. Studies to identify the BTV antigens involved in BTV-specific and protective CTL responses identified NS1 and VP2 as major and VP5, VP7 and NS3 as minor recognition antigens, with cross-reactivity to heterologous virus strains mostly directed against NS1 (Andrew *et al.*, 1995; Janardhana *et al.*, 1999). These studies suggest that BTV VP3, VP7 and particularly NS1 proteins could potentially be used as components of a cross-serotype reactive subunit vaccine.

The largest of the BTV non-structural proteins, NS1 forms tubules during viral infection. The NS1 protein is synthesized more abundantly than any other viral protein in mammalian cells and is relatively easy to purify (Ghosh *et al.*, 2002b). However, the role of the NS1 protein during BTV replication and in BTV-specific immune responses is unclear. NS1 expressed by recombinant baculoviruses also forms tubules, which have been used to carry epitope-peptides derived from other viruses (e.g. influenza and foot and mouth disease virus), generating a particulate vector system (Ghosh *et al.*, 2002a). Although an immune response was raised against the heterologous antigens, the NS1 tubules were also taken up efficiently by macrophages or dendritic cells and appear to induce CTL.

A comparative study also showed partial protection against recombinant vaccinia virus expressing the non-structural protein NS2 in mice, following intraperitoneal immunization with 20 µg of the recombinant NS2 protein (Jones *et al.*, 1997). However, no protection was observed when mice were immunized with the other recombinant non-structural proteins, NS1 or NS3, or minor structural proteins VP1, VP4 or VP6 of BTV-10. While part of the mechanism of protection could be attributed to the induction of NS2-specific CTL responses, the induction of recombinant antigen-specific CTLs did not necessarily correlate with protective immune responses in this system.

Overall, it is thought that the cellular immune response to BTV, of which CTLs are only one component (see Chapter 12), may be of major importance in the generation of a heterotypic immune response against different BTV strains, and may explain the partial protection observed in the absence of neutralizing antibodies (Jeggo *et al.*, 1985; Wade-Evans *et al.*, 1996). The nature of the cellular immune response to BTV is only partly understood, and consequently current vaccine preparations are mostly validated by their potential to induce neutralizing antibody response. The development of a vaccine which induces a broad heterotypic protection against several BTV serotypes will depend on further studies of cell-mediated protection against BTV.

Differentiating infected from vaccinated animals

Antibodies generated by infected animals against the conserved viral proteins, particularly the immunodominant core surface protein VP7 (Gumm and Newman, 1982), are detected in diagnostic assays (e.g. ELISA) to identify and distinguish BTV and related orbiviruses (Mecham and Wilson, 2004; see Chapter 17). Nucleic acid-based assays (real time or conventional RT-PCR) targeting the genes that encode conserved antigens are also increasingly being used as standard diagnostic assays for BTV [e.g. genome segment 1, encoding the viral polymerase VP1 (Shaw *et al.*, 2007); or genome segment 7, encoding VP7 (Anthony *et al.*, 2007)]. Antibodies to the serotype-specific BTV outer-capsid proteins, particularly VP2, are used to confirm diagnosis and serotype the virus involved in BTV infections. This can be carried out using conventional serum neutralization assays; however, recent sequencing studies of genome segment 2 from multiple isolates of different BTV serotypes from around the world (Maan *et al.*, 2007) have supported the development of BTV molecular epidemiology, and the development of BTV-serotype-specific RT-PCR assays (Mertens *et al.*, 2007).

Although effective diagnostic assays exist to identify and type BTV strains, differentiation between infected and vaccinated animals is more difficult, particularly for the live attenuated vaccines that cause an infection of the vaccinated animal. The use of VLP and CLP, subunit vaccines and purified (or partially purified) inactivated vaccines, may allow vaccinated and infected animals to be differentiated based on the detection of an immune responses against viral proteins that are either absent from the vaccine, or which only raise a low level of antibody response to a 'non-replicating' vaccine (e.g. the non-structural viral proteins – Anderson *et al.*, 1993).

Genome segment 2 of individual BTV strains varies significantly depending on both the strain and the geographical origin of the genetic lineage to which the virus belongs. On this basis, isolates of the BTV serotypes that have invaded Europe can be differentiated into eastern and western group viruses (see Chapter 7). These differences have made it possible to develop primers that can also be used in RT-PCR assays to distinguish some of the vaccine and field strains of BTV (Mertens *et al.*, 2007; see www.iah.bbsrc.ac.uk/dsRNA_virus_proteins/ReoID/rt-pcr-primers.htm). Previous studies have indicated that BTV nucleic acid can be detected in the blood of infected cattle and sheep for significant periods post viraemia (Bonneau *et al.*, 2002). It may therefore be possible to positively identify infections by field and vaccine strains of BTV for some time after infection. RT-PCR and sequence analyses of the resulting cDNA amplicons can be used to identify even small but significant differences in the nucleotide sequences of the wild type and vaccine virus strains derived from the same geographic origins (Breard *et al.*, 2003; Maan *et al.*, 2007; Mertens *et al.*, 2007). However, even this fine level of discrimination can be further complicated by the transmission and reassortment of the vaccine strains themselves in the field (Batten *et al.*, 2008a).

Active and accurate surveillance of BTV field strains is crucial. The development of safer non-replicating vaccines and the ability to identify and distinguish vaccinated and infected animals would clearly remove some of the understandable resistance to vaccination strategies.

Expression of viral proteins for use as vaccine components

Viral proteins can be expressed artificially using several different recombinant systems, including bacteria (usually *E. coli*), yeast, insect cells (using Baculovirus), or in eukaryotic cells by infection with appropriate viral vectors, by permanent stable transfection or by uptake of appropriate DNA constructs. Differences in the translation environment or post-translational modification, between bacterial, insect or mammalian cell-based expression systems can give altered protein structure and may therefore altered antibody recognition. Assembly of individual proteins into multimers and particles, of either the same or different proteins, can also influence both their structure and the presence of 'discontinuous' conformational epitopes that are constituted by the interactions of different protein chains. These factors may be particularly important in the antigenicity of expressed VP2 (the main BTV neutralization antigen), which is known to be highly conformationally dependent.

Expressed proteins can be secreted or may need to be extracted and purified, processes that can be simplified by engineering certain short amino acid sequences to either terminus of the expressed proteins. These 'tags' can help 'chaperone' the correct protein folding, can be recognized by specific antibodies and/or can be used to purify the expressed protein by affinity chromatography (e.g. nickel binding by His tags; Seleem *et al.*, 2007). Most 'vaccine-marker' strategies and assays to 'distinguish infected and vaccinated animals' (DIVA) are conventionally based on 'deletion' of at least one 'immunogen' from the vaccine, allowing the animals that have been infected [which will have an antibody response to the deleted antigen(s)] to be detected and differentiated from those that have only been vaccinated. However, the highly antigenic nature of certain tags and the availability of homologous high affinity and specificity antibodies raises the possibility that such tags could also be used in a marker-vaccine strategy, such that the detection of antibodies to the tag would indicate positively that an animal had been vaccinated.

Antigens for use in serological assay systems may be increasingly produced as synthetic peptides (Du Plessis *et al.*, 1994, 1995). However, this technology is most applicable to simple linear epitopes, and the importance of the antibody-mediated immune response for viral neutralization and the clear conformational dependence of VP2, suggest that discontinuous and highly conformational epitopes play a central role in its antigenicity. The full VP2 protein is therefore likely to be required for induction of a protective (neutralizing) response, and its relevant folding, assembly into trimers and/or association with other viral proteins (e.g. VP5 and VP7) are likely to increase its antigenic similarity to the native protein, and its ability to raise a protective (neutralizing) immune response (also see Subunit *Vaccine Candidates*, above).

Bacteria can be used for the synthesis of viral proteins, although this can result in mis-folding or modification. The highly conformation-dependent nature of the antigenicity of VP2 suggests that bacteria may not be the most suitable expression system, although this has not been widely explored for different BTV serotypes. The NS2 protein of BTV-1 has been successfully expressed in large amounts and in a native conformation, in bacteria (see Chapter 5), and has been used for structural studies by X-ray crystallography. High-level expression of the major BTV core protein VP7 and the non-structural protein NS3 of BTV serotype 1 has also been achieved in yeast cells (Martyn *et al.*, 1991). However, the baculovirus system has been more widely used for expression of single and multiple recombinant BTV proteins (e.g. for VLP and CLP). Recombinant baculovirus expression systems offer the potential for high-level protein expression in insect cells. The cloning of BTV cDNAs into a plasmid transfer vector, which are transfected into bacterial cells already containing the majority of the baculovirus genome (e.g. DH10Bac *E. coli* cells) followed by selection of successful clones, amplification of recombinant DNA and transfection of insect cells (e.g. *Spodoptera frugiperda* (typically ATCC No. CRL 1711 (Sf9), Sf21) or High-Five cells), allows subsequent harvesting of the recombinant baculovirus. It has frequently been shown that heterologous (non-baculovirus) genes can be expressed at high levels in this system, using the viral polyhedrin or P10 promoters. Expression in insect cells permits appropriate folding, post-translational modification and oligomerization that often appear identical to those that occur in other eukaryotic cells (e.g. mammalian cells). It may be significant in this context that BTV is an arbovirus and is therefore adapted to replicate successfully in insect cells [although these are usually derived from *Culicoides* species, it will also grow in mosquito (C6/36) cells (Mertens *et al.*, 1996)]. Insect cells do not require CO_2 for growth and can be readily adapted to high-density suspension culture for potential large-scale expression. High-Five cells do not require fetal calf serum and insect cells can also be grown at room temperature (ideally 25–27°C) and are considered safer than bacterial expression systems. Levels of expression of recombinant proteins have been cited as high as 300 mg/L of culture (expression of NS1 in *Sf* cells; Ghosh *et al.*, 2002a), suggesting that Baculovirus expression systems could potentially be used for large scale production of a low-cost and effective subunit vaccine for BTV or other orbiviruses.

Adjuvants

Adsorption of BTV particles to alum does not appear to affect their observed morphology using transmission electron microscopy (TEM) (Figure 18.1). Recent work has indicated that immune responses to BTV-1 particles can be enhanced in the guinea pig model using chitosan and aluminium phosphate as adjuvants for delivery (Somavarapu *et al.*, 2003a, b).

Other mechanisms that have been used in attempts to increase the immunogenicity of BTV VLP in other models, have included co-delivery of genes encoding a genetic adjuvant (such as interleukin-2 [Oh *et al.*, 2004]), conjugation or incorporation of potent adjuvants (such as cholera toxin subunit B (CTB) [Kang *et al.*, 2003]), as well as co-delivery of immunostimulating complexes (ISCOMs) (Nguyen *et al.*, 2003). It is also possible to increase certain aspects of immunogenicity by the use of adjuvants in conjunction with live attenuated viral vaccines (Luhrmann *et al.*, 2005), although much depends on the route of administration, and mostly, these approaches have not been assessed for BTV. Roy *et al.* (1990) reported that incomplete Freund's adjuvant enhanced the neutralizing antibody response generated against various recombinant proteins derived from BTV-10. In their study, VP5 enhanced the protective immune responses in sheep elicited by baculovirus expressed VP2 protein, mimicking the earlier results obtained by Huismans *et al.* (1987a, b) using native viral proteins purified from the virus capsid. It is not clear whether another more acceptable adjuvant could also enhance observed immune responses, but it appears likely that the efficacy of a subunit vaccine would be enhanced by the use of an appropriate adjuvant.

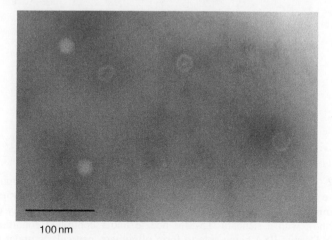

100 nm

Figure 18.1 TEM image of alum-adsorbed bluetongue virus particles. The bar represents 100 nm. (See color plate 30)

Synthetic and biodegradable delivery systems

Extensive research into particulate delivery systems has demonstrated the potential of liposomal and polymer-based carrier systems for the delivery of vaccine antigens (Alpar *et al.*, 2005; Bramwell and Perrie, 2005). The use of

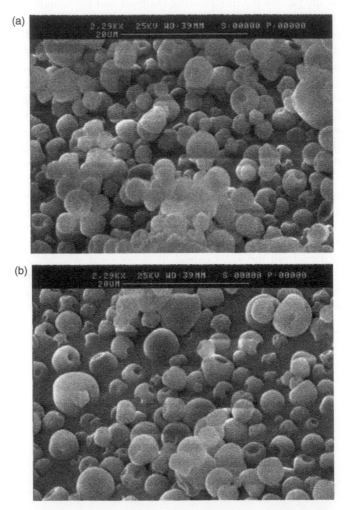

Figure 18.2 Scanning electron micrographs (SEM) of micro-particles formulated with chitosan base [high molecular weight (b)] or chitosan (b) and the model antigen bovine serum albumin (BSA) using concurrent spray drying with an SD-04 spray drier (Lab Plant, England), with a standard 0.5-Mm nozzle under controlled conditions (Alpar *et al.*, 2005). Despite BSA being able to form oligomers, particularly under the conditions employed during spray drying (i.e. Higher temperatures), SDS-PAGE analysis indicated that during the spray-drying process, the entrapped antigen remained stable.

co-adjuvants further enhances the promise of these particulate delivery systems and the recent delineation of specific Toll-like receptor (TLR) interaction for many adjuvants of microbial origin and their analogues has brought understanding to the mechanisms of innate immune activation and the specific molecular interactions involved. The mechanisms of adjuvant interaction ascribed to polymer and liposomal delivery systems include effective delivery to antigen-presenting cells, the depot effect of delayed or slow antigen release, with possible concomitant effects of increased antigen presentation and better antigen trafficking to lymph nodes and sites of immune induction. The use of an adjuvant and particulate combination can provide the basis for a potent vaccine delivery system.

Encapsulation of purified BTV particles (BTV-1) into chitosan nanoparticles has been used as a strategy for enhancement of immune responses against BTV (Somavarapu *et al.*, 2003a). Chitosans, in the form of derivatives soluble over increased ranges of pH, offer excellent potential as adjuvants for mucosal vaccine delivery (Alpar *et al.*, 2005). Their formulation into particulate delivery systems can be achieved using conventional double emulsion techniques or spray drying, similar to those used for the formulation of poly(lactide) (PLA), poly(lactide-co-glycolide) (PLGA) or other biodegradable polymers, depending upon the type of chitosan and co-polymer, if necessary (Figure 18.2). In the cited study (Somavarapu *et al.*, 2003a), chitosan nanoparticles were prepared as previously outlined (Calvo *et al.*, 1997) by the addition of a tri-polyphosphate solution to an aqueous solution of chitosan glutamate. The encapsulation of BTV in the nanoparticles was high (in excess of 80%), and the encapsulation did not result in any loss in viral infectivity.

Regulations and acceptability

Stringent regulations and guidelines apply to the development of vaccines as products for the veterinary market, such that no new veterinary-medicinal product may be authorized unless the applicant has demonstrated that the product fulfils the quality requirements, will be effective with regard to the claimed indications and does not present unacceptable risks for the environment, the target animals, consumers of food of animal origin or users of the product. A comprehensive list of standards is outlined at EudraLex (http://pharmacos.eudra.org); in 'The Rules Governing Medicinal Products in the European Union', Volumes 4–9: Good Manufacturing Practices, Medicinal Products for Human and Veterinary use; Pharmaceutical Legislation, Veterinary Medicinal Products; Notice to Applicants, Veterinary Medicinal Products; Guidelines, Veterinary Medicinal Products; Maximum residue limits, Veterinary Medicinal Products; and Pharmacovigilance, Medicinal Products for Human and Veterinary use.

The Guidelines are an extensive template for the required efficacy, safety, environmental acceptability and quality of animal vaccine products for the European market, and it should be noted also that there is a drive towards more flexibility to the European regulations concerning veterinary vaccines in the future. It is also clear that in the face of an outbreak, certain regulations for veterinary medical products can be suspended under emergency derogations.

Contamination of other vaccines by BTV (e.g. canine distemper vaccines caused by the use of contaminated bovine serum for growth of the vaccine in cell culture) has been reported to be responsible for problems seen following vaccination of pregnant bitches. Initial reports of disease and death in dogs associated with contaminated vaccine (Akita *et al.*, 1994; Wilbur *et al.*, 1994) were followed up with corroborative experimental reports. The experimental infection with BTV of pregnant bitches results in abortion and death (Brown *et al.*, 1996), and observed persistent infection of canine kidney cell cultures with different BTV serotypes indicates that more stringent screening of the cells used in the production of live vaccines for various contaminating viruses may be necessary (Ianconescu *et al.*, 1996). It is also suspected that BTV and African horse sickness virus can infect large carnivores in Africa, suggesting that similar problems could also occur in other species. Such observations certainly have implications for the propagation of attenuated BTV vaccines and raise the possibility for screening of other vaccines routinely for contamination.

Concluding remarks

The structural components of the outer-capsid layer are the most variable BTV proteins. VP2 and to a lesser extent VP5 vary in a manner that reflects virus serotype (Huismans and Bremer, 1981; Mertens *et al.*, 1987; Gould and Eaton, 1990; Singh *et al.*, 2004; Maan *et al.*, 2007, 2008). However, like many of the other BTV proteins, VP2 and VP5 also show significant sequence variations (within each serotype) that reflect the geographic origins of the virus lineage (VP2 and VP5 topotypes) (Maan *et al.*, 2007; Mertens *et al.*, 2007; Chapter 7). These differences in aa sequence (particularly those that are detected between serotypes) will inevitably have an impact on the development of vaccines that target the neutralizing antibody responses as a major tool of protective immunity. It is unclear if the significant aa sequence differences that exist between the different VP2 topotypes, within a single serotype, will also have an impact on the overall effectiveness of the neutralizing antibody response in protection for heterologous strains of the same serotype (particularly for subunit vaccines).

The lack of correlation sometimes observed between the neutralizing antibody response and protection suggests that T-cell determinants and CTL responses are also important in the broader clearance of BTV. The choice of the component antigens used in a mixed or subunit vaccine may therefore be crucial in overall development of effective immunological prophylaxis against BTV. Recombinant protein vaccine strategies have the potential for incorporation of marker properties that can facilitate differentiation of infected and vaccinated animals. Additionally, adjuvants and delivery systems offer considerable potential to enhance the protective immune response to BTV particles, subunit antigens and VLP or CLP vaccines. The justifiable concerns that exist about vaccine strain transmission, reassortment or adverse reactions associated with live vaccines could be answered by the use of an appropriate non-replicating vaccine/adjuvant combination. These could include individual expressed protein subunits, VLP or CLP, an inactivated virus, DNA-based vaccines (e.g. from bacterial of baculovirus sources), or the use of a heterologous replication-incompetent virus carriers (such as MVA or canary pox). An enhanced response may be generated by an appropriate combination of more than one of these technologies.

In view of the relatively low commercial value of the individual host animal (particularly sheep), the cost of the vaccine is a major driving factor in vaccine choice. Concerns associated with the use of the live vaccines currently available may be subordinate to the severe implications of allowing disease to spread. However, there is now clear evidence that these vaccine viruses can be transmitted in the field, and there is a significant risk that they will cause severe disease in European breeds, in their own right. It is therefore unlikely that these live attenuated vaccine strains will be used widely within Europe at any time in the future. The newer inactivated vaccines that are now coming onto the market for specific BTV serotypes clearly offer a much safer alternative, although as seen in 2007 with FMDV in the UK, there may still be risks associated with the production of any vaccine directly from live and virulent viruses. However, individually expressed BTV proteins cannot replicate and are therefore inherently 'safe' for use as vaccine components. More effective adjuvants and delivery strategies are now also available, some of which have been outlined here. Collectively these advances will lead to the development of 'next generation' BTV vaccines that avoid the problems with current vaccine strategies, offering affordable, realistic and effective alternatives.

Web sites

BTV type-specific primers. www.reoviridae.org/dsrna_virus_proteins//ReoID/rt-pcr-primers.htm
Standards for veterinary medicines at EudraLex. http://pharmacos.eudra.org
BTV reference collection. www.reoviridae.org/dsrna_virus_proteins//ReoID/BTV-isolates.htm
Data concerning BTV proteins and RNAs. www.reoviridae.org/dsrna_virus_proteins//BTV.htm
BTV serotype distribution. www.reoviridae.org/dsrna_virus_proteins//btv-serotype-distribution.htm

References

Akita, G.Y., Ianconescu, M., MacLachlan, N.J. and Osburn, B.I. (1994) Bluetongue disease in dogs associated with contaminated vaccine. *Vet. Rec.* **134**, 283–4.

Alpar, H.O., Somavarapu, S., Atuah, K.N. and Bramwell, V.W. (2005) Biodegradable mucoadhesive particulates for nasal and pulmonary antigen and DNA delivery. *Adv. Drug Deliv. Rev.* **57**, 411–30.

Anderson, J., Mertens, P.P.C. and Herniman, K.A. (1993) A competitive ELISA for the detection of anti-tubule antibodies using a monoclonal antibody against bluetongue virus non-structural protein NS1. *J Virol. Methods* **43**, 167–75.

Andrew, M., Whiteley, P., Janardhana, V., Lobato, Z., Gould, A. and Coupar, B. (1995) Antigen specificity of the ovine cytotoxic T lymphocyte response to bluetongue virus. *Vet. Immunol. Immunopathol.* **47**, 311–22.

Anthony, S., Jones, H., Darpel, K.E., Elliott, H., Maan, S., Samuel, A., Mellor, P.S. and Mertens, P.P.C. (2007) A duplex RT-PCR assay for detection of genome segment 7 (VP7 gene) from 24 BTV serotypes. *J. Virol. Methods* **141**, 188–97.

Appleton, J.A. and Letchworth, G.J. (1983) Monoclonal antibody analysis of serotype-restricted and unrestricted bluetongue viral antigenic determinants. *Virology* **124**, 286–99.

Batten, C.A, Maan, S. Maan, N. and Mertens, P.P.C. (2008b). Identification of the BTV-16 vaccine strain as the cause of the bluetongue outbreak on Sardinia 2004. *Trans. Dis.* (in Press).

Batten, C.A., Shaw, A, Maan, S., Maan, N. and Mertens, P.P.C. (2008a) Genome segment reassortment in the field between a European field strain of bluetongue virus serotype 16 and a serotype 2 vaccine strain. *Virus Res.* (in Press); [Epub ahead of print at: http://dx.doi.org/10.1016/j.virusres.2008.05.016].

Belyaev, A.S. and Roy, P. (1993) Development of Baculovirus triple and quadruple expression vectors: Co-expression of three or four bluetongue virus proteins and the synthesis of bluetongue virus-like particles in insect cells. *Nucleic Acids Res.* **21**, 1219–23.

Bonneau, K.R., DeMaula, C.D., Mullens, B.A., and MacLachlan, N.J. (2002) Duration of viraemia infectious to *Culicoides sonorensis* in bluetongue virus-infected cattle and sheep. *Vet. Microbiol.* **88**, 115–25.

Boone, J.D., Balasuriya, U.B., Karaca, K., Audonnet, J.C., Yao, J., He, L., Nordgren, R., Monaco, F., Savini, G., Gardner, I.A. and MacLachlan, N.J. (2007) Recombinant canarypox virus vaccine co-expressing genes encoding the VP2 and VP5 outer capsid proteins of bluetongue virus induces high level protection in sheep. *Vaccine* **25**, 672–8.

Bramwell, V.W. and Perrie, Y. (2005) Particulate delivery systems for vaccines. *Crit. Rev. Ther. Drug Carrier Syst.* **22**, 151–214.

Breard, E., Sailleau, C., Coupier, H., Mure-Ravaud, K., Hammoumi, S., Gicquel, B., Hamblin, C., Dubourget, P. and Zientara, S. (2003) Comparison of genome segments 2, 7 and 10 of bluetongue viruses serotype 2 for differentiation between field isolates and the vaccine strain. *Vet. Res.* **34**, 777–89.

Brown, C.C., Rhyan, J.C., Grubman, M.J. and Wilbur, L.A. (1996) Distribution of bluetongue virus in tissues of experimentally infected pregnant dogs as determined by in situ hybridization. *Vet Pathol.* **33**, 337–40.

Calvo, P., Remunan-Lopez, C., Vila-Jato, J.L. and Alonso, M.J. (1997) Chitosan and chitosan/ethylene oxide-propylene oxide block copolymer nanoparticles as novel carriers for proteins and vaccines, *Pharm. Res.* **14**, 1431–6.

Campbell, C.H. (1985) Immunogenicity of bluetongue virus inactivated by gamma irradiation. *Vaccine* **3**, 401–5.

Caporale, V., GIovannini, A., Patta, C., Calistri, P., Nannini, D. and Santucci, U. (2004) Vaccination in the control strategy of Bluetongue in Italy. Control of infectious animal diseases by vaccinatio. *Dev. Biol. Basel, Karger.* **119**, 113–27.

Chiam, R., Sharp, E., Maan, S., Rao, S., Mertens, P.P.C., Blacklaws, B., Davis-Poynter, N., Wood, J. and Castillo-Olivares, J. (2008) Induction of specific antibody responses to African horse sickness virus (AHSV) in ponies after vaccination with recombinant modified Vaccinia Ankara (MVA) viruses encoding AHS virus proteins. *Equine Vet. J.* (in Press).

Cowley, J.A. and Gorman, B.M. (1989) Cross-neutralization of genetic reassortants of bluetongue virus serotypes 20 and 21. *Vet. Microbiol.* **19**, 37–51.

Darpel, K.E. (2007) The bluetongue virus, ruminant host – Insect vector transmission cycle, the role of Culicoides saliva proteins in infection. Ph.D thesis, Royal Veterinary College, University of London.

Darpel, K.E., Batten, C.A., Veronesi, E., Shaw, A.E., Anthony, S., Bachanek-Bankowska, K., Kgosana, L., Bin-Tarif, A., Carpenter, S., Muller-Doblies, U.U., Takamatsu, H-H., Mellor, P.S., Mertens, P.P.C. and Oura, C.A. (2007) Clinical signs and pathology shown by British sheep and cattle infected with bluetongue virus serotype 8 derived from the 2006 outbreak in northern Europe. *Vet. Rec.* **161**, 253–61.

DeMaula, C.D., Bonneau, K.R. and MacLachlan, N.J. (2000) Changes in the outer capsid proteins of bluetongue virus serotype ten that abrogate neutralization by monoclonal antibodies. *Virus Res.* **67**, 59–66.

Du Plessis, D.H., Romito, M. and Jordaan, F. (1995) Identification of an antigenic peptide specific for bluetongue virus using phage display expression of NS1 sequences. *Immunotechnology* **1**, 221–30.

Du Plessis, D.H., Wang, L.F., Jordan, F.A. and Eaton, B.T. (1994) Fine Mapping of a Continuous Epitope on VP7 of bluetongue virus using overlapping synthetic peptides and a random epitope library. *Virology* **198**, 346–9.

Ferrari, G., Liberato, C.D., Scavia, G., Lorenzetti, R., Zini, M., Farina, F., Magliano, A., Cardeti, G., Scholl, F. and Guidoni, M. (2005) Active circulation of bluetongue vaccine virus serotype-2 among unvaccinated cattle in central Italy. *Prev. Vet. Med.* **68**, 103–13.

French, T.J. and Roy, P. (1990) Synthesis of bluetongue virus (BTV) core-like particles by a recombinant baculovirus expressing the two major structural core proteins of BTV. *J. Virol.* 64, 1530–6.

Ghosh, M.K., Borca, M.V. and Roy, P. (2002a) Virus-derived tubular structure displaying foreign sequences on the surface, elicit CD4 + Th cell and protective humoral responses. *Virology* **302**, 383–92.

Ghosh, M.K., Deriaud, E., Saron, M.F., Lo-Man, R., Henry, T., Jiao, X., Roy, P. and Leclerc, C. (2002b) Induction of protective antiviral cytotoxic T-cells by a tubular structure capable of carrying large foreign sequences. *Vaccine* **20**, 1369–77.

Giovannini, A., MacDiarmid, S., Calistri, P., Conte, A., Savini, L., Nannini, D. and Weber, S. (2004) The use of risk assessment to decide the control strategy for bluetongue in Italian ruminant populations. *Risk Anal.* **24**, 1737–53.

Gouet, P., Diprose, J.M., Grimes, J.M., Malby, R., Burroughs, J.N., Zientara, S., Stuart, D.I. and Mertens, P.P.C. (1999) The highly ordered double-stranded RNA genome of bluetongue virus revealed by crystallography. *Cell* **97**, 481–90.

Gould, A.R. and Eaton, B.T. (1990) The amino acid sequence of the outer coat protein VP2 of neutralizing monoclonal antibody-resistant, virulent and attenuated bluetongue viruses. *Virus Res.* **17**, 161–72.

Grieder, F.B. and Schultz, K.T. (1989) Conformationally dependent epitopes of bluetongue virus neutralizing antigen. *Viral. Immunol.* **2**, 17–24.

Grimes, J.M., Burroughs, J.N., Gouet, P., Diprose, J.M., Malby, R., Zientara, S., Mertens, P.P.C. and Stuart, D.I. (1998) The atomic structure of the bluetongue virus core. *Nature,* **395**, 470–8.

Gumm, I.D. and Newman, J.F.E. (1982) The preparation of purified bluetongue virus group antigen for use as a diagnostic reagent. *Arch. Virol.* **72**, 83–9.

Hassan, S.S. and Roy, P. (1999) Expression and functional characterization of bluetongue virus VP2 protein: Role in cell entry. *J. Virol.* **73**, 9832–42.

Hewat, E.A., Booth, T.F., Loudon, P.T. and Roy, P. (1992a) Three-dimensional reconstruction of Baculovirus expressed bluetongue virus core-like particles by cryo-electron microscopy. *Virology* **189**, 10–20.

Hewat, E.A., Booth, T.F. and Roy, P. (1992b) Structure of bluetongue virus particles by cryoelectron microscopy. *J. Struct. Biol.* **109**, 61–9.

Hewat, E.A., Booth, T.F. and Roy, P. (1994) Structure of correctly self-assembled bluetongue virus-like particles. *J. Struct. Biol.* **112**, 183–91.

Huismans, H. and Bremer, C.W. (1981) A comparison of an Australian bluetongue virus isolate (CSIRO 19) with other bluetongue virus serotypes by cross hybridization and cross immune precipitation. *Onderstepoort J. Vet. Res.* **48**, 59–67.

Huismans, H., Van Dijk, A.A. and Els, J.H. (1987a) Uncoating of parental bluetongue virus to core and subcore particles in infected L cells. *Virology* **157**, 180–8.

Huismans, H., Van Der Walt, N.T., Cloete, M. and Erasmus, B.J. (1987b) Isolation of a capsid protein of bluetongue virus that induces a protective immune response in sheep. *Virology* **61**, 3589–95.

Hwang, G.Y. and Li, J.K. (1993) Identification and localization of a serotypic neutralization determinant on the VP2 protein of bluetongue virus 13. *Virology,* **195**, 859–62.

Ianconescu, M., Akita, G.Y. and Osburn, B.I. (1996) Comparative susceptibility of a canine cell line and bluetongue virus susceptible cell lines to a bluetongue virus isolate pathogenic for dogs. *In Vitro Cell. Dev. Biol. Anim.,* **32**, 249–54.

Janardhana, V., Andrew, M.E., Lobato, Z.I. and Coupar, B.E. (1999) The ovine cytotoxic T lymphocyte responses to bluetongue virus. *Res. Vet. Sci.* **67**, 213–21.

Jeggo, M.H., Gumm, I.D. and Taylor, W.P. (1983a) Clinical and serological response of sheep to serial challenge with different bluetongue virus types. *Res. Vet. Sci.* **34**, 205–11.

Jeggo, M.H. and Wardley, R.C. (1982a) Generation of cross-reactive cytotoxic T lymphocytes following immunization of mice with various bluetongue virus types. *Immunology* **45**, 629–35.

Jeggo, M.H. and Wardley, R.C. (1982b) Production of murine cytotoxic T lymphocytes by bluetongue virus following various immunisation procedures. *Res. Vet. Sci.* **33**, 212–5.

Jeggo, M.H. and Wardley, R.C. (1982c) The induction of murine cytotoxic T lymphocytes by bluetongue virus. *Arch. Virol.* **71**, 197–206.

Jeggo, M.H., Wardley, R.C. and Brownlie, J. (1985) Importance of ovine cytotoxic T cells in protection against bluetongue virus infection. *Prog. Clin. Biol. Res.* **178**, 477–87.

Jeggo, M.H., Wardley, R.C., Brownlie, J. and Corteyn, A.H. (1986) Serial inoculation of sheep with two bluetongue virus types. *Res. Vet. Sci.* **40**, 386–92.

Jeggo, M.H., Wardley, R.C. and Taylor, W.P. (1983b). Host response to bluetongue virus In: R.W. Compans and D.H.L. Bishop (eds.), *Double Stranded RNA Viruses*. New York: Elsevier, pp. 353–9.

Jeggo, M.H., Wardley, R.C. and Taylor, W.P. (1984) Role of neutralizing antibody in passive immunity to bluetongue infection. *Res. Vet. Sci.* **36**, 81–5.

Jones, L.D., Williams, T., Bishop, D. and Roy, P. (1997) Baculovirus-expressed nonstructural protein NS2 of bluetongue virus induces a cytotoxic T-cell response in mice which affords partial protection. *Clin. Diagn. Lab. Immunol.* **4**, 297–301.

Julyr, A., Hochstein-Mintzel, V. and Stickl, H. (1975) Passage history, properties, and applicability of the attenuated vaccinia virus strain MVA. *Infection* **3**, 6–14.

Kang, S.M., Yao, Q., Guo, L. and Compans, R.W. (2003) Mucosal immunization with virus-like particles of Simian immunodeficiency virus conjugated with Cholera toxin subunit B. *J. Virol.,* **77**, 9823–30.

Kemeny, L. and Drehle, L.E. (1961) The use of tissue culture-propagated bluetongue virus for vaccine preparation. *Am. J. Vet. Res.* **22**, 921–5.

Lefèvre, P.-C. and Desoutter, D. (1988) La fièvre catarrhale du mouton (Bluetongue). Institut d'élevage et de médicine vétérinaire des pays tropicaux (IEMVT), Ètudes et synthèses, Montpellier p. 117.

Luedke, A.J. and Jochim, M.M. (1968) Clinical and serologic responses in vaccinated sheep given challenge inoculation with isolates of bluetongue virus. *Am. J. Vet. Res.* **29**, 841–51.

Luhrmann, A., Tschernig, T., Pabst, R. and Niewiesk, S. (2005) Improved intranasal immunization with live-attenuated measles virus after co-inoculation of the lipopeptide MALP-2. *Vaccine* **23**, 4721–6.

Maan, S, Maan, N.S., Ross-Smith, N., Batten, C.A., Shaw, A.E., Anthony, S.J., Samuel, A.R., Darpel, K.E., Veronesi, E., Oura, C.A.L., Singh, K.P., Nomikou, K., Potgieter, A.C., Attoui, H., van Rooij, E., van Rijn, P., De Clercq, K., Vandenbussche, F., Zientara, S., Bréard, E., Sailleau, C., Beer, M., Hoffman, B., Mellor, P.S. and Mertens, P.P.C. (2008) Sequence analysis of bluetongue virus serotype 8 from the Netherlands 2006 and comparison to other European strains. *Virology* **377**, 308–18.

Maan, S., Maan, N.S., Samuel, A.R., Rao, S., Attoui, H. and Mertens, P.P.C. (2007) Analysis and phylogenetic comparisons of full-length VP2 genes of the 24 bluetongue virus serotypes. *J. Gen. Virol.* **88**, 621–30.

Martyn, J.C., Gould, A.R. and Eaton, B.T. (1991) High level expression of the major core protein VP7 and the non-structural protein NS3 of bluetongue virus in yeast: Use of expressed VP7 as a diagnostic, group-reactive antigen in a blocking ELISA. *Virus Res.* **18**, 165–78.

McConnell, S.J.C., Morrill, J.C. and Livingston, C.W. (1985) Use of a quadrivalent modified-live bluetongue virus vaccine in wildlife species. *Prog. Clin. Biol. Res.* **178**, 631–8.

Mecham, J.O. and Wilson, W.C. (2004) Antigen capture competitive enzyme-linked immunosorbent assays using Baculovirus-expressed antigens for diagnosis of bluetongue virus and epizootic haemorrhagic disease virus. *J. Clin. Microbiol.* **42**, 518–23.

Mertens, P.P.C. (2004) The dsRNA viruses. *Virus Res.,* **101**, 3–13.

Mertens, P.P.C., Burroughs, J.N., Walton, A., Wellby, M.P., Fu, H., O'Hara, R.S., Brookes, S.M. and Mellor, P.S. (1996) Enhanced infectivity of modified bluetongue virus particles for two insect cell lines and for two *Culicoides* vector species. *Virology* **217**, 582–93.

Mertens, P.P.C. and Diprose, J. (2004) The bluetongue virus core: A nano-scale transcription machine. *Virus Res.* **101**, 29–43.

Mertens, P.P.C. Maan, N.S., Prasad, G., Samuel, A.R., Shaw, A.E., Potgieter, A.C., Anthony, S.J. and Maan, S. (2007) The design of primers and use of RT-PCR assays for typing European BTV isolates: Differentiation of field and vaccine strains. *J. Gen. Virol.* **88**, 2811–23.

Mertens, P.P.C., Maan, S., Samuel, A. and Attoui, H. (2005) Orbivirus, Reoviridae. In: M. Fauquet, M.A. Mayo, J. Maniloff, U. Desselberger, and L.A. Ball, (eds.), *Virus Taxonomy*, VIIIth Report of the ICTV, London: Elsevier/Academic Press, pp. 466–83.

Mertens, P.P.C., Pedley, S., Cowley, J. and Burroughs, J.N. (1987) A comparison of six different bluetongue virus isolates by cross-hybridization of the dsRNA genome segments. *Virology*, **161**, 438–47.

Mertens, P.P.C., Pedley, S., Cowley, J., Burroughs, J.N., Corteyn, A.H., Jeggo, M.H., Jennings, D.M. and Gorman, B.M. (1989) Analysis of the roles of bluetongue virus outer capsid proteins VP2 and VP5 in determination of virus serotype. *Virology*, **170**, 561–5.

Monaco, F, Camma, C, Serini, S, and Savini, G. (2006) Differentiation between field and vaccine strain of bluetongue virus serotype 16. *Vet. Microbiol.* **116**, 45–52.

Murray, P.K. and Eaton, B.T. (1996) Vaccines for bluetongue. *Aust. Vet. J.*, **73**, 207–10.

Nagesha, H.S., Wang, L.-F., Shiell, B., Beddome, G., White, J.R. and Irving, R.A. (2001) A single chain Fv antibody displayed on phage surface recognises conformational group-specific epitope of bluetongue virus. *J. Virol. Methods* **91**, 203–7.

Nguyen, T.V., Iosef, C., Jeong, K., Kim, Y., Chang, K-O., Lovgren-Bengtsson, K., Morein, B., Azevedo, M.S.P., Lewis, P. and Nielsen, P. (2003) Protection and antibody responses to oral priming by attenuated human rotavirus followed by oral boosting with 2/6-rotavirus-like particles with immunostimulating complexes in gnotobiotic pigs. *Vaccine*, **21**, 4059–70.

Noad, R. and Roy, P. (2003) Virus-like particles as immunogens. *Trends Microbiol.* **11**, 438–44.

Oberst, R.D., Squire, K.R., Stott, J.L., Chuang, R.Y. and Osburn, B.I. (1985) The coexistence of multiple bluetongue virus electropherotypes in individual cattle during natural infection. *J. Gen. Virol.* **66**, 1901–9.

Odeon, A.C., Gershwin, L.J. and Osburn, B.I. (1999) IgE responses to bluetongue virus (BTV) serotype 11 after immunization with inactivated BTV and challenge infection. *Comp. Immunol. Microbiol. Inf. Dis.* **22**, 145–62.

Oh, Y.K., Sohn, T., Park, J.S., Kang, M.J., Choi, H.G., Kim, J.A., Kim, W.K., Jae Ko, J. and Kim, C.K. (2004) Enhanced mucosal and systemic immunogenicity of human papilloma virus-like particles encapsulating interleukin-2 gene adjuvant. *Virology*, **328**, 266–73.

O'Hara, R.S., Meyer, A.J., Burroughs, J.N., Pullen, L., Martin, L.A. and Mertens, P.P.C. (1998) Development of a mouse model system, coding assignments and identification of the genome segments controlling virulence of African horse sickness virus serotypes 3 and 8. *Arch. Virol. Suppl.* **4**, 259–79.

OIE. (2007) Terrestrial Animal Health Code. *Bluetongue, Chapter 2.2.13.*, Paris: OIE, pp. 131–6.

Onderstepoort Biological Products (1996) Products Manual: National Department of Agriculture 1996. *OIE Manual of Standards for Diagnostic Tests and Vaccines* ISBN 062116142 X.

Parker, J., Herniman, K.A. and Gibbs, E.P. (1975) An experimental inactivated vaccine against bluetongue. *Vet. Rec.* **96**, 284–7.

Pritchard, L.I., Sendow, I., Lun, R., Hassan, S.H., Kattenbelt, J., Gould, A.R., Daniels, P.W. and Eaton, B.T. (2004) Genetic diversity of bluetongue viruses in south-east Asia. *Virus Res.* **101**, 193–201.

Purse, B.V., Mellor, P.S., Rogers, D.J., Samuel, A.R., Mertens, P.P.C. and Baylis, M. (2005) Climate change and the recent emergence of bluetongue in Europe. *Nat. Rev. Microbiol.* **3**, 171–81.

Roberts, D.H. (1990) Bluetongue: A review. *State Vet. J.* **44**, 66–80.

Roy, P. (2003) Nature and duration of protective immunity to bluetongue virus infection. *Dev. Biol. (Basel)* **114**, 169–83.

Roy, P., Urakawa, T., Van Dijk, A.A. and Erasmus, B.J. (1990) Recombinant virus vaccine for bluetongue disease in sheep. *J. Virol.* **64**, 1998–2003.

Samal, S.K., El-Hussein, A., Holbrookm, F.R., Beaty, B.J. and Ramig, R.F. (1987a) Mixed infection of *Culicoides variipennis* with bluetongue virus serotypes 10 and 17: Evidence for high frequency reassortment in the vector. *J. Gen. Virol.* **68**, 2319–29.

Samal, S.K., Livingston, C.W.Jr., McConnell, S. and Ramig, R.F. (1987b) Analysis of mixed infection of sheep with bluetongue virus serotypes 10 and 17: Evidence for genetic reassortment in the vertebrate host. *J. Virol.* **61**, 1086–91.

Savini, G., MacLachlan, N.J., Sanchez-Vizcaino, J.M. and Zientara, S. (2008) Vaccines against bluetongue in Europe. *Comp. Immunol. Microbiol. Infect. Dis.* **31**, 101–20.

Seleem, M., Ali, M., Al-Azeem, M.W., Boyle, S.M. and Sriranganathan, N. (2007) Enhanced expression, detection and purification of recombinant proteins using RNA stem loop and tandem fusion tags. *Appl. Microbiol. Biotechnol.* **75**, 1385–92.

Shaw, A.E., Monaghan, P., Alpar, H.O., Anthony, S., Darpel, K.E., Batten, C.A., Guercio, A., Alamena, G., Vitale, M., Bachanek-Bankowska, K., Carpenter, S., Jones, H., Oura, C.A.L., King, D.P., Elliot, H., Mellor, P.S. and Mertens, P.P.C. (2007) Development and validation of a real-time RT-PCR assay to detect genome bluetongue virus segment 1. *J. Virol. Methods* **145**, 115–26.

Singh, K.P., Maan, S., Samuel, A.R., Rao, S., Meyer, A. and Mertens, P.P.C. (2004) Phylogenetic analysis of bluetongue virus genome segment 6 (encoding VP5) from different serotypes. *Vet. Ital.* **40**, 479–83.

Somavarapu, S., Hamblin, C., Graham, S., Hawes, P., Mellor, P.S. and Alpar, H.O. (2003a) Encapsulation of purified bluetonge virus 1 (BTV-1) particles into chitosan nanoparticles. *J. Pharm. Pharmacol.* **55**, 72–3.

Somavarapu, S., Hamblin, C., Mertens, P.P.C. and Alpar, H.O. (2003b) Enhanced immune response to purified bluetongue virus (BTV-1) particles in guinea-pig model with chitosan-aluminium phosphate adjuvant. *J. Pharm. Pharmacol.,* **55**, 72.

Stott, J.L., Barber, T.L. and Osburn, B.I. (1985) Immunologic response of sheep to inactivated and virulent bluetongue virus. *Am. J. Vet. Res.* **46**, 1043–9.

Stott, J.L., Osburn, B.I. and Barber, T.L. (1982) Recovery of dual serotypes of bluetongue virus from infected sheep and cattle. *Vet. Microbiol.* **7**, 197–207.

Szmaragd, C., Wilson, A., Carpenter, S., Mertens, P.P.C., Mellor, P.S. and Gubbins, S. (2007) Mortality and case fatality during the 2007 BTV-8 re-occurrence. *Vet. Rec.* **161**, 571–2.

Takamatsu, H. and Jeggo, M.H. (1989) Cultivation of bluetongue virus-specific ovine T cells and their cross-reactivity with different serotype viruses. *Immunology* **66**, 258–63.

Takamatsu, H-H., Mellor, P.S., Mertens, P.P.C., Kirkham, P.A., Burroughs, J.N. and Parkhouse, R.M.E. (2003) A possible overwintering mechanism for bluetongue virus in the absence of the insect vector. *J. Gen. Virol.* **84**, 227–35.

Veronesi, E., Darpel, K., Carpenter, S., Elliott, H., Hamblin, C., Mellor, P.S., Takamatsu, H-H. and Mertens, P.P.C. (2008) Viraemia and pathology responses in Dorset Poll sheep after vaccination with live attenuated BTV vaccines serotype 16 and 4. *Vaccine* (in Press).

Veronesi, E., Hamblin, C. and Mellor, P.S. (2005) Live attenuated bluetongue vaccine viruses in Dorset Poll sheep, before and after passage in vector midges (Diptera: Ceratopogonidae). *Vaccine* **23**, 5509–16.

Verwoerd, D.W. and Erasmus, B.J. (1994) Bluetongue. In: J.A.W. Coetzer, G.R. Thomson, and R.C. Tustin, (eds.), *Infectious Diseases of Livestock with Special Reference to Southern Africa.* Cape Town: Oxford University Press, pp. 443–59.

Wade-Evans, A., Romero, C., Mellor, P.S., Takamatsu, H-H., Anderson, J., Thevasagayam, J., Fleming, M., Mertens, P.P.C. and Black, D. (1996) Expression of the major core structural protein (VP7) of bluetongue virus, by a recombinant Capripox virus, provides partial protection of sheep against a virulent heterotypic bluetongue virus challenge. *Virology* **220**, 227–31.

White, D.M., Wilson, W.C., Blair, C.D. and Beaty, B.J. (2005) Studies on overwintering of bluetongue viruses in insects. *J. Gen. Virol.* **86**, 453–62.

White, J.R. and Eaton, B.T. (1990) Conformation of the VP2 protein of bluetongue virus (BTV) determines the involvement in virus neutralization of highly conserved epitopes within the BTV serogroup. *J. Gen. Virol.* **71**, 1325–32.

Wilbur, L.A., Evermann, J.F., Levings, R.L., Stoll, I.R., Starling, D.E., Spillers, C.A., Gustafson, G.A. and McKeirnan, A.J. (1994) Abortion and death in pregnant bitches associated with a canine vaccine contaminated with bluetongue virus. *J. Am. Vet. Med. Assoc.* **204**, 1762–5.

Bluetongue control strategies

19

ORESTIS PAPADOPOULOS,* PHILIP S. MELLOR†
AND PETER P.C. MERTENS†

*Faculty of Veterinary Medicine, Aristotle University, Thessaloniki, Greece
†Institute for Animal Health, Pirbright Laboratory, Pirbright, Woking, Surrey, UK

Introduction

The strategy for the control and prevention of bluetongue (BT) follows the conventional steps of risk management decisions, based on current scientific knowledge, but it is also greatly influenced by the constraints of relevant national and international legislation and agreements.

BT virus (BTV) is transmitted between its ruminant hosts almost exclusively by certain species of biting midges of the genus *Culicoides*. It is not transmitted by contact or by fomites (i.e. it is not contagious), nor by animal products (e.g. meat or milk). Germ plasms are only rarely involved and are not thought to play an important role in the epidemiology of the disease. BTV does not infect humans.

The distribution, abundance, activity and competence of adults of vector *Culicoides* species are influenced by the weather and climate, particularly temperature (Purse *et al.*, 2005), but are unaffected by restrictions imposed on the movements of host animals. Geography and climate can also play critical roles in the movement of individual vector insects (which may be infected) and can therefore have a direct influence on the spread and transmission of the virus itself. Ambient temperature influences both the rate of virus replication (virogenesis) in the insect and the level of vector competence of the insects themselves, and is therefore a critical factor in virus transmission.

For the decision-maker, the first critical point is to set the objectives of the control measures to be considered and implemented. Morbidity, mortality and

ISBN-13: 978-0-12-369368-6

production loses caused by BT may be significant, depending upon the species and breeds of ruminants infected. Animal movement restrictions and export losses can also have significant cost implications. To this may be added the costs of surveillance, the laboratory testing of large numbers of samples and the possible implementation of a vaccination campaign. All of these options and actions should be set out in a national contingency plan. It is therefore clear that the precise strategy to be followed needs to be decided on a case-by-case basis following a risk/cost-benefit assessment.

In this review, the risk factors, methodology, control strategies, the international guidelines and cost-benefit assessment will be discussed, with examples being provided where appropriate.

Risk factors

Climate and environment

The climate, which can include dramatic seasonal variations in local weather conditions, affects the transmission of BTV in several ways and is thought to be one of the most important risk factors. Adult *Culicoides* are only active over a certain temperature range, which is characteristic for each species (Wittmann *et al.*, 2002). In tropical areas the presence and activity of adult vectors is continuous throughout the year, which means that the virus can also be transmitted throughout the year and hence in such regions BT may be endemic. Under these circumstances, the first objective of a control strategy is to reduce the severity and incidence of disease to tolerable levels, and if possible maintain them at those levels.

In temperate regions the abundance and activity of adult vectors is seasonal, which means that the disease incidence peaks at certain times of the year (in line with vector abundance and activity). Although these outbreaks usually subside and may even disappear completely during the 'vector-free' or 'low vector activity' season (usually in winter), they may reappear over one or more years (overwintering), depending to a great degree on the length of time between periods of vector abundance. In these situations eradication (i.e. the elimination of the virus from the host population or geographical area) could be considered.

In cooler climates any incursions that do occur are likely to result in more limited outbreaks (in terms of both area and time), as the local vectors will be active for a more limited period of the year and because BTV is transmitted increasingly poorly as temperature falls and not at all below approximately 15°C (see Chapter 16 – this volume). Under such conditions eradication should be less problematic or may occur naturally after a single or a brief series of transmission seasons.

Virtually all ruminant animals in areas where BTV is not endemic or frequently incursive will have no acquired immunity, and the ruminant species and breeds present may have little natural resistance to infection or disease. Consequently, outbreaks caused by the introduction of virus into such areas may be severe and the spread may be relatively rapid once the virus has become established.

Furthermore, the concept of global 'climate-change' is now generally accepted and appears to be progressive and on-going. The recent warming trend, as experienced in Europe, has already resulted in significant northerly extensions in the range of the Afro-Asiatic BTV vector, *Culicoides imicola*, which now occupies virtually the whole of the northern part of the Mediterranean Basin, thus dramatically expanding the areas at risk to BTV (Purse *et al.*, 2005). Current trends indicate that these temperature changes are likely to continue, suggesting that they will be a major influence in the future distribution and prevalence of BTV and its vectors.

However, microclimate conditions also need to be considered. Data from weather stations that are usually situated in only a limited number of locations (e.g. at airports) may not be representative of local conditions on the farms in the same region. Special attention should be given to 'protected' microenvironments that could become BTV foci and play an important role in overwintering of adult vectors and the virus.

The vectors

The presence of adults of competent vector species of *Culicoides* is essential for the transmission of BTV. Their distributions and seasonal variations in their numbers are therefore primary determinants of BT risk and must be considered. Each *Culicoides* species has a distribution and seasonal abundance that is determined by climate and a series of other environmental factors, including geography, soil type, moisture and breeding site availability. Temperature is particularly important as it increases the efficiency with which *Culicoides* transmit BTV so that species that had previously been considered non-vectors may become competent (see Chapter 16 – this volume). For example, in the epidemic of BT in NW Europe during 2006 and 2007, BTV-8 was transmitted efficiently by northern Palearctic *Culicoides* species (*Culicoides dewulfi* and *Culicoides obsoletus* complex) that had played little part in previous BTV incursions into southern Europe.

It is generally accepted that BTV is not transovarially transferred to the next generation of *Culicoides*. However, BTV nucleic acid sequences have been detected by reverse transcriptase-polymerase chain reaction (RT-PCR) in larvae and pupae of *Culicoides sonorensis* from an outbreak in North America. Although this could have significant implications as a potential overwintering mechanism should it be confirmed, it has not yet been possible to recover infectious virus from such insects (White *et al.*, 2005; also see Chapter 16 – this volume).

Vertebrate hosts

All ruminant species can become infected with BTV and are therefore thought to be epidemiologically important though improved breeds of sheep tend to exhibit more severe clinical signs than do cattle, goats or other ruminants. Such sheep are therefore far more useful in clinical surveillance, and it is unsurprising that the 2006 northern European outbreak was first identified by the recognition of typical clinical signs in infected sheep. However, many sheep breeds, particularly those that are native to endemic regions, are clinically more resistant – the recognition that a BTV incursion has occurred under such conditions can prove challenging.

Infection of the ruminant host is prolonged but does not appear to be persistent. BTV RNA can be detected in blood samples by RT-PCR for up to 100 days post infection (p.i.) in sheep and an average of 160 days in cattle (although a maximum of >220 days has been recorded for this species) (Bonneau *et al.*, 2002). These periods, which are thought to reflect the half life of the red blood cell (to which the virus becomes attached via its haema-glutinin-VP2), can vary significantly with virus strain (Singer *et al.*, 2001; Kirkland and Hawkes, 2004). In addition, such PCR-positive periods do not reflect the duration over which infectious virus is present in the blood (viraemia), which is critical for infection of feeding insects and transmission. In this respect cattle appear to be epidemiologically more significant, showing a longer viraemia averaging up to ~60 days p.i. (up to a detected maximum of about 100 days p.i.), compared with sheep and goats which appear to be viraemic for an average of only 21 days p.i. (up to a detected maximum of 54 days p.i.) (Koumbati *et al.*, 1999). The longer duration of viraemia in cattle suggests that this ruminant species could have significant involvement in overwintering of the virus.

The laboratory finding that BTV can persistently infect ovine γδ T lymphocytes could also provide a mechanism for virus persistence in the vertebrate host from one vector season to the next (Takamatsu *et al.*, 2003), although such a mechanism has not yet been recognized in the field.

Wild ruminant species may also be important as potential BTV hosts especially in situations where there is close contact between them and domestic livestock, making the application of comprehensive control measures in such cases extremely difficult. The role, if any, played by other exotic wildlife species (e.g. zoo animals, non-ruminant wildlife) in the epidemiology of BT is at present uncertain. There are a limited number of reports showing that some of the larger African carnivores have high titres of BTV-specific antibodies in their sera (Alexander *et al.*, 1994) but there is no information on whether such animals are able to mount a viraemia and at present they are not considered to play a part in the epidemiology of BT.

The BT virus

Twenty four distinct serotypes of BTV have been identified, based on the specificity of their reactions with the neutralizing antibodies generated by the mammalian host. The specificity of these reactions is determined primarily by protein components of the virus particle's outer surface, particularly outer-coat protein VP2 (Mertens *et al.*, 2007; Maan *et al.*, 2007a; see also Chapter 8 – this volume). However, VP2 and most of the other viral proteins and genes are also important as they show significant levels of variation that reflect the geographic origins of different isolates (Anthony *et al.*, 2007; Shaw *et al.*, 2007).

Different strains of BTV can show significant variations in pathogenicity and virulence for the host (usually expressed in sheep), although the genetic basis for these variations is poorly understood. Highly pathogenic strains of BTV can be responsible for significant direct loses due to mortality and morbidity but strains of low pathogenicity are also important economically due to the restrictions on animal movements and trade imposed when a BTV incursion is confirmed. In such situations, in the absence of overt clinical disease, diagnosis may also be delayed thereby allowing covert and wide-spread dissemination of the virus making eradication much more difficult.

Concurrent or successive epidemics caused by multiple BTV serotypes, as has been seen since 1998 in many parts of Mediterranean Europe (see Chapter 13 – this volume), including BTV serotypes 1, 4, 9 and 16 during the 1998–2001 epidemic in Greece, can also create particular diagnostic and control problems. Serological assays can often be used to identify the specificity of neutralising antibodies in sera from animals infected in the field. However, successive infections with multiple serotypes not only lead to production of antibodies against all of these types, but also generate increasing levels of cross-reactions with other serotypes (Jeggo *et al.*, 1984, 1986) making type identification difficult or impossible by these methods. In the past, serotype identification has therefore relied heavily upon virus isolation and typing by virus neutralisation assays, using a panel of reference antisera to the 24 BTV types. Although such virus neutralisation tests are still the gold standard for the identification of serotype, these methods are time consuming, are dependent upon the availability of high quality serological reagents which are not always available and can be temperamental in operation. Fortunately, recent advances in sequence analyses for dsRNA virus genomes (Maan *et al.*, 2007a) and serotype-specific RT-PCR methods have now made it possible to identify BTV to the serotype level in blood samples, more rapidly and more accurately than ever before (see *Molecular methods for surveillance*, below).

The relatively low levels of cross-neutralization and, therefore, protection between the different BTV serotypes make it vitally important to identify the outbreak serotype as quickly as possible so that the appropriate vaccine can be selected and deployed without delay. Similarly, the identification of the strain within the serotype may enable valuable information to be deduced on the origin and route of incursion.

Methods of surveillance and monitoring

Reliable diagnostic methods and assays are essential for use in BT monitoring and surveillance systems. These assays can be used to identify the virus type and strain, which may indicate its likely origins, providing vital epidemiological information concerning distribution, prevalence, routes of incursion and the recent or subsequent movements of individual viruses. These data form a basis to assess the risks of further spread, and for the design and implementation of appropriate control strategies by the national and international authorities. The standard diagnostic methodology is outlined in the World Organisation for Animal Health (OIE) Manual of Diagnostic Tests and Vaccines for Terrestrial Animals (OIE, 2004). The type of surveillance employed and the frequency of sampling will depend on the precise epidemiological questions being asked. The general and specific BT surveillance guidelines are detailed in the Appendices of the OIE Terrestrial Animal Health Code (OIE, 2007).

Clinical surveillance

Historically, clinical surveillance has been the standard method for the detection and monitoring of many diseases including BT. In order to be effective, clinical diagnosis requires a well-organized, properly trained veterinary service and an efficient clinical reporting system. Clinical diagnosis relies heavily on the experience of the individual veterinarian and with diseases such as BT, where many differential diagnoses are possible it may take some time before other common diseases are excluded, suggesting that it should not be a first or only method in use. For example, when BTV first occurred on the Greek island of Lesbos, in 1979, diagnosis was delayed because of the absence of the typical tongue lesions and lack of BT clinical experience in the local veterinarians. In one instance the presence of sub-maxillary oedema, combined with the coincidental post-mortem findings of heavy parasitism, diverted suspicion to parasitic causes (unpublished data).

In situations where outbreaks are infrequent or previously unknown, the clinical signs of BT, particularly in cattle, may be unrecognized or go unnoticed. Indeed, sub-clinical infections are the general rule in cattle, goats and resistant breeds of sheep which is one of the reasons why, in endemic regions, the clinical signs of BT may occur only in susceptible breeds of imported sheep. Also, as was seen in Greece during 1998–2001 and for reasons not well understood, clinical cases of BT may occur in one flock, whereas adjacent flocks or herds, although serologically positive, may exhibit no signs of the disease. In the 2006 northern European epidemic, it was estimated that in affected sheep flocks mean morbidity was 20% and mean mortality 5%. In affected cattle herds mean morbidity was 6.8% and mean mortality 0.3%. Approximately 7% of the affected sheep flocks did not show any clinical signs on clinical inspection, neither were clinical signs seen in goats (Elbers *et al.*, 2007). These observations show that an efficient laboratory diagnostic service is essential to confirm a clinical suspicion of BT. However, having said that, clinical diagnosis itself is vital for the initial recognition of the disease especially in new or unexpected locations where routine testing for BTV would not be expected. Indeed, the northern European outbreak of BTV-8 was first detected in August 2006 by recognition of typical clinical signs of the disease in an infected sheep.

Serological surveillance

The serological methods that are currently used for routine BT diagnosis and surveillance (e.g. ELISA) are specific for the members of the *BTV* species or serogroup, i.e. they detect antibodies to any and all of the strains and serotypes of the virus. The major serogroup-specific antigen of the virus is VP7 (Gumm and Newman, 1982), which is encoded by genome segment 7 and forms the surface layer of the virus-core particle (see chapter 8 – this volume). VP7 is highly conserved and, unlike the components of the outer capsid, shares many epitopes across the whole BTV species. However, VP7 is less conserved than the innermost core proteins. It shows little or no cross-reaction with the equivalent protein of other *Orbivirus* species. VP7 is also immuno-dominant, in many cases raising a much higher antibody response than other virus structural proteins (e.g. after a single inoculation with purified virus), making it particularly suitable for development of group-specific diagnostic assays (Afshar *et al.*, 1992).

Serotype-specific assays (e.g. serum neutralisation tests) can be used to differentiate between sera to each of the 24 serotypes of BTV to reveal evidence of past infection with a specific serotype, but it is not possible to determine the time interval since infection. Furthermore, these serological typing tests cannot be used to distinguish between different strains, lineages, topotypes or reassortants within a single BTV serotype and can be compromised by multiple or successive infections with different BTV serotypes (as described above).

Virus circulation during the vector season can be monitored using 'sentinel animals' or 'sentinel herds' that comprise carefully selected and tested seronegative animals, which may include young animals born after any previous outbreak, as long as these have been checked to confirm the absence of maternal antibodies. If these animals then become seropositive during subsequent and periodic testing, they must have been infected since the previous negative test, indicating active virus circulation in that time period taking into account the time interval required between infection and positive test (usually 1–3 weeks). Any animals that do seroconvert must be removed from the sentinel herd and ideally should be replaced by another seronegative individual if the sentinel herd is small. Such sentinel surveillance can be randomized or weighted towards areas where BTV transmission is considered most likely.

Virological surveillance

Virological surveillance is usually targeted towards clinically suspect cases or recently seroconverted sentinel animals (see above). However, viraemia is transient, with peak levels and therefore the highest chances of successful isolation covering a period of only 1–4 weeks after infection for each animal. Intravenous (i.v.) inoculation of 14-day-old embryonating hens' eggs with diluted and washed blood samples is the method of choice when attempting BTV isolation from the field, but inoculation of mammalian or insect cell cultures may also be used. The general rule when attempting to isolate viruses from the field is that the greater the range of assay systems employed the more virus isolations that will be made. Differentiation between 'wild' field strains and vaccine strains is a necessity, when vaccination with live vaccines is applied or their spread from vaccinated animals is suspected (Ferrari et al., 2004; Monaco et al., 2006).

Molecular methods for surveillance (RT-PCR and sequencing)

It is possible to identify BTV RNA using assays that are based on RT-PCR. These assays exist in a number of formats, including 'conventional' methodologies, where the targeted viral RNA sequence is copied and amplified, then detected by electrophoretic analyses of the cDNA products. A strong band of the expected size, particularly from multiple target regions, provides evidence of a positive result. These assays can be used to target sequences or genome segments that are conserved across the whole BTV species/serogroup, e.g. genome segment 7, which encodes the main serogroup antigen VP7 (Anthony et al., 2007). However, variations between isolates from different regions may require the use of multiplex (duplex) primer systems.

It is also possible to use conventional RT-PCR assays to identify virus serotype, by designing primers that specifically target genome segment 2 sequences from the appropriate type (Mertens et al., 2007; Maan et al., 2007b). These assays are more reliable and much faster than conventional serotyping methods, because they do not require virus isolation or access to stocks of highly characterized reference antisera for each of the 24 BTV serotypes, but rely primarily on specific oligonucleotides, which are inexpensive to buy and easy to obtain commercially. These methods also have significant advantages over serological typing assays as they can independently detect multiple virus types in a single blood sample, making them better suited for use in areas where multiple BTV serotypes are co-circulating or endemic. The amplified cDNA products can also be sequenced not only to confirm virus 'type' but also to identify different virus lineages and topotypes within each type, in a manner that is impossible by conventional serological typing methods (see Chapter 9 – this volume).

It is also possible to use real-time RT-PCR assays, which have been designed to detect conserved regions of the virus genome, e.g. genome segment 1 (Shaw *et al.*, 2007) and are therefore BTV species or serogroup specific. These assays detect fluorescence signals from an oligonucleotide probe that is activated during target amplification. The assay format is particularly well suited for routine diagnostic testing, as it uses a closed tube format rather than electrophoretic analyses to identify positive and negative samples, which significantly reduces the risks of cross-contamination and 'false-positive' results. The real-time RT-PCR is also faster to perform than gel-based RT-PCRs making it particularly well suited to high throughput automation. In research it can be used to provide sensitive quantitative measurements.

The persistence of viral RNA in blood, long after the infectious viraemia has subsided (e.g. for up to 220 days), also allows RT-PCR methods to be used to detect viral RNA and identify the BTV strain involved in an earlier infection, even if the animal is no longer infectious or viraemic. These molecular methods for virus detection and strain identification can also be used to explore the distribution and movement of individual virus strains, thereby providing data to underpin risk assessment and the design of appropriate control strategies. In addition, they can also be applied to differentiate between 'wild' and vaccine BTV strains (Ferrari *et al.*, 2004; Monaco *et al.*, 2006). The rapid nature of ELISA and RT-PCR assays and their potential for high-throughput automation are particularly valuable for large-scale surveillance by national and international reference laboratories.

When employing PCRs in surveillance, it must be remembered that they are designed to detect viral nucleic acids and so provide no information on the duration of infectivity in an animal; this information can only be acquired by the isolation or detection of live virus.

Entomological surveillance

The purpose of vector monitoring is to determine the spatial and temporal distribution of vector *Culicoides* spp. in a specific study area, which may be an affected farm, a breeding site, or even an entire region or country. In that case the region can be divided into smaller areas (quadrants), where a number of representative sampling points are selected. At each sampling point one or more insect-collecting light traps are set up and trapping collections are made at agreed time intervals throughout the year, according to an established trapping protocol. These studies can be used to show the presence, abundance, relative proportions and seasonal variations in the numbers of adult *Culicoides* of established, potential or newly identified vector species. Such data are vital in order to enable a comprehensive risk assessment for BTV infection and transmission and to determine the likelihood of the virus becoming established in a region.

Modelling

Using the virological, serological and entomological data collected as described above, mathematical models can be developed to describe the epidemiology of BT in time and space and to identify the climatic and other variables controlling BTV establishment and spread. Such analyses can be used to generate predictive maps that identify levels of risk in different regions and at different times of the year. A range of satellite-derived, proxy climatic variables (NDVI, LST, MIR) that have been shown to correlate with vector abundance, when combined with a detailed knowledge of the preferred habitats and previous distribution of known vector species and their virus transmission rates, can be used to predict geographical areas that could support an expansion of these insect populations and also of virus transmission (Baylis *et al.*, 2001; Purse *et al.*, 2005). Weather data, particularly those concerning temperature and wind speed/direction can also be used to predict the level of risk and likely direction of any potential spread, from an established outbreak site or region. Such models will enable veterinary and legislative authorities to monitor the levels of risk and to devise appropriate surveillance and control measures (Purse *et al.*, 2004; Torina *et al.*, 2004; Conte *et al.*, 2005; Guis *et al.*, 2007; Gloster *et al.*, 2007a, b).

Methods of control

Zoosanitary methods – Animal movement restrictions

As a general rule, animal movement restrictions apply to infected, possibly-infected and susceptible animals originating from or transiting through infected premises, defined risk zones or countries (see below). Infected ruminants may be viraemic and are deemed to be infectious for a maximum period of 60 days after infection, as currently defined by OIE (OIE, 2007). Infected animals become serologically (antibody) positive from the first to third week after infection, often for the duration of their lives. Such animals will be RT-PCR positive for approximately 3–100 days after infection in sheep, and 3–160 days in cattle (maximum known period in cattle – 220 days) (MacLachlan *et al.*, 1994; Singer *et al.*, 2001; Bonneau *et al.*, 2002).

Exemptions or derogations from the movement restrictions may be allowed under certain circumstances, e.g. movement to slaughter, movement during vector-free or low vector activity seasons, pre-movement isolation for set time periods plus protection from *Culicoides* vectors and testing with negative results, or positive serology but negative RT-PCR assay (see *legislation*). Similar conditions can also be applied to facilitate the safe movement of semen, ova and embryos from infected areas (Wrathall *et al.*, 2006). No BT-risk is associated with carcasses or animal products, or with the contamination of housing, equipment or fomites.

Husbandry modifications

The purpose of these measures is primarily to reduce biting by adult vector *Culicoides*, either by reducing their numbers through identification and destruction of their breeding sites, e.g. by draining wet areas and removing manure and straw, or by housing susceptible animals during times of maximum vector activity, e.g. the crepuscular periods, if the local vectors have been shown to be exophagic (Meiswinkel *et al.*, 2000). Susceptible animals could also be moved to areas where adult vector *Culicoides* are less common, e.g. high-altitude pastures or away from river valleys. In some countries where BTV occurs (e.g. Australia), clinical BT is rare or absent because the disease-susceptible species (sheep) are raised primarily in areas in the south where vector species are absent, so only cattle (which are largely asymptomatic) are present in northern areas where both vectors and BTV are present.

Vector abatement and/or control

In addition to husbandry modifications, attempts can also be made to reduce the local numbers of biting *Culicoides* by introducing vector control measures, although as eradication is considered impractical such measures should be regarded as mitigating the risks of transmission rather than preventing it (Mellor and Wittmann, 2002). These approaches rely on the use of insecticides against adult vector *Culicoides* on animals and buildings and larvicides on vector breeding sites. However, the widespread use of these toxic chemicals raises concerns in relation to collateral damage to non-target wildlife species (insects, fish, etc.) and persistence, particularly in animals or animal products that are likely to enter the human food chain (milk/meat). As a consequence, most authorities only recommend the targeted application of insecticides of known low mammalian toxicity (e.g. the synthetic pyrethroids). In cases where organic farmers are involved, insecticides are not permissible but insect repellents applied to individual animals may be allowed. Di-ethyl toluamide (DEET) is probably the best easily available repellent but protection rarely extends beyond 4–6 h (Braverman and Chizov-Ginsburg, 1997; Mellor and Wittmann, 2002).

Vaccination

Currently the choice of commercially available vaccines is limited. Attenuated (live) virus vaccines that were originally developed to protect animals in endemic areas in South Africa are available (Dungu et al., 2004; Savini et al., 2007a), but may be less appropriate for use in non-endemic areas such as Europe. Live BTV vaccines have been a subject of some controversy, as they may elicit a viraemia in vaccinated animals that is sufficient to allow vector *Culicoides* to become infected and transmit the virus (Venter et al., 2007); they can cause clinical signs of BT in some susceptible breeds of sheep (Veronesi et al., 2005), and the vaccine viruses can reassort with field strains of BTV to form 'new' viruses with uncertain virulence characteristics (Batten et al., 2008). They may also have teratogenic effects when used in pregnant animals (Flanagan and Johnson, 1995). Accordingly, live vaccine viruses should be selected and developed with a view to minimizing all of these potential hazards.

By contrast, inactivated vaccines should be safe and efficacious, provided that adequate antigen concentration and adjuvants are used, which may make their manufacturing more expensive than live virus vaccines. Partially purified inactivated vaccines are available for some BTV serotypes (2, 4 and 16) and have been provisionally authorized for emergency use in several European countries. These inactivated vaccines have been evaluated in sheep, or sheep and cattle, and appear to be efficacious in inducing protection from clinical disease and preventing or reducing development of a viraemia on subsequent challenge with the homologous virus serotype (Gerbier et al., 2004; Patta et al., 2004; Santi et al., 2004; Savini et al., 2007a, b). Other 'next generation' and subunit vaccines are currently under development in several laboratories. These include vaccines based on 'virus-like particles' (VLP), expressed BTV proteins, DNA-based vaccines and recombinant virus vectored vaccines (e.g. avipox or vaccinia virus) designed to express individual BTV protein components in the target animal. Critical features of these vaccines may include the use of appropriate delivery systems, adjuvants and micro-bead carriers.

The criteria for the use and authorizing of any BT vaccine include quality (application of good manufacturing practices, batch to batch consistency), safety for all species and categories of target animals (age, sex, reproductive status) and efficacy. The importance of neutralizing antibodies for protection against BTV suggests that vaccines need be the same serotype(s) as the outbreak virus(es), although some cross-protection may be observed after serial or multiple infections with several different serotypes (Jeggo et al., 1984, 1986). The onset and the duration of immunity after vaccination are extremely important parameters when assessing vaccine efficacy. It is also important that vaccinated and field-infected animals should be able to be distinguished – DIVA (differentiating infected from vaccinated animals) (see Chapter 20 – this volume). In the absence of fully authorized vaccines, the minimum data requirements for the authorization, under exceptional circumstances, for vaccines for emergency use against BT were set out in a reflexion paper by the European Medicines Agency (EMEA, 2007).

In essence, an ideal vaccine should provide enduring protection against the development of clinical signs of disease on field infection with the homologous serotype, should prevent the development of a viraemia in order to block onward transmission through the vector and should allow vaccinated and infected animals to be distinguished.

Stamping out

Stamping out has rarely been applied as a means of eradicating a BTV incursion. A major reason for this may be because BTV usually causes subclinical infections in a high proportion of infected animals so there is likely to be strong public and political resistance to the slaughter of apparently healthy animals. Also, because many cases of BTV infection are sub-clinical by the time an incursion has been identified, it is considered likely that virus will already be widely established in the local vector populations, which will not be affected by the slaughter of the host animals, and therefore transmission will continue.

However, there is benefit in the limited culling of infected and potentially infectious animals if there is sufficient reason to suspect that they represent the index case in an otherwise free zone, and such timely action may prevent virus dissemination into the local vector population. Such culling on an index premise should also be accompanied by the application of vector abatement measures to attempt to eliminate any vectors that may have been infected before they initiate onwards transmission. However, for this approach to be successful depends critically upon early detection and therefore the quality of clinical and/or other surveillance being operated at the time.

Control strategies

As already mentioned in the *Introduction*, the strategy for the control and prevention of BT follows the steps of risk assessment and management decisions, based on current scientific knowledge. It is therefore clear that the strategy to be followed will be decided on a case-by-case basis following a risk/cost-benefit assessment. However, the decision is also greatly influenced by the constraints of relevant national and international legislation and commitments.

Endemic situations

In tropical and subtropical environments, where vector activity continues throughout the year, BTV is usually endemic and more than one serotype is often present (IAH, 2007; WAHID, 2007). In these situations, protection against clinical disease is the first requirement, although such disease is often limited to exotic breeds of improved sheep. Live vaccines are often used because of their lower cost and potentially higher efficacy in inducing a long-lasting immune response and because inactivated vaccines are not generally available. These vaccines are often polyvalent (Dungu *et al.*, 2004). When there is a special need for export of live animals or semen/embryos the compartmental requirements of OIE should be followed (see below). To facilitate commercial exports, the creation of BTV-free zones could also be considered (Meiswinkel, 1997; Mogajane, 2004).

Incursion threats into BTV-free countries

The recent epidemic of BTV-8 in northern Europe (which started in 2006) demonstrated that few countries or regions can be considered completely safe from BT. There are three known ways of virus introduction: (1) importation of infected and infectious ruminants, semen or embryos; (2) importation of infected biologicals (e.g. BTV-contaminated vaccines or bovine serum) and (3) movement of infected *Culicoides*, on the wind or transported by other means (e.g. in lorries, planes). The arrival of the virus in one of these ways will only result in an outbreak if there are susceptible hosts and local populations of competent vectors already present (Giovannini *et al.*, 2004; Hoar *et al.*, 2004; MacLachlan and Osburn, 2006). Strict international rules exist to prevent transmission by (1) and (2), but such rules are rarely 100% effective as evidenced by the deaths of pregnant dogs in the USA after injection with a BTV-contaminated parvovirus vaccine (Wilbur *et al.*, 1994). In relation to (3) the movement of BTV-infected midges on the wind has been postulated by a number of workers (Sellers *et al.*, 1978, 1979). Although the risk of wind-borne vector incursions is higher in *countries neighbouring infected zones*, the introduction of the virus into Europe has occurred on several occasions from North Africa across the Mediterranean Sea and distances of over 200 km for such incursions have been suggested, based on virus introduction into the Balearic Islands, probably from Sardinia (Alba *et al.*, 2004). In such circumstances, prevention of incursions is impossible but recent work suggests that they may be able to be predicted in time and space thus aiding control (Gloster *et al.*, 2007a, b).

Epidemic situations (first or re-introductions)

BTV incursions may start insidiously, being totally sub-clinical, or may cause disease, depending on the species and breed of animals infected. In the first case, it is almost inevitable that the incursion will not be recognized until disease is evidenced, and identified as BT – by which time the virus may have spread widely. In the second case, unless a BTV incursion is anticipated from a known infected neighbouring country, for example the lack of diagnostic experience may lead to considerable delays before diagnosis is confirmed.

On recognition of a BTV incursion the most desirable option would be eradication. However, the measures to be applied will be decided on a case-by-case basis, following the outcome of a risk/benefit assessment, taking into account the risk factors and the probable efficiency and availability of the measures, and the costs involved. It is again emphasized that the options are greatly reduced by the restraints of international commitments and legislation. National contingency plans provide for the options and actions to be taken in BT emergency situations (see below).

In summary, the first action will be to isolate the index case and set out a protection zone, with restrictions in animal movements and application of vector control measures, aiming at reducing virus transmission. Around the protection zone, a surveillance zone will be established, aiming at the detection of further cases. If new cases are identified (as usually is the case), protection and surveillance zones are extended accordingly. Stamping out and vaccination may be considered (see below).

In temperate climatic zones, transmission is expected to be reduced and finally halted when temperatures fall below vector activity levels. New cases may continue to appear for a period, equal to the maximum incubation period of the disease, which for the purposes of OIE is set at 21 days (OIE, 2007).

A critical point will be overwintering of the virus. Following the vector-free (or low vector activity) season, foci may remain and create new infection cycles and spread in the following spring. This phenomenon depends on several factors, such as the duration of the vector-free season, the duration of viraemia (OIE maximum 60 days for cattle, but probably less for sheep and goats), survival of infected adult vectors and the presence of climatically protected areas. It is clear that the presence of a dense cattle population enhances the likelihood of overwintering. Serological surveillance carried out during the vector-free period, i.e. when virus is no longer being transmitted, will show the actual extent of the epidemic.

Eradication by stamping out of the infected and infective animals should only be considered if there is strong evidence to suggest that this is the first (index) case in an otherwise free zone, and with the expectancy that such a timely action may prevent virus dissemination into the local vector population (see *Stamping out*). An exception was the eradication programme that was attempted after the first BTV-4 epidemic on the Greek Island of Lesbos, in 1979. During the three following years the virus remained active in the cattle breeding region of the island, with seroconversions but no clinical disease. As virus survival was attributed to the short vector-free period and the longer viraemia in cattle, the control programme implemented consisted of slaughter of all seropositive cattle during the low vector activity season. In the first year (winter 1984), 605 of the 4525 cattle were found positive and were slaughtered. In the following year, 43 cattle were sero-converted and were slaughtered, and none afterwards. A sentinel bovine and ovine surveillance system was established, and in the absence of sero-conversions EU restrictions were lifted in 1991. After 1984, the seasonal movement of breeding sheep to continental Greece was allowed but only during winter and early spring and after a negative serological test and spraying with insecticide when leaving the island (Nomikou *et al.*, 2004, Unpublished data). No further evidence of BTV transmission was detected on the island until the new BTV incursions of 1999.

Emergency vaccination

Emergency vaccination, aiming at reducing clinical severity, and hopefully transmission, could be considered, depending on the availability of homologous vaccines and taking into account the time needed to achieve protection after completing the vaccination regime, the direction and speed of BTV spread and possible confusion in differentiating vaccinated from infected animals, for surveillance or for movement/exporting purposes. Such emergency vaccination is only likely to be successful if a vaccine that protects against clinical disease and reduces viraemia to a level below vector infectivity is deployed rapidly and used on an adequately high proportion of the susceptible ruminant population. However, all these requirements may not be easy to achieve in the field. 'Regular' vaccination using similarly efficacious vaccines could be considered at the end of the low vector activity period, in selected areas, with the aim of preventing a recrudescence of the epidemic. The duration of protection conferred by any of these vaccines should be known. Ideally, the duration should cover the whole period of the vector activity, otherwise, an appropriate re-vaccination date will need to be set and adhered to.

Vaccination, using live attenuated preparations, was used extensively against BTV serotypes 2, 4 and 16 in Italy, Corsica, Spain and Portugal, during the BT epidemic that started in 2000 and continued at least until 2006. Vaccination was reported as successful (Caporale *et al.*, 2004; Gerbier *et al.*, 2004; Patta *et al.*, 2004; Santi *et al.*, 2004; Savini *et al.*, 2007a, b) in reducing losses and spread of the BTV, but virus transmission was still being reported from some areas 5 years after the start of the vaccination campaign. By contrast, in continental Greece, where a series of epidemics involving BTV serotypes 1, 4, 9 and 16 occurred in 2000 and 2001, basic zoosanitary measures, such as movement restrictions and vector control were applied but not vaccination and yet virus transmission ceased after 2 years, as demonstrated by sentinel serological surveillance (Nomikou *et al.*, 2004). It is evident that the levels of protection conferred by the different control measures may be difficult to extrapolate from one location to another as the course of BT epidemics depends on many factors and its spread, incidence and clinical severity is often highly unpredictable.

Trade

Trade in live ruminants, their semen and embryos is affected by a declaration of BTV circulation in a country or zone but not of animal products, such as meat and milk. In this context BT is included in the World Trade Organisation's (WTO) Sanitary and Phytosanitary (SPS) agreement, with the technical details being dealt with by the World Organization for Animal Health (OIE) (see below: *legislation*). In brief, seronegative animals or their semen/embryos are accepted for trade provided that the period between sampling and movement is covered by quarantine for a set time period. Theoretically, seropositive animals are also safe after the viraemic period has ended. However, the possibility of a recent re-infection with another BTV serotype cannot always be excluded, and in such situations some free countries may refuse to accept the importation of BTV seropositive animals.

International legislation

The way that the above measures are applied, the tests to be used and the other responsibilities of the National Veterinary Authorities are described in the OIE Terrestrial Animal Health Code (OIE, 2007) and in the Manual of Diagnostic Tests and Vaccines for Terrestrial Animals (OIE, 2004). In line with OIE is the EU legislation, which sets out detailed procedures for BT control within the Union and is endorsed by the national legislations of all Member States. The WTO, under the mandate of its SPS Agreement, safeguards harmonization, equivalence and transparency of world trade by publishing health standards for international trade in animals and animal products. The SPS agreement's main intent is to provide guidelines and provisions to member countries designed to facilitate trade, while taking measures to protect human, animal and plant life, and health. The agreement dictates that all sanitary measures must be scientifically based and not more restrictive than required to avoid the identified risk. The agreement recommends the use of international standards as set out by OIE. If a country chooses to apply more restrictive measures than those recommended by the OIE, it has to justify its position through a risk analysis, thus avoiding the use of the sanitary measures as unjustified barriers to trade (WTO, 1995; Zepeda *et al.*, 2005).

Below, a summary of the current OIE and EU legislation is presented, with the provision that amendments are frequently adopted.

OIE guidelines

The OIE develops normative documents relating to rules that Member Countries can use to protect themselves from the introduction of diseases and pathogens, without setting up unjustified sanitary barriers. The main normative works produced by the OIE are the Terrestrial Animal Health Code (OIE, 2007) and the Manual of Diagnostic Tests and Vaccines for Terrestrial Animals (OIE, 2004), with specific chapters for BT, prepared by specialist working groups, and adopted by the International Committee (Schudel *et al.*, 2004). The Code is concerned mainly with the harmonization of criteria to support a declaration of freedom from BT of countries and zones, and sets out safety rules for trade in live ruminants, semen, ova and embryos.

For the purposes of the OIE Code, the global distribution of BTV is currently described as being between latitudes of approximately 53° N and 34° S, but is known to be expanding in the northern hemisphere. The infective period of BTV-infected animals is set at 60 days.

A BTV-infected country or zone is a clearly defined area where evidence of BTV has been reported during the past 2 years. A country or a zone may be considered free from BTV when BT is notifiable, and the whole country or zone lies wholly north of 53° N or south of 34° S and is not adjacent to a country or zone not having a free status; or a surveillance programme in accordance with the Code has demonstrated no evidence of BTV in the country or zone during the past 2 years; or a surveillance programme has demonstrated no evidence of *Culicoides* likely to be competent BTV vectors in the country or zone. A BTV seasonally free zone is a part of an infected country or zone for which for part of a year, surveillance demonstrates no evidence either of BTV transmission or of adult *Culicoides* likely to be competent BTV vectors.

The Code also sets out in detail the requirements to be certified by veterinary authorities for the importation of live ruminants, semen, ova and embryos from BTV-free, seasonally free or infected countries or zones, including potential risk management strategies for protecting animals from *Culicoides* attack. The specific guidelines for the surveillance for BT are set out in detail in an Appendix and include the data required for declaring freedom from BTV.

Current epidemiological information is available on the OIE website (http://www.oie.int). The WAHID Interface provides access to all data held within OIE's World Animal Health Information System (WAHID, 2007). It replaces and extends the former web interface named Handistatus II System. The database includes immediate notifications and follow-up reports submitted by Member Countries in response to exceptional disease events occurring in those countries, as well as follow-up reports about these events, 6-monthly reports describing the OIE-listed disease situations in each country and annual reports providing further background information on animal health, on laboratory and vaccine production facilities, etc.

European Union legislation

In the European Union (EU), the outbreaks of BT at the end of the 1990s demonstrated that the Community legislation then in force did not deal with the specific features of the disease. In view of this, specific provisions for the control and eradication of BT were set out by Council Directive 2000/75/EC (EU Council, 2000), which lays down measures to control and eradicate BT, and were incorporated into the National Law of all Member States. In this legislation no-action is not an option. Suspected or confirmed cases of BT must be notified to the competent authority of the Member State concerned. Suspicion and confirmation of the disease each trigger a certain number of measures for prevention and control. Where the presence of the disease is suspected in a flock, the official veterinarian is to place the holding(s) concerned under surveillance and undertake a certain number of investigative measures (inventory of the animals and the premises, clinical and epidemiological surveillance) and protective measures (ban on movements of animals, housing, treatment of animals and habitats with insecticide, destruction and disposal of animal carcases). For epidemiological reasons, the measures may be extended to other holdings. If the presence of BT is confirmed, the official veterinarian shall extend the measures introduced on suspicion, to all holdings within a 20-km radius around the infected holding(s) and may proceed, informing the Commission thereof, with the slaughter deemed necessary to prevent extension of the epidemic.

At the same time, the competent authority must establish a protection zone and a surveillance zone. The protection zone will extend over a radius of at least 100 km around the infected holding(s). Within that zone, all holdings with animals must be identified and the animals may not leave the zone. The competent authority must also establish an epidemiological surveillance programme, including sentinel animals and vector studies. A vaccination programme may also be set up within the protection zone (including the 20-km zone) after informing the Commission. The surveillance zone will extend for at least 50 km beyond the protection zone. Within that zone, the same measures applicable within the protection zone must be implemented. However, the vaccination of animals against BT is forbidden, as it could confound the results from the surveillance programme which is intended, in part, to detect evidence of any virus extensions from the infected holdings.

The Commission, assisted by the Standing Committee on the Food Chain and Animal Health, may adopt supplementary measures or derogations. Commission Decision 2005/393/EC (as amended) sets out current requirements in relation to the protection and surveillance zones corresponding to the specific epidemiological situations for each country or region, and the conditions for granting derogations from bans on the movement of animals within and out of those zones. In particular, provisions may be adopted to allow animals, sperm, ova and embryos to leave a restricted zone under certain conditions e.g. post-vaccination, after the seasonal cessation of vector activity, direct movement to slaughterhouse, after serological and/or virological testing (EU Commission, 2005).

Each Member State should draw up a contingency plan indicating the means by which it applies the measures laid down in the Directive; several are available online (e.g. Defra, 2007).

The Directive also designates the national reference laboratories and fixes the Community Reference Laboratory (CRL). At present the CRL for BT is the Institute for Animal Health, Pirbright Laboratory, UK. The main functions and duties include:

1. To co-ordinate, in consultation with the European Commission, the methods employed in the Member States for diagnosis of specified animal diseases.
2. To make the necessary arrangements for training or retraining of experts in laboratory diagnosis.

Sources of information

OIE home page: http://www.oie.int/
OIE WAHID database: http://www.oie.int/wahid-prod/public.php?page=home
EU disease information: http://ec.europa.eu/food/animal/diseases/adns/index_en.htm
dsRNA virus web pages: http://www.iah.bbsrc.ac.uk/dsRNA_virus_proteins/
EU reference laboratory: http://www.iah.bbsrc.ac.uk
EU EFSA web page: http://www.efsa.europa.eu/en/in_focus/bluetongue.html

References

Afshar, A., Eaton, B.T., Wright, P.F., Pearson, J.E., Anderson, J., Jeggo, M. and Trotter, H.C. (1992) Competitive ELISA for serodiagnosis of bluetongue: evaluation of group-specific monoclonal antibodies and expressed VP7 antigen. *J. Vet. Diagn. Invest.* **4**, 231–7.

Alba, A., Casal, J. and Domingo, M., (2004) Possible introduction of bluetongue into the Balearic Islands, Spain, in 2000, via air streams. *Vet. Rec.* **155**, 460–1.

Alexander, K.A., MacLachlan, N.J., Kat, P.W., House, C., O'Brien, S.J., Lerche, N.W., Sawyer, M., Frank, L.G., Holekamp, K., Smale, L. et al. (1994) Evidence of natural bluetongue virus infection among African carnivores. *Am. J. Trop. Med. Hyg.* **51**, 568–76.

Anthony, S., Jones, H., Darpel, K.E., Elliott, H., Maan, S., Samuel, A., Mellor, P.S. and Mertens, P.P.C. (2007) A duplex RT-PCR assay for detection of genome segment 7 (VP7 gene) from 24 BTV serotypes. *J. Virol. Methods* **141**, 188–97.

Batten, C.A., Maan, S., Shaw, A., Mann, N.S. and Mertens, P.P.C. (2008) A European field strain of bluetongue virus derived from two parental vaccine strains by genome segment Reassortment. *Virus Res.* (in press, doi:10.1016/j.virusres.2008.05.016).

Baylis, M., Mellor, P.S., Wittmann, E.J. and Rogers, D.J. (2001) Bluetongue distribution around the Mediterranean – identification of areas at risk by predictive modelling of the vector distribution. *Vet. Rec.* **149**, 639–43.

Bonneau, K.R., DeMaula, C.D., Mullens, B.A. and MacLachlan, N.J. (2002) Duration of viraemia infectious to *Culicoides sonorensis* in bluetongue virus-infected cattle and sheep. *Vet. Microbiol.* **88**, 115–25.

Braverman, Y. and Chizov-Ginzburg, A. (1997) Repellency of synthetic and plant-derived preparations for *Culicoides imicola. Med. Vet. Entomol.* **11**, 355–60.

Caporale, V., Giovannini, A., Patta, C., Calistri, P., Nannini, D. and Santucci, U. (2004) Vaccination in the control strategy of bluetongue in Italy. *Dev. Biol. (Basel)* **119**, 113–27.

Conte, A., Colangeli, P., Ippoliti, C., Paladini, C., Ambrosini, M., Savini, L., Dall'Acqua, F. and Callistri, P. (2005) The use if web-based interactive geographical information system for the surveillance of bluetongue in Italy. *Rev. Sci. Tech.* **24**, 857–68.

Defra (2007) Bluetongue disease contingency plans for Great Britain (draft). Available online at http://www.defra.gov.uk/animalh/diseases/pdf/bluetongue-contplan.pdf

Dungu, B., Potgieter, C., Von Teichman, B. and Smit, T. (2004) Vaccination in the control of bluetongue in endemic regions: the South African experience. *Dev. Biol. (Basel)* **119**, 463–72.

Elbers, A.R.W., Mintiens, K., Staubach, C., Gerbier, G., Meroc, E., Ekker, H.M., Conraths, F.J., van der Spek, A.N. and Backx, A. (2007) Epidemiological analysis of the 2006 bluetongue virus serotype 8 epidemic in north-western Europe – Nature and severity of disease in sheep and cattle. EFSA Report of the Epidemiology Working Group. Appendix 2. Available online at http://www.efsa.europa.eu/en/in_focus/bluetongue/bluetongue_report_s8.html

EMEA (2007) Reflexion paper on data requirements for an authorisation under exceptional circumstances for vaccines for emergency use against bluetongue. European Medicines Agency, London, UK. Available online at: http://www.emea.europa.eu/pdfs/vet/iwp/10500807en.pdf

EU Commission (2005) Commission Decision 2005/393/EC (2005) of 23 May 2005, on protection and surveillance zones in relation to bluetongue and conditions applying to movements from or though these zones. Official Journal of the European Union L130, pp. 22–28.

EU Council (2000) Council Directive 2000/75/EC of 20 November 2000, laying down specific provisions for the control and eradication of bluetongue. *Off. J.Eur. Commun.* **L327**, 74–83.

Ferrari, G., De Liberato, C., Scavia, G., Lorenzetti, R., Zini, M., Farina, F., Magliano, A., Cardeti, G., Scholl, F., Guidoni, M., Scicluna, M.T., Amaddeo, D., Scaramozzino, P. and Autorino, G.L. (2004) Active circulation of bluetongue vaccine virus serotype-2 among unvaccinated cattle in central Italy. *Prev. Vet. Med.* **68**, 103–113.

Flanagan, M. and Johnson, S.J. (1995) The effects of vaccination of Merino ewes with an attenuated Australian bluetongue virus serotype 23 at different stages of gestation. *Aust. Vet. J.* **72**, 455–7.

Gerbier, G., Hendrikx, P., Roger, F., Zientara, S., Biteau-Coroller, F., Grillet, C., Baldet, T. and Albina, E. (2004) Bluetongue control using vaccines: experience of the Mediterranean islands. *Vet. Ital.* **40**, 611–15.

Giovannini, A., Conte, P., Calistri, P., Di Francesco, C. and Caporale, V. (2004) Risk analysis on the introduction into free territories of vaccinated animals from restricted zones. *Vet. Ital.* **40**, 697–702.

Gloster, J., Manning, A., Hort, M., Webster, H. and Mellor, P.S. (2007a) Assessing the risk of windborne spread of bluetongue in the 2006 outbreak of the disease in northern Europe. *Vet. Rec.* **160**, 54–6.

Gloster, J., Mellor, P.S., Burgin, L., Sanders, C. and Carpenter, S (2007b) Will bluetongue come on the wind to the United Kingdom in 2007? *Vet. Rec.* **160**, 422–6.

Guis, H., Tran, A., de la Rocque, S., Baldet, T., Gerbier, G., Barragué, B., Biteau-Coroller, F., Roger, F., Viel, J-F. and Mauny, F. (2007) Use of high spatial resolution satellite imagery to characterize landscapes at risk for bluetongue. *Vet. Res.* **38**, 669–83.

Gumm, I.D. and Newman, J.F.E. (1982) The preparation of purified bluetongue virus group antigen for use as a diagnostic reagent. *Arch. Virol.* **72**, 83–9.

Hoar, B.R., Carpenter, T.E., Singer, R.S. and Gardner, I.A. (2004) Probability of introduction of exotic strains of bluetongue virus into the US and into California through importation of infected cattlet. *Prev. Vet. Med.* **66**, 79–91.

IAH (2007) http://www.iah.bbsrc.ac.uk/dsRNA_virus_proteins/btv-serotype-distribution.htm

Jeggo, M.H., Wardley, R.C., Brownlie, J. and Corteyn, A.H. (1986) Serial inoculation of sheep with two bluetongue virus types. *Res. Vet. Sci.* **40**, 386–92.

Jeggo, M.H., Wardley, R.C. and Taylor, W.P. (1984) Clinical and serological outcome following the simultaneous inoculation of three bluetongue virus types into sheep. *Res. Vet. Sci.* **37**, 368–70.

Kirkland, P.D. and Hawkes, R.A. (2004) A comparison of laboratory and 'wild' strains of bluetongue virus – is there any difference and does it matter? *Vet. Ital.* **40**, 448–55.

Koumbati, M., Mangana, O., Nomikou, K., Mellor, P.S. and Papadopoulos, O. (1999) Duration of bluetongue viraemia and serological responses in experimentally infected European breeds of sheep and goats. *Vet. Microbiol.* **64**, 277–85.

Maan, S., Maan, N.S, Samuel, A.R., Rao, S, Attoui, H. and Mertens, P.P.C. (2007a) Analysis and phylogenetic comparisons of full-length VP2 genes of the twenty-four bluetongue virus serotypes. *J. Gen. Virol.* **88**, 21–30.

Maan, S., Rao, S., Maan, N.S., Anthony, S., Attoui, H., Samuel, A.R. and Mertens, P.P.C. (2007b) Rapid cDNA synthesis and sequencing techniques for the genetic study of bluetongue and other dsRNA viruses. *J. Virol. Methods* **143**, 132–9.

MacLachlan, N.J., Nunamaker, R.A., Katz, J.B., Sawer, M.M., Akita, G.Y., Osburn, B.I. and Tabachnick, W.J. (1994) Detection of bluetongue virus in the blood of inoculated calves: comparison of virus isolation, PCR assay, and in vitro feeding of *Culicoides variipennis*. *Arch. Virol.* **136**, 1–8.

MacLachlan, N.J. and Osburn, B.I. (2006) Impact of bluetongue virus infection on the international movement and trade of ruminants. *J. Am. Vet. Med. Assoc.* **228**, 1346–9.

Meiswinkel, R. (1997) Discovery of a *Culicoides imicola*-free zone in South Africa: Preliminary notes and potential significance. *Onderstepoort J. Vet. Res.* **1997**, 81–6.

Meiswinkel, R., Baylis, M and Labuschagne, K (2000) Stabling and the protection of horses from *Culicoides bolitinos* (Diptera: Ceratopogonidae), a recently identified vector of African horse sickness. *Bull. Entomol. Res.* **90**, 509–15.

Mertens, P.P.C., Maan, N.S., Prasad, G., Samuel, A.R., Shaw, A., Potgieter, A.C., Anthony, S.J. and Maan, S. (2007) The design of primers and use of RT-PCR assays for typing European BTV isolates: Differentiation of field and vaccine strains. *J. Gen. Virol.* **88**, 2811–23.

Mellor, P.S. and Wittmann, E.J. (2002) Bluetongue virus in the Mediterranean Basin 1998–2001. *Vet. J.* **164**, 20–37.

Mogajane, M.E. (2004) Trade implications of bluetongue in Africa. *Vet. Ital.* **40**, 691–2.

Monaco, F., Camma, C., Serini, S. and Savini, G. (2006) Differentiation between field and vaccine strain of bluetongue virus serotype 16. *Vet. Microbiol.* **116**, 45–2.

Nomikou, K., Mangana-Vougiouka, O. and Panagiotatos, D.E. (2004) Overview of bluetongue in Greece. *Vet. Ital.* **40**, 108–15.

OIE (2004) Manual of Diagnostic Tests and Vaccines for Terrestrial Animals, 12th edn., World Organisation for Animal Health, Paris. Available online at http://www.oie.int

OIE (2007) Terrestrial Animal Health Code, 16th edn., World Organisation for Animal Health, Paris. Available online at http://www.oie.int

Patta, C., Giovannini, S., Rolesu, S., Nannini, D., Savini, G., Calistri, P., Santucci, U. and Caporale, V. (2004) Bluetongue vaccination in Europe: the Italian experience. *Vet. Ital.* **40**, 601–10.

Purse, B.V., Baylis, M., Tatem, A.J., Rogers, D.J., Mellor, P.S., Van Ham, M., Chizov-Ginzburg, A. and Braverman, Y. (2004) Predicting the risk of bluetongue through time: climate models of temporal patterns of outbreaks in Israel. *Rev. Sci. Tech.* **23**, 761–75.

Purse, B.V., Mellor, P.S., Rogers, D.J., Samuel, A.R., Mertens, P.P.C. and Baylis, M. (2005) Climate change and the recent emergence of bluetongue in Europe. *Nat. Rev. Microbiol.* **3**, 171–81.

Santi, A., Loli Piccolomini, L., Viappiani, P., Tamba, M., Calabrese, R. and Massirio, I. (2004) Bluetongue control using vaccines: the experience of Emilia Romagna, Italy. *Vet. Ital.* **40**, 623–6.

Savini, G., MacLachlan, N.J., Sanchez-Vizcaino, J-M. and Zientara, S. (2007a) Vaccines against bluetongue in Europe. *Comp. Immunol. Microbiol. Infect. Dis.* **31**, 101–20.

Savini, G., Ronchi, F., Luis, T.M., Vaz, A., Duarte, M., Henriques, M., Cruz, B. and Fevereiro, M. (2007b) An inactivated vaccine for the control of bluetongue virus serotype 16 infection in sheep in Italy. *Vet. Microbiol.* **124**, 140–6.

Schudel, A., Wilson, D. and Pearson, J.E. (2004) Office International des pizooties international standards for bluetongue. *Vet. Ital.* **40**, 676–81.

Sellers, R.F., Pedgley, D.E. and Tucker, M.R. (1978) Possible windborne spread of bluetongue to Portugal, June–July 1956. *J. Hyg. Camb.* **81**, 189–96.

Sellers, R.F., Gibbs, E.P.J., Herniman, K.A.J., Pedgley, D.C. and Tucker, M.R. (1979) Possible origin of the bluetongue epidemic in Cyprus, August 1977. *J. Hyg. Camb.* **83**, 547–55.

Shaw, A.E., Monaghan, P., Alpar, H.O., Anthony, S., Darpel, K.E., Batten, C.A., Carpenter, S., Jones, H., Oura, C.A.L., King, D.P., Elliot, H., Mellor, P.S. and Mertens, P.P.C. (2007) Development and validation of a real-time RT-PCR assay to detect genome bluetongue virus segment 1. *J. Virol. Methods* **145**, 115–26.

Singer, R.S., MacLachlan, N.J. and Carpenter, T.E. (2001) Maximal predicted duration of viraemia in bluetongue virus-infected cattle. *J. Vet. Diagn. Invest.* **13**, 43–49.

Takamatsu, H., Mellor, P.S., Mertens, P.P.C., Kirkham, P.A., Burroughs, J.N. and Parkhouse, R.M.E. (2003) A possible overwintering mechanism for bluetongue virus in the absence of the insect vector. *J. Gen. Virol.* **84**, 227–35.

Torina, A., Caracappa, S., Mellor, P.S., Baylis, M. and Purse, B.V. (2004) Spatial distribution of bluetongue virus and its *Culicoides* vectors in Sicily. *Med. Vet. Entomol.* **18**, 81–89.

Venter, G.J., Mellor, P.S., Wright, I. and Paweska, J.T. (2007) Replication of live-attenuated vaccine strains of bluetongue virus in orally infected South African *Culicoides* species. *Med. Vet. Entomol.* **21**, 239–47.

Veronesi, E., Hamblin, C. and Mellor, P.S. (2005) Live attenuated bluetongue vaccine viruses in Dorset Poll sheep, before and after passage in vector midges (Diptera: Ceratopogonidae). *Vaccine* **23**, 509–16.

WAHID (2007) OIE's World Animal Health Information Database interface. World Organisation for Animal Health, Paris. Available online at: www.oie.int/wahid-prod/public.php?page=home

White, D.M., Wilson, W.C., Blair, C.D. and Beaty, B.J. (2005) Studies on overwintering of bluetongue viruses in insects. *J. Gen. Virol.* **86**, 453–62.

Wilbur, L.A., Evermann, J.F., Levings, R.L., Stoll, L.R, Starling, D.E., Spillers, C.A., Gustafson, G.A. and McKeirnan, A.J. (1994) Abortion and death in pregnant bitches associated with a canine vaccine contaminated with bluetongue virus. *J. Am. Vet. Med. Assoc.* **204**, 1762–5.

Wittmann, E.J., Mellor, P.S. and Baylis, M. (2002) Effect of temperature on the transmission of orbiviruses by the biting midge, *Culicoides sonorensis. Med. Vet. Entomol.* **16**, 147–56.

Wrathall, A.E., Simmons, H.A. and Van Soom, A. (2006) Evaluation of risks of viral transmission to recipients of bovine embryos arising from fertilisation with virus-infected semen. *Theriogenology* **65**, 247–74.

WTO (1995) The World Trade Organization (WTO) agreement on the application of sanitary and phytosanitary measures (SPS agreement). WTO, Geneva. Available online at http://www.wto.org/English/tratop_e/sps_e/spsagr_e.htm, accessed on 30 October 2007

Zepeda, C., Salman, M., Thiermann, A., Kellar, J., Rojas, H. and Willeberg, P. (2005) The role of veterinary epidemiology and veterinary services in complying with the World Trade Organization SPS agreement. *Prev. Vet. Med.* **67**, 125–40.

Conclusions

20

PHILIP S. MELLOR,* PETER P.C. MERTENS* AND MATTHEW BAYLIS†

*Institute for Animal Health, Pirbright Laboratory, Pirbright, Woking, Surrey, UK
†Veterinary Clinical Science, University of Liverpool, Neston, Cheshire, UK

It can be hypothesised that, historically, the primary epizootiological cycle of bluetongue virus (BTV) was between species of African antelope and vector *Culicoides*. The disease, bluetongue (BT), became evident only when susceptible European breeds of sheep were imported into Africa.

With the agricultural development in large parts of Africa and the introduction of many domestic ruminants, the traditional epizootiological role of wild animals has now largely been taken over by cattle. In many regions such as southern Africa, the infection seems to progress in a 'cattle–midge cycle' during spring and early summer, and once a certain level of infection is reached, it spills over to sheep. This generally occurs in late summer or in autumn. Sheep, therefore, seem to be involved mainly in a secondary epizootiological cycle (see Chapter 2).

Bluetongue first achieved prominence outside Africa during the 1950s and 1960s when it caused major outbreaks of disease in SW Europe, killing over 179 000 sheep. Its identification at around the same time in the Americas and SE Asia confirmed in the minds of many farmers, scientists and legislators that it was an emerging disease that could devastate the livestock industries of the more advanced farming nations in Europe, the Americas and Australia. For these reasons it was allocated the status of a List 'A' disease by OIE.

However, during the 1970s and 1980s in the absence of major outbreaks of disease the perception that BTV had reached its potential to emerge and should be down-graded in importance and removed from the OIE List 'A' category grew. More recently, possibly as a result of climate-change or other

ISBN-13: 978-0-12-369368-6

environmental changes, BT has once again captured the attention of much of the developed world, with extensions of new serotypes into North America, Australia and Europe and with the most devastating outbreak of BT ever recorded in Europe stretching as far as 53° N, the furthest north the virus has ever been recorded anywhere in the world.

Spurred on by these new outbreaks, since the late 1990s, there has been an enormous increase in scientific work focussing on BTV, and great strides have been made in our understanding of the distribution of the different BTV serotypes and also in the fields of epidemiology, molecular epidemiology, biochemistry, virus structure and function, transmission, vector identification and ecology, control including vaccination, and predictive modelling. The contents of this book describe the advances that have recently been achieved in each of these areas.

Several chapters deal primarily with the current distribution of the serotypes and strains of BTV, in time and space, in locations across the world from the Americas to Europe, southeast Asia and Australasia, and also describe recent variations in distribution in the light on ongoing environmental changes.

Much has also been learned about the assembly and replication of BTV and other orbiviruses and of their inter-relationships and taxonomy. Such studies have benefited greatly from the detailed structural work on the viral core and through a synthesis of the information derived from structural, biochemical and cell biology studies.

Our knowledge of the mechanisms involved in causing and controlling cellular and tissue damage is rapidly increasing, providing a better understanding of the responses of different species (vertebrate and invertebrate) and different tissues to BTV infection. However, further work is necessary to determine the mechanisms that lead to the death of most BTV-infected mammalian cells but to persistence in insect cells and possibly some mammalian cells.

Advances in molecular epidemiology now mean that the geographical movements of BTV serotypes and strains can be closely tracked in time and space to help identify the origins of virus incursions, determine their routes of entry into new territories and even elucidate movement between individual farms, thus potentially enabling control responses to be better targeted. These studies also enable reassortments between BTV types and strains (field or live vaccine) to be quickly identified and the consequences of such interactions to be determined. Such forensic work requires not only the application of powerful new molecular analytical tools, such as SNPS analysis, but also the availability and maintenance of comprehensive and up-to-date sequence databases that are easily and rapidly accessible to scientists, veterinary authorities and those involved in control.

Bluetongue virus infection of its mammalian and insect hosts involves complex host–pathogen–vector–environment interactions. In relation to the insect vector, it has been shown that at higher temperatures, a greater proportion of a vector population is able to transmit BTV, transmission occurs more quickly and additional species of *Culicoides* may be recruited as novel vectors. It has also been shown that a single bite from an infectious vector *Culicoides* is sufficient to infect a susceptible ruminant host. Such findings may help explain why, for the first time, BTV has been able to enter and be transmitted by Palearctic species of *Culicoides* in northern Europe – the virus invading the area at a time when the temperature was some 6.2°C higher than ever before (Gloster *et al.*, 2007). In a time of climate change, further examples of pathogens being transmitted beyond their normal range and by novel vectors may be expected.

Until recently it was supposed that only ruminant hosts with a high viraemia represent a risk for virus transmission to the insect vector. However, this is clearly not the case as recent work has shown that vector insects can be infected by feeding on infected hosts at very low viraemia titres or even when no viraemia at all is detectable (Bonneau *et al.*, 2002). It is not yet known whether these represent chance events whereby individual vectors may occasionally ingest an infectious unit of BTV even at very low viral titres or whether, since midges feed on skin capillary seepage and tissue fluid, the virus titre as measured from sampling the major blood vessels is less relevant. In this context, it has recently been shown that virus may sometimes be present in the skin of a ruminant host even when it is aviraemic (Darpel, 2007).

A fundamental question that has long vexed scientists is why virulent strains of BTV often produce disease in sheep but not in cattle. It has now been shown that BTV infection of bovine endothelial cells results in endothelial activation with the increased transcription of cell surface adhesion molecules, while a similar infection of sheep results in minimal activation of endothelial cells. Furthermore, the ratio of thromboxane to prostacyclin, which is indicative of enhanced coagulation and possibly consumptive coagulopathy, is significantly greater in sheep than in cattle (DeMaula *et al.*, 2001, 2002a, 2002b).

Rapid and accurate diagnosis of infection is of ever increasing importance in the fight against BT. However, during the current European BTV outbreak, most northern authorities operated only a passive clinical surveillance system. As the proportion of clinical cases, in the principal vertebrate host cattle, is small in comparison to infections (<5%), this has been insufficient to determine, accurately, the distribution of the virus and has rendered control and eradication challenging. Nevertheless, accurate and rapid diagnostic tests for confirmation of a clinical infection and for the identification of subclinical infections, at both BTV group and serotype-specific levels, have now been developed and are widely available (see Chapter 17).

Bluetongue virus control strategies are changing rapidly. Concerns on the use of live virus vaccines, as some have been shown to cause disease in susceptible breeds of sheep, some can be transmitted by vector *Culicoides* and reassortment has been recorded between vaccine and field strains of virus in the field, have meant that a range of inactivated vaccines is now being marketed. These have none of the potential drawbacks listed above and are a timely addition to our control armoury. However, the cost per dose is higher than for live vaccines and level of efficacy, duration of protection and occurrence of colostral immunity remain to be fully determined.

Other methods of BT control – animal movement restrictions, vector control and housing of livestock to protect from vector attack – have all been applied across northern Europe during 2006–2007 but in the absence of vaccination have failed to prevent the spread of disease. At the time of writing, few studies have been undertaken specifically to assess the efficacy of insecticides in controlling adult *Culicoides*, and as the breeding sites of most of the vector species are poorly defined, these are difficult or impossible to treat. Furthermore, the degree of protection from *Culicoides* attack afforded by housing stock is uncertain and depends upon the species of vector present, the level of enclosure, the time of the year and the periods of vector activity. Consequently, such methods of control are unlikely to be completely effective in the absence of a comprehensive vaccination campaign and are best regarded as mitigation measures of uncertain efficacy, at least until more detailed studies have been carried out.

The ability of BTV to successfully overwinter in northern Europe from 2006 to 2007 and from 2007 to 2008, in regions where there is a virtual vector-free period of >60 days and a transmission-free period of >90 days (Wilson *et al.*, 2007; Wilson, Carpenter and Mellor, Personal Communication), has been one of the most startling observations in the recent history of BT. As the maximum infectious period for BTV, in its ruminant hosts, has been set at 60 days by OIE, the virus should not be able to overwinter under such conditions but it did so easily, especially in regions where intense transmission had occurred in the previous year. Investigations to determine how this might happen have shown that BTV-8 from northern Europe is able to pass transplacentally from bovine dams infected during pregnancy to the foetus, many of which look perfectly healthy at birth but are strongly PCR-positive (De Clercq *et al.*, 2008; Menzies *et al.*, 2008; Darpel *et al.*, Personal Communication). Live virus has also been isolated from some of these calves. As the bovine gestation period is of 9 months duration, this mechanism should easily enable BTV-8 to overwinter. On the face of it then, this would seem to be the most likely way in which BTV-8 has overwintered. However, there is a problem; although live virus has been recovered from a small proportion of PCR-positive calves, usually those sampled pre-colostrum, this has not been possible from the vast majority. The reasons for this difficulty are uncertain at the time of writing, though it may have to do with the formation of virus–antibody complexes in calves that have ingested colostrum from their antibody- positive mothers. Clearly, if live virus can be recovered only from a very small proportion of transplacentally infected animals, this mechanism is not likely to be the main way in which BTV has overwintered so easily, and other mechanisms must also be involved. The identity of these is uncertain. Bluetongue virus has provided many surprises over its history, but despite the advances in our understanding, it seems that the surprises are set to continue.

References

Bonneau, K.R., DeMaula, C.D., Mullens, B.A. and MacLachlan, N.J. (2002) Duration of viraemia infectious to *Culicoides sonorensis* in bluetongue virus-infected cattle and sheep. *Vet. Microbiol.* **88**, 115–25.

Darpel, K.E. (2007) The bluetongue virus 'ruminant host – insect vector' transmission cycle; the role of *Culicoides* saliva proteins in infection. A thesis submitted in partial fulfilment of the requirements of the University of London for the degree of Doctor of Philosophy.

De Clercq, K., Vandenbussche, F., Vandemeulebroucke, E., Vanbinst, T., De Leeuw, I., Verheyden, B., Goris, N., Mintiens, K., Meroc, E., Herr, C., Hooybergs, J., Houdart, P., Sustronck, B., De Deyken, R., Maquet, G., Burghin, J., Saulmont, M., Lebrun, M., Bertels, G and Miry, C. (2008) Transplacental bluetongue infection in cattle. *Vet. Rec.* **162**, 564.

DeMaula, C.D., Jutila, M.A., Wilson, D.W. and MacLachlan, N.J. (2001) Infection kinetics, prostacyclin release, and cytokine-mediated modulation of the mechanism of cell death during bluetongue virus infection of cultured ovine and bovine pulmonary artery and lung microvascular endothelial cells. *J. Gen. Virol.* **82**, 787–94.

DeMaula, C.D., Leutennegger, C.M., Bonneau, K.R. and MacLachlan, N.J. (2002b) The role of endothelial cell-derived inflammatory and vasoactive mediators in the pathogenesis of bluetongue. *Virology,* **296**, 330–7.

DeMaula, C.D., Leutennegger, C.M., Jutila, M.A. and MacLachlan, N.J. (2002a) Bluetongue virus induced activation of primary bovine lung microvascular endothelial cells. *Vet. Immunol. Immunopathol.* **86**, 147–57.

Gloster, J., Mellor, P.S., Burgin, L., Sanders, C and Carpenter, S. (2007) Will bluetongue come on the wind to the United Kingdom in 2007? *Vet. Rec.* **160**, 422–6.

Menzies, F.D., McCullough, S. D.J., McKeown, I.M., Forster, J.L., Jess, S., Batten, C., Murchie, A.K., Gloster, J., Fallows, J.G., Pelgrim, W., Mellor, P.S. and Oura, C.A.L. (2008) Evidence for both transplacental and contact transmission of bluetongue virus in cattle. *Vet. Rec.* (in press).

Wilson, A., Carpenter, S., Gloster, J. and Mellor, P.S. (2007) Re-emergence of bluetongue in northern Europe in 2007. *Vet Rec.* **161**, 487–9.

provided many surprises over its history, but despite the advances in our understanding it seems that the surprises are set to continue.

References

[references illegible]

Glossary

Acronym	Meaning
ADCC	Antibody-dependent cell-mediated cytotoxicity
AdoHCy	*S*-adenosyl homocysteine
AdoMet	*S*-adenosyl methionine
ADP	Adenosine diphosphate
AGE	Agarose gel electrophoresis
AGID	Agar gel immunodiffusion
AHS	African horse sickness
AHSV	African horse sickness virus
AINP	All India Network Project
AMP	Adenosine monophosphate
APHIS	Animal and Plant Health Inspection Service
ATP	Adenosine triphosphate
BBSRC	Biotechnology and Biological Sciences Research Council
BEI	Binary ethylenimine
BHK	Baby hamster kidney
bp	Base pair
BRDV	Broadhaven virus
BSR	cell line derived by sub-cloning from BHK cells
BT	Bluetongue
BTV	Bluetongue virus
C6-36	*Aedes albopictus* cells
CaP	Capping enzyme
CD	Cluster of differentiation
cDNA	Complementary DNA
CEE	Central European encephalitis
CF	Complement fixation
CK2	Casein kinase II
CLP	Core-like particle
CORV	Corriparta virus
CPAE	Calf pulmonary artery endothelial cells

Acronym	Meaning
CPE	Cytopathic effect
CPV	Capripoxvirus
CRL	Community Reference Laboratory
CSBF	Central Sheep Breeding Farm
CSWRI	Central Sheep and Wool Research Institute
Ct	Cycle-threshold
CTB	Cholera toxin subunit B
CTL	Cytotoxic T-cell
CTP	Cytidine triphosphate
DAHDF	Department of Animal Husbandry, Dairying and Fisheries
DEFRA	Department for Environment, Food and Rural Affairs
DIA	Dot immunobinding assays
DIVA	Differentiating Infected from Vaccinated Animals
DNA	Deoxyribonucleic acid
dpi	Days post-infection
dsRNA	Double stranded RNA
DTT	Dithiothreitol
ECE	Embyonated chicken eggs
EDTA	Ethylene-diamine-tetra-acetic acid
EE	Equine encephalosis
EEV	Equine encephalosis virus
EFSA	European Food Safety Authority
EHD	Epizootic haemorrhagic disease
EHDV	Epizootic haemorrhagic disease virus
EIP	Extrinsic incubation period
ELD	Egg lethal dose
ELISA	Enzyme-linked immunosorbent assay
EM	Electron microscopy
EMEA	European Medicines Agency
EU	European Union
FAT	Fluorescent antibody test
FAO	Food and Agriculture Organisation
FMD	Foot and mouth disease
FMDV	Foot and mouth disease virus
FRET	Fluorescence resonance energy transfer
FYR	Former Yugoslav republic
GDP	Guanosine diphosphate
GFP	Green fluorescent protein
GIV	Great Island virus
GMP	Guanosine monophosphate
GT	Guanylyltransferase
GTP	Guanosine triphosphate
HCl	Hydrochloric acid
Hel	Helicase

Acronym	Meaning
HeLa	Henrietta Lacks
HIT	Histidine triad
hsc	Heat shock protein
IAH	Institute for Animal Health
ICAR	Indian Council of Agricultural Research
ICTV	International Committee for the Taxonomy of Viruses
Il	Interleukin
IR	Infection rate
ISCOM	Immunostimulating complex
ISVP	Infectious subviral particle
JAM	Junction-adhesion molecule
KC	KC cells
kDa	kilodalton
KL	Kinase-like domain
LBV	Lebombo virus
LST	Land surface temperature
MCFV	Malignant catarrhal fever virus
MEB	Midgut escape barrier
MEM	Minimal Essential Medium
MER	Molecular evolutionary rate
MEVP	Membrane enveloped virus particles
MHC	Major histocompatibility complex
MIB	Midgut infection barrier
MIR	Middle infrared
MPOV	Middle Point orbivirus
mRNA	Messenger RNA
MS	Member State (of the European Union)
MVA	Modified Vaccinia Ankara
NCRs	Non-coding region
NDP	Nucleotide diphosphate
NDVI	Normalised difference vegetation index
NO	Nitric oxide
NS/NSP	Non-structural protein
NTP	Nucleotide triphosphate
NVSL	National Veterinary Services Laboratories
OBP	Onderstepoort Biological Products
OIE	Office International des Epizooties
ORF	Open reading frame
ORV	Orungo virus
ORV	Orthoreovirus
PABP	Poly(A)-binding protein
PAGE	Polyacrylamide agarose gel electrophoresis
PALV	Palyam virus
PBMC	Peripheral blood mononuclear cells

Acronym	Meaning
PBS	Phosphate Buffered Saline
PCR	Polymerase chain reaction
PDP	Protein database
PFU	Plaque forming units
PHSV	Peruvian horse sickness virus
Pi	Inorganic phosphate
PLA	Poly(lactide)
PLGA	Poly(lactide-co-glycolide)
Pol	Polymerase
PP	Pyrophosphate
PPi	Pyrophosphate
PPR	Peste des petits ruminants
RdRp	RNA dependent RNA polymerase
REPA	Restriction enzyme profile analysis
RNA	Ribonucleic acid
RT-PCR	Real-time PCR
RTP	RNA triphosphate
RV	Rotavirus
RVFV	Rift Valley Fever Virus
SALT	Skin associated lymphatic tissue
SAM	S-adenosyl methionine
ScFV	Single chain variable fragment
SCRV	St Croix River virus
SDS PAGE	Sodium Dodecyl Sulfate PAGE
SEM	Scanning electron microscopy
siRNA	Small interfering RNA
SNP	Single nucleotide polymorphism
SNT	Serum neutralisation test
SPS	Sanitary and Phytosanitary
SR	Susceptibility rate
ssRNA	Single stranded RNA
SYBR	Synergy Brands
T	Triangulation number
TC	Transcriptase complex
TCA	Trichloracetic acid
TCID	Tissue culture infective dose
TEM	Transmission electron microscopy
TLD	Thoracic duct lymphocytes
TLR	Toll-like receptor
TOT	Transovarian transmission
TSE	Transmissible spongiform encephalopathies
TuP	Tubule protein
USDA	United States Department of Agriculture
UTP	Uridine triphosphate

Acronym	Meaning
VIB	Viral inclusion body
VIP	Viral inclusion protein
VLP	Virus like particle
VNT	Virus neutralisation test
VP	Viral protein
VT	Vertical transmission
WAHID	World Animal Health Information Database
WGV	Wongorr virus
WTO	World Trade Organization
YUOV	Yunnan orbivirus

Index

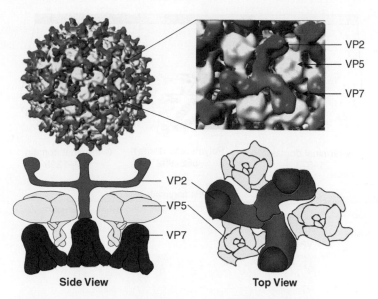

Plate 1 Arrangement of the proteins in the outer capsid of bluetongue virus (BTV). (Top) Cryo-EM reconstructions highlighting the relative position of the triskelion protein (VP2) and the globular protein (VP5). (Below) Cartoon showing the arrangement of VP2, VP5 and VP7 relative to each other (See Figure 4.4 in p. 56).

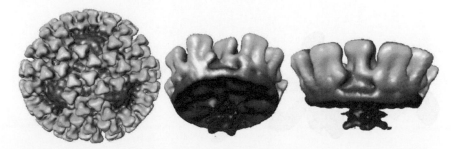

Plate 2 Cryo-EM reconstruction of core-like particles (CLPs) incorporating VP1 (polymerase) and VP4 (capping enzyme). (Left) CLPs in which the VP7 trimer at the five-fold axis has been lost. (Centre and right) Position and shape of the transcriptase complex (blue) relative to VP3 (pink) and VP7 (green) at the five-fold axes (See Figure 4.5 in p. 58).

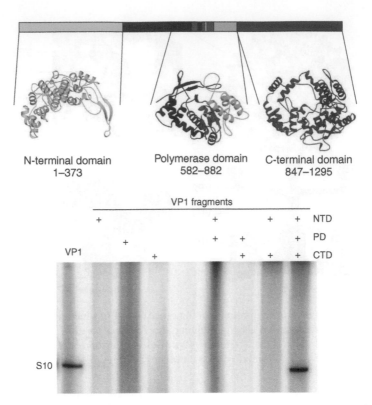

Plate 3 (Top) Predicted structure of domains of VP1 based on homology with other viral RNA-dependent RNA polymerases. (Bottom) The assignment of domains of VP1 was validated by the demonstration that polymerase activity could be reconstituted by mixing expressed N-terminal domain (NTD), polymerase domain (PD) and C-terminal domain (CTD) in a standard polymerase assay for the intact VP1 protein (See Figure 4.7 in p. 60).

Plate 4 Steps in the assembly of the VP3 layer (subcore) of bluetongue virus (BTV). VP3 adopts two different conformations in the mature core structure. Mutagenesis studies have revealed that the first stable structural intermediate in the assembly of the subcore is a decamer and that these decamers are the building blocks of subcore assembly (See Figure 4.9 in p. 62).

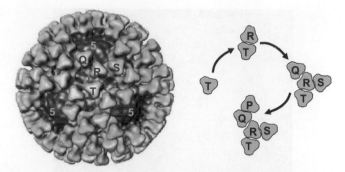

Plate 5 Combined mutagenesis and cryo-EM studies have revealed the order of assembly of the VP7 layer onto the VP3 subcore. If VP7 trimers in core-like particle (CLPs) are labelled P, Q, R, S, T with the P timer positioned at the five-fold axis and the T trimer at the three-fold axis of the VP3 layer, the P trimer has the poorest contacts with the VP3 layer and is easily removed and the T trimer has the most favourable contacts. (Left) Cryo-EM reconstruction of CLP in which the VP7 P trimer has been removed in a CsCl gradient and the other trimers and the visible five-fold axes are labelled. (Right) Cartoon showing order of assembly of VP7 trimers on the VP3 subcore. The T trimer nucleates assembly at the three-fold axis and other trimers are sequentially added from this point (See Figure 4.10 in p. 63).

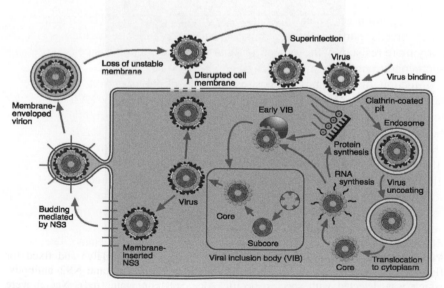

Plate 6 The bluetongue virus replication cycle. (See Figure 5.1 in p. 80).

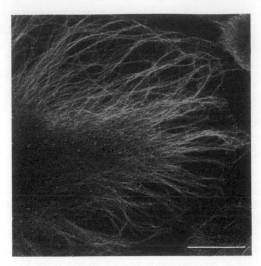

Plate 7   PAEC cell infected with bluetongue virus (BTV) and fixed for fluorescence microscopy at 20 hpi. The microtubules were labelled with mouse anti-tubulin detected with species-specific Alexa488 conjugate (green) and NS2 was labelled with specific antibody detected with species-specific Alexa568 conjugate (red). The majority of small viral inclusion antibodies (VIBs) are associated with the microtubules. Nuclei were stained with DAPI (blue). Scale bar = 10 μm (See Figure 5.6 in p. 84).

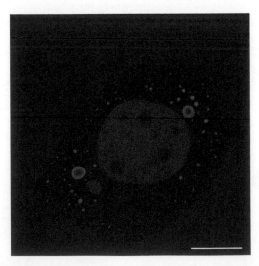

Plate 8   BHK-21 cells were infected with bluetongue virus (BTV) and fixed for fluorescence microscopy at 18 hpi. Cells were labelled with anti-NS2 antibody, which was detected with species-specific Alexa568 conjugate (red). Nuclei were stained with DAPI (blue). Scale bar = 10 μm (See Figure 5.7 in p. 85).

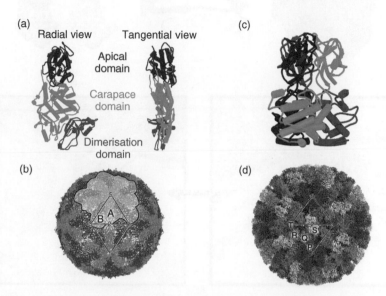

Plate 9 (a) Cartoon showing the domain structure of VP3(T2). The radial view shows the molecule as if it were in the subcore when viewed from outside the particle looking towards the centre. The tangential view is rotated 90° away from the radial view around the long axis of the molecule. Note how thin the molecule appears in tangential compared to radial profile showing just how thin the subcore shell is compared to the volume it encloses. (b) The BTV subcore made up of 120 copies of VP3(T2). The conformationally distinct A and B molecules have been coloured green and red, respectively. The icosahedral 5-fold, 3-fold and 2-fold symmetry axes have been marked. One decamer, a putative assembly intermediate, has been highlighted. (c) Cartoon showing the structure of the VP7 trimer. The three monomers have been coloured differently. The trimer sits on the exterior surface of the subcore, making contact through the bottom surface in this view. Inter-trimer contacts are mediated through the sides of the lower half of the molecule. (d) Cartoon showing the core structure. The icosahedral 5-fold, 3-fold and 2-fold axes are labelled and the icosa-hedrally independent trimers P (purple), Q (orange), R (green), S (yellow) and T (blue) are identified (Grimes *et al.*, 1998). (See Figure 6.2 in p. 104).

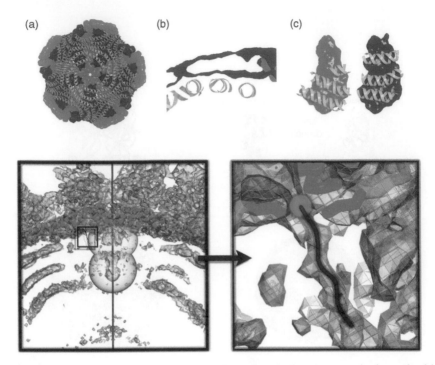

(a)   (b)   (c)

Plate 10 (Top) Cartoon showing dsRNA following shallow grooves on the underside of the VP3(T2) layer. (a) The top layer of dsRNA is modelled as a spiral around the 5-fold axis. The strands shown in orange are related to the strands shown in blue by icosahedral symmetry. (b) A slice through the VP3(T2) layer (molecular surface shown in red and green) showing shallow grooves on the underside of the layer. The strands appear to follow these grooves. (c) The grooves are in roughly the same place on both the A (green) and B (red) molecules. When assembled into the subcore layer, A and B are offset, and the grooves on one molecule feed into a groove further from the 5-fold on the next, giving rise to the spiral pattern (Gouet *et al.*, 1999). (Bottom) Cartoon showing the putative location of VP1(Pol) and VP4(Cap) within the subcore. VP3(T2) is shown as red and green worms, the observed electron density is shown as a grey semi-transparent surface, the icosahedral 5-fold axis is shown as a purple bar. VP1(Pol) and VP4(Cap) are thought to lie directly under the subcore layer on the icosahedral 5-fold axis, with VP1(Pol) being uppermost. The red and purple semi-transparent surfaces have been calculated to have volumes equal to those estimated for VP1(Pol) and VP4(Cap) and indicate their likely location. Shown in the right-hand panel is some weak density that may indicate that the disordered N-terminus of VP3(T2)A actually leads down into the core to make contact with VP1(Pol) and/or VP4(Cap). The possible path of the protein backbone is shown in blue (Gouet *et al.*, 1999) (See Figure 6.3 in p. 107).

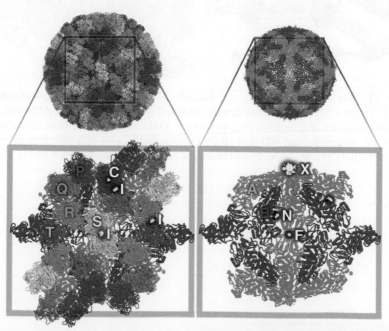

Plate 11 Cartoon showing substrate/product-binding sites on the bluetongue virus (BTV) core particle. The left panel shows the VP7(T13) layer (trimers P, Q, R, S, T are labelled) and associated binding sites (I, C). The right-hand panel shows the VP3(T2) layer (molecules A and B are labelled) and associated sites (X, N and F). The inter-trimer site I binds NTP and NDP and is found on the local 2-fold dyad between the top domains of trimers S–S', P–Q and R–T. The cation-binding site C is found within all of the trimers but shown only for the P trimer for simplicity. Site X, on the icosahedral 5-fold axis, is the likely site of export of the nascent mRNA. Site N was shown to bind NTP and NDP and is a putative pore allowing NTP into the particle (see Figure 6.12 for more detail). Site F was shown to bind phosphate and may be a route out of the core for waste phosphate (Diprose *et al.*, 2001) (See Figure 6.5 in p. 112).

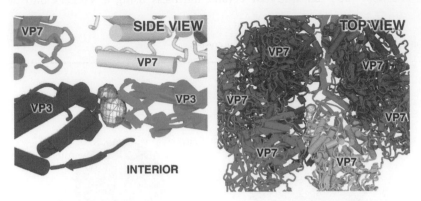

Plate 12 Site N, the NTP entry site through the outer core VP7(T13) and subcore VP3(T2) layers. The difference in density observed when comparing an ADP soak to a reference is shown as a semi-transparent surface. This site, accessible from both sides of the capsid, is seen to bind NTP and NDP. This suggests a role as a pore allowing NTP into the particle and waste NDP out (Diprose *et al.*, 2001) (See Figure 6.6 in p. 114).

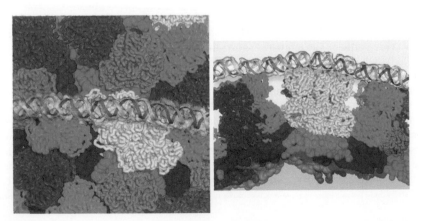

Plate 13 Cartoon showing bound dsRNA on the surface of core. The VP3(T2) layer is shown as balls, the VP7(T13) layer is shown as worms, the dsRNA is shown as ribbons and the observed additional electron density is shown as a semi-transparent surface. The RNA makes extensive contacts with the top surface of the VP7(T13) trimers (Diprose *et al.*, 2002) (See Figure 6.7 in p. 116).

(b)

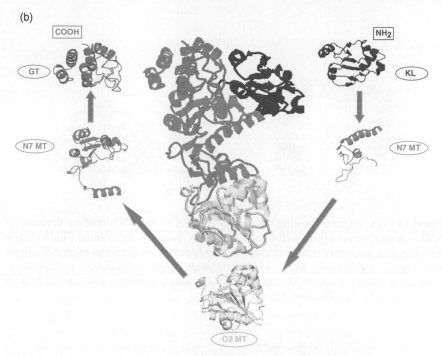

Plate 14   (b) The atomic structure of VP4 coloured by domain. From the N-terminus to the C-terminus: the kinase-like (KL, in red) domain, which is thought to interact with the polymerase; the N7-methyltransferase (N7MT, in cyan); the O2-methyltransferase (O2MT, in yellow) and the guanylyltransferase (GTase, in green) (See Figure 6.9 in p. 119).

(a)   (b)

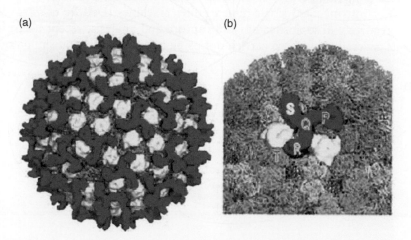

Plate 15   The outer capsid proteins VP2 and VP5. (a) Arrangement of outer capsid proteins VP2 and VP5 on the core surface. The 60 trimers of the VP2 triskelion are shown in *red* and the 120 copies of the globular VP5 density is shown in *yellow*. (b) The arrangement of the icosahedral asymmetric portion of VP2 and VP5 (in *grey*) on the core. The positions of the icosahedrally unique trimers of VP7(T13) are labelled (*P–T*) to help clarify the positions of VP2 and VP5 (See Figure 6.10 in p. 121).

(c)   (d)

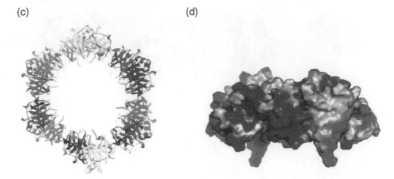

Plate 16 (c) Diagram showing the helical packing of the RNA-binding domains of NS2 observed in the crystals (Butan *et al.* 2004). Subunits are coloured individually and drawn as secondary structural cartoons. (d) Electrostatic potential of an NS2 dimer showing the amphipathic nature of the molecule (negative potential: *red*, positive: *blue*, on an arbitrary scale). The putative RNA-binding domain is towards the bottom of the molecule (See Figure 6.12 in p. 124).

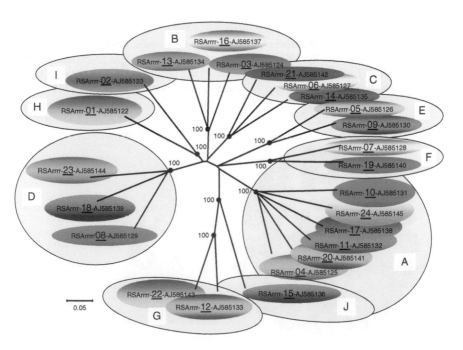

Plate 17 Phylogenetic tree for Seg-2 of the 24 BTV serotypes. This neighbour-joining tree was constructed using MEGA2 with the default parameters and the full-length Seg-2 (*VP2* gene) sequences of the reference strains of the 24 BTV serotypes. Nine evolutionary branching points are indicated by black dots (along with their bootstrap values) on the tree, which correlate with the ten Seg-2 'nucleotypes' designated A–J (Maan *et al.*, 2007a). (See Figure 7.1 in p. 138).

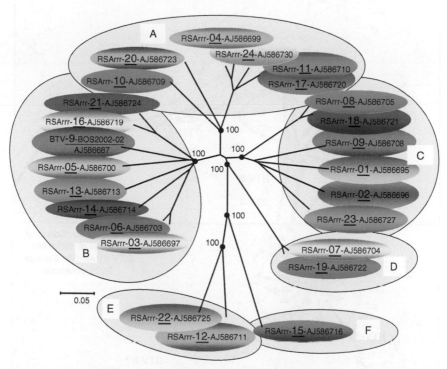

Plate 18 Phylogenetic tree for Seg-6 of the 24 BTV serotypes. Unrooted neighbour-joining tree showing relationships between the nucleotide sequences of Seg-6 from reference strains of the 24 BTV serotypes. The tree was constructed using MEGA2 with the default parameters and the full-length genome Seg-6 (*VP5* gene) sequences of the reference strains of the 24 BTV serotypes. Six evolutionary branching points are indicated by black dots (along with their bootstrap values) on the tree, which correlate with the six Seg-6 'nucleotypes', designated A–F. (See Figure 7.4 in p. 141).

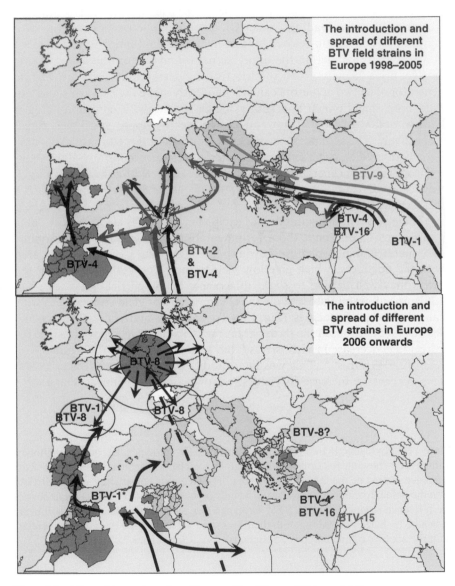

Plate 19 Map of incursions into Europe (See Figure 7.7 in p. 144).

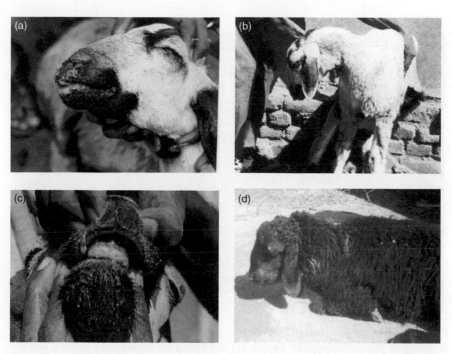

Plate 20 (a) Bluetongue (BT)-affected Nellore Palla sheep showing thick mucopurulent nasal discharge; (b) BT-affected Nellore Palla sheep showing muscle stiffness and torticollis; (c) BT-affected Nellore Palla sheep showing ulceration and erosions in the mucocutaneous borders and bleeding lips. (d) BT-affected Deccani sheep showing oedema of head and recumbency (See Figure 8.3 in p. 176).

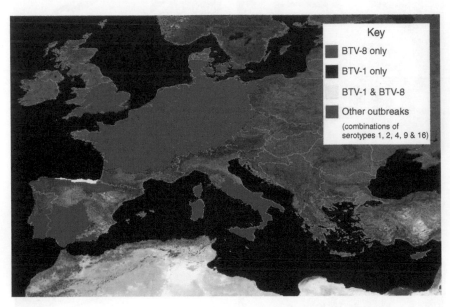

Plate 21 Bluetongue virus (BTV) restriction zones in the European Union at the time of going to press, showing the current BTV-8 epizootic in Northern Europe and the recent incursion of BTV-1 into Northern Spain and SW France (Adapted from: http://ec.europa.eu/food/animal/diseases/controlmeasures/bluetongue_en.htm) (See Figure 11.5 in p. 253).

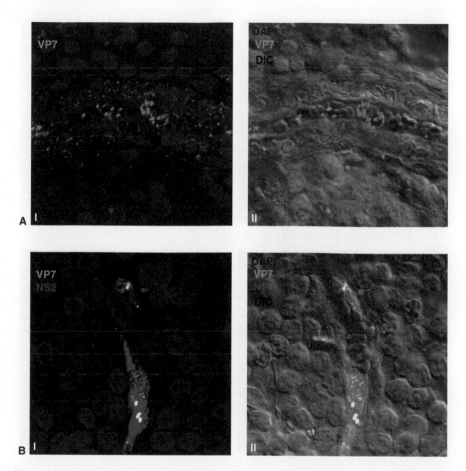

Plate 22 Bluetongue virus (BTV) protein in capillaries of lymph nodes. A paraformaldehyde-fixed, microtome cut thick section of lymph nodes from BTV-2 (RSA1971/03) infected sheep labelled with different anti-BTV antibodies and analysed via confocal microscopy. In both panels A and B, cellular nuclei are stained in blue (DAPI) and tissue morphology is shown in II using differential interference contrast (DIC). Viral proteins are stained either red or green as colour coded in the pictures, sometimes using double labelling for different proteins. (Panel A) BTV protein VP7 in the lumen of a small blood vessel in the prescapular lymph node at 3 days post-infection (d.p.i.). (Panel B) BTV proteins NS2 and VP7 in the endothelium of a capillary in the mandibular lymph node at 8 d.p.i. (See Figure 12.1 in p. 269)

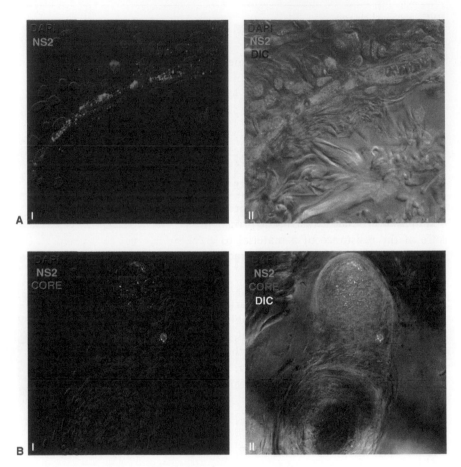

Plate 23 Bluetongue virus (BTV) proteins in cellular components of the skin. A paraformaldehyde-fixed microtome cut thick section of lymph nodes from BTV-infected sheep labelled with different anti-BTV antibodies and analysed via confocal microscopy. In both panels A and B, cellular nuclei are stained in blue (DAPI) and tissue morphology is shown in II using differential interference contrast (DIC). Viral proteins are stained either red or green as colour coded in the pictures, sometimes using double labelling for different proteins. (Panel A) BTV protein NS2 in the endothelium of a small blood vessel of the dermis of the skin at 3 days post-infection (d.p.i) with BTV-2 (RSA1971/03). (Panel B) BTV core proteins and NS2 in a glandular structure branching of a hair at 8 d.p.i. with BTV-8 (NET2006/01) (See Figure 12.2 in p. 271).

Plate 24   Ulcers in the dental pad and haemorrhages in the palate of a sheep with BT (See Figure 13.1 in p. 286).

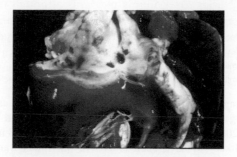

Plate 25   Sub-intimal haemorrhage of the pulmonary artery of a sheep with BT (See Figure 13.2 in p. 287).

Plate 26   Acute necrosis and haemorrhage in the myocardium (papillary muscle of the left ventricle) of a sheep with BT (See Figure 13.3 in p. 287).

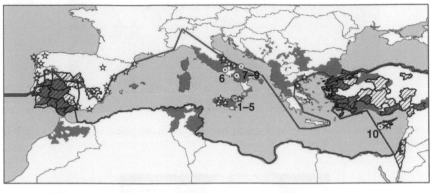

▨ Areas affected by BTV before 1998	Sites where BTV has been isolated from
▦ Areas affected by BTV 1998–2005	⊙ C. obsoletus group
—— Known limit of C. imicola to 2005	⊚ C. pulicaris s.s.
—— Known limit of C. imicola to 1997	☆ Sites without C. imicola before 1998

Plate 27 The changed distribution of bluetongue (BTV) and its vectors in Europe: Map showing the distribution of BTV prior to 1998 and that of BTV since 1998 (up to October 2004). The distribution of BTV prior to 1998 in North Africa and the Middle East is likely to have been extensive but much transmission in these endemic areas occurs silently in disease-resistant host animals. The reported outbreaks mapped here therefore vastly underestimate the extent of historical transmission in these fringe areas. Lines indicate the known northern range limit of the major Old World vector midge species *Culicoides imicola* up to 1998 (blue) and up to the present day (red). Dotted circles indicate sites where BTV has been isolated from wild–caught, non-engorged individuals from Palearctic vectors groups – from *C. pulicaris s.s.* in green and from *C. obsoletus* group in yellow. All these isolations were made during 2002 [Sites 1–5 (Caracappa *et al.*, 2003ʼ), Site 6 (De Liberato *et al.*, 2005ʼ), Sites 7–9 (Savini *et al.*, 2003, 2005)] except for site 10 in Cyprus (Mellor and Pitzolis, 1979). Stars indicate sites where *C. imicola* was found to be absent before 1998 (Mellor *et al.*, 1984; Gallo *et al.*, 1984; Gloukova *et al.*, 1991; Dilovski *et al.*, 1992; Scaramozzino *et al.*, 1996; Rawlings *et al.*, 1997; Ortega *et al.*, 1998); sites in Bulgaria were surveyed but not georeferenced, so are not shown (See Figure 16.1 in p. 345).

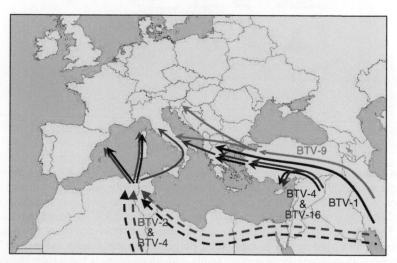

Plate 28 The molecular epidemiology of bluetongue virus (BTV) in Europe. Sequence analysis of the five European BTV serotypes has identified six lineages, which have arrived from at least two sources (Mertens et al., this volume). The European strain of BTV-1 (Greece2001/01) belongs to an eastern group of viruses and is similar to viruses isolated in India. The European strain of BTV-2, which first appeared in Tunisia in 1998, belongs to a western group of viruses and is similar to strains from South Africa, Nigeria, Sudan and the United States and probably entered Europe from the south. Both European BTV-9 and -16 belong to eastern groups while the European type 4, initially isolated in Greece (in 2000), is very similar to viruses periodically isolated in the region since 1969. This suggests that it may have been circulating in the fringes of Europe for many years. However, in late 2003, a new strain of BTV-4 arrived in Corsica and the Balearics, which is distinct from that seen in Greece and Turkey, and therefore may have arrived from North Africa. Importantly, distinct strains are still entering Europe on an annual/biannual basis (See Figure 16.2 in p. 346).

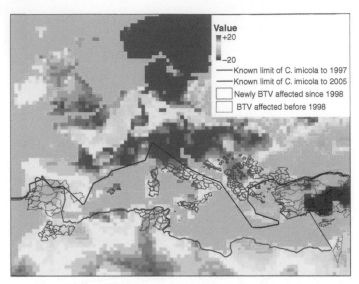

Plate 29 Spatial variation in recent climate change in Europe. These images show the changes in annual minimum temperatures between the 1990s and 1980s for each 0.5° square of longitude and latitude on a sliding colour scale ranging from a reduction of 2.0 °C (dark blue and −20 on inset legend) to an increase of 2.0 °C (dark red and +20 on inset legend). Temperature increases are most marked in both central (Italy, Corsica, the Balearic Islands) and eastern Europe (West Bulgaria, northern Greece, Albania, F.Y.R. Macedonia, Bosnia and Herzegovina, Serbia and Montenegro, Croatia) whilst central Iberia and the zone around the border between northern Morocco and Algeria have cooled. This is overlaid with the areas historically affected by BT in grey outline and the newly affected areas in red outline and the distribution limits of *C. imicola* from Figure 16.1. (This image was produced by temporal Fourier processing the raw time series of data and reconstituting it by summing the annual, biannual and triannual harmonics – essentially smoothing the data. The minimum values here are the minima of the reconstituted series for the period in question.) (See Figure 16.3 in p. 350)

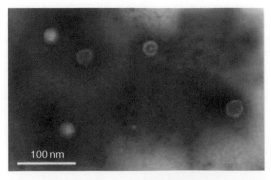

Plate 30 TEM image of alum-adsorbed bluetongue virus particles. The bar represents 100 nm. (See Figure 18.1 in p. 417)

Printed and bound by CPI Group (UK) Ltd, Croydon, CR0 4YY

03/10/2024

01040412-0018